PHYSICOCHEMICAL AND BIOMIMETIC PROPERTIES IN DRUG DISCOVERY

PHYSICOCHEMICAL AND BIOMIMETIC PROPERTIES IN DRUG DISCOVERY

Chromatographic Techniques for Lead Optimization

Klara Valko

CSC-Analytical Chemistry
Platform Technology & Sciences
GlaxoSmithKline, United Kingdom

and

Department of Pharmaceutical and Biological Chemistry
UCL, School of Pharmacy
London, United Kingdom

WILEY

Published by John Wiley & Sons, Inc., Hoboken, New Jersey
Published simultaneously in Canada

Library of Congress Cataloging-in-Publication Data:

Valko, Klara
Physicochemical and biomimetic properties in drug discovery : chromatographic techniques for lead optimization / Klara Valko, CSC-Analytical Chemistry, Platform Technology & Sciences, GlaxoSmithKline, United Kingdom, and Department of Pharmaceutical and Biological Chemistry, UCL, School of Pharmacy, London, United Kingdom.
pages cm
Includes bibliographical references and index.
ISBN 978-1-118-15212-6 (cloth)
1. Drug development–Statistical methods. 2. Chromatographic analysis. 3. Clinical trials–Statistical methods. 4. Pharmaceutical biotechnology. I. Title.
RM301.25.V35 2014
615.1′9–dc23

2013027975

Printed in the United States of America
10 9 8 7 6 5 4 3 2 1

CONTENTS

PREFACE

During the past 10 years I have been teaching the Physchem/ADME (absorption, distribution, metabolism, and elimination) module for "Drug Discovery M.Sc." students at the School of Pharmacy, University of London (University College London since 2012). The module covers ADME and the underlying physicochemical properties of drugs. This book is intended to summarize the course material, providing detailed explanations of the physicochemical aspects of drug absorption and distribution *in vivo*. It is well recognized now that drug molecules occupy a relatively small physicochemical property space in comparison to the huge number of possible physicochemical entities. Lipophilicity, solubility, permeability, and the charge state of molecules are the most important properties that influence absorption and *in vivo* distribution. Thus, the measurement and calculation of lipophilicity, solubility, permeability, and the charge state are essential early on in the drug discovery process in order to select compounds for further studies. Compounds selected in this way have the best possible chance to make it to development and eventually help patients recover or at least improve their quality of life. The principles of the measurement of physicochemical and biomimetic properties are explained in more detail. Special emphasis is given to the chromatographic measurements of various physicochemical and biomimetic properties of drugs. It will be shown how to interpret the chromatographic retention data by putting them into various models for the estimation of *in vivo* distribution behavior of compounds. Other techniques are also mentioned, and further references are provided but not discussed in detail.

The chromatographic dynamic equilibrium process provides an excellent model to describe a compound's *in vivo* distribution between the moving plasma/blood compartment and the stationary tissue compartments. The reader will learn that chromatography is a powerful technique not only for the separation of closely related compounds but it is also very useful to measure a compound's interactions with various biomimetic stationary phases covered with alkyl chain, proteins, and phospholipids. The obtained binding data are suitable for deriving quantitative structure–physicochemical property relationships (QSPR) that can be used in drug design. Mathematical models can be constructed for the estimation of a compound's *in vivo* distribution. Active transporters and other specific binding of the molecules to various types of proteins and phospholipids can, of course, modulate the estimated distribution, and these potential interactions need to be taken into account during the lead optimization process. When significant differences are observed between the *in vivo* measured and estimated distribution behavior of a compound based on physicochemical properties, it indicates to the drug discovery scientist that the compound undergoes some active transport mechanism. The active transport process can push the compound concentration from the thermodynamic equilibrium, but it requires constant energy investment from the body via biochemical processes.

This book contains previously published ideas, concepts, and methodologies. This literature review is far from comprehensive; it is rather critical and selective in order to reveal various ways to approach problems. However, the book provides simple descriptions and explanations of the essence of selected publications that usually help students understand the methods, results, and conclusions of scientific papers. A chapter dedicated to chromatography describes the basic principles and practical considerations that are needed to set up and run the measurements that are discussed in later chapters. Examples are shown how to use experimental data in various models of absorption, tissue binding, volume of distribution, and other *in vivo* ADME characteristics of the drugs. The property data obtained by fully automated chromatographic measurements can be used to select and prioritize compounds for further *in vitro* and *in vivo* studies at early stages of the drug discovery process. A special chapter is dedicated for the interpretation of the data, the interrelationship between the physicochemical properties, and some structure–property relationships in the hope that medicinal chemists can use the information for designing new potential drug molecules.

Although there are several books that cover similar topics, this book is unique in many ways. First, the chromatographic technique is normally described as an analytical separation method, and it is rarely discussed as a tool for property measurements of drug discovery compounds. There are hundreds of research papers and a few reviews that demonstrate the usefulness of the technique for such purposes but the general explanations together with a deeper understanding has not been published in a textbook yet. Sufficient detail is provided to enable the reproduction of chromatographic measurements in any laboratory equipped with HPLC. Second, this book is unique in the sense that it contains some new insight into the interdisciplinary knowledge needed for designing efficient drugs with minimal side effects. The intention is to reveal the relationships between various disciplines such as physical chemistry, analytical chemistry, biology, and pharmacokinetics/pharmacodynamics by the underlying basic thermodynamic laws.

Communication between scientists from different disciplines is hindered not only by differences in vocabulary but also many times by the use of different units. For example, the first measurements of activity of a compound is usually expressed as a pIC_{50}, which means a quantity of the compound expressed as the negative logarithm of the molar concentration that causes 50% of the maximum inhibition on a particular target or enzyme. Then we measure the solubility in millimolar or milligram per milliliter units as the maximum soluble concentration of the drug. Experts in drug metabolism and pharmacokinetic (DMPK) measure the quantity of drug absorbed as a percentage, the plasma protein binding is measured as unbound fraction, and volume of distribution as liters per kilogram. Finally, the patient takes a quantity of drug (dose) expressed as an *X* mg tablets three times a day. In all of the above instances, we would like to express the amount or concentration of drug molecules. Using different units in various phases of the drug discovery may hinder our understanding. I do not think we can change this practice in the near future. However, at present, we can learn to convert the units and understand the relationships between various measurements. The mathematical and physical–chemical rules serve as links between disciplines in this book. It is demonstrated that mathematical approaches (paying special attention to the units of the measurements) help interdisciplinary thinking, converting, and translating knowledge between scientists working in different fields. In this way, we can maximize our understanding of the huge amount of data that are normally generated during the drug discovery process.

Chapter 12 presents examples, mostly using historical data of successful known drug molecules or published project examples from several pharmaceutical companies. At the end of each chapter, there is a short summary containing the conclusions and help revision for students taking examinations of this material. Finally, some typical examination questions are added at the end of each chapter to be able to test the reader's comprehension of the material. The answers are provided in Appendix A.

Some of the conclusions have been made on the basis of the author's personal view and experience in the subject. These do not always agree with the accepted and widely used theories in big pharmaceutical companies. Ideas for further research areas, unanswered questions, and hypotheses are presented *in italics* in order to encourage and motivate readers to form their own opinion. The book contains several questions that were raised by the students during the lessons and the answers still have to be found by the next generation of scientists. Therefore, I hope that the text raises the motivation and enthusiasm of scientists who are willing to engage themselves in drug discovery for a long period.

This book provides the essential experimental details for the determination of physicochemical and biomimetic properties that can be carried out in any analytical laboratories equipped with HPLC. Medicinal chemists will be able to understand these properties and the structure–property relationships described here and successfully use them in drug design and lead optimization. DMPK and ADME scientists will be able to interpret the results of their *in vitro* and *in vivo* experiments after thorough comprehension of the content, and it will help them in designing the necessary experiments to support the drug discovery process. I am aware that several drug discovery institutions and universities have already set up the chromatography-based determination of lipophilicity, protein binding, and phospholipid binding. I hope that with the help of this book, they will be able to use these properties in their full potential for the estimation of *in vivo* properties, thus reducing the need for animal experiments.

Finally, I would like to thank all those who have been of tremendous help during the preparation of this book. I am grateful to my past and present colleagues at GlaxoSmithKline, especially to Derek Reynolds, Chris Bevan, and Alan Hill with whom I gained experience in the field of physicochemistry and its application in drug discovery and who have supported me in many ways. I would like to thank my colleagues Shenaz Bunally, Elisabetta Chiarparin, and Paul Leeson, who read the manuscript and gave me excellent advice to improve it. I would like to acknowledge the help I have received from the colleagues at the School of Pharmacy, University College London, Professor David Thurston, Professor Simon Gibbons, Professor Anne Stevenson, Dr. Michael Munday, Dr. Mire Zloh, and Dr. Rosemary Smyth.

I have received the greatest inspiration for the preparation of this book from my students. Many of them are extremely talented, providing me with the hope that they will continue to improve drug discovery and will help millions of patients who desperately need cure for their disease. Drug Discovery M.Sc. students from year 2011 to 2012 have gone through the manuscript and provided excellent feedback where I need to explain the material in more detail or more clearly. I would like to thank especially Godfrey Mayoka, Samar Youshif, and Hajir Azam for carefully reading the manuscript. I have used and referenced the project works of Manju Kalyani and Andriana Rapti, who carefully investigated and validated the methodology for solubility determination and α-1-acid glycoprotein binding measurements during their M.Sc. studies.

I am very grateful for the well-known experts who provided valuable suggestions and corrections for specific chapters. Professor Krisztina Takács-Novák, School of Pharmacy, Semmelweis University, has reviewed chapters on solubility and pK_a; Dr. Alex Avdeef, In Adme, Ltd., has reviewed the chapter on permeability; Professor Michael Abraham, University College London; has reviewed the chapter on lipophilicity. They are recognized experts in these fields and their valuable corrections and suggestions are appreciated.

I would like to thank my family and friends for their emotional support and understanding, especially to my son Adam Valko who coped very well with his studies and his life on his own during the preparation of this book.

1

THE DRUG DISCOVERY PROCESS

The way we discover drugs now is very different from the process followed 30–50 years ago [1–3]. In the past, the drug discovery process was based on the structure of known active compounds. The structure of the active compound, which was either endogenous (e.g., acetylcholine, adrenaline, and steroid hormones) or a natural product (e.g., morphine, papaverine, cocaine, atropine, and digitalis glycosides), was identified and modified by a trial-and-error method. Scientists isolated the active material, revealed the structure and the mechanism of the pharmacological action, and based on this knowledge they would modify the chemical structures with the aim of improving the activity, decreasing the toxicity, and increasing the duration of action. Using this approach, they discovered, for example, a series of cholinesterase inhibitors, a series of adrenaline analogs for sympatomimetic or sympatolytic compounds, semisynthetic steroid hormones, and many more. Another example is pethidine containing the pharmacophores of morphine. The ethoxy derivative of papaverine called *No-Spa* has a longer half-life and is a stronger spasmolytic agent. Figure 1.1 shows a few analog-based drug molecules designed from natural compounds. There are numerous examples of naturally occurring molecules or their derivatives being used as drugs, such as warfarin (anticoagulant derived from dicumarol found in sweet clover, hirudin (anticoagulant from leeches), statins (a fungal metabolite that reduces plasma

Physicochemical and Biomimetic Properties in Drug Discovery: Chromatographic Techniques for Lead Optimization, First Edition. Klara Valko.

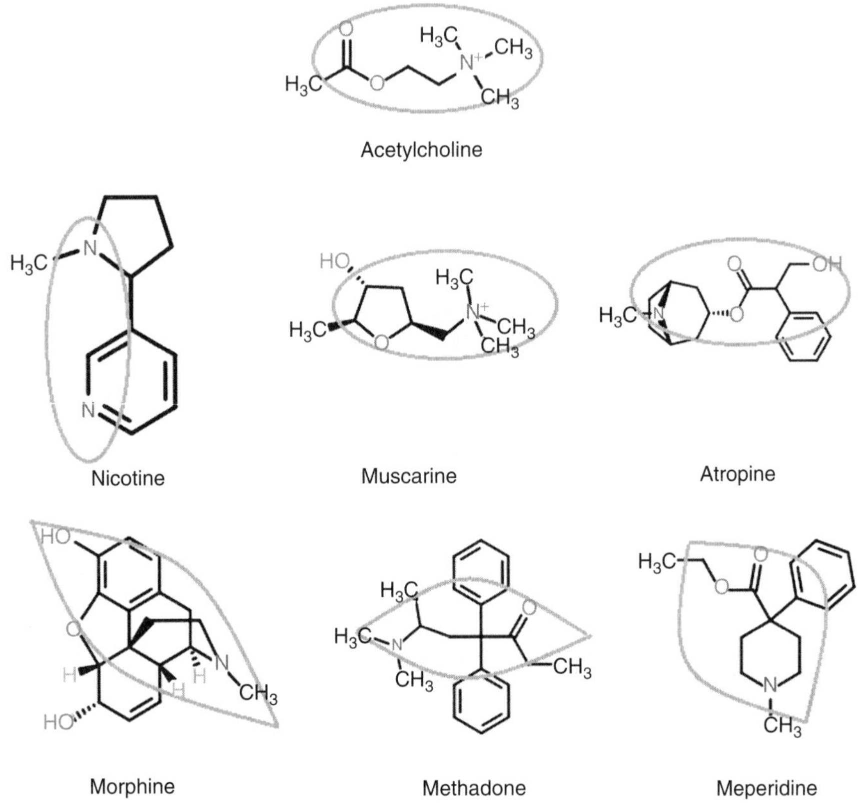

Figure 1.1 Drug molecules designed as analogs of natural active molecules containing the active pharmacophore.

cholesterol), vinca alkaloids (anticancer drugs isolated from periwinkle plant families), and antibiotics (derived from fungal metabolites).

Today, drug discovery utilizes recent advances in biochemistry and genetics. Drug discovery usually starts with discovering a so-called "target" that can be enzymes or receptors, such as G protein-coupled receptors (GPCR) and enzyme targets, such as kinases, ion channels, and hormone receptors.

The medicinal chemists and computational chemists work closely together to generate ideas about potential active molecules that would fit to the "target" molecule. They develop a so-called high throughput screening (HTS) method that enables the big pharmaceutical companies to screen large number of compounds against a particular target. Compounds that show activity in HTS are called *hits*. The activity of the "hits" is further evaluated by repeated experiments for obtaining the standard concentration–potency curve. Figure 1.2 shows a typical concentration–response/potency profile during hit generation and confirmation.

This phase of the drug discovery process is often called *hit generation*. Synthetic chemists then synthesize a whole library of molecules, making numerous

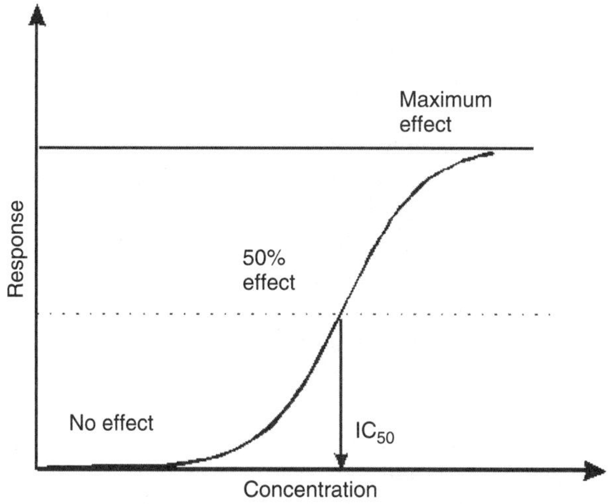

Figure 1.2 Typical concentration versus potency profile used to derive the pIC_{50} values that refer to a concentration of the compound producing 50% of the maximum measurable response in a particular screen.

combinations of substituents on an active "skeleton" or scaffold. This part of the process is often called *lead generation*. The "lead" molecule usually shows high activity on a particular target. It is advisable to test the selectivity of the lead molecule to avoid so-called promiscuous binders that produce a hit on a variety of targets; that is, the lead molecule should show activity preferably only on one target. The "lead" molecule also should show reasonable solubility and bioavailability during the first *in vivo* animal experiments.

Sometimes it is possible to get the three-dimensional structure of the targets using X-ray crystallography or nuclear magnetic resonance (NMR) and the binding sites are identified. Computational chemists, using various docking methods, find or design small molecules that fit to the active site of the target [4]. The primary role of a docking program is to identify biologically active compounds that may act as ligands to a target binding site based on the calculated intermolecular interaction energy. When the X-ray crystallographic structure is not available, there are various computational methods that can be used to assess the binding site of the target protein and design molecules that will bind to it.

The drug discovery process moves into the lead optimization phase, where molecules then undergo a series of structural modifications, which helps to establish a so-called SAR (structure–activity relationship) for the series. It allows the chemists to establish which part of the molecule is essential for activity. During this process, they often find much more active analogs than the original lead molecule. The activity measurements are repeated using cell-based assays or in the presence of plasma proteins. Usually that is the point where potential liabilities may be observed. The reduced cellular activity could be due to low permeability or solubility or strong nonspecific binding to cellular components.

This is the time when they approach the physicochemists to measure solubility, permeability, and protein binding in order to determine the cause of the loss of cellular activity. Usually the activity is not lost completely but only weakens. This is called *drop off* of the activity. The lead optimization process often takes a year or two, maybe even considerably longer. It includes various *in vitro* assays for the assessment of selectivity and *in vitro* microsomal stability. For a smaller number of potentially promising compounds, animal studies, bioavailability measurements, and pharmacokinetic studies are also carried out before candidate selection. At this stage, scientists try to establish the dose necessary for effective *in vivo* pharmacological activity. It is important to observe *in vivo* efficacy and receptor or target engagement assays to prove that the observed pharmacological activity is a consequence of the molecule binding to the target. The first toxicological studies are also performed at this stage.

Chemists utilize information from other disciplines to establish structure–property relationships and modify the structures accordingly. If successful, at the end of this process, the molecule that shows the desired potency, bioavailability, *in vivo* receptor occupancy, and acceptable pharmacokinetic properties will be declared as the "candidate" and will then enter a long development process. When a molecule is identified as a candidate, the research process ends and the selected molecule is passed over to development scientists, who then carry out preclinical studies, extensive toxicity studies using much higher doses than are needed for efficacy, formulation studies, large-scale synthesis, and preparation for the first clinical trial "first time in human." Usually, a "backup" compound is also selected in case the candidate molecule fails during the development process.

During this process, we also need a disease model to prove the concept that the compound acting on the selected target actually can cure the disease. It is called *proof of concept* and target validation. Figure 1.3 summarizes the major phases of the early drug discovery process. The high throughput chemistry is often replaced nowadays by lower throughput synthesis of smaller number of compounds as a designed library. The HTS campaign is very expensive; therefore, it is used when hit molecules are not available from other sources.

However, traditional HTS has not delivered on its promise of increasing the numbers and quality of new drugs entering clinical trials. The typical hit rate of a screening campaign is usually less than 1%. This lack of success is due in part to the complexity and the relatively large size of the compounds routinely being screened. On the basis of the mathematical probability, the ability of finding perfect binding, including four to six interactions at the binding site, is very low [5].

Recently, the so-called fragment approach has been used more often. It has been found that the probability of binding decreases rapidly as the complexity of the ligand increases. However, there is much higher probability of finding one or two binding interactions at a time and then mapping the other possible interactions at the binding site one by one. Therefore, instead of screening a very large number of molecules to find the one that shows affinity to the receptor, there is much higher probability of success in finding molecular fragments that have only one or two binding interactions. From the small, so-called "fragment" molecules, we can

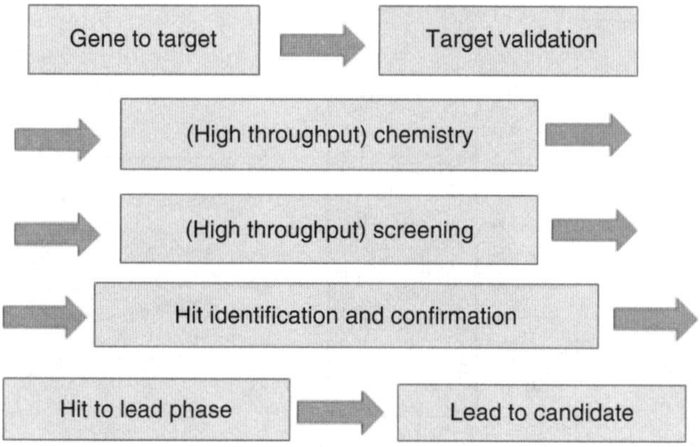

Figure 1.3 Major steps of the modern drug discovery process including both the biological and chemical aspects.

build bigger molecules by combining several fragments into a single molecule. In this way, we can increase the likelihood of finding a highly potent larger molecule. The effective use of the fragment-based approach is dependent on the use of biophysical techniques such as X-ray crystallography or NMR measurements to identify how and where the fragment binds to the target. Traditional bioassays used in HTS are generally unable to detect such small drug fragments because of their low potency binding to the protein target. The approach requires screening of molecules at higher concentration (millimolar and not micromolar or nanomolar) to be able to detect the weaker interactions. The fragment-based drug design approaches are adopted by many big and small pharmaceutical companies. The typical hit rate is usually an order of magnitude higher than the HTS hit rate (around 10%). The typical binding interactions between such small molecules and the target are hydrogen bonding (by donating or accepting proton), hydrophobic, and π–π interactions. The small molecule has to fit spatially well to the binding sites, so that the interacting molecular forces can produce energetically stable binding. Ideally, fragment molecules should allow chemists to attach other functional groups and grow the molecules. It means that the fragments should be chemically tractable for modification.

The early drug discovery process sometimes produces several thousands of potent molecules, still only a few reach the market. The attrition is very high. Although this part of the drug discovery research is expensive, it costs only around 5% of the total cost of getting a drug molecule to the market, which is estimated to be around $800 million. Unfortunately, only 1 in 10 candidate molecules obtain approval from the Food and Drug Administration (FDA). This huge attrition of the compounds is due to so-called "developability" issues. If a candidate is to fail, the earlier the better, as it saves a lot of effort and investment for the company. The most common reasons for stopping the candidates from further development are disadvantageous pharmacokinetics, lack of efficacy in the clinic, toxicity,

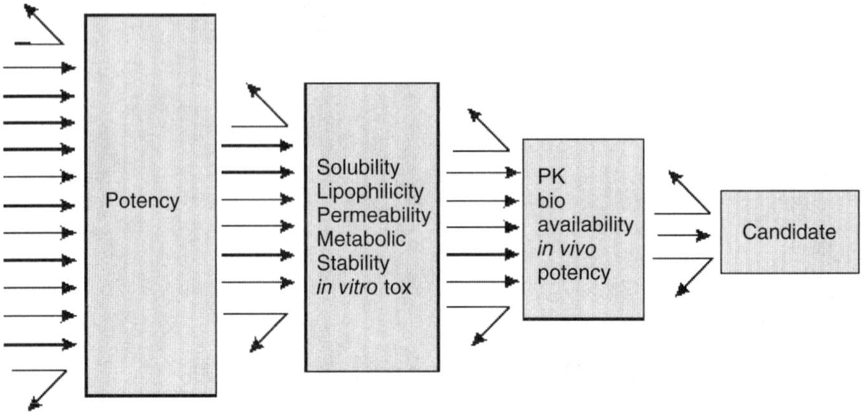

Figure 1.4 Developability considerations and physicochemical properties should be taken early in the drug discovery process.

and formulation problems. Sometimes development is stopped for commercial reasons, as other compounds by competitors appear on the market. The potential disadvantageous pharmacokinetics includes poor oral exposure, high clearance, drug–drug interaction, too short or too long half-life, etc. The physicochemical properties of the molecules have a crucial impact on these failures [6–8]; therefore, it is vital to include these measurements early in the drug discovery process. Figure 1.4 shows how the developability considerations and physicochemical properties are built into the candidate selection process.

For example, poor solubility of the discovery compounds may compromise the results of the screening. It may produce false-positive or false-negative potency values. Very lipophilic compounds often bind to many different types of proteins and to various nonspecific binding sites as well, thus reducing the actual available free concentration of the compound and shifting the pIC_{50} values. Ionized compounds can have various ionic interactions; they can easily change the physicochemical properties such as solubility and lipophilicity because of small changes in the pH of the environment. The most widely known description of optimum physicochemical property space of orally absorbed marketed drug molecules is the Lipinski rule of five [9].

1. Lipophilicity should be less than 5 (calculated by clog P).
2. Number of H-bond donor groups (–NH, –OH) should be less than 5.
3. Number of H-bond acceptor groups should be less than 10.
4. Molecular weight is less than 500 Da.
5. Molecules that are undergoing active transport are exceptions.

Since the introduction of the seminal work of Lipinski [9], a great number of publications have highlighted the importance of simple physicochemical properties such as size, lipophilicity, H-bond acidity and basicity, and polar surface area that characterizes the drug-like property space. Thus, it was found that marketed drugs fall within a certain size, lipophilicity, and polarity range regardless of the pharmacological class [10–11]. As compounds progress through the lead optimization process, they tend to increase in size and lipophilicity, so-called "molecular obesity" [12], and fall outside the desired drug-like property space.

The aim of this book is to explain how to measure and use these physicochemical properties during the lead optimization process. We shall discuss how these properties can be used for compound selection and prioritization for further studies. At this stage, we need to measure solubility, lipophilicity, permeability, and other biomimetic properties of a large number of compounds in a short period of time. Compounds are usually not available in large quantities at this stage and often not available in a well-characterized solid form either. The measurements should be automated; compound dispensing and handling are being robotized. The physicochemical measurements are often pragmatic and limited to the experimental conditions used. Therefore, these measurements are quite different from the physicochemical measurements that are carried out on selected candidate molecules at the development stage.

SUMMARY

The drug discovery process has gone through major changes in the past decade. The discovery of defined human molecular targets based on genomics provides the possibility to develop HTS methods. In spite of finding numerous potent molecules, the number of drug approvals has not increased proportionally as drug candidates often fail at later stages in the drug development process. The failure is most often due to the lack of efficacy, poor bioavailability, unfavorable pharmacokinetics, and toxic side effects. The undesirable outcome in the clinic is often due to the physicochemical properties of the molecules, such as size, lipophilicity, permeability, and solubility. The scope of this book is to investigate the measurement and calculation of these properties at early stages of the drug discovery process, reveal general structure/property relationships, and suggest appropriate structural modifications during the lead optimization. The aim is to provide insight and explain why these physicochemical properties are important to take into consideration for reducing late-stage attrition of potent candidate molecules.

QUESTION FOR REVIEW

Q1.1 What are the major steps of the drug discovery process?

REFERENCES

1. Drug Discovery and Development, Ed. Rang, H. P. (2006) Elsevier, London.

2. Kerns, E. H., Di, L. (2008) Drug-like Properties: Concepts, Structure Design and Methods. Academic Press, Elsevier, Amsterdam.
3. Analogue Based Drug Discovery II, Eds Fischer, J., Ganellin, C. R. (2010) Wiley-VCH, Weinheim.
4. Balius, T. E., Mukherjee, S., Rizzo, R. C. (2011) Implementation and evaluation of a docking-rescoring method using molecular footprint comparisons. *Journal of Computational Chemistry*, *32*, 2273–2289.
5. Leach, A. R., Hann, M. M. (2011) Molecular complexity and fragment-based drug discovery: Ten years on. *Current Opinion in Chemical Biology*, *15*, 1–8.
6. Valkó, K., Reynolds, D. P. (2005) High throughput physico-chemical and *in vitro* ADMET screening: A role in pharmaceutical profiling, *American Journal of Drug Delivery*, *3*, 83–100.
7. Van de Waterbeemd, H. (2009) Improving compound quality through *in vitro* and *in silico* physicochemical profiling, *Chemistry and Biodiversity*, *6*, 1760–1766.
8. Gleeson, M. P., Hersey, A., Montanari, D., Overington, J. (2011) Probing the links between *in vitro* potency, ADMET and physicochemical parameters. *Nature Reviews; Drug Discovery*, *10*, 197–208.
9. Lipinski, C. A., Lombardo, F. B., Dominy, W., Feeney, P. J. (1997) Experimental and computational approaches to estimate solubility and permeability in drug discovery and development settings. *Advanced Drug Delivery Reviews*, *23*, 3–25.
10. Lipinski, C. A. (2004) Lead- and drug-like compounds: The rule-of-five revolution. *Drug Discovery Today: Technologies*, *1*, 337–341.
11. Vieth, M., Siegel, M. G., Higgs, R. E., Watson, I. A., Robertson, D. H., Savin, K. A., Durst, G. L., Hipskind, P. A. (2004) Characteristic physical properties and structural fragments of marketed oral drugs. *Journal of Medicinal Chemistry*, *47*, 224–232.
12. Hann, M. M. (2011) Molecular obesity, potency and other addictions in drug discovery. *Medicinal Chemistry Communications*, *2*, 349–355.

2

DRUG-LIKENESS AND PHYSICOCHEMICAL PROPERTY SPACE OF KNOWN DRUGS

There has been a lot of interest in finding common features in drug molecules that are used for the treatment of various diseases. The potent molecules usually contain a so-called pharmacophore group that is responsible for the receptor binding, enabling it to produce the desired pharmacological activity. However, it has long been recognized that it is not enough to make an active molecule in order to produce a successful drug. Many highly potent compounds failed to reach the market because of poor absorption, poor distribution, safety concerns, and so on. On the basis of the study by Kola and Landis [1], only around 5% of selected candidate molecules reach patients as drugs. A pioneering study about the desirable properties of drugs was published by Lipinski et al. [2]. The Lipinski "rule of five" has been mentioned in Chapter 1. The rule has been established on the basis of analysis of the physicochemical properties of orally administered drugs. Their publication has been cited over 10,000 times, and many new modified rules have been proposed [3–11].

Other physicochemical properties have also been considered in subsequent publications such as the polar surface area, number of rotatable bonds, and number of heavy atoms. More recently, the importance of the number of aromatic rings has been emphasized [12–14]. Increasing the number of aromatic rings in molecules increases the chance of poor solubility. Lipinski [15] introduced the term *drugability*, emphasizing that the chemical space, where the drug molecules can be

Physicochemical and Biomimetic Properties in Drug Discovery: Chromatographic Techniques for Lead Optimization, First Edition. Klara Valko.

found, occupies only a small region of the chemical space of all possible types of molecules. Van de Waterbeemd et al. [16] introduced the new phase of drug discovery, where the property-based design replaces the structure-based design of new drug molecules. Kerns and Di [17] published a book in 2008 in which they discuss the drug-like properties, their measurements, and applications in the drug discovery. The essence of these publications is that drug molecules occupy a reasonably well-defined physicochemical property space. It was found that size, lipophilicity, solubility, permeability, polar surface area, and H-bond acidity and basicity play an important role in the description of the physicochemical property space. Gleeson et al. [18] published various *in silico* models for estimating absorption, plasma protein binding, metabolism, volume of distribution, central nervous system (CNS) penetration, and so on. These models highlighted the most important physicochemical properties that influence these absorption, distribution, metabolism, and elimination (ADME) characteristics. Lipophilicity and size were the two most dominant descriptors in estimating almost all of the ADME properties. For oral absorption, the compound should be easily dissolved in biorelevant media and should be permeable through the intestinal wall. Solubility and permeability can be measured *in vitro*. However, measurements are not sufficient for designing drug molecules; we need to understand how these properties relate to each other and also how they relate to the chemical structure of the molecules. Both solubility and permeability can be related to compound lipophilicity or hydrophobicity. The more lipophilic the compound, the less likely it is to be soluble in water. On the other hand, the molecule has to be reasonably lipophilic to be able to go through the lipophilic membrane bilayers. Hydrophilic compounds might contain H-bond donor and acceptor groups that can form hydrogen bonds in crystal form, which would reduce solubility and dissolution rate due to the increased crystal lattice energy. Solubility and permeability are both dependent on lipophilicity in the opposite direction and we need to find the optimum lipophilicity in order to achieve acceptable solubility and permeability. Also, when introducing a solubilizing group into a molecule, we increase the size of the molecule too, which can have a detrimental effect on the permeability. Figure 2.1 illustrates a hypothetical multidimensional property space projected onto a two-dimensional space using the first two principal components as axes. We need to be in the right chemical property space to achieve good potency, good absorption, good safety profile, good metabolic stability, and so on. We do it by making compounds and measuring its properties. Figure 2.1 shows a desirable situation when there is a small space that overlaps with all the required property spaces. This is usually not known to the researcher at the start of the chemistry program. Figure 2.2 demonstrates a hypothetical situation when potent compounds are outside the desired property space for good absorption and metabolic stability. Therefore, where possible, it is very important to start a research program with several series of leads and try to map the boundaries of the desired property space. Several research programs have ended without success as, for example, all the soluble and potent compounds were metabolically unstable, and only the insoluble potent compounds had the desired metabolic stability. In such cases, it is better to search for a new chemical starting point using different scaffolds. An in-depth

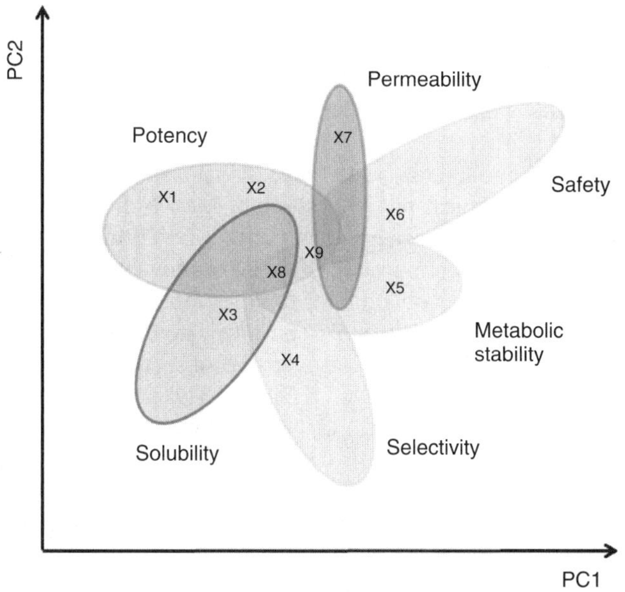

Figure 2.1 Illustration of the small overlapping property space (plotted using the first two principal components PC1 and PC2) that covers all the required important attributes of a drug molecule.

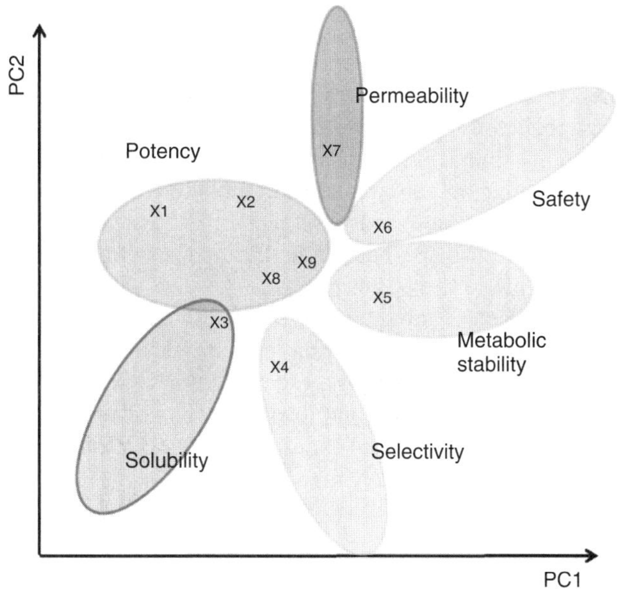

Figure 2.2 Illustration of the lack of overlapping property space (plotted using the first two principal components PC1 and PC2) that fulfills the requirement of a drug molecule.

lead optimization program should start only when a high probability exists that the optimization process is possible. As the number of newly discovered targets is increasing and no previous example of potent drug molecule exists, there is no guarantee that it is possible to find molecules that have the overlapping desired property space of a drug.

There is a debate in literature as to make molecules that are outside the boundaries of the physicochemical property space of known drug molecules [11]. The analysis of lipophilicity data suggests that only a very narrow lipophilicity range (from 1 to 3 log *D*, logarithmic value of the octanol/water distribution coefficient at pH 7.4) provides a high probability of finding a good drug molecule. The paper discusses in detail what the risks are when a research molecule is outside of the optimum lipophilicity range. There are examples of successful drug molecules that do not obey the Lipinski "rule of five." There are always exceptions to the general rules. As has been shown by Bade et al. [10], many natural products, for example, erythromycin, digitalis glycosides, do not fit the rule of five as the molecular weight is over 1000 Da and the number of H-bond donor and acceptor groups are also over the limit and still they are orally absorbed (most often by active transport). In drug discovery, we need to have a sense of urgency not only due to considerations of the cost but also the benefit of finding new drug molecules that can help patients live longer and healthier lives. In modern drug discovery, we need to focus on the probability of the success and need to increase our knowledge and understanding of why some molecules fail during the development process.

These rules and boundaries have to be used with caution. Intuition and experience can also increase the probability of success of a research program. The conclusion from the studies of the physicochemical properties of drugs, research compounds, and failed drug candidates is that we can define physicochemical property boundaries into which the majority of known drug molecules can fit. The most important physicochemical properties that we can measure are the lipophilicity, solubility, permeability, and acid/base character. Size is also important and is usually described by molecular weight or calculated molar refraction (cMR). These measurable properties are not independent of each other; so, it is difficult to alter only one property at a time. We also need to understand the effect of the conditions of the measurements (pH, solvents, etc.).

The following chapters provide deeper insight into these important physicochemical properties and how we can measure them using chromatographic techniques. It is demonstrated through examples that describe how we can apply measured data to develop general and local models to estimate a compound's *in vivo* behavior in the early stages of the drug discovery process and increase the probability of success.

SUMMARY

Several physicochemical properties of known drug molecules have been analyzed. It has been found that absorption, distribution, metabolism, and other pharmacokinetic parameters, such as volume of distribution and half-life can be related to the physicochemical properties of drugs. The majority of known drug molecules fit into a narrow range of physicochemical properties that are termed *drug-like* properties. When drug discovery compounds show these "drug-like" properties, they have a higher probability of becoming successful drugs. Therefore, it is important to identify, understand, and measure these properties early in the drug discovery process. The most important physicochemical properties that show correlations to the absorption, distribution, and pharmacokinetic properties are lipophilicity, solubility, permeability, and acid/base character. The size of the molecule is also very important and can be easily and precisely characterized by the molecular weight or calculated molar refractivity. These physicochemical parameters are not independent. The measured values are highly dependent on the conditions of the measurements. Therefore, it is very important to gain a better understanding of how to measure them and how to relate the measured values to the measurement conditions and to the chemical structure.

QUESTIONS FOR REVIEW

Q2.1 What are the most important physicochemical properties that can be related to the ADME properties?

Q2.2 What is "drug-likeness?"

Q2.3 Why is there a need to measure solubility, lipophilicity, and permeability in the early stages of the drug discovery process?

Q2.4 Which physicochemical properties are the most important in terms of drug-like property space?

REFERENCES

1. Kola, I., Landis, J. (2004) Can the pharmaceutical industry reduce attrition rates? *Nature Reviews Drug Discovery*, *3*, 711–716.
2. Lipinski, C. A., Lombardo, F., Dominy, B. W., Feeney, P. J. (2001) Experimental and computational approaches to estimate solubility and permeability in drug discovery and development settings. *Advanced Drug Delivery Reviews*, *46*, 3–26.
3. Oprea, T. I., Allu, T., Fara, D., Rad, R., Ostopovici, L., Bologa, C. (2007) Lead-like, drug-like or "Pub-like": How different are they? *Journal of Computer Aided Molecular Design*, *21*, 113–119.
4. Gleeson, M. P. (2008) Generation of a set of simple, interpretable ADME rules of thumb. *Journal of Medicinal Chemistry*, *51*, 817–834.
5. Hou, T. J., Wang, J. M., Zhang, W., Xu, X. J. (2007) ADME evaluation in drug discovery. 6. Can the oral bioavailability in humans be effectively predicted by simple molecular property-based rules? *Journal of Chemical Information and Computational Sciences*, *47*, 460–463.
6. Veber, D. F., Johnson, S. R., Cheng, H. Y., Smith, B. R., Ward, K. W., Kopple, K. D. (2002) Molecular properties that influence the oral bioavailability of drug candidates. *Journal of Medicinal Chemistry*, *45*, 2615–2623.

7. Leeson, P. D., Empfield, J. R. (2010) Reducing the risk of drug attrition associated with physicochemical properties. *Annual Reports in Medicinal Chemistry*, *45*, 393–407.
8. Hughes, J. D., Blagg, J., Price, D. A., Bailey, S., DeCrescenzo, G. A., Devraj, R. V., Ellsworth, E., Fobian, Y. M., Gibbs, M. E., Gilles, R. W., Greene, N., Huang, E., Krieger-Burke, T., Loesel, J., Wager, T., Whiteley, L., Zhang, Y. (2008) Physicochemical drug properties associated with *in vivo* toxicological outcomes. *Bioorganic and Medicinal Chemistry Letters*, *18*, 4872–4875.
9. Keserü, Gy., Makara, G. (2008) The influence of lead discovery strategies on the properties of drug candidates. *Nature Reviews Drug Discovery*, *8*, 2003–212.
10. Bade, R., Chan, H.-F., Reynisson, J. (2010) Characteristics of known drug space. Natural products, their derivatives and synthetic drugs. *European Journal of Medicinal Chemistry*, *45*, 5646–5652.
11. Waring, M. J. (2010) Lipophilicity in drug discovery. *Expert Opinion in Drug Discovery*, *5(3)*, 235–248.
12. Lovering, F., Bikker, J., Humblet, C. (2009) Escape from flatland: Increasing saturation as an approach to improving clinical success. *Journal of Medicinal Chemistry*, *52(21)*, 6752–6756.
13. Ritchie, T. J., MacDonald, S. J. F. (2009) The impact of aromatic ring count on compound developability—Are too many aromatic rings a liability in drug design? *Drug Discovery Today*, *14*, 1011–1020.
14. Hill, A. P. Young, R. J. (2010) Getting physical in drug discovery: A contemporary perspective on solubility and hydrophobicity. *Drug Discovery Today*, *15*, 648–655.
15. Lipinski, C. A. (2000) Drug like properties and the causes of poor solubility and permeability. *Journal of Pharmacological and Toxicological Methods*, *44*, 235–249.
16. Van de Waterbeemd, H., Smith, D. A., Beaumont, K., Walker, D. K. (2001) Property-based design: Optimization of drug absorption and pharmacokinetics, *Journal of Medicinal Chemistry*, *44*, 1313–1333.
17. Kerns, E. H., Di, L. (2008) Drug-like Properties: Concepts, Structure Design and Methods. Elsevier, Academic Press, San Diego, California, USA.
18. Gleeson, M. P., Hersey, A., Hannongbua, S. (2011) In-silico ADME models: A general assessment of their utility in drug discovery applications. *Current Topics in Medicinal Chemistry*, *11*, 358–381.

3

BASIC PHARMACOKINETIC PROPERTIES

There are several useful monographs [1–3] about the topics of absorption, distribution, metabolism and elimination; this has been mainly referred to as *ADME* characteristics, pharmacokinetic (*PK*), and pharmacodynamic (*PD*) parameters that are measured *in vivo* during the drug discovery process. When a drug molecule is absorbed, it gets into the blood circulation and distributes into various organs and tissues. Basic PK measurements include the concentration determination of the drug in plasma at various time intervals. Drug plasma concentration decreases with time as the body eliminates the drug molecule via metabolism and clearance. PD studies reveal the relationship between the free concentration of the drug at the site of action, causing a certain level of receptor occupancy, and it describes how it affects the pharmacological efficacy. It is very important to know where the drug molecules partition, how they are transformed in the body, and to determine the metabolites and the rate of elimination.

DMPK (distribution, metabolism, and pharmacokinetic) scientists carry out these measurements. In this chapter, the definition of the basic PK terms and their measurements are described in order to understand how these processes depend on the physicochemical properties of the drug molecules. These measurements are based on administering the compound *in vivo* using oral or intravenous routes and measuring the concentration of the compound in the plasma over time. From the plasma concentration–time profile, important PK behavior of the compound

Physicochemical and Biomimetic Properties in Drug Discovery: Chromatographic Techniques for Lead Optimization, First Edition. Klara Valko.

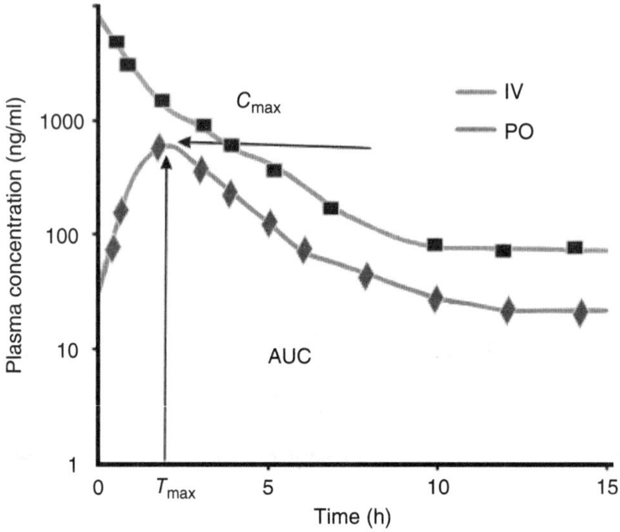

Figure 3.1 Typical plasma concentrations–time profile after oral (PO) and intravenous (IV) administration. (c_{max} is the maximum plasma concentration at T_{max} time point).

can be calculated, such as percentage absorption, elimination rate, volume of distribution, and half-life. Figure 3.1 shows a typical plasma concentration versus time profile, where plasma concentration is plotted against the time from the point of administration. The maximum plasma concentration (c) is reached at t_{max} time after the administration. The half-life ($t_{1/2}$) of the drug is the time when the maximum plasma concentration is halved. It depends on the absorption and disposition rates as well as clearance. The absorption, distribution, and elimination processes are governed by the laws of thermodynamic equilibrium processes, such as passive diffusion, partition, and binding equilibrium constants. According to the most simplistic model, the body can be divided into two compartments, the moving plasma compartment and the stationary tissue compartment in which the compound distributes. The drug molecule interacts and binds to various specific and nonspecific binding sites, most often by a reversible equilibrium process.

The metabolizing enzymes turn over and eliminate the compounds via enzymatic reactions. Therefore, these processes can be described on the basis of physicochemical principles. Figure 3.2 shows the simplistic view of the fate of drug molecules *in vivo*. The body can be viewed as primarily composed of a series of membrane barriers dividing a series of compartments that contain proteins and nucleic acids. Rate processes, such as absorption, depend on the ability of the molecule to cross the membrane. Drug distribution equilibrium depends also on the affinity of the drug to membrane as well as to proteins. There are specific active transport processes as well that result in the distortion of the thermodynamic equilibrium processes, which requires energy investment from various biochemical processes.

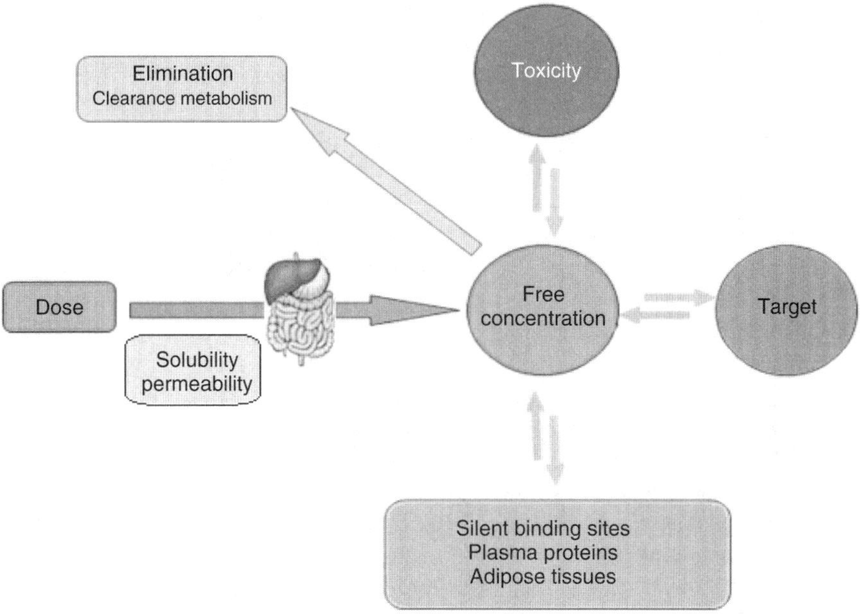

Figure 3.2 A simplistic physicochemical basis of the distribution of drugs *in vivo*.

The drug molecule concentration may vary in certain compartments because of permeability barriers or active transport/efflux. In spite of these active biological processes, it is always important to know what to expect from a molecule based only on the physicochemical principle. It helps to identify the active processes when the expected and measured *in vivo* behavior shows large discrepancies. In this chapter, we review the ADME properties that are measured during the lead optimization processes and that can be modeled or estimated on the basis of physicochemical properties of the molecules. Therefore, the description of the parameters is simplified. Scientists who are engaged in the field of ADME and PK–PD studies are encouraged to read the relevant suggested monographs.

ABSORPTION

There are several ways by which a drug can be administered, such as intravenous, subcutaneous, topical, inhaled, intranasal, and most commonly oral routes. When we speak about absorption, normally we are considering oral drug administration. Intestinal absorption is the process that a drug molecule undergoes following oral administration to its appearance in the bloodstream. It covers the process that a drug molecule has to go through, starting from dissolution in the gastrointestinal tract and subsequent permeation through the gastrointestinal wall. It is usually expressed by the percentage or fraction of the dose that can be found in the bloodstream. However, it should not be used interchangeably with oral bioavailability.

Bioavailability means the proportion of the dose of drug that is able to reach the target *in vivo* to exert the pharmacological activity. It is considered as 100% when administering the drug via intravenous route. Oral bioavailability is usually less than 100%, as a fraction of the dose can be lost during the absorption of the drug from the intestine and by the first-pass metabolism in the liver. Metabolism may start in the gut as metabolizing enzymes can be found there. Thus, bioavailability can be equal to or less than the fraction absorbed depending on the extent of metabolism and loss of the drug during the absorption process. Figure 3.3 shows a schematic representation of the processes involved in the oral bioavailability in the small intestine. We can investigate the underlying physicochemical processes of absorption. Suppose the drug molecule is administered in some sort of a solid form. It has to go into solution first to be able to cross the intestinal wall; this is determined by the solubility and the dissolution rate. From the stomach to the large intestine, the pH of the gut lumen can change from very acidic (pH 1.5) to alkaline (pH 10). It means that ionizable molecules can go through various ionization states. We have to keep in mind that the ionized species are more soluble than the nonionized species. Therefore, acidic molecules that lose ionization at low pH are less soluble in the stomach than basic compounds that get ionized. There are several other molecules, micelles, and enzymes in the intestinal lumen that may help solubilization. The most important micelles are taurocholate and phosphatidyl choline micelles. Lipophilic drug molecules can partition into the micelles; this usually accelerates the dissolution rate. When the drug molecule is in solution, there is a chance that it will go through the intestinal wall. The passive absorption process is governed by the rate of permeation, which follows Fick's law of diffusion that describes the flux of molecules

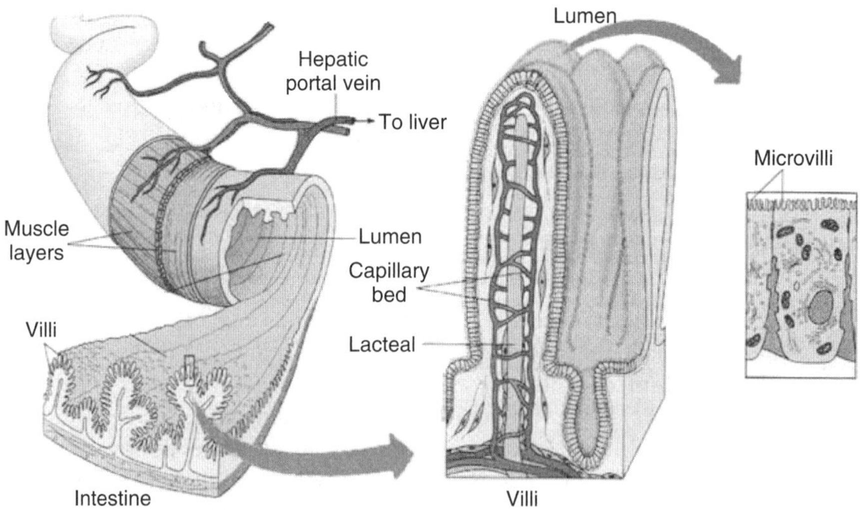

Figure 3.3 The schematic representation of the gut where oral absorption of drug molecules takes place.

from higher concentration to lower concentration. The major physicochemical driving force of absorption is the concentration difference between the intestinal lumen and the other side of the intestinal wall. The intestinal wall is densely networked with blood vessels. The continuously moving plasma reduces the concentration of the compounds and drives the permeability of compounds through the intestinal wall. Small molecules can go through the paracellular path. These small passages are between the intestinal cells. Larger molecules have to diffuse through the cells, which means they have to go through several phospholipid bilayers.

There are also active transport processes that carry molecules through the intestinal wall at a higher rate than would be estimated from passive diffusion alone. There are active efflux processes as well, which protect the body from the absorption of potentially harmful exogenous compounds. There are metabolizing enzymes that break down large nutrient molecules to aid absorption, and they can metabolize the drug molecules also. Besides the physicochemical aspects of absorption, there are several other important physiological factors. For example, the effect of food intake influences the pH, the amount of metabolizing enzymes, and the motility of the gut. There are large individual variations in the size of the absorption surface in the gut, depending on age, gender, disease state, etc. The usual way to determine absorption during the drug discovery and development is to measure the plasma concentration after oral and intravenous administration and compare the two. The first-pass metabolism in the liver can be determined in a separate *in vitro* experiment using liver microsomes or cytochrome P450 enzyme assay. The absorption can be estimated from the bioavailability and first-pass metabolism that occur in the liver. Compounds that go through the intestinal wall are first taken to the liver via the portal vein. The liver protects the systemic circulation from harmful exogenous material that might be absorbed and metabolizes the compounds to less harmful (most often more polar) compounds that can easily be eliminated either back to the gut or via the kidney. When the first-pass metabolism is significant, the bioavailability of the compound can be significantly less than the absorption. At early drug discovery, it is easier to measure bioavailability than absorption. Absorption can be measured by *in situ* intestinal perfusion in rodents. It is not always successful to estimate the human absorption and bioavailability from rodents. For example, Cao et al. [4] have found no correlation between rat and human bioavailability. The discrepancy is mainly due to the significant differences of metabolizing enzymes.

The bioavailability can also be determined from the AUC (area under the curve) values, obtained by plotting the time and the plasma concentration and the dose. Bioavailability of the drug can be calculated by comparing the AUC values after oral and intravenous administrations and is usually expressed as a percentage ($F\%$). As the absorption process involves complex physicochemical and biological processes, it is difficult to predict solely from the chemical structure [5]. One of the most comprehensive descriptions of the physicochemical basis of absorption is described by Avdeef [6].

PLASMA PROTEIN BINDING

The absorbed drug molecules get into the general circulation, which helps to distribute the compound in the body. The blood can be divided into the plasma and the cellular components (red blood cells and white blood cells). The plasma contains proteins, small molecules, and inorganic ions. Approximately 60% of the plasma protein is serum albumin, 3% is α-1-acidglycoprotein, and the rest is immunoglobulin. The plasma is not a clear solution; moreover, owing to the presence of proteins, we can consider the plasma as a colloid solution containing two aqueous compartments similar to aqueous biphasic systems. The small molecules including the absorbed drugs can partition between the clear aqueous and the protein phase. Some proteins in plasma such as albumin and the α-1-acidglycoprotein bind drug molecules strongly. Albumin is considered a transport protein as albumin binding accelerates the transport of small molecules such as short-chain fatty acids. There are a few other proteins that are considered as transport proteins and bind specific small molecules very strongly, such as vitamin D-binding protein, cortisol-binding globulin, and sex-hormone-binding protein. However, drug molecules that bind to proteins are not available as "free" to bind to the biological target. It is also well known that repeating some biological assays in plasma or blood usually shifts the measured potency (pIC_{50}) values, that is, we need to have higher drug concentration in plasma to be able to achieve the same level of biological activity as without plasma. This is the so-called shift assay. Therefore, it is common to measure the fraction of unbound drug in plasma. The total concentration of the drug in plasma (c_{total}) is calculated from the amount of drug added to 1 ml of plasma and from the measured free concentration (c_{free}). After equilibration, the proteins together with the bound drug are separated from the aqueous compartment, and the free concentration of the drug is determined using an analytical technique, such as high performance liquid chromatography (HPLC). The drug binding is calculated by dividing the free concentration of the drug by the total concentration added to the plasma before equilibration and is expressed as an unbound fraction (f_u) according to Equation 3.1.

$$f_u = \frac{c_{free}}{c_{total}} \tag{3.1}$$

The unbound fraction (f_u) is necessary for the calculation of the unbound volume of distribution; we can calculate the unbound AUC and the unbound c_{max}. Some authors express the plasma protein binding as percentage bound (%PPB), which can be calculated from f_u using Equation 3.2.

$$\%PPB = 100 \times (1 - f_u) \tag{3.2}$$

There are several opinions as to when and how to use this information in the drug discovery process [7–9]. We need to understand on a physicochemical basis what the unbound fraction means. If we want to use the plasma protein binding to estimate the free concentration of the drug in plasma, we need to know the total amount of drug. It is also useful to express the fraction unbound (f_u) as a binding constant, assuming one-to-one binding of a compound to a protein. Equation 3.3 shows the expression of the binding affinity constant from the free drug and protein, such as human serum albumin (HSA).

$$K_{aff} = \frac{[\text{DrugHSA}]}{[\text{Drug}][\text{HSA}]} \tag{3.3}$$

where [DrugHSA] is the concentration of the drug and HSA complex, and [Drug] and [HSA] are the free drug and HSA concentrations, respectively. The equation represents the situation when the binding is one to one. The real situation is usually much more complex. First of all, the albumin concentration in plasma is much higher than the drug concentration; so, the HSA concentration can be considered constant; secondly, many drugs bind nonspecifically to several plasma proteins depending on their lipophilicity. Therefore, it is more accurate to express the binding strength as a partition coefficient, which means the concentration ratio of the drug in the free and protein phases is as described by Equation 3.4.

$$K = \frac{[\text{Drug bound}]}{[\text{Drug free}]} \tag{3.4}$$

Equation 3.4 shows that the K is constant regardless of the total amount of drug. This condition also applies to the unbound fraction expression, showing that it is considered as constant. When compounds bind to saturable binding sites in plasma, it may happen that the measured unbound fraction varies with the concentration of the drug.

When we describe the plasma-protein-binding phenomenon as the drug partitioning between the free and protein phases, we need to know the volume of the protein phase instead of the concentration of the protein in plasma. In practice, it is difficult to measure or estimate the volume of the free and protein phases, as plasma is a colloid solution when both phases contain water. However, it should be proportional to the total plasma protein concentration. The total plasma protein concentration in humans should be the same unless certain disease conditions changes this significantly in individuals.

The relationship between the two types of measures of the plasma protein binding is described by Equation 3.5 and Equation 3.6.

$$K = \frac{1 - f_u}{f_u} - \log\ [\text{HSA}] \tag{3.5}$$

$$K = \frac{\%\text{HSA bound}}{100 - \%\text{HSA bound}} \tag{3.6}$$

The logarithmic value of the partition coefficient or the binding constant is used most often as it shows a linear relationship to the free energy of binding. When we plot the logarithmic value of the plasma protein (HSA) binding constants as a function of the percentage binding, the relationship is sigmoid as shown in Figure 3.4.

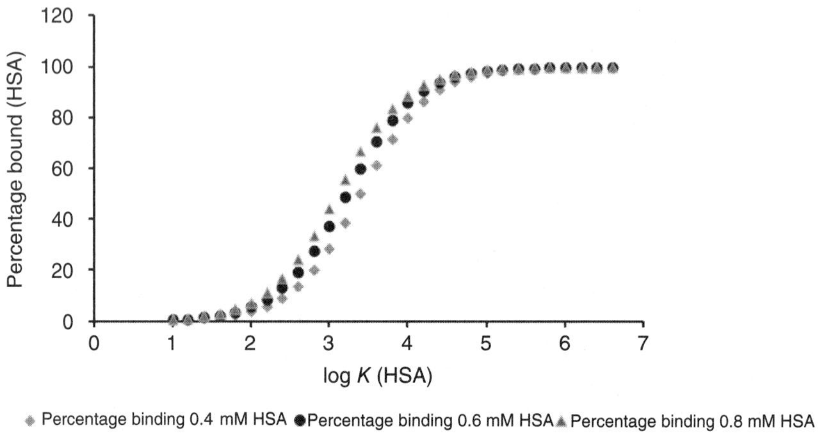

Figure 3.4 The sigmoidal relationship between the %HSA binding and the logarithmic value of the HSA binding constant (*K*) expressed as logarithmic partition coefficient at various HSA concentrations that can be found in human plasma.

Figure 3.4 demonstrates two important aspects of using the different measures of plasma protein or HSA binding. First, we can see that smaller changes in the percentage binding at the higher end represent larger changes in the log *K* values. Second, the percentage binding depends on the amount of protein present in the plasma in which the measurements are made. Therefore, it is more appropriate to use log *K* values when we want to relate this property to the chemical structure of the compound.

When the unbound fraction (f_u) or %PPB depends on the dose (i.e., the amount of drug in plasma), then we speak about dose-dependent plasma protein binding. Dose-dependent plasma protein binding may result in nonlinear PK. A possible explanation for this phenomenon could be that the drug molecule specifically binds to one particular binding site of one particular plasma protein. In these situations, Equation 3.3 should be applied. When two drugs are administered simultaneously and they bind to the same albumin binding site, they can cause potential drug–drug interactions.

The stronger binding drug replaces the weaker binder from the albumin binding site and thus may increase the free concentration of the weaker binder, which may reach toxic levels. The most well-known example is warfarin (anticoagulant) administered together with ibuprofen or other nonsteroid anti-inflammatory drugs that bind to the same binding site of albumin.

DISTRIBUTION

After absorption, the compound reaches the blood and distributes in the body. From the moving bloodstream, it partitions into the tissues and the clearance process also starts in the liver or kidneys. Various PK models can describe the distribution of the drug as a function of time. The body can be considered as consisting of several compartments. In the simplest example, only two compartments are considered; the plasma compartment and the tissue compartment. Sometimes the PK profile is more complex and to fit the plasma concentration versus time curves, we need more complicated equations that describe the body as a multicompartment system. If a compound distributes more slowly to tissues that have low blood supply such as fat, muscle, or skin, then multicompartment models have to be applied to fit the curve obtained from the plasma concentration–time profile. These PK models can help understand the drug distribution, calculate various PK parameters, such as half-life and apparent volume of distribution.

VOLUME OF DISTRIBUTION

There are several experimental approaches that are used to characterize the *in vivo* distribution of drug molecules and used to describe their PK and PD behaviors. The simplest model divides the body into two compartments, the plasma compartment and the tissue compartment. The volume of distribution (V_d) was used as a hypothetical volume that reveals how extensive the compound distribution is into tissues from the plasma. It can be defined in two different ways and can be obtained from the given intravenous dose and the plasma concentration (c_{plasma}), as described in Equation 3.7.

$$V_d = \frac{\text{Dose}}{c_{plasma}} \tag{3.7}$$

As the plasma concentration is continuously changing with time as the compound distributes and then starts to eliminate, the so-called steady-state volume of distribution (V_{ss}) is introduced, which considers a repeated dose that results in a steady-state plasma concentration. Using the steady-state volume of distribution and knowing the volume of the plasma compartment (V_p), we can rearrange Equation 3.7 to Equation 3.8.

$$V_{ss} = \frac{\text{Dose} \times V_p}{A_{plasma}} \tag{3.8}$$

In Equation 3.8, A_{plasma} is the amount of drug in the plasma compartment. We can define an equilibrium constant ($K_{tissue/plasma}$) for the drug partitioning between the tissue and the plasma compartments as the tissue concentration is divided by the plasma concentration, assuming that the drug distribution between these compartments are independent from the dose as shown by Equation 3.9.

$$K_{tissue/plasma} = \frac{\text{Dose} - A_p/V_t}{A_p/V_p} \tag{3.9}$$

where V_t is the volume of the tissue compartment. It can be supposed that the amount of drug in the tissue compartment is the difference between the dose and the amount of drug in the plasma compartment. Combining Equation 3.8 and Equation 3.9, Equation 3.10 is obtained, which expresses the drug's partition coefficient between the tissue and plasma compartments as a function of the steady-state volume of distribution.

$$K = \frac{V_{ss} - V_p}{V_t} \tag{3.10}$$

Equation 3.10 shows that the *in vivo* steady-state volume of distribution is proportional to the drug's partition coefficient between the tissue and the plasma compartments. It means that although the volume of distribution is usually given as volume, often expressed as liter per kilogram body weight, it can easily be converted to a dimensionless partition coefficient when we include the average plasma and tissue volume of a human (or animal). The schematic illustration of the drug's partitioning *in vivo* is shown in Figure 3.5.

We talk of apparent volume of distribution when the plasma concentration is taken as a hypothetical average concentration. It cannot be lower than the plasma volume, which is 3 l for an average human weighing 70 kg, that is, 0.04 l/kg. It means that the total amount of drug (dose) is found in the plasma; so, the drug did not leave the plasma compartment. If the drug distributes to the tissue and only a minimal amount can be found in the plasma the volume of distribution can be very high, above 20 l.

When we speak of the steady-state volume of distribution (Vd_{ss}), we put an average plasma concentration into the equation, which is obtained after continuous infusion of a drug when the rate of infusion is equivalent to the rate of elimination. So, we achieve a quasi-equilibrium distribution of the drug between the plasma and the tissue compartments. The concept of the steady-state volume of distribution by Benet et al. [10, 11] has been introduced to have a volume parameter independent of the clearance of the compound. The steady-state volume of distribution can be regarded as the product term of clearance and mean residence time (MRT).

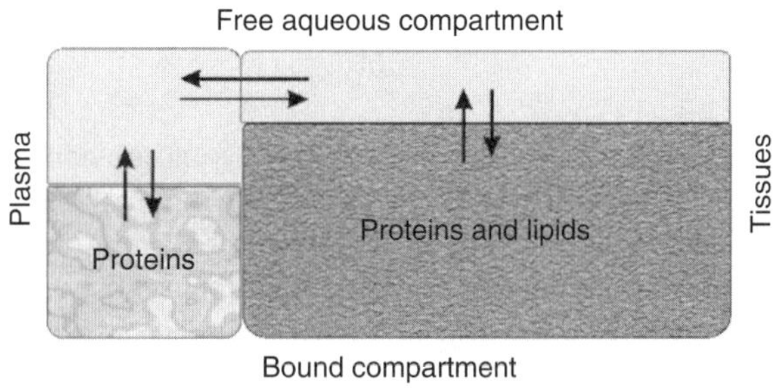

Figure 3.5 Schematic illustration of the drug's partitioning *in vivo* between plasma and tissue compartments as well as the free and bound compartments in plasma and tissues.

It can be derived that the volume of distribution is basically proportional to the drug distribution coefficient between the plasma and the tissue compartments (Eq. 3.10). The more the drug distributes into the tissues, the larger the volume of distribution becomes. Depending on where the target enzyme or receptor is, various drug molecules may have a wide range of volume distribution. For certain therapeutic areas, the volume of distribution should be low (target is in the plasma compartment), and for others, it should be high (the target is in one of the tissue compartments, such as in the brain) depending on where the target is located in the body.

The steady-state volume of distribution can also be expressed as the product term of systemic clearance and MRT. It is mathematically equivalent to the equations traditionally used, describing the relation of apparent volume of distribution (V_d) to systemic clearance (Cl_s) and half-life ($t_{1/2}$) using a simple one-compartment model shown in the following equation.

$$V_d = \frac{Cl_s}{k} \quad \text{where } k = \frac{0.693}{t_{1/2}} \quad \text{so } V_d = Cl_s \times \frac{t_{1/2}}{0.693} \tag{3.11}$$

This means that the rate of elimination determines how frequently we need to administer the given dose to achieve the steady-state plasma concentration and consequently reach the steady-state volume of distribution. When the clearance is high, the half-life or MRT is short, when the clearance is low, the half-life or MRT is longer. The product term of the two parameters is proportional to the volume distribution. However, the steady-state volume of distribution will depend only on the compound distribution between the plasma and tissue compartments. Figure 3.6 demonstrates the situation for two hypothetical molecules that have different clearance and half-life but the same steady-state volume of distribution.

Figure 3.7 and Figure 3.8 show the plot of logarithmic values of steady-state volume distribution of known drug molecules against the clearance and half-life, respectively. The literature data [12] of the volume of distribution, half-life, and clearance are shown in Table 3.1. It can be seen that neither the logarithmic value of

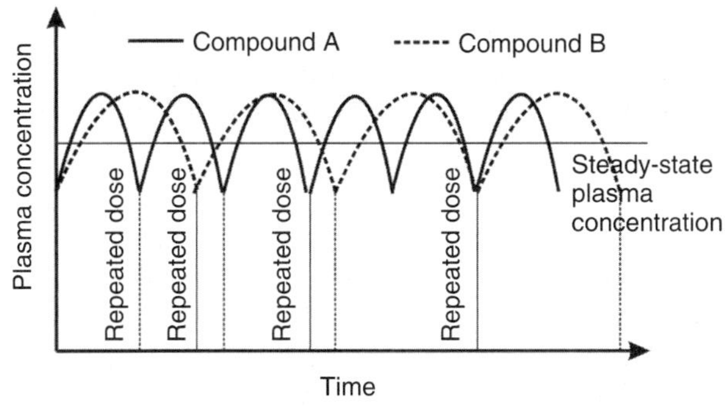

Figure 3.6 Two hypothetical drug molecules having the same steady-state volume of distribution but different clearance and half-lives.

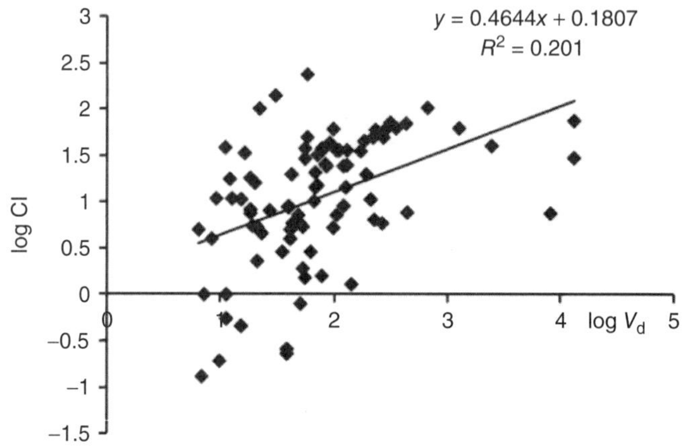

Figure 3.7 The plot of logarithmic value of steady-state volume of distribution (log Vd_{ss}) and logarithmic value of systemic clearance (log Cl) of known drug molecules (poor correlation).

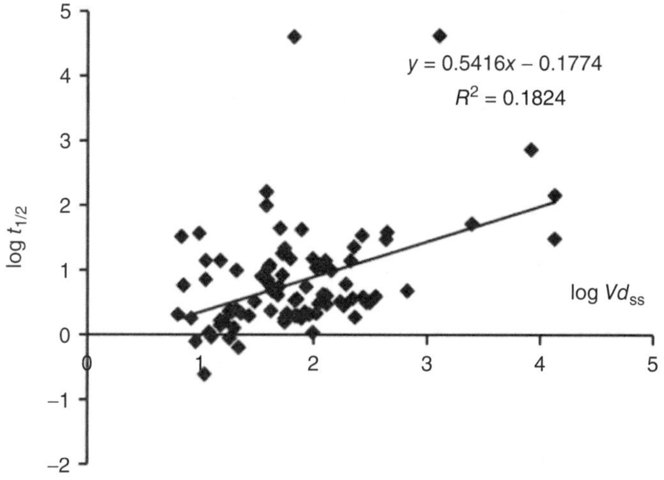

Figure 3.8 The plot of logarithmic value of steady-state volume of distribution (log Vd_{ss}) and logarithmic value of half-life (log $t_{1/2}$) of known drug molecules (poor correlation).

the clearance nor the logarithmic value of the half-life on its own shows correlation to the logarithmic value of the volume distribution. However, the product term of the two or the sum of the logarithmic value of clearance and half-life shows excellent correlation to the logarithmic value of the steady-state volume of distribution, as is shown in Figure 3.9. The fit is not perfect as it is based on *in vivo* measured human clinical data that shows individual variations and measurement error, but the trend is clear.

Table 3.1 The literature data of human volume of distribution, clearance, and half-life of diverse set of known drugs obtained from various databases

Drug	Volume of Distribution (l/70 kg)	log Vd_{ss}	Clearance (l/h/70 kg)	log Cl	Half-Life (h) $t_{1/2}$	log $t_{1/2}$
Acetaminophen	67	1.83	21	1.32	2	0.3
Acyclovir	42	1.62	19.8	1.3	2.4	0.38
Amikacin	19	1.28	5.5	0.74	2.3	0.36
Amoxicillin	15	1.18	10.8	1.03	1.7	0.23
Amphotericin	53	1.72	1.9	0.28	18	1.26
Ampicillin	20	1.3	16.2	1.21	1.3	0.11
Aspirin	11	1.04	39	1.59	0.25	−0.6
Atropine	120	2.08	24.6	1.39	4.3	0.63
Azathioprine	57.4	1.76	239.4	2.38	0.16	−0.8
Bumetanide	9.1	0.96	10.9	1.04	0.8	−0.1
Caffeine	43	1.63	5.9	0.77	4.9	0.69
Captopril	57	1.76	50.4	1.7	2.2	0.34
Carbamazepine	98	1.99	5.3	0.72	15	1.18
Cefazolin	8.4	0.92	4	0.6	1.8	0.26
Cefoperazone	6.3	0.8	5	0.7	2.1	0.32
Cefoxitin	21.7	1.34	103	2.01	0.65	−0.19
Ceftazidime	16.1	1.21	34	1.53	1.6	0.2
Ceftriaxone	11.2	1.05	1	0	7.3	0.86
Cephalexin	18	1.26	18	1.26	0.9	−0.05
Chlordiazepoxide	21	1.32	2.3	0.36	10	1
Chloroquine	8050	3.91	7.6	0.88	738	2.87
Chlorpropamide	6.8	0.83	0.13	−0.89	33	1.52
Cimetidine	70	1.85	32.4	1.51	2	0.3
Ciprofloxacin	130	2.11	25.2	1.4	4.1	0.61
Clonazepam	224	2.35	6.5	0.81	23	1.36
Codeine	182	2.26	46.2	1.66	2.9	0.46
Cyclosporine	85	1.93	24.6	1.39	5.6	0.75
Diazepam	77	1.89	1.6	0.2	43	1.63
Diclofenac	11.9	1.08	17.6	1.25	1.1	0.04
Digitoxin	38	1.58	0.23	−0.64	161	2.21
Digoxin	440	2.64	7.8	0.89	39	1.59
Diltiazem	220	2.34	50.4	1.7	3.7	0.57
Disopyramide	41	1.61	5	0.7	6	0.78
Doxorubicin	425	2.63	71.4	1.85	30	1.48
Enalapril	40	1.6	9	0.95	11	1.04
Erythromycin	55	1.74	38.2	1.58	1.6	0.2
Ethambutol	110	2.04	36	1.56	3.1	0.49
Ethosuximide	50	1.7	0.8	−0.1	45	1.65
Flucytosine	48	1.68	7.2	0.86	4.2	0.62
Fluoxetine	2450	3.39	40.3	1.61	53	1.72
Foscamet	30.1	1.48	140	2.15	3.3	0.52
Gentamicin	21	1.32	5.4	0.73	2.5	0.4

(continued)

Table 3.1 (*Continued*)

Drug	Volume of Distribution (l/70 kg)	log Vd_{ss}	Clearance (l/h/70 kg)	log Cl	Half-Life (h) $t_{1/2}$	log $t_{1/2}$
Indomethacin	18	1.26	8.4	0.92	2.4	0.38
Ketoconazole	168	2.23	35.3	1.55	3.3	0.52
Labetalol	660	2.82	105	2.02	4.9	0.69
Lidocaine	77	1.89	38.4	1.58	1.8	0.26
Lithium	55	1.74	1.5	0.18	22	1.34
Meperidine	310	2.49	72	1.86	3.2	0.51
Methadone	266	2.42	5.9	0.77	35	1.54
Methotrexate	39	1.59	9	0.95	7.2	0.86
Methylprednisolone	84	1.92	26	1.41	2.3	0.36
Metoprolol	290	2.46	63	1.8	3.2	0.51
Metronidazole	52	1.72	5.4	0.73	8.5	0.93
Morphine	230	2.36	60	1.78	1.9	0.28
Naproxen	11.2	1.05	0.55	−0.26	14	1.15
Nifedipine	55	1.74	29.4	1.47	1.8	0.26
Nortriptyline	13,300	4.12	30	1.48	31	1.49
Pancuronium	18.2	1.26	7.6	0.88	2.3	0.36
Pentamidine	13,300	4.12	75.6	1.88	144	2.16
Phenobarbital	38	1.58	0.26	−0.59	99	2
Piperacillin	12.6	1.1	10.9	1.04	0.93	−0.03
Prednisolone	105	2.02	36.5	1.56	2.2	0.34
Prednisone	70	1.85	15.1	1.18	3.6	0.56
Primidone	41	1.61	4	0.6	12	1.08
Procainamide	130	2.11	36	1.56	3	0.48
Propranolol	270	2.43	50.4	1.7	3.9	0.59
Pyridostigmine	77	1.89	36	1.56	1.9	0.28
Quinidine	190	2.28	19.8	1.3	6.2	0.79
Ranitidine	91	1.96	43.8	1.64	2.1	0.32
Rifampin	68	1.83	14.4	1.16	3.5	0.54
Sulfamethoxazole	141	2.15	1.3	0.11	10.1	1
Tacrolimus	62	1.79	2.9	0.46	15	1.18
Terbutaline	125	2.1	14.4	1.16	14	1.15
Tetracycline	105	2.02	7.2	0.86	11	1.04
Theophyline	35	1.54	2.9	0.46	8.1	0.91
Tobramycin	23.1	1.36	4.6	0.66	2.2	0.34
Tocainide	210	2.32	10.8	1.03	14	1.15
Tolbutamide	7	0.85	1	0	5.9	0.77
Trimethoprim	119	2.08	9.2	0.96	11	1.04
Tubocurarine	27	1.43	8.1	0.91	2	0.3
Valproic acid	15	1.18	0.46	−0.34	14	1.15
Vancomycin	49	1.69	5.9	0.77	5.6	0.75
Verapamil	350	2.54	63	1.8	4	0.6
Warfarin	9.8	0.99	0.19	−0.72	37	1.57
Zidovudine	98	1.99	61.8	1.79	1.1	0.04

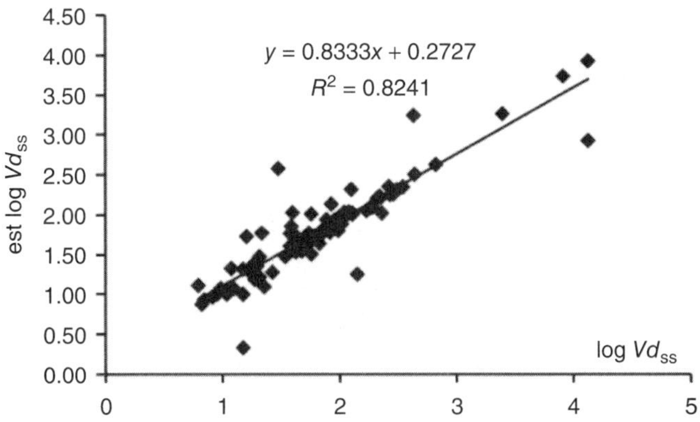

Figure 3.9 The plot of the logarithmic value of steady-state volume of distribution (log Vd_{ss}) and the sum of the logarithmic value of systemic clearance (log Cl) and half-life (log $t_{1/2}$) (significant correlation).

UNBOUND VOLUME OF DISTRIBUTION

The unbound volume of distribution is a relatively new concept. It can be derived from the volume of distribution by taking into account the plasma protein binding. The unbound volume of distribution (Vd_u) can be expressed by Equation 3.12.

$$Vd_u = \frac{\text{Dose}}{c_{\text{plasma free}}} \tag{3.12}$$

The difference with Vd_{ss} is that instead of the plasma concentration we use the free plasma concentration in the equation. Therefore, (Vd_u) can also be calculated from Vd_{ss} when the unbound fraction of the drug (f_u) is known in plasma by Equation 3.13.

$$Vd_u = \frac{V_{ss}}{f_u} \tag{3.13}$$

We shall see later in Chapter 11 when we discuss the drug efficiency concept how important the unbound volume distribution is for ranking research compounds for further studies. In general, a low unbound volume of distribution is desirable as it means we have higher free concentration in plasma from a lower dosage. This is advantageous as we do not have to administer a large amount of compound that distributes everywhere in the body with possible side effects or uses the natural metabolism of the body to clear. If we have high free plasma concentration, we can assume high free concentration of the drug near the active site, which results in higher receptor occupancy.

HALF-LIFE

The half-life ($t_{1/2}$) is the time when the drug concentration drops to half of the maximum concentration in the plasma after one single dose and it is estimated from the drug plasma concentration–time profile. The half-life can be short because of the high rate of clearance/metabolism. Sometimes it is advantageous to have a short half-life (short-term narcotic drugs). When the drug needs to be taken for long periods of time, a longer half-life is advantageous, as the patient has to take the drug less frequently to maintain the therapeutic drug level in the plasma. As we have discussed earlier, the half-life and the clearance together can be used to calculate the steady-state volume of distribution. The half-life ($t_{1/2}$) depends on the elimination rate constant (k) according to Equation 3.14.

$$t_{1/2} = \frac{0.693}{k} \quad (3.14)$$

Equation 3.14 is held only when we suppose the first-order elimination rate constant. When we consider linear PKs, we suppose that this parameter is independent of the dose, and so is the steady-state volume of distribution, plasma protein binding, etc.

METABOLISM AND CLEARANCE

Clearance is a measure of the ability of the body or an organ to eliminate a drug molecule from the systemic circulation. Systemic clearance (Cl) relates to the whole body, while hepatic or renal clearance refers to the particular organ—liver or kidney. The clearance shows the elimination rate of the drug from the entire body relative to the drug concentration in plasma. Clearance is usually measured by the apparent volume of a reference fluid (usually the plasma or blood) that is cleared of drug during a unit time. So, it means the volume of blood or plasma that is cleared by the drug in 1 min. Similar to the volume of distribution, clearance is also related to the body mass and expressed as milliliter per minute per kilogram. Clearance can be calculated from the dose and the AUC values obtained from the plasma concentration–time profile. The organ clearance of the drug is dependent on the blood flow rate in the clearing organ (liver or kidney), and it also depends on the free plasma concentration. Clearance cannot be higher than the blood flow. During clearance, several metabolizing enzymes make chemical modification/transformation of the drug molecules. Normally, the metabolite is more polar and more water soluble than the original drug. The major metabolizing enzymes in the liver are the cytochrome P450 family. This family of enzymes contains several identified species such as 2C9, 3A, and 2C9D. These enzymes have slightly different active sites and they carry out different modifications. The most typical metabolic transformation reactions are N-demethylation and para-hydroxylation of phenyl rings. Some of the enzymes perform conjugation of the drug molecules with endogenous polar molecules, such as glucuronic acid and glutathione. Although the metabolic transformations are specific reactions, the P450 enzyme family has similar attractions toward lipophilic, basic drugs. We shall discuss the relationships between basic physicochemical properties and metabolism in more detail and show that some basic physicochemical properties correlate well with the rate of metabolic transformations in spite of the chemical specificity of the reaction. In this part of the book, the kinetic aspects of clearance are not discussed, the principles are simplified, and the detailed mathematical formulae are avoided in order to get a basic understanding of how and why these PK parameters can be related to physicochemical properties.

FREE DRUG HYPOTHESIS

Brodie [13] describes the concept that only the free drug is available for interaction with the pharmacological target and produces the desired pharmacological effect. The free drug hypothesis also means that the free concentration of the drug in all compartments is the same. Thus, the free concentration of the drug in plasma is the same as the free concentration of the drug in the tissue when the steady-state equilibrium is reached. This hypothesis is true when no active transport process or permeability barrier exists, which works against the thermodynamic equilibrium process in steady state. When we measure various physicochemical properties of molecules *in vitro* using directly biological material such as plasma, enzyme, receptor, tissue homogenate, phospholipid membrane, or simple octanol/water partition and use the data for extrapolation to the *in vivo* situation, we ignore the active biological transport processes. We use the free drug hypothesis to estimate the free concentration of the drug at the site of action in two ways. From the *in vitro* potency data, we know the desired concentration of the compound needed to produce a certain level of biological response. Knowing the percentage absorption, plasma protein binding, and tissue binding, we can estimate the required dose. The other way of applying the concept is to estimate the free concentration at the site of action for a given dose. Smith et al. [8] summarized the exceptions to the free drug hypothesis. The free concentration is not equal on both sides of the membrane when active efflux or influx processes take place, for example, by P-glycoprotein (PgP). The free concentration of the drug at the site of action is not proportional to the pharmacological effect when the drug binds to the receptor irreversibly, or the drug causes biological activation or an inhibition effect that is independent of the occupied proportion of the receptor. Otherwise, the free drug hypothesis is a valid assumption.

SUMMARY

The distribution of drug molecules *in vivo* is governed by basic thermodynamic rules. The body can be considered as containing several compartments separated by a membrane. Each compartment contains a free aqueous phase and another phase containing proteins and phospholipids. The continuously moving bloodstream helps to develop the steady-state equilibrium process. The drug molecule administered orally need to be absorbed from the intestine and reach the general circulation. It partitions in the plasma (plasma protein binding) and in the tissues. When it reaches the liver or the kidney, it is eliminated. The typical way of measuring the drug distribution process is by monitoring the plasma concentration as a function of time starting from the drug administration. Several PK parameters can be derived from the plasma concentration–time profile. The t_{max} is the time needed to reach the maximum plasma concentration, while the half-life ($t_{1/2}$) is the time when the plasma concentration drops to half of the maximum because of systemic clearance. The clearance (Cl) is expressed as the volume of the plasma that is cleared of the drug in a unit time. The AUC of the plasma concentration–time profile is proportional to the drug exposure. Bioavailability, expressed as a $\%F$, can be calculated from the proportion of the AUC after oral administration relative to the intravenous administration. The steady-state volume of distribution (V_{ss}) is defined as the dose divided by the steady-state plasma concentration. It can also be calculated from the product of the clearance and the half-life. It has been demonstrated that the steady-state volume of distribution is proportional to the compound distribution coefficient between the plasma and the tissue compartments. Plasma protein binding, expressed as an unbound fraction (f_u), percentage bound (%PPB), or as a partition coefficient (K) between the free aqueous and the protein phases is an important property of the drug. Using the unbound fraction (f_u) value, we can calculate the unbound volume of distribution (Vd_u) as $Vd_u = V_{ss}/f_u$. It is defined as the dose divided by the free plasma concentration (c_{free}). The AUC can also be expressed as unbound (AUC_u) divided by f_u. The unbound volume of distribution can be used to estimate the dose that produces the desired free plasma concentration for the pharmacological activity based on the free drug hypothesis. On the basis of the free drug hypothesis, we can assume that the free concentration of the drug is the same in all compartments (plasma, tissue, and at the site of action). This assumption is not true when there is a permeability barrier or active transport does not allow the drug to achieve the same equilibrium concentration on both sides of a membrane. The free drug concentration is sometimes not proportional to the pharmacological effect when the drug irreversibly binds to the receptor or induces active series biological processes that are not related to the receptor occupancy of the drug.

QUESTIONS FOR REVIEW

Q3.1 Define the following parameters: fraction absorbed, unbound fraction in plasma, volume of distribution, clearance, half-life, and unbound volume of distribution.

Q3.2 What is AUC, and how it can be measured?

Q3.3 Why is plasma protein binding important?

Q3.4 What is the "free drug hypothesis" and when are these assumptions not true?

REFERENCES

1. Smith, D. A., Van de Waterbeemd, H., Walker, D. K. (2006) Pharmacokinetics and Metabolism in Drug Design. Wiley-VCH, Darmstadt, Germany.
2. Kwon, Y. (2001) Handbook of Essential Pharmacokinetics, Pharmacodynamics and Drug Metabolism for Industrial Scientists. Springer-Verlag, New York, USA.
3. Schoenwald, R. D. (2002) Pharmacokinetics in Drug discovery and Development. CRC Press, Boca Raton, NJ, USA.
4. Cao, X., Gibbs, S. T., Fang, L., Miller, H. A., Landowski, C. P., Shin, H-C., Lennernas, H., Zhong, Y., Amidon, G. L., Yu, L. X., Sun, D. (2006) Why is it challenging to predict intestinal drug absorption and oral bioavailability in human using rat model? *Pharmaceutical Research*, *23*, 1675–1686.
5. Burton, P. S., Goodwin, J. T., Vidmar, T. J., Amore, B. M. (2002) Predicting drug absorption: How nature made it a difficult problem. *Journal of Pharmacology and Experimental Therapeutics*, *303*, 889–895.
6. Avdeef, A. (2003) Absorption and Drug Development. Wiley, Hoboken, NJ, USA.
7. Trainor, G. L. (2007) The importance of plasma protein binding in drug discovery, *Expert Opinion in Drug Discovery*, *2*, 51–64.
8. Smith, D. A., Di, L., Kerns, E. H. (2010) The effect of plasma protein binding on *in vivo* efficacy: Misconceptions in drug discovery. *Nature Reviews Drug Discovery*, *9*, 929–939.
9. Liu, X., Chen, C., Hop, C. E. C. A. (2011) Do we need to optimize plasma protein binding and tissue binding in drug discovery? *Current Topics in Medicinal Chemistry*, *11*, 450–466.
10. Benet, L. Z., Galeazzi, R. L. (1979) Noncompartmental determination of the volume of distribution in steady-state. *Journal of Pharmaceutical Sciences*, *68*, 1071–1074.
11. Benet, L. Z. (2010) Clearance (nèe Rowland) concepts: A downdate and an update. *Pharmacokinetic and Pharmacodynamic*, *37*, 529–539.
12. Obach, R. S., Lombardo, F., Waters, N. J. (2008) Trend analysis of a database of intravenous pharmacokinetic parameters in humans for 670 drug compounds. *Drug Metabolism and Disposition*, *36*, 1385–1405.
13. Brodie, B. B., Kurtz, H., Shanker, L. J. (1960) The importance of dissociation constants in lipid solubility on influencing the passage of drugs into CSF. *Journal of Pharmacology and Experimental Therapy*, *130*, 20–25.

4

PRINCIPLES AND METHODS OF CHROMATOGRAPHY FOR THE APPLICATION OF PROPERTY MEASUREMENTS

In 1905, a Russian botanist Michael Tsvet wrote his thesis about the separation of plant pigments, chlorophylls, in a glass column filled with calcium carbonate. The plant extraction was placed onto the top of the column and he observed separation of components with various colors by pouring a petrol ether and ethanol mixture on it. He coined the name chromatography, which means "color writing," and recognized the potential of the method for separating the components of complex mixtures. It took another 50 years until a Nobel Prize was awarded to Martin and Synge for developing a gas chromatography instrument and establishing the basic theoretical principles describing the chromatographic separation process. Chromatography, as a part of separation science, is a rapidly growing field even today. We can improve the separation by developing new stationary phases (chiral phases for chiral separations, tailor-made phases for biopolymer separation, etc.) or by improving the instrumentation (scale-up or miniaturize), in order to widen the application field. There are still numerous potential applications of chromatography to be discovered. The application of separation science with biochemistry, molecular biology, pharmacology, pharmacognosy, or pharmaceutics increases the possibility of new and significant developments in these disciplines. HPLC has emerged as a computer-controlled liquid chromatographic instrumentation with high performance stationary phases. HPLC is nowadays an important tool in pharmaceutical analysis, quality control procedures, stability investigation, metabolite

Physicochemical and Biomimetic Properties in Drug Discovery: Chromatographic Techniques for Lead Optimization, First Edition. Klara Valko.

research, and drug design. Several research journals, such as *Journal of Chromatography*, *Journal of Liquid Chromatography*, *Journal of Chromatographic Sciences*, and *Chromatographia*, regularly publish new theories and applications of HPLC. There are several books published for beginners in the field, citing only a few here [1–4] and directing the readers if they need more detailed knowledge about the principles of HPLC and descriptions of running and troubleshooting HPLC experiments.

THEORETICAL BACKGROUND OF CHROMATOGRAPHY

Chromatography is a common name for methods for separation of substances by differences in rates of migration that arise from their different distribution ratios between a flowing mobile phase (gas or liquid) and a stationary phase (liquid or solid). HPLC is a controlled chromatographic procedure where the compounds are injected into a chromatographic system consisting of a liquid mobile phase and a solid stationary phase. Compounds are separated as a result of differential migration in the chromatographic system depending on their interactions between the stationary phase and the mobile phase. Components of a mixture that is injected into the system will move with the mobile phase, but the speed of the component zones is different for each type of molecule, depending on their average distribution between the mobile and stationary phases. Figure 4.1 shows the schematic principles of chromatographic separation.

The mobile phase content is monitored at the end of the chromatographic column by a suitable detector, usually UV. The detector signal is recorded as a function of time measured from the injection of the sample, which takes place at the top of the chromatographic column filled with small stationary phase particles. The recorded signal as a function of time is called a *chromatogram*. The typical information content of an HPLC chromatogram is illustrated in Figure 4.2.

A chromatogram may show several peaks depending on the number of components in the mixture injected onto the top of the column. The peak retention time, the time elapsed from the injection until the appearance of the peak maximum, is characteristic of quality of component, while the peak area is proportional to the injected quantity of the component. Thus, a chromatogram provides information about both the quantity and the quality of the mixture. The retention time depends on the chromatographic conditions such as the type and composition of the stationary and mobile phases, the mobile phase flow rate, and column dimensions. Under a set of conditions, the same compound should always show the same characteristic retention time. When a known standard compound is analyzed under the same conditions, the retention time of the standard can be used for identification of an unknown sample. Similarly, for quantification, we need to inject a known amount of a known compound under the same condition to calibrate the detector signal and compare the peak area of a known quantity to a peak area of an unknown quantity of components in the samples.

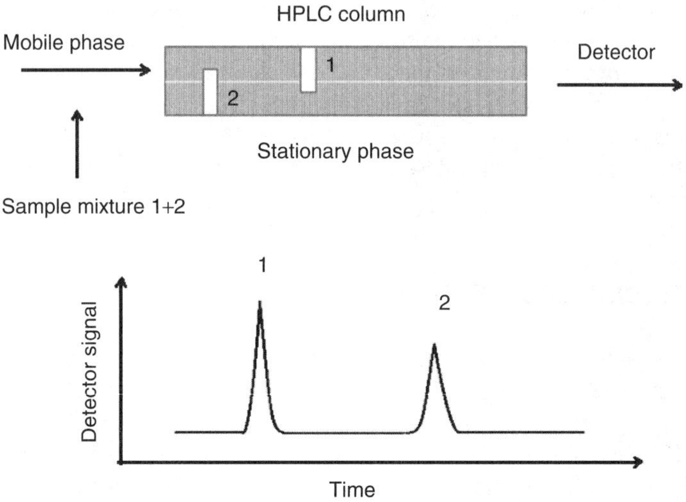

Figure 4.1 The schematic principles of the chromatographic separation process.

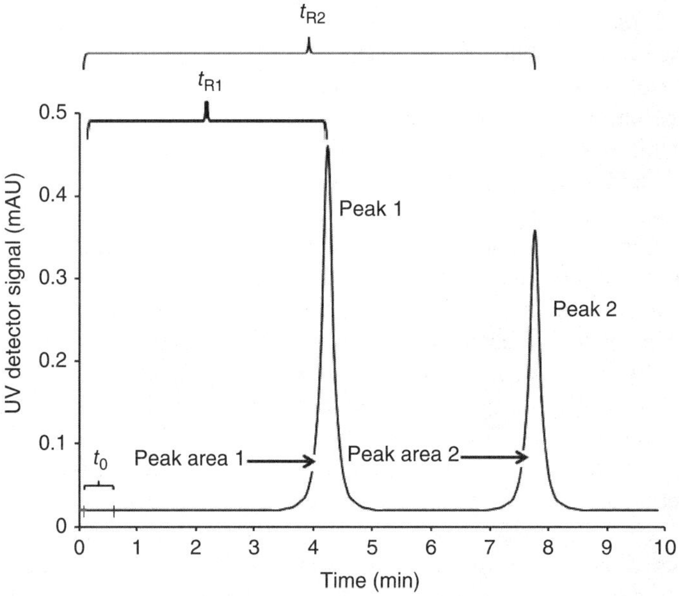

Figure 4.2 Key factors of a chromatogram: t_0 is the dead time, time necessary for a nonretained mobile phase component to travel from the injector to the detector, t_{R1} and t_{R2} are the retention times of components 1 and 2 represented by peaks 1 and 2. The peak areas 1 and 2 are the areas under the peak of the two components. Retention time is characteristic for the type of compound, while the peak area is a measure of the quantity of the compound in the mixture.

Nowadays, typical column dimensions and the particle size of the stationary phase are reduced to 30–50 mm in length with diameter 2.1–3 mm and 2–5 μm spherical particles. The mobile phase flow rate is less than 1 ml/min.

The chromatogram can be characterized by three features that are important in describing the resulting separation. First, each compound leaves the column in the form of more or less symmetrical peaks. The second characteristic feature is the difference in retention times of adjacent peaks, for example, t_{R2} for compound 2 minus t_{R1} for compound 1. The larger this difference, the easier is the separation of the two compounds. Finally, each band is characterized by a peak width, t_w. Tangents are drawn to each side of the peak and extended to touch the baseline (the detector signal for zero sample concentration). The separation efficiency can be regarded as better for the narrower peaks and thus smaller t_w values.

RETENTION FACTOR AND ITS RELATION TO THE DISTRIBUTION CONSTANT BETWEEN THE MOBILE AND THE STATIONARY PHASES

The speed of movement of various components during the chromatographic separation depends on the solute distribution between the mobile and the stationary phases. It means that the larger the proportion of molecules is in the mobile phase, the faster their movement is. If we consider the average velocity of the solvent molecules S within the column (u, cm/s) and some sample component band X as u_x, the speed of the sample band movement can be described by Equation 4.1:

$$u_x = uR \tag{4.1}$$

where R stands for the fraction of sample component present in the mobile phase. If the fraction of the X sample molecule in the mobile phase is 0 ($R = 0$), no migration can occur and u_x is zero. If the fraction of X sample molecules in the moving phase is 1 (i.e., all molecules are in the mobile phase, $R = 1$), then the sample molecules move through the column at the same rate as the solvent molecules and $u_x = u$. Thus, R can be considered as the relative migration rate of the X compound and expressed by the chromatographic retention parameter (k), which in the past was called *capacity ratio*. The k is equal to the proportion of the total moles of X molecule in the stationary phase (n_s) divided by the total mole of the molecules in the mobile phase (n_m); so, $k = n_s/n_m$. We can set up Equation 4.2 from which Equation 4.3 can be derived for describing the relationship between the retention factor and R.

$$k + 1 = \frac{n_s}{n_m} + \frac{n_m}{n_m} = \frac{n_s + n_m}{n_m} \tag{4.2}$$

$$R = \frac{n_m}{n_s + n_m} = \frac{1}{1 + k} \tag{4.3}$$

Combining Equation 4.1 and Equation 4.3, the band velocity can be expressed by the retention factor according to Equation 4.4.

$$u_x = \frac{u}{1+k} \tag{4.4}$$

The quantity of u_x can be related to the retention time t_R and column length (L) according to Equation 4.5.

$$t_R = \frac{L}{u_x} \tag{4.5}$$

The mobile phase velocity can be expressed by the length of the column L divided by the time, which is needed to reach the end of the column for the mobile phase component. This time, t_0 called *dead time*, can be expressed by Equation 4.6.

$$t_0 = \frac{L}{u} \tag{4.6}$$

Eliminating L between these last two equations gives Equation 4.7.

$$t_R = \frac{ut_0}{u_x} \tag{4.7}$$

Considering Equation 4.1 for the relationship of R to u and Equation 4.3 for the relationship of R to k Equation 4.8 can be set up.

$$t_R = t_0(1+k) \tag{4.8}$$

The rearrangement of Equation 4.8 gives the expression for k, which can be easily calculated from the experimentally measured chromatographic retention times.

$$k = \frac{t_R - t_0}{t_0} = \frac{n_s}{n_m} \tag{4.9}$$

From the above-mentioned consideration, it is clear that the velocity of the separated components in the chromatographic system depends on the distribution of the components between the mobile and the stationary phases. The greater the portion of the molecules in the mobile phase is, the faster the movement of that component through the column. The retention factor (k) is the proportion of the average number of molecules in the stationary phase (n_s) divided by the mobile phase (n_m). This means that the higher the retention factor, the greater number of molecules is in the stationary phase, that is, the compound is more retained. The retention factor can be calculated from the measured retention time by relating it to the dead time (t_0), the time needed for the nonretained molecules to go through the column. It can be seen that the retention factor is independent of the column length and the mobile phase velocity (L and u can be eliminated from the equation).

The important column parameter, t_0, can be measured in various ways. In most cases, it is at the center of the first band or after a baseline disturbance (following a sample injection). If there is any doubt about the position of t_0 on the chromatogram, a weaker solvent (or other nonretained compound) can be injected as sample, and its "retention time" will be the dead time.

Retention in liquid chromatography is sometimes measured in volume units (ml), rather than in time units. Thus, the retention volume V_R is the total volume of the mobile phase, which is required to elute at the center of a given band. It means the volume of the solvent that passed through the system from the time of sample injection to the appearance of the band at the detector. The retention volume V_R is equal to the retention time (t_R) multiplied by the flow rate (F), expressed usually in milliliter per minute, according to Equation 4.10.

$$V_R = t_R F \tag{4.10}$$

Similarly, the total volume of the solvent within the column, V_m, or, in other words, void volume is equal to F multiplied by t_0.

$$V_m = t_0 F \tag{4.11}$$

Elimination of F from Equation 4.10 and Equation 4.11 gives Equation 4.12:

$$V_R = V_m \frac{t_R}{t_0} = V_m(1 + k) \tag{4.12}$$

So far, we have ignored the effect of the volume of the stationary phase, V_s, on the retention volume. The quantity of k equals n_s/n_m, according to the previously described definition, where n_s and n_m can be expressed by the mobile phase concentration of the molecules in the stationary phase (X_s) and (X_m) as $n_s = X_s V_s$ and $n_m = X_m V_m$. Thus, k can be expressed by Equation 4.13.

$$k = \frac{X_s V_s}{X_m V_m} = \frac{K V_s}{V_m} \tag{4.13}$$

where $K = X_s/X_m$ is the well-known distribution constant, which is the equilibrium constant of the compound between the mobile and the stationary phases.

Taking the logarithmic values on both sides of Equation 4.13, we obtain the basic linear free-energy-related chromatographic retention parameter, log k, widely used in quantitative structure–retention investigations.

$$\log k = \log K + \log\left(\frac{V_s}{V_m}\right) \tag{4.14}$$

Equation 4.14 means that the logarithmic value of the retention factor is in linear correlation with the logarithmic value of the distribution coefficients of the compounds. The value of $\log(V_s/V_m)$ can be regarded constant in a given chromatographic system.

MEASURE OF SEPARATION EFFICIENCY

We can easily imagine that if we can obtain very narrow peaks on the chromatogram, more peaks could be detected during the same time frame than would be present if the peaks were wide. The efficiency of the separation therefore can be related somehow to the peak width. When we look at any chromatogram, it is noticeable that the longer the retention time, the wider will be the peaks, which means that the longer the sample component stays in the chromatographic system, the wider the peaks will be. The band width or peak width, t_w, is commonly expressed in terms of theoretical plate numbers, N, of the column. The plate number was used to characterize the efficiency of fraction distillation columns and the same expression is adopted for the characterization of the separation efficiency of a chromatographic column.

$$N = 16\left(\frac{t_R}{t_w}\right)^2 \tag{4.15}$$

N is approximately constant for different bands in a chromatogram for a given set of operating conditions. Therefore, N is a useful measure of column efficiency, which is the relative ability of a given column to provide narrow peaks with longer retention times and therefore more efficient separation. As N can be considered constant for every peak, Equation 4.15 can predict the peak width for every retention time. Minor exceptions to the constancy of N for different peaks exist. In gradient elution chromatography, all peaks show approximately the same peak width, as the later eluting peaks usually start their movement later from the top of the column, when the mobile phase composition is such that it is able to elute them.

Because of the widening of the chromatographic peaks as the retention time increases, later eluting bands show a corresponding reduction in peak height (as the peak area is proportional to the concentration) and eventually disappear into the baseline. It also means that under standard analytical HPLC conditions without gradient elution, the measurable peaks elute within less than an hour. Later eluting peaks are so wide that they cannot be detected and they merge with the baseline.

N is proportional to column length L, so that (if other factors are the same) an increase in L results in an increase in N and better separation. This proportionality of N and L can be expressed in terms of 4.16.

$$N = \frac{L}{H} \tag{4.16}$$

where H is the so-called height equivalent of a theoretical plate (plate height) or HETP value. The H measures the efficiency of a given column per unit length. Small H values mean more efficient columns and larger N values. A central goal in HPLC practice is the attainment of small H values for maximum N and highest column efficiency. In general, H is smaller for small particles of column packing and for low mobile phase flow rates. H is smaller for less viscous mobile phases and at higher separation temperatures. Large N values and improved separation are usually favored by long columns packed with small particles, by nonviscous mobile phases, flowing relatively slowly through the column, and by higher separation temperatures. The separation of macromolecules such as proteins is often more difficult than the separation of compounds with molecular weights below 1000 Da. The dependence of column H values on the mobile phase velocity (u) is of great importance. Knowledge of this relationship is the key to understand and control the separation efficiency besides the selection of the column and mobile phase. It is possible to describe the dependence of H on u by means of the van Deempter equation:

$$H = A + \frac{B}{u} + Cu \tag{4.17}$$

The constants A, B, and C vary from one column to another and also depend to some extent on the sample, mobile phase, and separation temperature. The resulting plots of H versus u exhibit a characteristic shape shown in Figure 4.3. The constant A is a characteristic constant for the shape and distribution of the stationary phase particles in the column. The peak broadening due to the uneven length of routes between stationary phase particles is called the *eddy diffusion effect*. It can be decreased using smaller particles with spherical shape and better packing procedure. The constant B is a characteristic constant for the longitudinal diffusion of the compounds, and C is a characteristic constant for the speed of the mass transfer.

The form of the H versus u plot in Figure 4.3 has a diagnostic value and can be used to determine whether a given column is operating properly. Therefore, it is useful to construct such a plot for every new column before the column is placed for use. An H versus u plot also provides a reference point against which the column can be compared at some later time if it is believed that the column

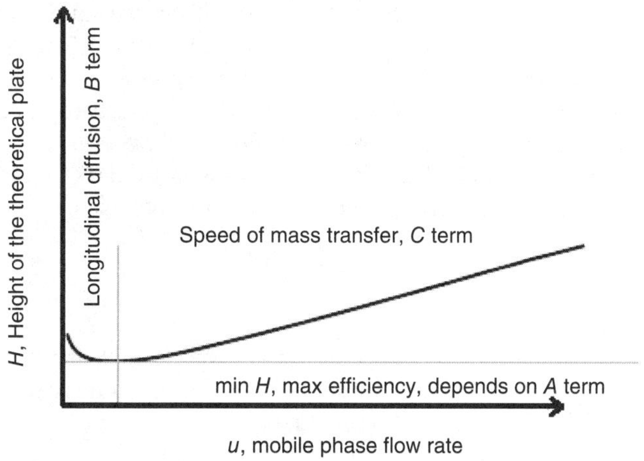

Figure 4.3 Typical plot of H (inversely related to efficiency) versus u (mobile phase velocity) in HPLC.

performance has changed. The column efficiency can be checked using a certain mobile phase flow rate and a test mixture. The column manufacturers usually provide a test chromatogram together with the new column. It shows the components of the test mixtures, the HPLC conditions, and the observed theoretical plate number (N) of the column.

The efficiency of the column can deteriorate by a continuous loss of the chemically bonded moiety from the surface, or by pushing the particles tightly together, forming a hole at the top of the column, or breaking the particles. All of these events decrease the lifetime of the column and are manifested in the decrease of the efficiency (smaller N values). The appearance of a shoulder, tailing, or leading on the peaks, resulting in wider than normal peaks, is a typical sign of a hole at the top or the end of the column.

RESOLUTION AND SEPARATION TIME

The usual goal in HPLC is the adequate separation of a given sample mixture. In approaching this goal, we must have some quantitative measure of the relative separation or resolution achieved. The resolution (R_S) of two adjacent peaks 1 and 2 is defined by the distance between the two peaks maxima divided by the average peak width, as shown in Figure 4.4 and Equation 4.18.

$$R_s = \frac{t_{R2} - t_{R1}}{0.5\,(t_{w1} + t_{w2})} \tag{4.18}$$

The quantities t_{R1} and t_{R2} refer the retention times of peaks 1 and 2, and t_{w1} and t_{w2} are their respective peak width values. When $R_s = 1$, the two peaks are reasonably well separated; that is, only 2% of one band overlaps with the other. Larger values of R_s mean better separation and smaller values of R_s mean poorer separation. The R_s parameter defines the quality of the separation. To control the resolution, we must know how R_s varies with experimental parameters such as k and N. Next we derive such a relationship for two closely eluted compounds. Equation 4.9 gives $t_{R1} = t_0(1 + k_1)$ and $t_{R2} = t_0(1 + k_2)$, where k_1 and k_2 are the k values (retention factor) for the two components. If we consider t_{w1} and t_{w2} to be approximately equal and N assumed to be constant for peaks 1 and 2, inserting these values into Equation 4.18, Equation 4.19 can be obtained.

$$R_s = \frac{t_0(k_2 - k_1)}{t_{w1}} \tag{4.19}$$

From Equation 4.15, the peak width can be expressed by the theoretical plate number as $t_{w1} = 4t_1/\sqrt{N} = 4t_0\ (1 + k_1)/\sqrt{N}$. By inserting this expression for t_{w1} into Equation 4.19, Equation 4.20 is obtained.

$$R_s = \frac{(k_2 - k_1)\sqrt{N}}{4(1 + k_1)} = \frac{1}{4}k_2(k_2 - 1)\sqrt{N}\frac{k_1}{1 + k_1} \tag{4.20}$$

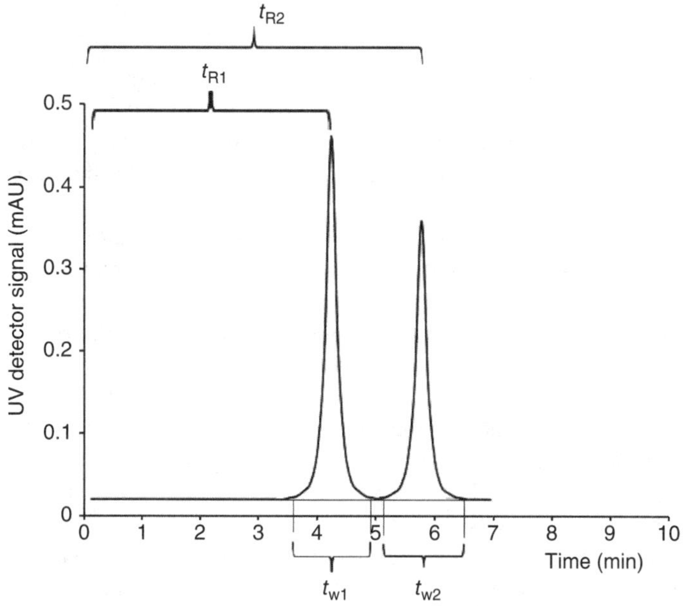

Figure 4.4 Resolution of two adjacent peaks (R_s). t_{R1} and t_{R2} are the retention times, and t_{w1} and t_{w2} are the peak widths. The resolution can be calculated according to Equation 4.18.

If we define the separation factor $\alpha = k_2/k_1$ for peaks 1 and 2 and recognize that k_1 is approximately equal to k_2, average k can be introduced in Equation 4.20 and Equation 4.21 can be obtained.

$$R_s = \frac{1}{4(\alpha - 1)}\sqrt{N}\frac{k}{1+k} \tag{4.21}$$

Equation 4.21 describes the peak resolution as a function of several experimental parameters. It can be seen that increasing separation selectivity (α) can increase the resolution. Separation selectivity can be increased by changing the mobile phase composition. The column efficiency (N) has only small effect as it is implied by the square root term.

In general, the column efficiency and resolution are important parameters when chromatography is used as a separation technique, and the aim is to achieve the best possible separation of two (or more) components. However, it is also important when using chromatography for measurement of molecular properties of the compounds. When the chromatographic system has high efficiency and high resolution power, it means that we can measure small differences in molecular properties of different compounds. Therefore, it is important to be aware of the separation efficiency of the chromatographic system that is used for property measurements.

GRADIENT ELUTION

Sometimes the mixtures to be separated contain components that are very different in polarity, so that no isocratic conditions result in the retention factor (k) of all components being within the range 0.5–20 (the normal practically sensible time window). In these cases, it is necessary to change the strength of the mobile phase during the chromatographic run. By using a mobile phase with strong elution power, all components elute quickly without separation. Using a mobile phase with weak elution strength does not elute the strongly retained components. Therefore, it is common to start the chromatography with a weak mobile phase composition and continuously increase the concentration of the stronger component of the mobile phase during the chromatographic run. This is called *gradient elution*. The most commonly used gradient profile is linear, that is, continuously increasing the concentration of the stronger solvent within a given time interval. Typical chromatograms of a homolog mixture using isocratic and gradient elutions are shown in Figure 4.5. It can be noticed that when applying gradient elution, the later eluting peaks do not necessarily get wider. Also, when separating homolog mixtures, the difference in retention times by adding one CH_2 group to the molecules does not increase as it is observed under isocratic conditions. This is caused by increasing migration velocities of the bands along the column during gradient elution. Therefore, the previously described calculations of resolution and column efficiency with the theoretical plate number are not applicable. It is not because the rules are not valid. It is because each component starts its migration at different times and not when the injection is made. The velocity of the band movement is not constant, but accelerates as the mobile phase composition changes. Therefore, a so-called peak capacity is used to describe the efficiency in gradient

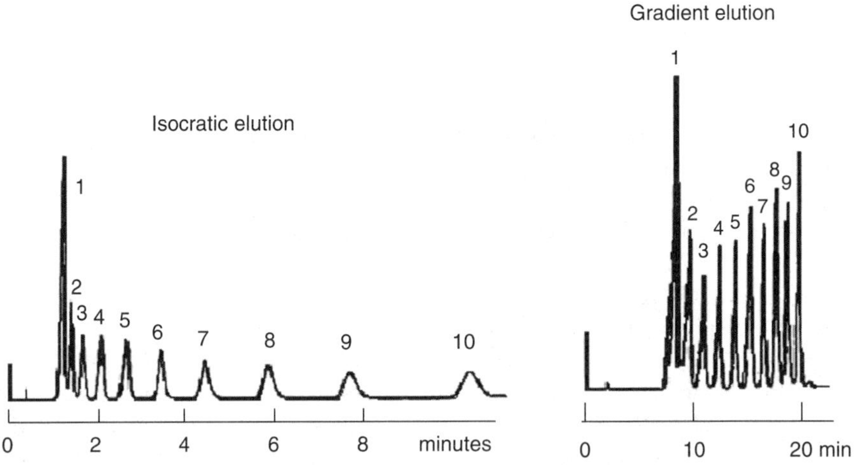

Figure 4.5 Elution of a homolog series by isocratic and gradient modes. (Reproduced with permission from Reference [5].)

mode. The peak capacity equals the number of peaks that can be fully separated in a given retention time window, for example, in 1 min. The peak capacity can be calculated as 60 divided by the average peak width (expressed in seconds), as described by Equation 4.22.

$$n = 1 + \frac{(t_g - t_0)}{w} \tag{4.22}$$

where n is the peak capacity (number of peaks), t_g is the gradient retention time, t_0 is the dead time, and w is the average peak width at baseline.

Figure 4.6 demonstrates how the peak capacity depends on the flow rate and gradient run time based on the work carried out by Neue [6]. A peak capacity above 15 can be considered as good separation efficiency. In gradient mode, the peak capacity depends on the mobile phase flow rate much more than on the length of the column.

Figure 4.7 shows the gradient separation of a mixture using longer column with slower mobile phase and shorter column with faster mobile phase flow [8]. Applying the faster flow rate significantly reduces the analysis time without decreasing the separation efficiency (peak capacity). This example clearly shows that the factors that affect separation efficiency are quite different in gradient mode than in

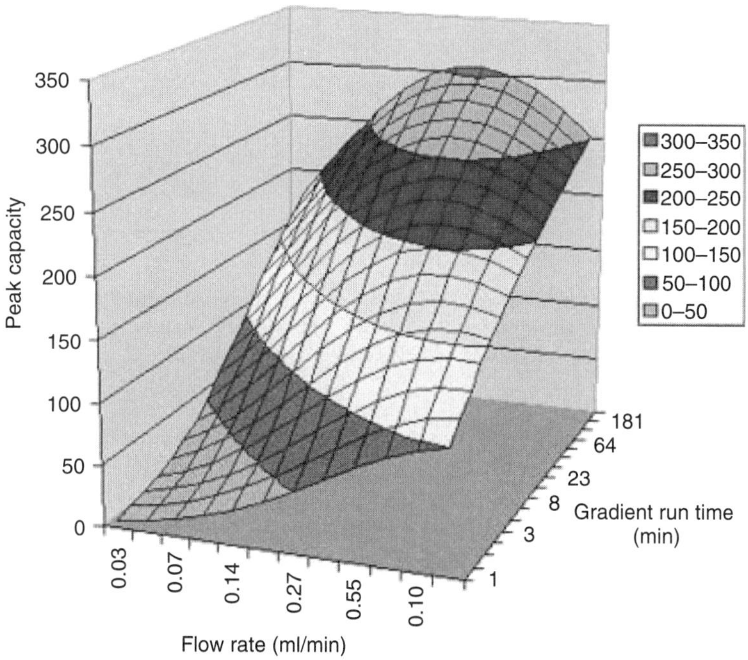

Figure 4.6 The interdependence of peak capacity, gradient time, and flow rate. (Reproduced with permission from Reference [7].)

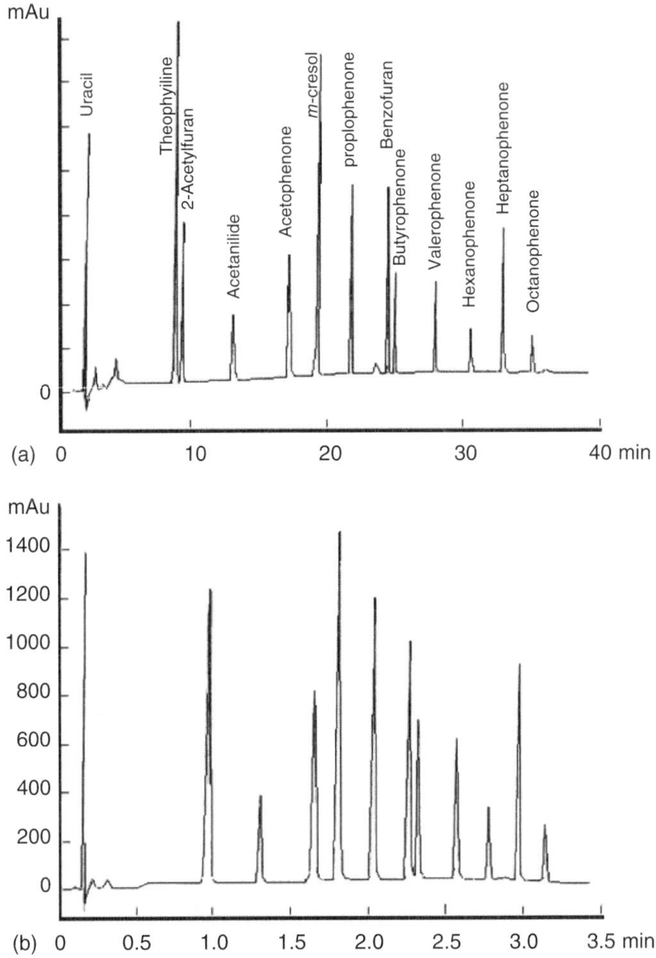

Figure 4.7 Comparison of the gradient reversed phase separations of a mixture using (a) longer column and slower flow rate and (b) with a shorter column and faster flow rate HPLC conditions. (Reproduced with permission from Reference [8].)

isocratic elution. The higher the mobile phase flow rate also means faster gradient (i.e., shorter time from 0% to 100% mobile phase B, usually acetonitrile, with higher flow rate means less steep gradient). Basically short columns, high flow rate, and short gradient can increase the gradient separation efficiency. This finding is very significant in developing fast generic gradient chromatographic methods that enabled to develop open-access generic HPLC systems available for chemists for the quick analysis of large number of compounds [7]. The generic gradient chromatographic method was developed using reversed phase stationary phase and aqueous acetonitrile as mobile phase. The aqueous phase usually has a low pH and the acetonitrile concentration ranges from 5% to 95%. The only limitation is the increased backpressure of the column. A more detailed explanation of the theoretical backgrounds of gradient chromatography is described by Mutton [8] and Jandera [5].

The application of gradient chromatographic analysis increased the throughput of physicochemical measurements as well. The gradient chromatographic method can be used for concentration determination during solubility and partition coefficient measurements without the need for the development of HPLC analytical procedures for each sample. Usage of a generic fast gradient HPLC analysis [8] has increased the measurement capacity of physicochemical laboratories and dramatically reduced the method development and analysis time.

APPLICABILITY OF CHROMATOGRAPHY FOR MEASUREMENTS OF MOLECULAR PROPERTIES

The chromatographic technique has great potential to determine the strength of a compound's interaction with the stationary phases and measure its partition coefficients between the mobile and stationary phases. As has been shown, the retention time can be used to derive the retention factor, which is directly related to the compound's distribution between the mobile and the stationary phases. The retention factor (k) determined from the retention time (t_R) and dead time (t_0), as we have derived earlier, $(t_R - t_0)/t_{0,}$ is equal to the ratio of the average number of analyte molecules in the stationary phase to the average number of molecules in the mobile phase (Equation 4.13) during the elution process. The only constant that we need to know is the value of the volume ratio of the stationary and mobile phases (V_s/V_m) to be able to obtain the absolute value of the chromatographic partition coefficient. It is difficult to measure the volume ratio of the mobile and stationary phases; therefore, a set of reference compounds are needed to derive the absolute values of the partition/distribution coefficients of compounds between the mobile and the stationary phases. In reversed phase chromatography, the compound's retention is governed by hydrophobic forces, while retention on protein stationary phases depends on the compound's interaction with the protein phase. Thus, partition coefficients and the compound's protein binding can be obtained from time measurements instead of concentration determination. The retention time is independent of the compound concentration/amount injected into the chromatographic system, and it is the main indicator of partition process. Impurities do not affect the measurements as they are usually separated from the main component. The solvent used to dissolve the compound also separates, and thus low solubility does not affect the measurements. The above-mentioned advantages are important in the early drug discovery stage, when compounds for further optimization have to be selected from combinatorial libraries containing thousands of compounds. The other advantage of using the chromatographic technique is that various stationary phases can be used to provide information about the physicochemical properties of compounds in a wide variety of conditions. For example, normal paraffin hydrocarbons, immobilized octanol, and biomimetic phases such as immobilized artificial membrane (IAM), HSA, and α-acid-glycoprotein (AGP) stationary phases can be used, on which the retention time will reveal the compound's interaction/partition into the particular phase. The pH and polarity of the mobile phase can also be altered, which will affect compound interaction with the stationary phase, too. The technique, therefore, is easily applied to the measurement of more than one type of compound interaction. Since the early publications [9–11] in the beginning of the 1970s, several hundred papers have been published regarding the applicability of reversed phase liquid chromatography for the determination of lipophilicity. Several review papers summarized the major approaches [11–14].

Dorsey and Khaledi [15] have discussed the theoretical differences between a chromatographic lipophilicity measure and partitioning between bulk solvents. They emphasize that the chromatographic distribution process is more entropy driven than the bulk liquid–liquid distribution driven by enthalpy. They have also discussed the thermodynamic basis for possible failures in comparing partitioning of small molecules between bulk solvents and chromatographic mobile and stationary phases, as well as partitioning of solutes into lipid layers and biological membranes. Biological partitioning of compounds in aqueous biphase systems, such as blood and various tissues, certainly happens on a large area of contact surfaces between the partitioning phases. This is similar to the situation in the chromatographic partition process. Significant differences between bulk solvent partition and partition involving large surface may be observed mainly for surface-active compounds, such as amphiphilic and charged molecules with large hydrophobic moieties.

The chromatographic distribution process takes place on a large surface in a dynamic way, which is very similar to the in vivo distribution of the compound. The in vivo distribution is also a dynamic process, which means that there is a continuous change in compound distribution that results in "steady-state" equilibrium. Therefore, the chromatographic process can be considered as being very similar to the in vivo process, where the moving plasma/blood contains the drug molecules that are interacting with the "stationary" tissue compartment.

In spite of the obvious advantage of the HPLC technique and the possibility of using biomimetic phases that may prove to be better models for biological partitions, the collation of HPLC-based lipophilicity and protein binding data is not widespread. The major problem in providing a more easily measured alternative to other methods such as octanol/water partition measurements for lipophilicity or equilibrium dialysis method for protein binding lies in the lack of standardization of various stationary phases, which makes interlaboratory comparison of the data very difficult. To overcome these difficulties, the chromatographic system needs to be calibrated using known standards that also allow the chromatographic retention data to be converted into biomimetic partition coefficients. The reproducibility of current commercially available stationary phases, the automation, and reliability of modern HPLC instrumentation now make it possible to fully exploit the advantages of the technique and allow collation of a large amount of lipophilicity and biomimetic binding data in a reproducible way. However, it also requires a consensus for standardization and validated models for the applicability of chromatographic techniques to model the biological distribution process. If the modeling is successful, the chromatographic technique is suitable to replace animal experiments for the estimation of a compound's *in vivo* distribution in humans.

Another future potential of using chromatography for estimating a compound's in vivo distribution is to use the peak width and peak shape for the determination of the kinetic aspects of the binding. As we have seen from the van Deempter plot, the C term refers to the speed of the mass transfer. When we have slow onset and offset rates during the binding, it can cause peak broadening. When the offset rate is much slower than the onset rate, it can cause peak tailing, that is, asymmetrical peaks. There are only a few methods that can be used for measuring binding kinetics, and HPLC has not been explored yet in this field, although, in principle, it provides an easy way of using the information from the peak width versus temperature or peak width versus flow rate profiles. We would need a standard calibration set of compounds for which the rate constants are available in order to exclude the peak broadening effect from the eddy diffusion (term A in the van Deempter equation) and from the longitudinal diffusion. Only a handful of publications that demonstrate the applicability of the chromatographic methodology for such purposes are available in the literature [16, 17].

SUMMARY

Chromatographic separation is based on the different velocities of compounds in a chromatographic system consisting of a mobile and a stationary phase. The difference in velocity is governed by the compound's different strengths of interactions with the stationary phase. The retention time is the time a compound takes to get from the injection to the detector. A compound that has no interaction with the stationary phase moves with the speed of the mobile phase and appears in the detector with the so-called dead time. The retention factor, which is the retention time expressed relative to the dead time, is directly proportional to the quotient of the average number of moles of the compound in the stationary and mobile phases; thus, it can be used to measure the compound's dynamic equilibrium distribution between the stationary and mobile phases. This provides the theoretical basis of using the chromatographic principle for the measurements of a compound's interaction with a wide variety of stationary phases under a wide variety of mobile phase composition and pH. The chromatographic separation process has very high efficiency and selectivity. The efficiency of the separation can be measured by the theoretical plate number per unit of column length. The narrower the chromatographic peaks are with longer retention time, the more efficient is the chromatographic separation. The efficiency depends on the mobile phase flow rate, which can be described by the van Deempter equation in isocratic elution. The column packing (particle size and shape of particles) causes the eddy diffusion effect on peak broadening, which is independent of the mobile phase flow rate. The longitudinal diffusion that causes peak broadening, and thus reduces efficiency, is greater at low flow rates, while the mass transfer resistance, which increases the peak broadening, is greater at high flow rates. We need to use highly efficient chromatographic systems to be able to measure small differences in compound properties. Gradient elution is used when compounds with a wide range of properties are to be separated, which means constant change of the mobile phase composition of solvents from the weaker to the stronger elution strength. Separation efficiency and retention are very different in gradient mode. Shorter columns and higher flow rates provide more efficient separation, which is measured by the peak capacity. To be able to derive the absolute values of physicochemical properties we need to use a set of compounds with known data from other methodologies, which are measured under the same chromatographic conditions as the unknown compounds.

QUESTIONS FOR REVIEW

Q4.1 Explain how chromatographic separation takes place.

Q4.2 What is the retention factor and how does it relate to a compound's distribution coefficient between the mobile and the stationary phases?

Q4.3 How can we measure the efficiency of a chromatographic system and how does it relate to the mobile phase flow rate in isocratic and gradient elutions?

Q4.4 Explain the differences between isocratic and gradient elution.

REFERENCES

1. Scott, R. P. W. (1995) Techniques and Practices of Chromatography. Marcel Dekker, New York.
2. Meyer, V. R. (1994) Practical High-performance Liquid Chromatography (2nd Edition) Wiley, Hoboken, New York.
3. Bidlingmeyer, B. A. (1992) Practical HPLC Methodology and Applications. Wiley, Hoboken, New York.
4. Snyder, L. R., Kirkland, J. J. (1979) Introduction to Modern Liquid Chromatography, Wiley, New York.
5. Jandera, P. (2000). Comparison of Various Modes and Phase Systems for Analytical HPLC. Chapter 1. In: Separation Methods in Drug Synthesis and Purification, Ed. Valkó, K., Elsevier, Amsterdam, pp. 1–69.
6. Neue, U. D. (2005) Theory of peak capacity in gradient elution. *Journal of Chromatography A*, *1079*, 153–164.
7. Lane, S. (2000). Coupled Chromatography—Mass Spectrometry Techniques for the Analysis of Combinatorial Libraries. Chapter 5. In: Separation Methods in Drug Synthesis and Purification, Ed. Valkó, K., Elsevier, Amsterdam, pp. 127–160.
8. Mutton, I. (2000). Fast Gradient HPLC Methods. Chapter 2. In: Separation Methods in Drug Synthesis and Purification, Ed. Valkó, K., Elsevier, Amsterdam, pp. 73–85.
9. Call, J. M. (1975) Liquid-liquid partition coefficients by high-pressure liquid chromatography. *Journal of Medicinal Chemistry*, *18*, 549–552.

10. Brent, D. A., Sabatka, J., Minick, D. J., Nehry, D. J. (1983) A simplified high-pressure liquid chromatography method for determining lipophilicity for structure-activity relationships. *Journal of Medicinal Chemistry*, *26*, 1015–1020.
11. Minick, D. J., Brent, D. A., Frenz, J. (1986) Modeling octanol-water partition coefficients by reversed-phase liquid chromatography. *Journal of Chromatography A*, *461*, 177–191.
12. Tomlinson, E. (1975) Chromatographic hydrophobicity parameters in correlation analysis of structure–activity relationships. *Journal of Chromatography A*, *113*, 1–45.
13. Braumann, Th., Weber, G., Grimme, L. H. (1983) Quantitative structure–activity relationships for herbicides: Reversed-phase liquid chromatographic retention parameter, log k_w versus liquid-liquid partition coefficient as model of the hydrophobicity of phenylureas, s-triazines and phenoxycarbonic acid derivatives. *Journal of Chromatography*, *261*, 329–343.
14. Kaliszan, R. (1992) Quantitative structure–retention relationships. *Analytical Chemistry*, *64*, 619A–631A.
15. Dorsey, J. G., Khaledi, M. G. (1993) Hydrophobicity estimations by reversed-phase liquid chromatography: Implication for biological partition processes. *Journal of Chromatography A*, *656*, 485–499.
16. Talbert, A. M., Tranter, G. E., Holmes, E., Francis, P. L. (2002) Determination of drug-plasma protein binding kinetics and equilibria by chromatographic profiling: Exemplification of the method using L-tryptophan and albumin. *Analytical Chemistry*, *74*, 446–452.
17. Meng, Q. C., Johansson, J. S., Eckenhoff, R. G. (2002) Chromatographic approach for determining the relative membrane permeability of drugs. *Journal of Chromatography B*, *774*, 89–95.

5

MOLECULAR PHYSICOCHEMICAL PROPERTIES THAT INFLUENCE ABSORPTION AND DISTRIBUTION—LIPOPHILICITY

Lipophilicity is one of the most important physicochemical properties of molecules that determine how they distribute in the body. As the body consists of several lipid and aqueous compartments, molecules will have a preference for one compartment to another based on their lipophilic nature. Lipophilic molecules will prefer the lipid environment, while hydrophilic molecules will prefer the aqueous environment. There is a slight difference between lipophilicity and hydrophobicity as defined by the International Union of Pure and Applied Chemistry (IUPAC) committee. "Lipophilicity represents the affinity of a molecule or a moiety for a lipophilic environment, while hydrophobicity measures the association of non-polar groups or molecules in an aqueous environment, which arises from the tendency of water to exclude non-polar molecules."

PARTITION COEFFICIENT

Lipophilicity of a compound is usually determined and expressed as partition coefficient. Partition coefficient is the concentration ratio of a compound that is measured after the compound is equilibrated in two immiscible solvents (one of them is water or buffered water to adjust the pH), while the other is a nonpolar organic liquid. After the equilibrium has been established, the concentration of the compound

Physicochemical and Biomimetic Properties in Drug Discovery: Chromatographic Techniques for Lead Optimization, First Edition. Klara Valko.

is determined in both the solvents. If the compound prefers to be in the organic solvent instead of the water, it shows that it prefers the nonpolar environment, that is, it is lipophilic. The concentration ratio of a compound in two immiscible solvents is constant. However, we need to avoid high compound concentrations in either liquid to avoid potential self-association of the molecules that might change their lipophilic character. The partition coefficient is a thermodynamic constant (and as such is temperature dependent). It is usually expressed in logarithmic term and referred to as $\log K$ ($\text{solvent}_1/\text{solvent}_2$). The most often used partition system in drug discovery is the octanol/water partition coefficient since the pioneering work of Hansch and Leo [1] in 1964. They found that octanol is a very good model for the lipophilic environment in the body as it is not only nonpolar but also contains a polar moiety, the hydroxyl group, which is able to accept and donate a proton, thus forming hydrogen bonds, similar to the protein environment in the body. Octanol is also a "wet" solvent as it contains around 4% water when it is equilibrated with water. The molar solubility of water in octanol is 0.234, which means that every fourth molecule is water. The proteins also contain structural water molecules hydrating the proteins and reinforcing their secondary and tertiary structures in their native forms. The logarithmic value of the octanol/water partition coefficient is also commonly noted as $\log P$. Collander [2] and Leo [3] studied the partition coefficients of compounds in several immiscible solvent pairs. Collander found that the partition coefficient measured in one solvent pair (1/2) correlates with the partition coefficients measured in another solvent pair (3/4) according to Equation 5.1.

$$\log K_{1/2} = a \log K_{3/4} + b \tag{5.1}$$

where $K_{1/2}$ is the partition coefficient between solvents 1 and 2, while $K_{3/4}$ is the partition coefficient of the same compound in solvents 3 and 4; a and b are constants. The Collander equation is a linear-free energy equation as the logarithmic value of a partition coefficient is linearly related to ΔG free energy. The Collander equation suggests that if a compound shows high partition coefficient in one solvent pair, in other words, it is lipophilic; it shows a high partition coefficient in another solvent pair. It makes sense for if the compound does not like to be in water (hydrophobic), it will partition to a nonaqueous environment. However, Leo [3] has pointed out that the Collander type of correlation does not hold when proton donor and acceptor compounds are considered together and the organic solvents in the two partition systems differed from each other in terms of H-bonding abilities. He has studied more than 300 compounds in more than 30 partition systems (such as octanol/water, chloroform/water, hexane/water, and diethylether/water). He found that the Collander type of correlation between the partition coefficients holds true only if the compounds are similar in structure (all proton donors or all proton acceptors) and the partitioning solvents are similar in their H-bonding ability. It also means that we can expect that the octanol/water partition coefficients will not always be good models for biological partition processes. For example, as we shall discuss later, the blood–brain barrier distribution cannot be modeled well by octanol/water partition coefficients.

The measurements of octanol/water partition coefficients for drug discovery compounds are very important. As we discussed earlier, the *in vivo* distribution of drugs between the plasma compartment and various tissue compartments is predominantly dependent on the compound's lipophilicity. The octanol/water partition proposed to be a good model for biological partition has been studied extensively, and several experimental protocols have been developed and published for its measurement. In principle, the following procedure can be followed.

Step 1. Prepare an aqueous buffer solution with the required pH. For the determination of the partition coefficient (P) of the unionized form of the compound, the pH of the aqueous buffer should be such that the compound is in its nonionized form (i.e., basic compounds are measured at high pH, and acidic compounds are measured at low pH).

Step 2. Equilibrate the octanol with the above-mentioned buffer solution and the buffer with the octanol. Mix and leave it for 24 h at room temperature. Then separate the two layers. (Note: water also dissolves around 2.3 g/l of octanol!)

Step 3. Dissolve the compound in a known amount of octanol-saturated buffer or the buffer-saturated octanol depending on its expected lipophilicity. Put a known amount of water-saturated octanol and octanol-saturated water with the dissolved compound into a shake-flask device. Shake and equilibrate (temperature can be controlled for more precise measurements).

Step 4. Separate the two phases (centrifuging the layers can help) and determine the compound concentration in the two layers using standard concentration determination techniques (e.g., UV spectrophotometry and HPLC).

It is worth mentioning that for practical considerations, the volumes of the octanol and water phases can be altered, so that about equal amount of compound can be found in the two phases. This makes the quantitative determination of the concentration ratio more accurate. Figure 5.1 illustrates the schematic principles of the partition coefficient determination.

The above-mentioned basic principles of the octanol/water partition coefficient determination have to be applied in every method; however, the actual details of a protocol make it possible to carry out these measurements in a high throughput manner for a large number of compounds in an automated manner. The first problem is to establish the "appropriate" volume of the two solvents. When the compound is very lipophilic, it means that 1000 to 10,000 times higher concentration of the compound is in octanol relative to the aqueous phase. Even in this situation, we need to be able to detect the compound in the water, and probably the octanol phase needs to be diluted for the concentration determination. Also, if we want to automate the procedure, we need to determine several (preferably 96, in a 96-well format plate or tubes) compounds at the same time. It requires reducing the volumes of the partitioning solvents and also reducing the amount of

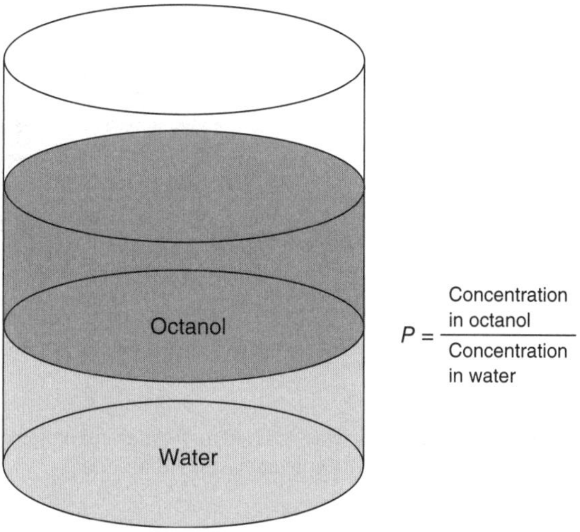

Figure 5.1 Schematic principles of the partition coefficient measurements.

compound needed for the measurements. For example, a method has been developed at GSK by the Discovery Physicochemical Characterization Group [4] using HPLC autosampler vials that can hold approximately 2 ml of solvent. To avoid the preparation of a standard solution with a known amount of compound for the concentration determination, the following procedure has been applied. Approximately 0.5 mg of research compound is dissolved in 10 ml of buffer (pH 7.4 phosphate buffer is used in order to imitate the physiological blood pH). Although the precise concentration of the stock solution is not known, it allows the use of this solution for the relative concentration determination, which is needed for the partition coefficient. One milliliter of the stock solution was dispensed in four autosampler vials. No octanol was added to the first vial, and 20, 200, and 1000 µl octanol were added to the remaining three vials, respectively. Then the vials are placed in a roller machine for equilibration (Figure 5.2).

After the equilibrium process and a slight centrifugation of the vials, the aqueous phase is analyzed by HPLC. We expect to obtain four peaks with the same retention times but different peak area values from the analysis of four vials containing the same compounds. A generic reversed phase gradient procedure can be used to determine the peak areas from the aqueous solutions (see section "Gradient Elution" in Chapter 4 [7,8]). Figure 5.3 shows the chromatograms obtained for a compound that was partitioned using various amounts of octanol. The peak area is proportional to the injected amount of the compound when no partitioning to the octanol occurred (A_0). In the second vial, when the compound partitioned into the octanol, the concentration of the aqueous phase is reduced; therefore, the obtained peak area (A_1) is less than A_0. The partition coefficient can be calculated by taking into account the volumes (V) of the octanol and aqueous phases (supposing the same injection volumes have been applied from the octanol and the water phases),

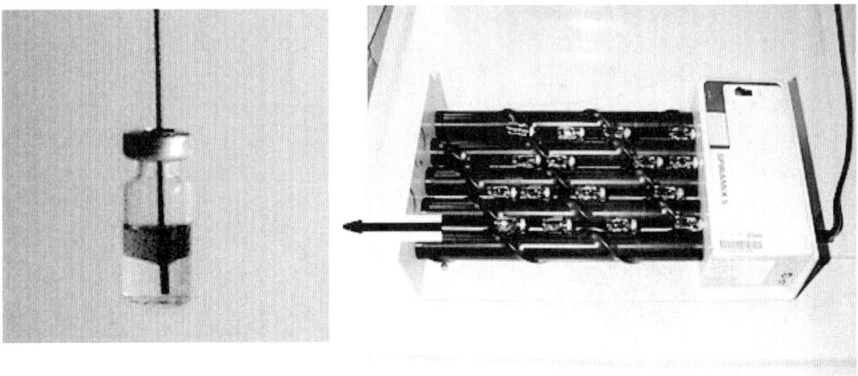

Figure 5.2 The semiautomated "roller-vial" method for the determination of the octanol/water partition coefficient. (Reproduced with permission from Reference [4].)

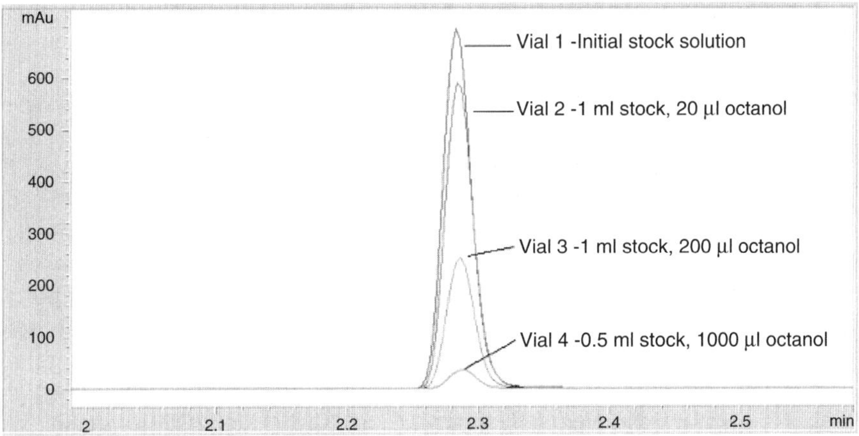

Figure 5.3 Example of chromatograms obtained for a research compound after analyzing the aqueous phase of the four vials with the "roller-vial" method of octanol/water partition coefficient determination. (With permission from Reference [4].)

as shown by Equation 5.2:

$$P = \frac{A_0 - A_1}{A_1} \times \frac{V_{\text{water}}}{V_{\text{octanol}}} \tag{5.2}$$

The partition coefficient of the same compound can be calculated in a similar way from the results obtained for the third and fourth vials. In an ideal case, we can obtain three parallel determinations of the partition coefficients, which can be averaged, and the measurement error can also be estimated. In practice, many times the peak areas are too small or very close to the peak area of the original stock solution because of the partition coefficient of the compound being too high or too low; therefore, the partition coefficients are obtained only from one or two chromatograms. However, the advantage of the method is that the concentration determination can be carried out automatically overnight, and the calculations can be programmed. Without changing the experimental protocol, the log P values from −1 to 3.5 to 4 can be determined. Several new automated experimental procedures have been developed using only one vial and one partition experiment. With a special dilutor program, it is possible to inject the compound from the octanol phase by diluting the octanol with methanol, before introducing an aqueous mobile phase where the neat octanol would not be miscible. Figure 5.4 shows two chromatograms obtained by injecting a compound after equilibrium distribution from the aqueous and the octanol phases.

To further automate the procedures, a 96-well plate-based assay can also be used. The two phases are dispensed into a 96-well plate together with the compounds in reduced volume of octanol and water. By shaking the plates, equilibration can take place, and then the separation of the two layers can be carried out by centrifugation. The concentration ratio of the compound in the two layers can be determined again by injecting the samples from the two layers to HPLC or to chemiluminescence nitrogen detector (CLND) equipment when the concentration of the compound can be determined by its nitrogen content.

One of the most accurate methods for the determination of octanol/water partition coefficients was introduced by Sirius Analytical Ltd (http://www.sirius-analytical.com/) with its GLpKa and T3 instruments that provide a highly accurate determination of octanol/water partition coefficient under a "good laboratory practice" (GLP) environment. The instrument is based on a proprietary Ag/AgCl double-junction electrode for accurate pH measurement. It carries out two potentiometric titrations of the compound with and without the presence of octanol. The partition coefficients are calculated using the Bjerrum plots to estimate the pK_a shift caused by the compound partitioning into the octanol phase [6]. From the two titration plots, the pK_a and the log P values are calculated. Sirius pH-metric methods can measure log P values of ionizable compounds between −1 and +7, which is the widest range of any log P measurement method. Unlike indirect chromatographic methods, they study the partition directly between water and octanol and dodecane or other partition solvents. They do not suffer from problems of microemulsion formation and lack of dynamic range that affect shake-flask methods. Figure 5.5 shows the GLpKa instrument; a new version T3 instrument, which is even more automated, uses smaller amount of samples and is suitable also for solubility determination.

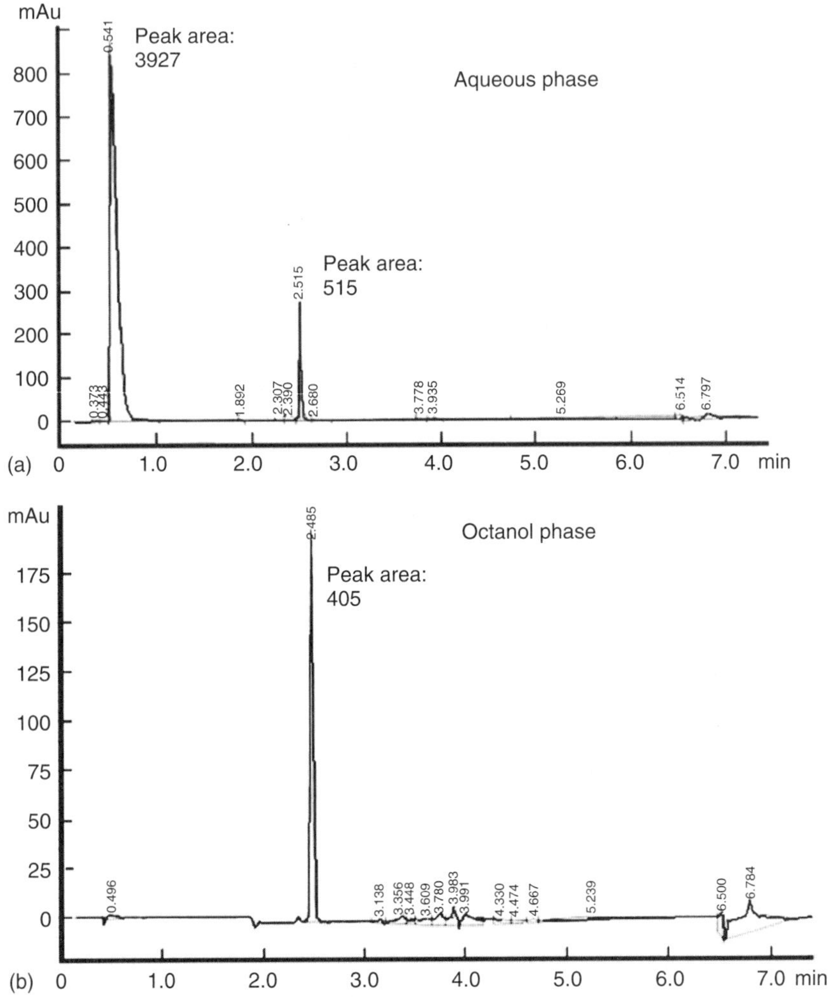

Figure 5.4 Chromatograms obtained for a compound after equilibrating in 1 ml of octanol-saturated water and 0.2 ml of octanol. (a) Chromatogram and peak area obtained from the octanol phase injecting 1 μl and (b) from the aqueous phase 25 μl. Chromatograms were obtained using generic acetonitrile gradient reversed phase chromatography with 7.5 min cycle time.

LIPOPHILICITY MEASUREMENTS BY REVERSED PHASE CHROMATOGRAPHY WITH ISOCRATIC ELUTION

The solvophobic theory of reversed phase liquid chromatography described by Horvath et al. [7] relates the reversed phase retention to solvophobic forces that cause the solute retention on the nonpolar stationary phase surface. When the mobile phase contains water, the "solvophobic" forces essentially mean hydrophobicity.

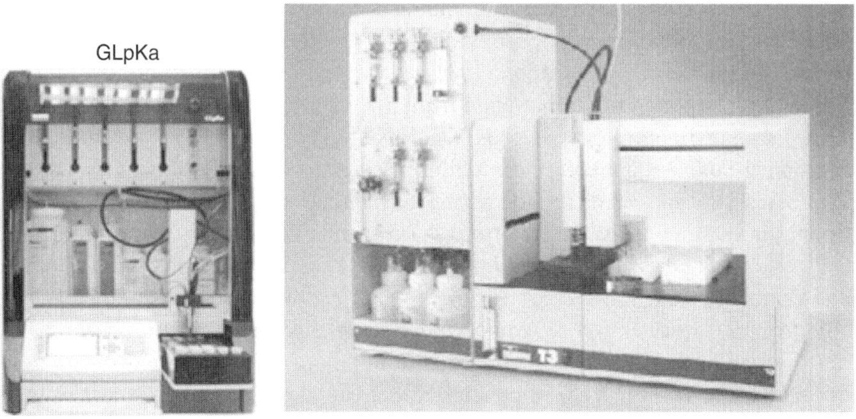

Figure 5.5 The Sirius GlpKa and the T3 instruments are able to measure pK_a, octanol/water partition coefficients (log *P*/*D*), and solubility based on automated acid/base titration process, which are monitored by potentiometric or dip probe UV spectrophotometric detection. (Picture from www.sirius-analytical.com with permission.)

Therefore, reversed phase chromatographic retention can be used, in principle, for hydrophobicity determination. It means that the more hydrophobic the compound is, the more likely it will be excluded from the hydro-organic mobile phase and stick to the lipophilic stationary phase, resulting in longer retention times. The typical characteristic of reversed phase chromatography is that the mobile phase contains water (and some other water-miscible organic solvent). A typical reversed phase stationary phase contains a chemically bonded alkyl-silica surface, such as C-8 or C-18 alkyl chain. Reversed phase HPLC is usually performed with high efficiency bonded reversed phase columns, and today the commercially available columns are both robust and reproducible. Solute partitioning from aqueous/organic mobile phase into standard reversed phase (often C-18) HPLC stationary phase can be used as a direct measurement of hydrophobicity. The retention time of a compound in reversed phase chromatography depends on its interaction/partition to the stationary phase, that is, its lipophilicity. Thus, from the retention time, the compound's partition coefficient between the aqueous mobile phase and the lipophilic stationary phase can be derived as described in more detail in Chapter 4 by Equation 4.14. As we have seen in Chapter 4, the retention factor (k) can be obtained from the retention time (t_R) of the compound and the dead time (t_0) (the time needed for a nonretained compound to go through the chromatographic system). The logarithmic value of the retention factor is linearly related to the compound's logarithmic partition coefficient (log K) in the dynamic chromatographic partition system (Equation 4.14). The $\log(V_s/V_m)$ parameter, which is the volume ratio of the stationary and mobile phases in the chromatographic columns, is constant for a given system. As it is difficult to determine this constant, the chromatographic systems have to be "calibrated" by measuring the retention times of compounds with known lipophilicity data under the same conditions, which are used for the lipophilicity measurements of the unknown compounds. Compounds that show big differences in their lipophilicity cannot be measured under the same reversed phase condition as a less lipophilic compound shows too short retention time, while a very lipophilic compound might not elute from the stationary phase. In order to cover a wide range of lipophilicity, various concentrations of the organic solvent (usually acetonitrile or methanol) in the mobile phase must be used. Higher organic solvent concentrations in the mobile phase reduce the retention times and even highly lipophilic compounds will elute from the column. Thus, it requires a preliminary estimation of the expected lipophilicity of the compound in order to choose the appropriate mobile phase composition for the retention time measurements. To be able to compare retention factors obtained under different organic phase concentrations, we have to extrapolate the retention data to the same condition, which is normally zero organic solvent concentration. Equation 5.3 is used most often for such extrapolation.

$$\log k = S\varphi + \log k_{\mathrm{w}} \tag{5.3}$$

Equation 5.3 shows the linear relationship between the organic solvent concentration (φ) and the logarithmic retention factor. The relationship is not linear for the full range of organic solvent concentrations, but is a good approximation within the working limits of $-0.5 < \log k < 1.5$. Valkó et al. [5] studied in detail the relationship, which is described by Equation 5.3, and the meaning of the S (slope) and $\log k_{\mathrm{w}}$ (intercept). The $\log k_{\mathrm{w}}$ (intercept) is the extrapolated retention factor to zero organic phase concentration. It means a hypothetical logarithmic retention factor that the compound would have when 100% water is used as mobile phase, as is shown in Figure 5.6.

These extrapolated $\log k_{\mathrm{w}}$ values are recommended in the Organization for Economic Co-operation and Development (OECD) guidelines [8] as chromatographic lipophilicity measures after calibration using octanol/water $\log P$ values. Since 1975 when Tomlinson [9] and McCall [10] published the use of reversed phase retention to measure a compound's lipophilicity, the number of publications in this field has reached over 300. There are some milestone papers and review articles [11,12] that the reader is referred to. Several publications are available, showing the search for the most suitable stationary and mobile phases to measure compound lipophilicity that closely mimics the lipophilicity obtained by the octanol/water partition coefficients [13–17]. They tried to use various types of reversed phase stationary phases and applied acetonitrile, methanol, and even a small amount of octanol in the mobile phase and searched for correlation with the octanol/water partition coefficients. The adjustment of the pH of the mobile phase causes a problem when the lipophilicity of the neutral form of basic drug molecules has to be determined. It requires a high pH mobile phase, which can be detrimental to silica-based hydrocarbon stationary phases, as the silica dissolves at high pH. The secondary interactions of the free unmodified silanol groups caused deviations from the good correlation between reversed phase retention and the

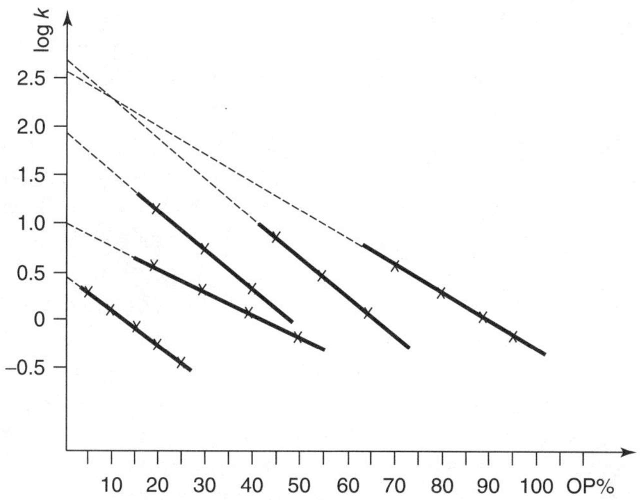

Figure 5.6 The determination of the log k_w values, the extrapolated retention factors to 100% water mobile phase. The extrapolation is necessary when the retention factors for a set of compounds cannot be determined under the same mobile phase composition because of retention times being too short or too long.

octanol/water lipophilicity, especially when a high organic phase concentration has to be used to elute lipophilic compounds.

The isocratic chromatographic methods for the determination of lipophilicity have the advantages that quantitative determination of the compound is not required, no long equilibration time is needed, and impurities do not disturb the lipophilicity measurements. However, the problem with the HPLC methodology is that various types of stationary and mobile phases can be used, all of them providing different measures of compound interaction. The lack of standardization of the stationary and mobile phases causes large variability of the measured data. Also, even in the same laboratory, the correlation between the reversed phase retention data and octanol/water lipophilicity can be broken when new chemical series with different numbers of H-bond donor groups are included [18]. As long as we use different chromatographic partition systems from the octanol and water, we measure different partition properties of the compounds. In other words, we can search for new stationary phases and mobile phase compositions, but they will never be exactly the same as octanol/water partition. There are publications that describe dynamically coated octanol stationary phase [16,19], which means that the octanol is adsorbed (not chemically bonded) in a lipophilic stationary phase. As the mobile phase flow continuously removes the octanol, it has to be added to the mobile phase. The advantage of this methodology is that in this chromatographic system, the retention will be proportional to the octanol/water distribution coefficient of the compound. However, there are several disadvantages of this type of approach. The volume ratio of the stationary and mobile phases should be constant, which is very difficult to achieve and control. The separation efficiency of such a system is usually very low because of very wide peak and distorted peak shapes. It is very difficult to determine the retention time accurately when the peaks are wide. The interface between the mobile and stationary phases are very large, which is essential for efficient chromatographic separation, but it makes the chromatographic partition system different from the octanol/water partition system, where the molecules are not measured at the interface. Discrepancies can be expected especially for surface-active, amphiphilic compounds. The question remains as to which is the best solution. Lombardo et al. [20] from Pfizer published a method for high throughput determination of lipophilicity of drug discovery compounds using the principle of extrapolating the retention time for the 0% organic concentration and tuning the chromatographic system to model the octanol/water partition system as close as possible, as is demonstrated in Figure 5.7. The authors have used commercially available HPLC columns (Supelcosil C18-ABZ) that were developed to produce improved peak shape (efficiency) for acidic (A), basic (B), and zwitterionic (Z) compounds, introducing charged moiety deep into the silica surface among the C-18 chains.

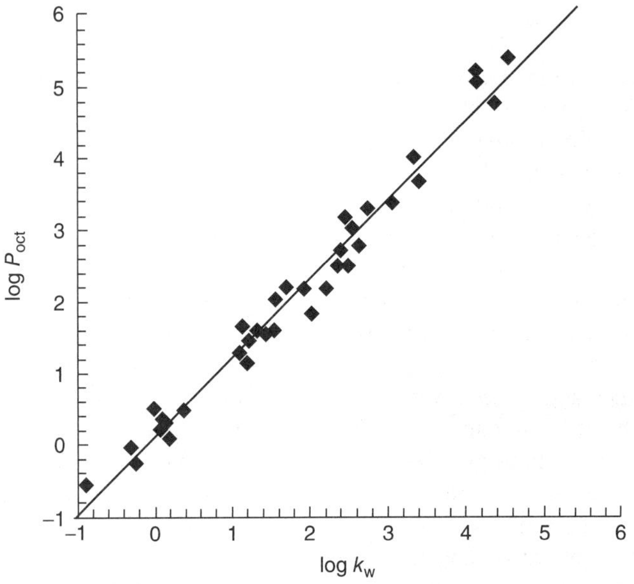

Figure 5.7 The "Pfizer" method of using HPLC for the determination of octanol/water partition coefficients [20]. The log k_w is the extrapolated logarithmic retention factor to the 100% water (buffer) calculated from the fitted straight lines obtained from at least three retention time measurements at various methanol concentrations in the mobile phase. The log P_{oct} values were calculated using log $P = 1.1 \times \log k_w + 0.13$ formula. The mobile phase was saturated with octanol; the reversed phase stationary phase was Supelcosil LC-ABZ. (Reproduced with permission from Reference [20].)

They used methanol and a minute amount of octanol in the mobile phase in order to adjust the mobile phase conditions for the best fit for octanol/water partition.

They have developed "in house" software, which, based on the expected lipophilicity of compounds, programmed the HPLC pumps to mix the appropriate concentration of organic phase in the mobile phase. They have determined the retention factor of each compound using three different mobile phase compositions, applying a higher concentration of methanol for more lipophilic compounds. From the three retention factors, the extrapolated log k_w was calculated by fitting a straight line. The so-called $E\log P$ values were calculated using the slope and intercept values of the straight line that was obtained for a set of structurally nonrelated compounds measured with octanol/water $\log D$ and the HPLC-based log k_w data.

Sirius Analytical Ltd [21] has developed an instrument using a very similar approach to the $E\log P$ using chromatographic principles. They have patented a proprietary stationary phase that mimics the octanol. They have developed software to apply the appropriate method for the retention determination based on the expected lipophilicity of the compound. However, they have measured the retention of the compound using only one condition. They did not change the mobile phase composition, but rather changed the column length and mobile phase flow rate based on the lipophilicity of the compound. This reduced the number of measurements for each compound from 3 to 1. For the retention measurement of highly lipophilic compounds, they used a shorter column and/or a higher mobile phase flow rate. For each condition, they have established a relationship between the retention and the shake-flask octanol/water partition data. The principles of the measurements are demonstrated in Figure 5.8.

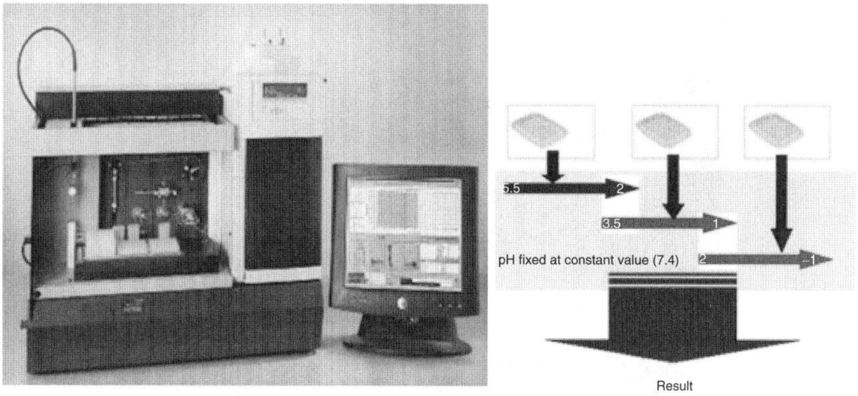

Figure 5.8 The Sirius profiler for the determination of octanol/water partition coefficients using chromatographic principles. Instead of applying organic modifier to elute lipophilic compounds, the flow rate and the column lengths have been adjusted to cover the $\log D$ range from −1 to 5.5. The preselection of the method is based on the calculated $\log D$ values. (With permission from www.syrius-analytical.com.)

The methodology, which involves several isocratic measurements using several organic phase concentrations and extrapolating to the 100% water condition, is not a strictly correct approach for two reasons. First, the log k versus organic phase concentration plot is not linear in the full range when the log k_w values are derived using a different set of mobile phase compositions (even if the same column and instrument are used); therefore, different log k_w values can be obtained for the same compound. This is because we are trying to fit a straight line to a curve. Depending on which part of the curve we do the fitting, the slope and the intercept of the straight line are different. Also, different log k_w values are obtained when acetonitrile or methanol is used as the organic modifier. Second, the reversed phase chromatographic system does not model the octanol/water system for structurally diverse compounds successfully. It has also been demonstrated [22] that the straight lines obtained by Equation 5.3 can cross each other, as is shown in Figure 5.9; hence, different lipophilicity rankings are obtained at different organic solvent concentrations.

The best correlation with the octanol/water partition coefficients is not necessarily achieved at the zero organic phase concentration; therefore, the S value in Equation 5.3 can be used as a second independent variable together with the log k_w value, as shown by Equation 5.4.

$$\log P = aS + b \log k_w + c \tag{5.4}$$

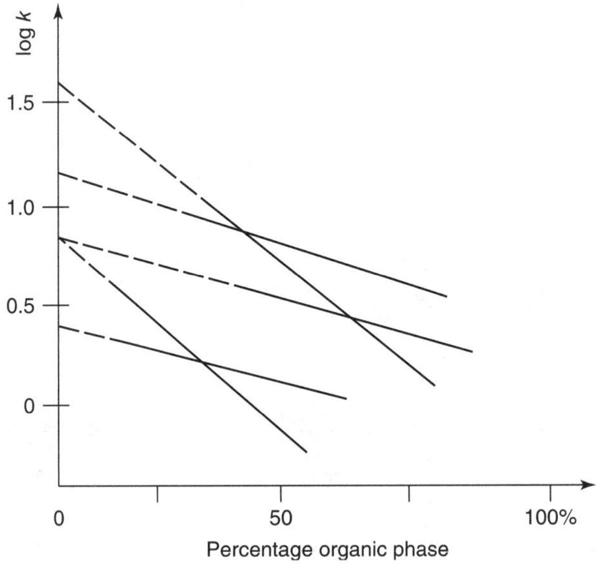

Figure 5.9 The hypothetical straight lines obtained for five compounds by plotting the logarithmic retention factor (log k) against the organic phase concentration in the mobile phase (%). The dotted lines show the extrapolation to zero organic phase concentration (log k_w).

where a, b, and c are regression coefficients. The ratio of a to b gives an organic modifier concentration value to which the log k values should be extrapolated to give the best statistical correlation with the log P values. In this way, the chromatographic partition can be tuned to model the octanol/water partition better. An interesting example was published by Valkó et al. [23], where the offset rate of several azidomorphine derivatives was studied using a guinea pig ileum preparation. Figure 5.10 shows the chemical structure of the investigated compounds. Two compounds could not be washed off from the ileum at all by applying Ringer solution (isotonic aqueous solution), while the other compounds could be washed off and the natural contraction of the ileum reappeared after a certain period of time, which was recorded. The question was whether it was caused only by the high lipophilicity of the two compounds (both had two lipophilic substituents at R_1 and R_2 positions, $-OC_2H_5$ and $-CH_2CH_2$–phenyl or $-CH_2CH_2$–cyclopropyl) or some irreversible specific interactions took place. The reversed phase logarithmic retention factors of the compounds (log k) were determined using various acetonitrile concentrations, and the slope values and the extrapolated log k_w values were calculated. They did not show good correlation, as is shown in Figure 5.11. When the offset rate of the compounds was correlated with the slope and log k_w as is shown in Equation 5.4, Equation 5.5 was obtained.

$$\log t_{1/2} = 33.4 \times \text{slope} + 0.86 \times \log k_w + 0.22 \quad (5.5)$$

where log $t_{1/2}$ was the time that was needed for washing in order to return to the half of the original magnitude of the guinea pig ileum contractions. When we divide the coefficients of the slope and the log k_w parameters, we get approximately 40, which means that the chromatographic system with 40% acetonitrile concentration in the mobile phase would provide the best model for the partition behavior of the compounds on the guinea pig ileum preparation. It has been demonstrated that when isocratic elution of the compounds was attempted, all compounds provided a measurable retention except the two highly lipophilic compounds. The two compounds that could not be washed from the ileum preparation could not be eluted from the

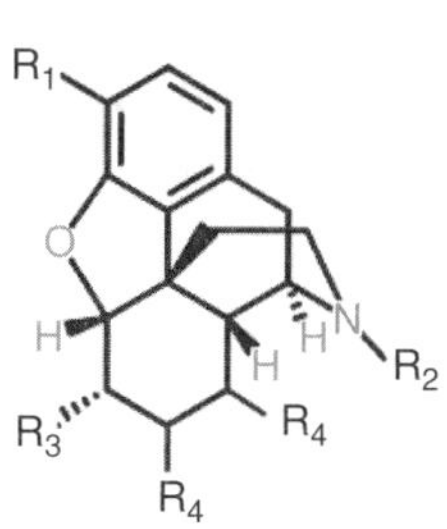

Figure 5.10 The azidomorphine derivatives for which the offset rate from guinea pig ileum was modeled by reversed phase liquid chromatography [23] using the slope and the log k_w values. $R_1 = -OH, -OCH_3, -OC_2H_5, -OCOCH_3$; $R_2 = -CH_3, -CH_2CH_2$–phenyl, $-CH_2CH_2$–cyclopropyl; $R_3 = -N_3, -OH$; R_4 = double bound, –H.

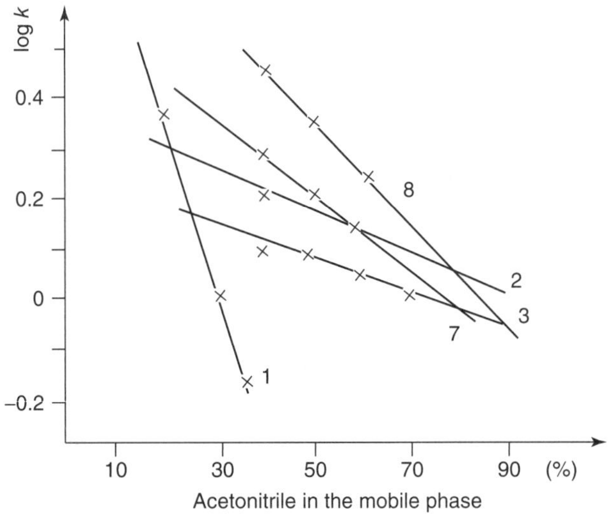

Figure 5.11 The crossing log *k* versus acetonitrile concentration lines obtained for the azidomorphine derivatives. (Reproduced with permission from Reference [23].)

reversed phase column. So, the answer to the question is that it is very likely that the lipophilic interactions were so strong between the tissue and the lipophilic compounds that it explains the seemingly irreversible binding to the biological matrix. This example demonstrates that the chromatographic system can model directly the biological partition/distribution processes.

In this, the chromatographic system was adjusted in such a systematic manner that it was able to model directly the biological partition processes. The optimization of the chromatographic mobile phase using the slope and intercept method has never become widespread. The main reason for this is that the reversed phase stationary phases have gone through a significant development using procedures that masked the adsorption effect of the silanol groups. With these improvements in stationary phases, the slope and intercept values (S and $\log k_w$) show good correlation, which means that the same retention order can be obtained at various organic phase concentrations even for structurally unrelated compounds. Thus, the ability to tune the reversed phase chromatographic system to model the octanol/water or any other biological system is greatly reduced.

It has been found that the quotient of the intercept and slope in Equation 5.4 ($\log k_w/S = \varphi_0$) also shows better correlation with the log P values than $\log k_w$ [23], as is shown by Equation 5.6.

$$\log P = d\varphi_0 + e \tag{5.6}$$

where d and e are regression coefficients. The φ_0 is the x-coordinate of the intersection of the log k versus φ lines, as demonstrated in Figure 5.12.

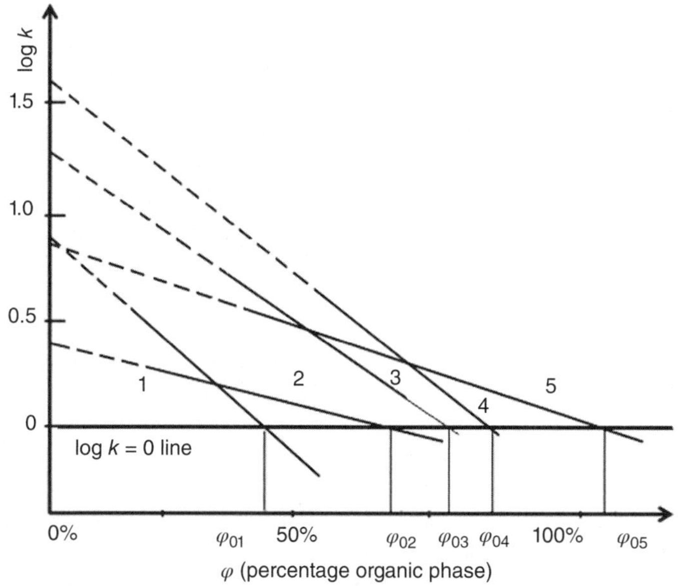

Figure 5.12 The intersection points of the log*k* versus φ lines for hypothetical compounds. The *x*-axis provides the φ_0 values of each compound, which are defined as the isocratic Chromatographic Hydrophobicity Index.

The quotient (φ_0) is equivalent to the organic solvent concentration in the mobile phase, which is required to get the compound retention time exactly twice that of the dead time, that is, $\log k = 0$. The φ_0 value of a compound is defined as the Chromatographic Hydrophobicity Index (CHI) [24]. It ranges from 0% to 100%, which indicates also an appropriate organic phase concentration by which the compound can be eluted from the column with a reasonably short time (retention time is double the dead time). On the basis of the data for over 500 compounds [24], an approximation can be made for the organic phase concentration in the mobile phase using acetonitrile and methanol based on calculated octanol/water partition coefficients (log *P*) of a compound using Equation 5.7 and Equation 5.8.

$$\varphi_{0(\mathrm{AcN})} = 16 \log P + 26 \tag{5.7}$$

$$\varphi_{0(\mathrm{MeOH})} = 13 \log P + 42 \tag{5.8}$$

The advantage of φ_0 is that it does not involve an extrapolation to outside a measurable range of retention as is often required to obtain the log k_w values. An important aspect of the isocratic CHI is that it has been derived from the retention data of compounds that were analyzed at a mobile phase pH when the compounds were not ionized and the octanol/water partition coefficients also were obtained for the neutral form of the compounds. The pH dependence of lipophilicity and the reversed phase chromatographic retention are discussed later in this chapter.

The other advantage of this approach is that direct linear relationship could be established between the log *P* values and the chromatographic hydrophobicity indices (φ_0), allowing the design of HPLC conditions for the separation of compounds with known structure and calculated log *P* values.

The major application of the CHI (or φ_0) was in gradient mode. It has been shown that the gradient retention times of the compounds showed acceptable correlation with the φ_0 values; so, it could be converted to the CHI values, which are the gradient chromatographic equivalent of the isocratic φ_0 values, as is described below.

LIPOPHILICITY MEASUREMENTS BY REVERSED PHASE CHROMATOGRAPHY WITH GRADIENT ELUTION

A linear gradient increase in the organic solvent concentration of the mobile phase during a chromatographic run helps elute compounds over a wide range of lipophilicity. In isocratic mode, several retention time measurements of the same compound are required to establish the CHI, while one measurement in gradient mode can cover compounds with approximately five orders of magnitude of difference in partition coefficient. In order to maintain the resolution of the separation, slow organic phase gradient methods are used. Mutton [25] has pointed out that by increasing the flow rate and using short columns, it is possible to reduce the gradient time without the loss of separation efficiency and resolution. The analysis time per compound can be reduced to 5 min. During a gradient run, any point of the run time is equivalent to a particular organic solvent concentration. By knowing the dead volume and the dwell volume (the volume of the organic solvent needed to reach the top of the column from the pump) of the HPLC system, it is possible to estimate the organic phase concentration at the time point when the compound elutes from the column. During a fast organic phase gradient, the solvent slope parameter *S* (Equation 5.3) has a negligible influence on the gradient retention time. When the appropriate organic phase concentration reaches the column, each compound will travel at approximately the mobile phase velocity. The retention time in a fast gradient run should be linearly related to the isocratic φ_0 values. It has been shown that the gradient retention times are in good correlation with the isocratically obtained φ_0 values using experimental data from 76 diverse drug molecules obtained from both isocratic and gradient elutions [26]. The gradient retention times of a calibration set of compounds give a straight line when plotted against the φ_0 values. The slope and the intercept of the calibration line can be used to convert the gradient retention times to a gradient CHI, which is suitable for interlaboratory comparison and database building [26]. The CHI measurement can take only 5 min using standard generic fast reversed phase gradient HPLC conditions. The conditions used cover a 5–6 log *P* unit range of lipophilicity, and simple data processing can be used to convert the gradient retention times to CHI values. The CHI values can also be projected to the widely used logarithmic value of partition coefficient (log *P*) by *linear conversion* [27].

It is important to note that the gradient retention times show a linear relationship with the logarithmic value of the isocratic retention factor. Equation 5.9 shows the CHI conversion to CHI log D [27], while Equation 5.10 shows the CHI conversion to Chrom log D [28].

$$\text{CHI } \log D = 0.054 \text{ CHI} - 1.467 \tag{5.9}$$

$$\text{Chrom } \log D = 0.0654 \text{ CHI} - 2 \tag{5.10}$$

The constants for the conversion of CHI log D from CHI have been obtained by plotting measured CHI values for 98 known drug molecules and their measured logarithmic octanol/water partition coefficient (log P). The constants for the Chrom log D conversion have been derived from the plot of CHI values and calculated log P values (c log P) for over 20,000 research compounds.

Kerns et al. [29] used the gradient retention times of a calibration set of compounds with known log D octanol/water partition coefficients and converted the gradient retention times directly to octanol/water log D values. The gradient retention times correlated approximately with the log D values using a Polaris C-18 column and acetonitrile gradient. Donovan and Pescatore [30] have reported the use of gradient methods to obtain a wide range of measured lipophilicity values from gradient retention times, but they used an octadecyl polymer column and a methanol gradient. They claim their values to be close to octanol/water partition coefficients. The starting mobile phase pH is 7.4. The biggest advantage of this methodology is the provision of a lipophilicity parameter directly from the liquid chromatography-mass spectrometry (LC-MS) compound integrity/purity measurements. In general, by using gradient HPLC and C-18 phases, reproducible lipophilicity values can be obtained in a high throughput way, while the perfect match with the traditionally used octanol/water partition coefficients is lost. It should be noted that there are significant differences between log D values derived from gradient retention times and the octanol/water log D values. For neutral molecules, H-bond donor compounds generally show a lower chromatographic log P than the octanol/water log P. Charged molecules, however, tend to give higher log D values in the chromatographic system than octanol/water log D values (as is discussed later). So, the conversion is pragmatic and it is used only for expressing the CHI scale (normally 0–100) as a log P scale (normally ranging from −1 to 5). CHI values of more than 200,000 research compounds have been collected at GlaxoSmithKline and proved to be useful in numerous lead optimization projects. Camurri and Zaramella [31] have adapted the methodology for mass spectrometry (MS) detection of compounds during the gradient run. The MS detection makes it possible to determine CHI values for mixtures of compounds this way, further reducing the analysis time. Equivalent selectivity for the fast gradient and the isocratic systems has been demonstrated by Du et al. [32]. The connection between the gradient and isocratic retention times has been discussed extensively by Snyder [33] and utilized in DryLab software, which computes isocratic method development conditions based on a few carefully selected gradient measurements.

Typical Measurement Condition for Lipophilicity Determination by Reversed Phase Gradient HPLC

A typical experimental procedure with a suggested set of compounds for calibrating the gradient retention times is described in this section.The experimental condition for CHI lipophilicity determination on C-18 stationary phases [34] using three different mobile phase pH values is as follows:

HPLC column: Luna C(18)2 5 μm 50 mm × 3 mm Phenomenex Ltd, United Kingdom

Mobile phase A1 (pH 7.4): 50 mM ammonium acetate pH 7.4

Mobile phase A2 (pH 10.5): 50 mM ammonium acetate plus concentrated ammonia solution to adjust the pH to 10.5

Mobile phase A3 (pH 2): 0.01 M phosphoric acid

Mobile phase B: acetonitrile (HPLC grade)

Flow rate: 1.0 ml/min

Gradient: 0–2.5 min 0%–70% B

2.5–3 min 70% B

3.0–3.2 min 0% B

3.2–5 min 0% B

Any HPLC instrument that is able to produce fast gradient analysis with a lowest possible dwell volume is suitable. UV detection of the compound can be used when a UV chromophore is available in the analyte. As was highlighted previously, the calibration of the retention time is essential. Any set of compounds that covers a wide range of lipophilicity and chemical diversity can be used for calibration. However, for accurate data that is useful for interlaboratory comparison, the constants for the gradient system should be determined by isocratic measurements. Alternatively, one can use the calibration set of compounds presented in Table 5.1 and use the data given for retention time calibration. The gradient retention times of the compounds are converted to CHI values using the obtained slope and intercept values of the calibration plots. The CHI values can be converted to the octanol/water partition coefficient scale (log D) using the formula given in Equation 5.9. Table 5.1 shows the calibration set of compounds and their CHI values at three different pH values obtained from isocratic measurements with typical gradient retention time data.

Figure 5.13 shows a typical calibration plot obtained between gradient retention times and CHI values for the calibration set of compounds listed in Table 5.1. The correlation coefficient of the fitted straight line is expected to be above 0.99. When the straight line fit is not as good, it shows a sign of some discrepancies in the chromatographic system used. It is possible that the column may not have been equilibrated properly with the starting mobile phase, or the mobile phase pH is different from the pH used to obtain the CHI values. It is important to use HPLC systems, in which the dwell volume is very small, which means that the actual

Table 5.1 The calibration set of compounds for the CHI lipophilicity determination at three different pH values and typical gradient retention times obtained under the conditions described in Reference [34]

Name	t_R pH 2	CHI 2	t_R 7.4	CHI 7.4	t_R 10.5	CHI 10.5
Theophylline	1.38	17.9	1.15	18.4	1	4.97
Phenyltetrazole	1.78	42.2	1.24	23.6	1.29	15.98
Benzimidazole	1.3	6.3	1.41	34.3	1.58	30.61
Colchicine	1.83	43.9	1.67	42.0	1.70	43.86
Phenyltheophylline	1.97	51.7	1.79	51.2	1.39	21.48
Acetophenone	2.12	64.1	1.98	65.1	2.05	64.12
Indole	2.29	72.1	2.15	71.5	2.19	72.07
Propiophenone	2.37	77.4	2.23	77.4	2.29	77.42
Butyrophenone	2.55	87.3	2.43	87.5	2.46	87.33
Valerophenone	2.72	96.4	2.60	96.2	2.61	96.36

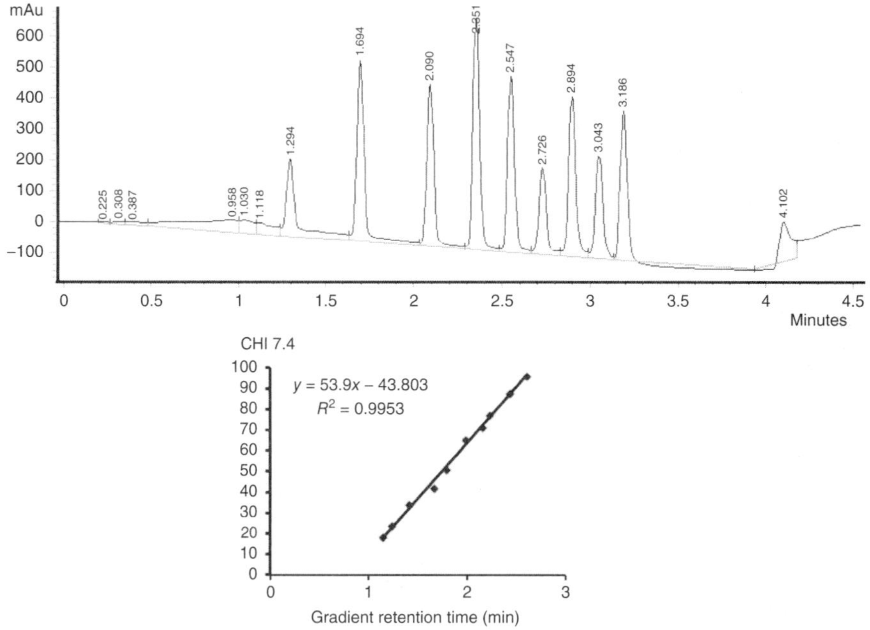

Figure 5.13 Typical calibration plot of gradient retention times and CHI values for the calibration set of compounds listed in Table 5.1. By fitting a straight line, the obtained slope and intercept values can be used to convert the gradient retention time of a new compound to CHI values.

organic phase concentration on the top of the column is very close to the nominal value at which the pumps mix the aqueous and the organic solvents in the mobile phase. When all these conditions are met and the straight line fit is still very poor, it suggests that the HPLC column has a different selectivity from the usual C-18

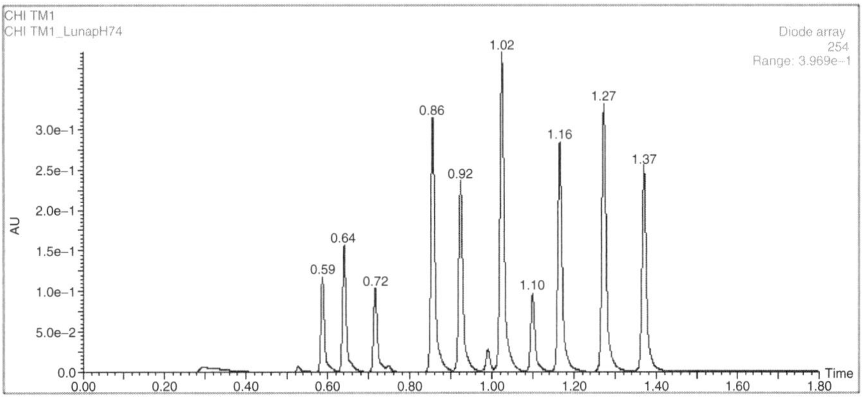

Figure 5.14 Typical chromatogram of the CHI calibration set of compounds obtained Waters ACQUITY UPLC system with LunaC-18 column (2.1 mm × 3.0 mm) fast acetonitrile gradient. (Courtesy of Ms Shenaz Bunally, GSK.)

reversed phase columns. Then it is advisable to determine the φ_0 values of the compounds from isocratic measurements using the same column and fit the gradient retention times with the obtained φ_0 values.

In principle, any C-18 column can be used with acetonitrile gradient. By reducing the column diameter, the mobile phase flow rate can also be reduced proportionally to keep the linear velocity of the mobile phase flow approximately the same. For the reduction of analysis time, the mobile phase flow rate has to be increased proportionally in order to keep the gradient slope similar. It is advisable to inject the calibration set of compounds into a mixture and check that the baseline separation of all components has been achieved. Thus, it is possible to use narrow bore columns (0.2–0.1 mm diameter) packed with ultrahigh performance stationary phase (particle size around 1 μm) in an Ultra Performance Liquid Chromatography (UPLC) system. In such systems, the lipophilicity measurement for a compound can take less than 2 min. An example is shown in Figure 5.14.

LIPOPHILICITY OF CHARGED MOLECULES—pH DEPENDENCE OF LIPOPHILICITY

Drug molecules often contain ionizable groups; this means that their ionization state depends on the pH of the environment. When the compound is ionized, the lipophilicity drops dramatically and so does the octanol/water distribution coefficient. The logarithmic value of the octanol/water partition coefficient of the unionized molecule is denoted as log P, while the distribution coefficient of the ionized and unionized species at a given pH is denoted as log D. The partition of ionized species into the octanol phase is very small unless a suitable neutral ion pair is formed. The Henderson–Hasselbalch equation describes the relationship between pH, the acid dissociation constant (pK_a), and percentage ionization of compounds. It forms the basis of the equation, which describes the pH dependence of distribution coefficients ($\log D$), as shown by Equation 5.11a for acids and Equation 5.11b for bases:

$$\log D_{\mathrm{pH}} = \log(P \times 10^{(\mathrm{pH}-\mathrm{p}K_a)} + D_{\text{fully ionized}}) - \log(1 + 10^{(\mathrm{pH}-\mathrm{p}K_a)}) \quad (5.11a)$$

$$\log D_{\mathrm{pH}} = \log(P \times 10^{(\mathrm{p}K_a-\mathrm{pH})} + D_{\text{fully ionized}}) - \log(1 + 10^{(\mathrm{p}K_a-\mathrm{pH})}) \quad (5.11b)$$

Figure 5.15 shows a typical log D versus pH profile for an acid with $pK_a = 8$, while Figure 5.16 shows the log D versus pH profile for a base with $pK_a = 8$.

The partition coefficient of the fully ionized form of the molecules is approximately three to four orders of magnitude smaller than the neutral form's partition

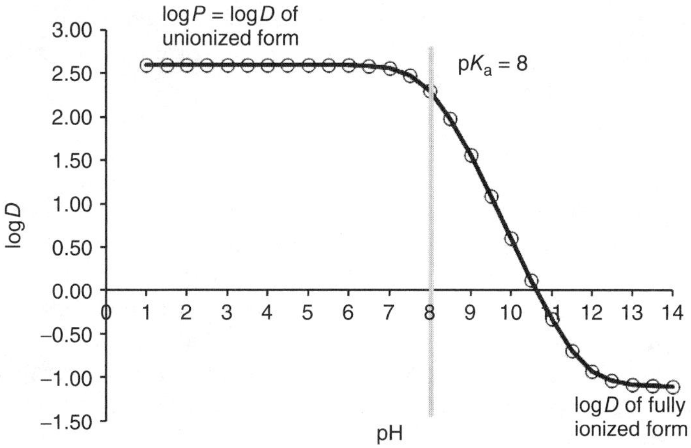

Figure 5.15 Typical log D–pH profile for an acid ($pK_a = 8$).

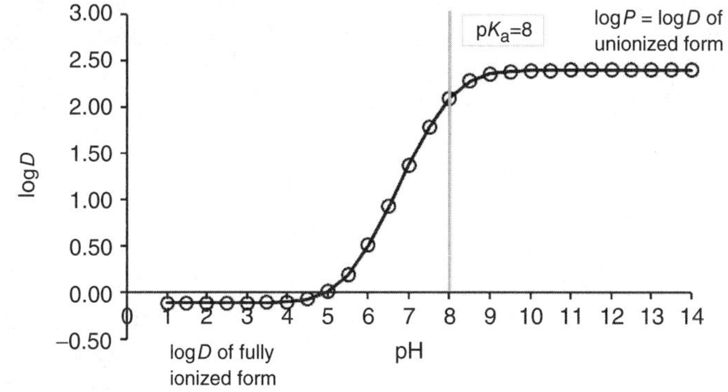

Figure 5.16 Typical log D–pH profile for a base ($pK_a = 8$).

coefficient (depending on the compound). Note that when 50% ionization occurs (pH = 8), the lipophilicity of the compound is only 0.3 (log 2) less than the lipophilicity of the neutral form.

Reversed phase retention depends on the pH of the aqueous mobile phase. The presence of charge on a compound decreases its reversed phase retention [35] as its lipophilicity decreases, similar to the octanol/water partition. However, the effect of charge on chromatography is different from that observed in solvent–solvent partitions. The reason for this is that the chromatographic distribution takes place on a surface and the stationary phase molecules do not solvate the solute molecule completely. It means that large lipophilic molecules can be retained on the surface of the lipophilic stationary phase even if they have a positive or negative charge, which would prevent their solvation in a lipophilic solvent. Such amphiphilic molecules would accumulate in the interface in a solvent–solvent partitioning system. It means that charged molecules might show higher lipophilicity in a reversed phase chromatographic system than in a solvent–solvent partition system. Therefore, we may observe significant discrepancies between bulk solvent partition and the chromatographic partition. As the biological partition processes also involve large surfaces, for example, drug molecules are attached to the hydrophobic surface of protein or membrane, the chromatographic system might be a better model.

Canals et al. [36] investigated the gradient retention time versus pH profile using acetonitrile gradients. A complication is that the mobile phase pH is not constant during gradient elution. It is influenced by the acetonitrile concentration. Canals et al. [36] have reported that the pH 2 starting mobile phase increased up to nearly pH 5 when the acetonitrile concentration reached 95%. The high pH starting mobile phase pH dropped from pH 10.5 to pH 8.5 when the acetonitrile concentration changed from 0% to 95%. Interestingly, the neutral pH (pH 7.4) remained stable when ammonium acetate buffer and acetonitrile gradient were used, as shown in Figure 5.17.

However, the compound's pK_a values are also affected by the presence of the organic phase in the aqueous buffer. Figure 5.18 shows the log D versus pH profile of 4-*tert*-butylpyridine obtained from the gradient CHI expressed as CHIlog D, and the pH profile expected from the octanol/water partition system. The result of the study is that an empirical constant needs to be introduced for weak acids and weak bases into Equation 5.11a and Equation 5.11b to fit the gradient retention time versus pH profile obtained experimentally, as shown by Equation 5.12.

$$gt_R = \frac{\left(gt_{R(HA)}\ 10^{s(pK_a - pH)} + gt_{R(A)}\right)}{\left(10^{s(pK_a - pH)} + 1\right)} \tag{5.12}$$

Fuguet et al. [37] investigated the gradient retention time (CHI) variations for ionizable solutes using acetonitrile and methanol gradient elutions at various pH values. They found that the empirical s parameter in Equation 5.12 depends on how similar the solute pK_a and the solvent pH change during gradient elution.

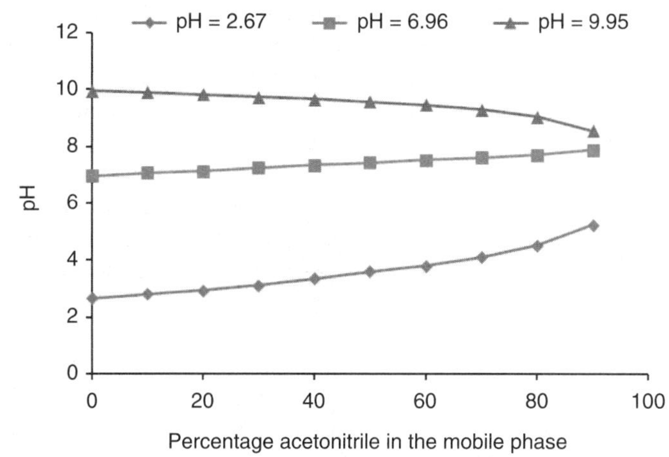

Figure 5.17 The change of the buffer pH by mixing with various amounts of acetonitrile based on the data from Reference [36].

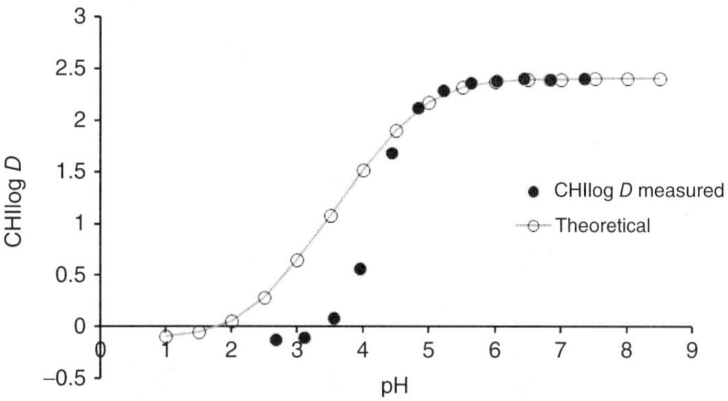

Figure 5.18 The theoretical CHIlog *D* versus pH profile (open circle by Equation 5.11b) and the experimental CHIlog *D* versus pH profile (full circles by Equation 5.12) for a weak base.

They found that that there was no difference in s values for acids in methanol or acetonitrile, while basic compounds had slightly lower s values in methanol. These differences could be explained by the different effects of the mobile phase pH change and the pK_a change of compounds.

Thus, it is possible to measure the gradient retention time (gt_R) of a compound using several starting mobile phase pH values and fitting Equation 5.12 to the gt_R versus pH plot and calculate the compound's pK_a. The precision of the pK_a determination in this way depends on the number of gradient measurements using various starting mobile phase pH values. It means that even if we use a fast generic pH gradient that takes only 4 min, it would take an hour to measure the pK_a of a compound using 15 experiments at different pH values. In order to have an approximate idea of the compounds' acid/base character, in practice, we use only three gradient retention time measurements, low pH (pH 2–3), physiological pH (pH 7.4), and high pH (pH 9–11). We calculate the CHI at three different pH values and by applying a simple algorithm, the compounds can be grouped into six categories, namely, acid, base, neutral, weak acid, weak base, and amphoteric (Figure 5.19). Zwitterionic compounds that have the same proportion of ionized molecules at all pH values appear as neutral compounds; so, the algorithm is not able to differentiate them from neutral compounds. Therefore, it is advisable to check the chemical structure of the molecules too.

This simple algorithm is based on the comparison of obtained CHI values at the three different pH values. If the CHI values are the same (within 5 CHI units) at all three pH values, the compound is considered to be neutral at physiological pH. We have to be careful, however, as zwitterionic molecules that are always ionized (either positive or negative or both) could behave as neutral. When the CHI value is significantly greater (more than 5 CHI units) at low pH, but they are the same (within 5 CHI units) at pH 7.4 and high pH, the compound is considered as acidic, and we can be sure that it contains negatively charged molecules at physiological pH. When the same CHI values are obtained at low pH and pH 7.4, but it is significantly lower (more than 5 CHI units) at basic pH, then the compound is assigned as weak acid and it is considered neutral (unionized) at pH 7.4. When a compound has the same CHI value at pH 7.4 and basic pH, but the CHI is significantly lower at acidic pH, the compound is assigned as weak base, and it is considered as neutral (unionized) at pH 7.4. Basic compounds show significantly higher CHI values

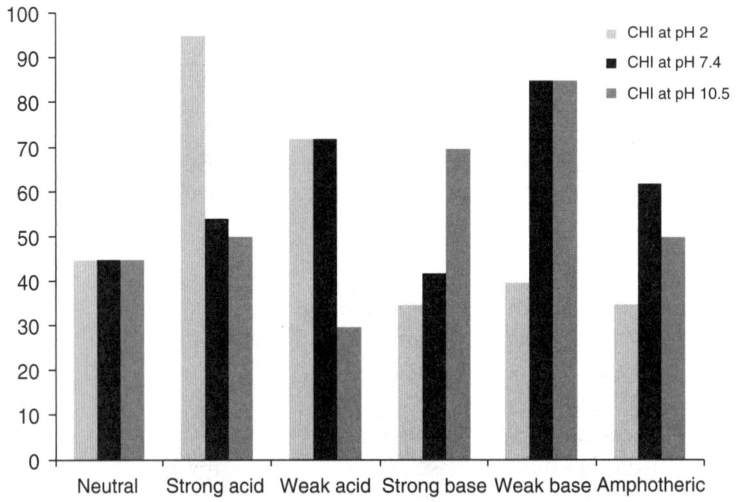

Figure 5.19 Categorizing compounds according to their acid/base character using the Chromatographic Hydrophobicity Index at three pH values (acidic, neutral, and basic).

(more than 5 units) at higher pH values, and are positively charged at physiological pH. In order to increase the throughput of HPLC-based lipophilicity determination of compounds, several instrument vendors came up with an idea of parallel HPLC equipment. Injecting the compound into several chromatographic columns with different mobile phases at the same time not only decreases the time for compound characterization but also reduces the cost of instrumentation. This is because several modules of the instrument can serve all of the combined HPLC systems, for example, one injector, one data acquisition system, and multichannel detector. Figure 5.20 shows an Agilent four-way parallel HPLC system that has been built using a CTC autosampler that can inject into four HPLC systems simultaneously; four binary gradient pumps, four column holders with temperature control, four multiwavelength UV detectors, and one computer with HPCHEM software control the system and collect data into one file from the four detectors.

A typical four-channel chromatogram of a basic compound is illustrated in Figure 5.21. From the retention time, one can easily assess the compound's lipophilicity—pH profile and membrane partition data using an IAM stationary phase as one of the channels, which is discussed later. The retention times obtained at acidic, neutral, and basic mobile phase pH values show the increase of retention as the compound loses the positive charge with increasing pH. It is also evident that the compound is still partially protonated (positively charged) at pH 7.4 as the retention time is shorter than at pH 10.5. For acids, the retention time order would be the opposite. The longest retention would have been observed at low pH, while the retention time decreases as the pH increases. Although the pK_a values cannot be determined using the retention data at three pH values only, it is enough to assess the ionization state of the compound at physiological pH (pH 7.4).

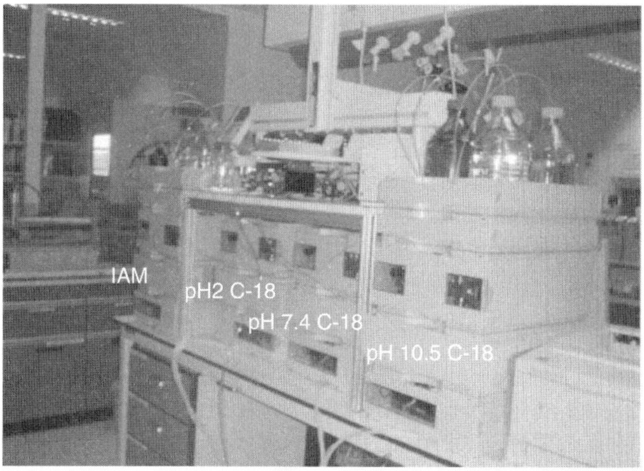

Figure 5.20 An example of a four-way Agilent HPLC system that has a CTC autosampler injecting the same sample into four columns; one computer controls the four HPLC gradient pumps and collects data from four detectors. (*See insert for color representation of this figure.*)

Kaliszan et al. [38,39] have published another approach for the determination of the acid dissociation constant (pK_a) and lipophilicity of compounds. They have proposed the use of only two gradient reversed phase chromatographic measurements. The first measurement should be carried out using a starting mobile phase pH at which the compound is not ionized (basic compounds high pH and acidic compounds low pH) and using an organic phase (methanol or acetonitrile) gradient. On the basis of the compound's retention time in the first experiment, the organic phase concentration that results in a retention factor greater than 10 should be calculated. The second experiment is a pH gradient that starts at high pH for basic compounds and reduces to low pH, and starts at low pH for acidic compounds and increases to high pH. The organic phase concentration should be kept constant. From the retention time of the second experiment, the compound's pK_a can be calculated.

The question may arise as to how it is possible that we see only one chromatographic peak of a compound, which is present both in ionized and unionized forms during the chromatographic process. If the two forms have very different lipophilicities, there should be a significant difference in their chromatographic retention time as well. This is a valid question. One possible answer is that the dissociation process is much faster than the compound partitioning equilibrium into the stationary phase. Therefore, there is no time to separate the fully ionized form of the molecules from the uncharged form as the protonation or dissociation is so fast that in the moment they are separated, the equilibrium reestablishes. However, it is very rare to see two peaks from one compound that is partially ionized. It happens mostly for basic compounds when the protonation occurs between two basic nitrogen atoms and the charge is delocalized. The protonation equilibrium is very slow and the separation is faster. The two peaks are not fully separated even in that case, and the peaks are much wider than one could expect from the efficiency of the separation system.

BIOMIMETIC LIPOPHILICITY MEASUREMENTS BY CHROMATOGRAPHY

One of the great advantages of using HPLC technology for measuring properties of compounds is that various types of stationary phases can be applied including ones that mimic biologically important constituents such as membranes and proteins.

Immobilized Human Serum Albumin (HSA) Stationary Phase

Protein stationary phases chemically bonded to silica particles were originally developed for chiral separations [40]. There are two commercially available protein-bonded stationary phases, the HSA and α-1-acid glycoprotein (AGP). They are important not only for chiral separation but also for measuring compound interaction with these proteins. HSA and AGP can be found in human and animal plasma. While plasma proteins contain around 60% HSA, the AGP concentration is

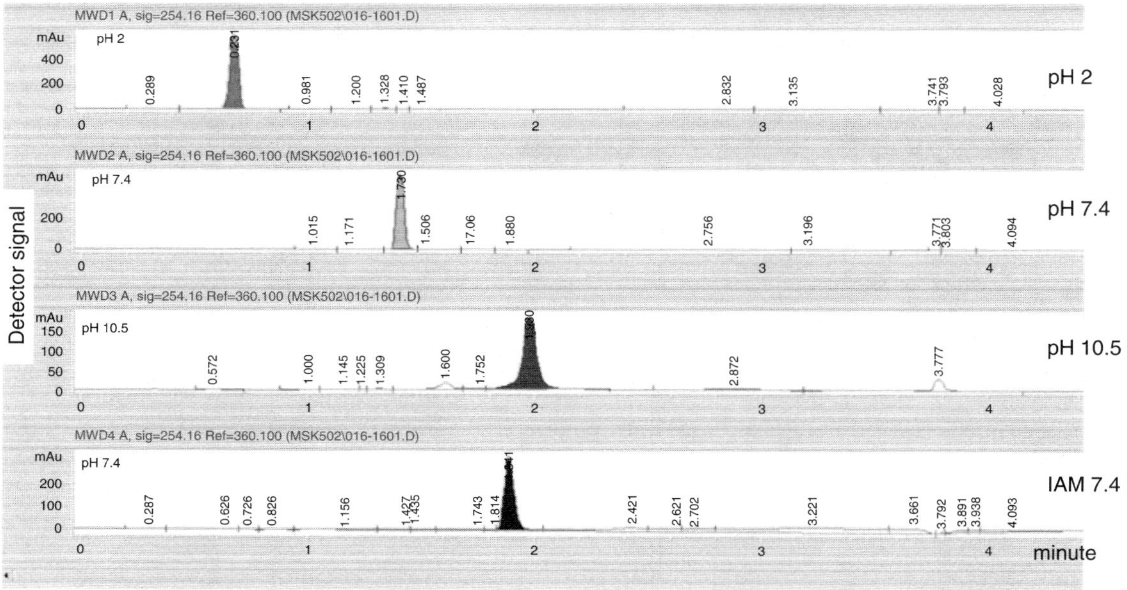

Figure 5.21 Typical four-way chromatogram of a basic compound. The retention times obtained at acidic, neutral, and basic mobile phase pH values show the increase in retention as the compound loses a proton and becomes less positively charged.

much lower (1%–3%). Although the AGP concentration in plasma is relatively low, it has great significance as it can bind basic (positively charged) drug molecules very strongly, while HSA binds the lipophilic and negatively charged compounds. The other major components of the plasma proteins are immunoglobulins, which bind compounds based on their lipophilicity regardless of charge. Several authors have presented good correlations between the logarithmic retention factors measured on the HSA stationary phase and plasma protein binding of drugs obtained by equilibrium dialysis or ultrafiltration methods [41–44]. The principles of using immobilized target biopolymers to measure drug affinity by HPLC have been reviewed by Bertucci et al. [45]. However, Buchholz et al. [46] reported a lack of correlation between plasma protein binding and HPLC retention measured on HSA columns. There are two important aspects that should be taken into account when comparing the HPLC method with the equilibrium dialysis method for measuring compound binding to plasma proteins. First, the unit of the binding measure has to be comparable. Plasma protein binding is most often expressed as percentage bound or fraction unbound, while in HPLC, the logarithmic retention factor measures the partition coefficient of the compound between the aqueous mobile phase and the protein stationary phase. The two measures are in sigmoid relationship with each other, as can be seen in Figure 5.22. As is shown, the relatively big changes in log k result in minor changes in percentage bound at the high binding region. Both measures of binding can be calculated from each other using Equation 5.13 and Equation 5.14.

$$\%\text{HSA bound} = 100\left(\frac{k}{k+1}\right) \tag{5.13}$$

$$\log K(\text{HSA}) = \log\left(\frac{\%\text{HSA bound}}{100 - \%\text{HSA bound}}\right) - \log[\text{HSA}] \tag{5.14}$$

where k is the chromatographic retention factor on HSA stationary phase and K is the binding affinity constant. However, this equation is valid only for nonspecific binding, when compound retention is independent of the injected amount and the overall binding can be modeled by partition. The chromatographic retention factor k is equivalent to the albumin partition coefficient only if the volumes of the stationary and mobile phases are equal (V_s/V_m is unity, see Equation 4.14).The affinity constant also depends on the HSA concentration, which may vary. Figure 5.22 also shows that the same binding affinity constant may represent a different value when it is expressed as percentage bound depending on various physiological (or pathological) concentrations of HSA in the plasma. The second typical mistake is to compare the value of the albumin binding to that of plasma protein binding. Plasma contains only 60% HSA; therefore, it is very likely that the compound will bind to a certain extent to other plasma proteins. The deviations are most often observed with basic compounds that are positively charged at physiological pH values and bind strongly to AGP, for example. Therefore, for comparing various methodologies, it is important to compare the data that refers only to the albumin binding as Kratochwil et al. [47] rightfully did when they compared literature HSA binding

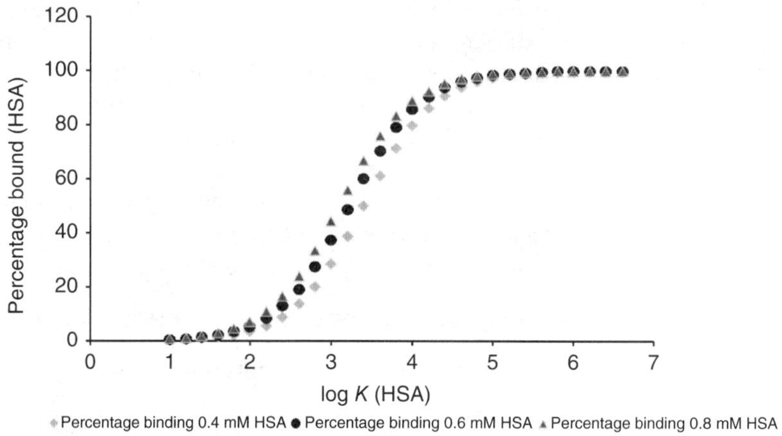

Figure 5.22 The nonlinear relationship between the logarithmic retention factor (log *k*) on HSA stationary phase and percentage-bound HSA.

data from various methodologies. It is also important to keep in mind that the chromatographic method measures the proportion of the average number of moles of the drug in the mobile and the stationary phases during the dynamic equilibrium process. Thus, it measures not only the compounds binding to one particular binding site of the albumin, but many other nonspecific binding sites as well. We cannot therefore assume a one-to-one binding between the albumin and drug molecules.

From a methodological point of view, it is also important to imitate the physiological condition during the chromatographic retention measurements. The best approach is to measure the binding at the same pH as the plasma (pH 7.4) and use isocratic conditions preferably without using an organic modifier in the mobile phase As it was found by Tiller et al. [42], the $k/(k+1)$ proportion derived from the retention factor multiplied by 100 gave very close values to the reported plasma protein binding percentage for a variety of drug molecules. They used small amount of iso-propanol (3%) and pH 7.4 buffer in the mobile phase, which provided good agreement with the binding condition in the plasma. They have obtained similar binding data from the mixture to the data obtained from individual compounds, which suggested that compounds did not interfere with each other during the binding measurement. Valkó et al. [44] applied a fast iso-propanol gradient during the retention time measurements in order to reduce the analysis time for strongly bound compounds. The calibrated logarithmic retention time values showed acceptable correlation with the literature plasma protein binding data. This was endorsed by the study of Kratochwil et al. [47], reporting good correlation ($r^2 = 0.86$) when they compared chromatographic binding data with binding data obtained by equilibrium dialysis. However, the correlation using the data obtained by Colmenarejo et al. [48] who used acetonitrile as an organic modifier in the mobile phase was not as good. This is possibly because the acetonitrile changes the natural three-dimensional structure of the albumin and changes its binding characteristics. The practical recommendation for the HSA HPLC columns for chiral separation is to use a small percentage of iso-propanol as organic modifier. This supports the hypothesis that under such conditions, the naturally chiral-selective binding sites are intact. Therefore, it is advisable to use an enantiomeric mixture of warfarin that binds to the major binding site of albumin to check the enantioselectivity of the stationary phase as it often deteriorates after extensive use of the column. On the basis of the comparative data and explanations, the following general procedure is suggested for HSA binding measurements by HPLC.

Typical Measurement Conditions of HSA Binding by HPLC Retention

The conditions for generic HSA binding measurements [44] are as follows:

HPLC column: immobilized HSA 50 mm × 3 mm (Chromtech or Chiral Technologies)
Mobile phase A: 50 mM ammonium acetate pH 7.4
Mobile phase B: 2-propanol (HPLC grade)
Flow rate: 1.8 ml/min
Gradient: 0–3.0 min 0%–30% B
3.0–10.0 min 30% B
10.0–10.5 min 0% B
10.5–15.0 min 0% B

Any HPLC instrument that is able to deliver the high flow rate and operate up to 600 bar backpressure can be used. Elevated temperatures up to 40 °C can be used to reduce the backpressure. The iso-propanol is a viscous solvent and when the 30% gradient is applied, the backpressure can be close to the operational limit of the HPLC instrument. It is possible to apply a lower flow rate, but the gradient slope should be kept the same, and the analysis time should be increased proportionally. UV detection is suitable for UV-active compounds. The application of mass spectrometric detection is very useful as isomers and enantiomers often produce multiple peaks for which the identification is important. Column performance checks and calibration should be performed before the analysis. The calibration set of compounds can be dissolved separately in 0.5 mg/ml concentration in 50% 2-propanol and 50% pH 7.4 ammonium acetate solutions. The calibration set of compounds and their literature percentage plasma protein binding data are listed in Table 5.2. For calibrating the retention time, we need to use the linear conversion (log K_{lit}) of the percentage plasma protein binding data. The typical retention times for the calibration set of compounds and their logarithmic values are also listed in Table 5.2. Figure 5.23 shows a typical chromatogram of warfarin. If the racemates are separated, it is considered that the warfarin binding site of the HSA is intact. Figure 5.24 shows a typical calibration plot. The logarithmic value of the gradient retention times (gt_{R}) obtained from the HPLC experiments was plotted against

Table 5.2 Calibration set of compounds with their literature and typical measured chromatographic data obtained with the HSA column[a]

	Literature Percentage of PPB	Linearized Percentage of PPB (log K_{lit})	gt_R	log t_R	log K from Calibration	Percentage of HSA Binding (From Calibration)
Nizatidine	35	−0.28	0.937	−0.03	−0.25	36.27
Bromazepam	60	0.17	1.2	0.08	0.09	55.95
Carbamazepine	75	0.46	1.63	0.21	0.52	77.66
Budesonide	88	0.83	1.837	0.26	0.69	83.85
Nicardipine	95	1.20	3.01	0.48	1.38	96.95
Indometacin	99	1.69	3.845	0.58	1.72	99.12
Piroxicam	94.5	1.16	2.821	0.45	1.29	96.06
Dichlofenac	99.8	1.92	4.2	0.62	1.85	99.58
Flubriprofen	99.96	1.98	4.3	0.63	1.88	99.68

[a]Literature data were obtained from Reference [44] with permission.

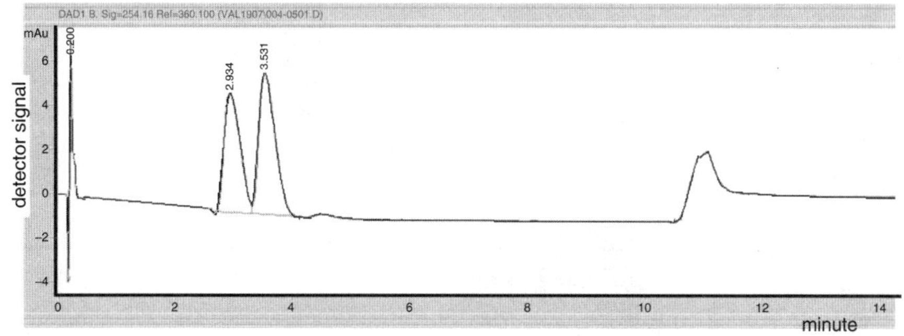

Figure 5.23 Typical chromatogram of racemic mixture of warfarin. Chromatographic conditions are described in the experimental section.

the linearized values of the percentage PPB (i.e., log K). The slope and the intercept were then used to convert the gradient retention time values to log K values for a new compound. From the so-obtained log K values, the estimated percentage protein binding was obtained by Equation 5.15.

$$\%\text{HSA bound} = \frac{101 \times 10^{\log K}}{1 + 10^{\log K}} \tag{5.15}$$

Typical chromatograms of the calibration set of compounds (without the flurbiprofen) are shown in Figure 5.25. The day-to-day reproducibility of the retention time is excellent. We can see that we have several minutes after eluting the diclofenac, which is the most highly bound compound in this mixture. It means that we can easily distinguish and rank very highly bound compounds.

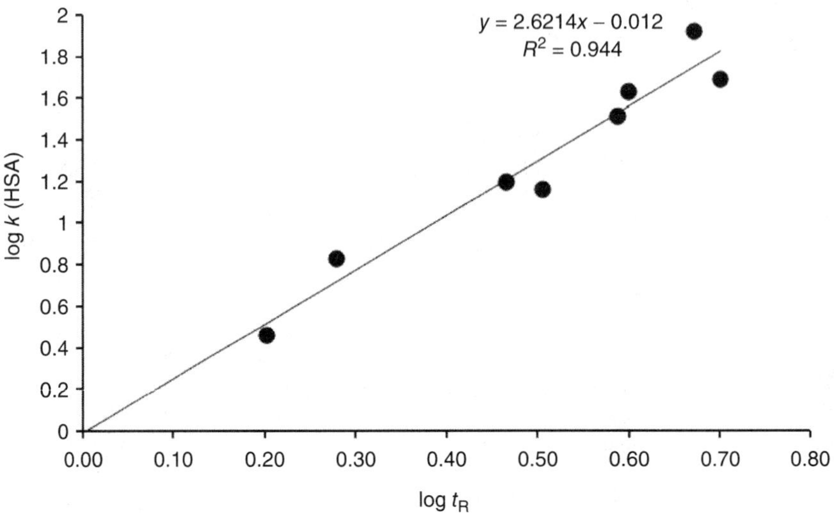

Figure 5.24 Typical calibration plot obtained on the HSA column.

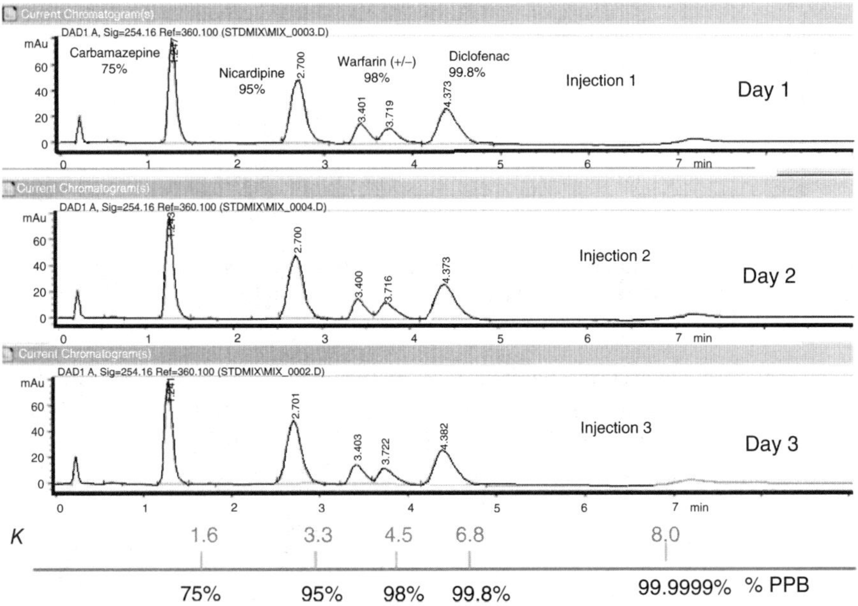

Figure 5.25 Typical chromatograms obtained for the calibration set of compounds using chemically bonded HSA as stationary phase and fast 2-propanol gradient with pH 7.4 ammonium acetate buffer.

Immobilized α-1-Acid-glycoprotein (AGP) Stationary Phase

Like the HSA stationary phase, the AGP stationary phase has been developed for chiral separations. The AGP stationary phase was studied by Jewell et al. [49] who concluded that the retention factors obtained on AGP columns did not show correlation with AGP binding data obtained by *in vitro* displacement experiments. This is not surprising as the displacement experiments measure compounds binding to a particular binding site of the protein where the probe molecule binds. The chromatographic approach based on retention time measurements indicates the compound's overall partitioning to the protein phase and, as such, includes specific and nonspecific binding. Good agreement has been found between the AGP binding of known drugs obtained by ultrafiltration and chromatographic methods [50], as is demonstrated in Figure 5.26.

Some of the positively charged compounds showed stronger binding by the HPLC-based measurements than by ultrafiltration. A possible explanation could be that under the ultrafiltration conditions the positively charged compounds saturated the negatively charged specific binding site because of the higher proportion of the drug concentration relative to the protein. The proportion of the drug concentration relative to the protein concentration in chromatography is much closer to the physiological condition. Kaliszan et al. [51] studied the binding behavior of positively charged drugs on the immobilized AGP stationary phase, and proposed a funnel-shaped binding for positively charged compounds being negatively charged at the narrow end of the funnel. As the carbohydrate content of AGP represents 45% of the molecular weight attached in the form of five to six highly sialylated

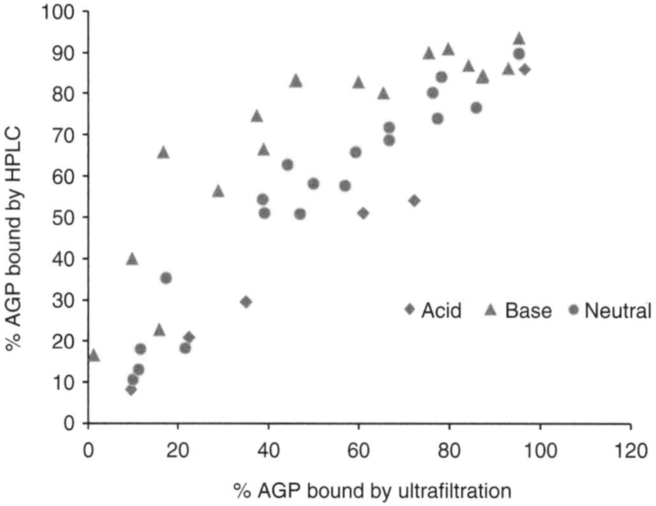

Figure 5.26 Comparison of AGP binding obtained by ultrafiltration method and chromatographic retention time measurements on immobilized AGP HPLC stationary phase. (From Reference [50] with permission.)

complex-type-*N*-linked glycans [52], it could explain the strong affinity to basic compounds.

There are other chromatographic methods such as frontal chromatography and displacement chromatography that can be used to probe a compound's interactions with proteins, as reviewed by Oravcova et al. [53]. For a quick assessment and ranking of compound binding to plasma proteins, the HPLC-based retention time measurements on immobilized protein stationary phases are recommended. It is advisable to apply a calibration set of compounds for which the retention time measurements are frequently repeated in order to check the column performance and suitability and to increase the day-to-day and instrument-to-instrument reproducibility of the data.

Typical Measurement Conditions of AGP Binding by HPLC Retention

The general experimental conditions for the AGP binding measurements are as follows:

Column: immobilized AGP 50 mm × 3 mm (Chromtech, UK or Chiral Technologies, France)

Mobile phase A: 50 mM ammonium acetate pH 7.4

Mobile phase B: 2-propanol (HPLC grade)

Flow rate: 1.8 ml/min

Gradient: 0–2.5 min 0%–25% B

2.5–4.5 min 25% B

4.5–4.6 min 0% B

4.6–6 min 0% B

Instrumentation and detection techniques are the same as described earlier. Column performance checks and calibration should be performed before the analysis. The compounds used for the column calibrations were dissolved separately in 0.5 mg/ml concentration in 50% 2-propanol and 50% pH 7.4 ammonium acetate solutions. The calibration set of compounds, their literature percentage AGP binding, and their linear conversion value using Equation 5.14 (log K_{lit}), as well as typical retention times are listed in Table 5.3 with their logarithmic values. Figure 5.27 shows a typical calibration plot.

It is important to observe that for the calibration during the protein binding measurements, we used the logarithmic values of the gradient retention time, unlike the CHI measurements. This is because the iso-propanol gradient is very shallow and does not differ significantly from an isocratic retention time measurement. The calibration of the chemically bonded protein column and the HPLC system is carried out to obtain directly a logarithmic binding constant derived from the percentage binding data obtained by equilibrium dialysis method.

Table 5.3 The compounds and their literature and experimental AGP binding data

Compound	Percentage Binding (Literature)	log K_{lit} AGP	t_R	log t_R	log K from Calibration	Percentage AGP form Calibration
Nizatidine	36	−0.257	1.248	0.10	−0.25	36.33
Bromazepam	50	−0.009	1.601	0.20	0.03	52.23
Warfarin	88	0.831	2.877	0.46	0.76	86.01
Propranolol	86	0.758	2.997	0.48	0.81	87.59
Imipramine	91	0.959	3.167	0.50	0.87	89.02
Nicardipine	92	1.010	3.349	0.52	0.93	90.32
Chlorpromazine	92	1.010	3.975	0.60	1.15	94.34

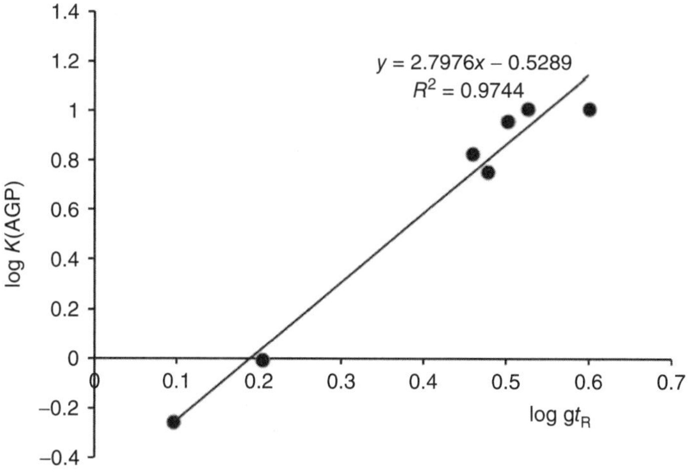

Figure 5.27 Typical calibration plot for AGP binding measurements obtained by plotting the literature log K(AGP) binding data as a function of the logarithm of gradient retention time (log gt_R).

Immobilized Artificial Membrane (IAM) Stationary Phase

Pidgeon and Venkataram [54] patented a method for immobilizing phospholipids on HPLC-grade silica stationary phases. The immobilized phospholipids mimic the lipid environment of a fluid cell membrane on a solid matrix (see Figure 5.28 for illustration). Nowadays, various IAM HPLC stationary phases are commercially available. The IAM stationary phases are prepared by covalently bonding the phospholipids to amino-propyl silica stationary phases. The remaining amino-propyl groups and free silanol groups are treated to reduce their secondary interactions with the compounds. The most frequently used IAM stationary phase contains covalently bonded phosphatidyl choline covering the silica particles at monolayer density. There are excellent reviews about the various commercially available IAM stationary phases and their applications for determination of

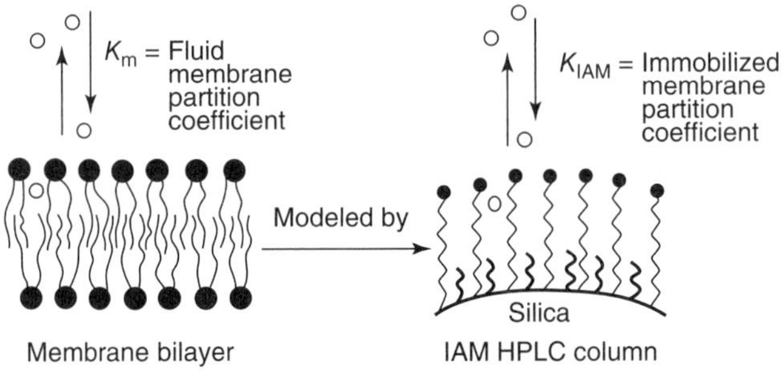

Figure 5.28 The similarity between the IAM stationary phase and a phospholipid bilayer.

compound binding to phospholipids by Taillardat-Bertschinger et al. [55,56] and Giaginis and Tsantili-Kakoulidou [57]. Stewart et al. [58] reviewed the application of IAM chromatography for modeling drug transport. Usually, isocratic methods are suggested, incorporating a very low volume (15%) of acetonitrile in the mobile phase. The logarithmic retention factor or its extrapolated value to the 0% organic concentration is suggested for derivation of membrane partition data from isocratic retention factors. Taillardat-Bertschinger et al. [56] reviewed the application of IAM chromatography in drug discovery in great detail. A gradient method has been developed and published [59], which applies acetonitrile gradient on IAM HPLC columns and provides a CHI(IAM). The gradient retention times are calibrated with the data obtained from isocratic measurements. The measurements and ranking of the phospholipid binding of drug discovery compounds can be used for characterizing and prioritizing compounds for progression.

Typical Measurement Conditions of IAM Binding by HPLC Retention

The following conditions can be applied for the phospholipid binding measurements [59].

Column: IAM column (Regis Analytical Ltd) 150 mm × 4.6 mm
Mobile phase A: 50 mM ammonium acetate pH 7.4
Mobile phase B: acetonitrile (HPLC grade)
Flow rate: 2.0 ml/min
Gradient: 0–2.5 min 0%–70% B
  2.5–3 min 70% B
  3.0–3.2 min 0% B
  3.2–5 min 0% B

Instrumentation and detection are as described earlier. The column performance check and the calibration should be performed before the analysis. The compounds

Table 5.4 The calibration set of compounds and their literature [59] CHI IAM values

Compound	Typical Retention Time t_R	CHI IAM	log k IAM
Octanophenone	3.18	49.4	2.69
Heptanophenone	3.07	45.7	2.52
Hexanophenone	2.94	41.8	2.34
Valerophenone	2.79	37.3	2.14
Butirophenone	2.58	32	1.89
Propiophenone	2.35	25.9	1.61
Acetophenone	2.04	17.2	1.21
Acetanilide	1.85	11.5	0.95
Paracetamol	1.62	2.9	0.55

used for the column calibrations are dissolved separately in 0.5 mg/ml concentration in 50% acetonitrile and 50% pH 7.4 ammonium acetate solution mixtures. The CHI (IAM) values for the calibration set of compounds are measured isocratically [59] and listed in Table 5.4. A typical calibration plot is shown in Figure 5.29.

The CHI (IAM) values were converted to the isocratically equivalent log k IAM. The formula for the conversion (Equation 5.16) was obtained by plotting the isocratic log k values (extrapolated for 100% aqueous mobile phase) against the CHI (IAM) values [59].

$$\log k(\text{IAM}) = 0.046 \times \text{CHI(IAM)} + 0.42 \tag{5.16}$$

We shall discuss later how useful this methodology is for highlighting potential developability issues with drug discovery compounds. The biomimetic HPLC binding data are suitable for deriving structure-binding relationships, helping the

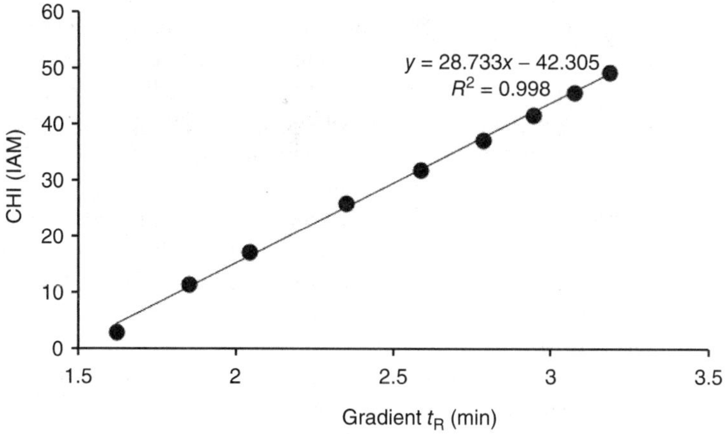

Figure 5.29 Typical calibration curve for CHI (IAM) measurement.

chemists to modify the structure in order to reduce the strong, often specific binding of the molecules. The biomimetic partition coefficients obtained by chromatography are the sum of all the specific and nonspecific binding interactions of the compound with the stationary phase. The predominant force of compound's binding is the hydrophobic force. If the dominant binding is specific to a particular binding site, the retention time of the compound would depend on the injected amount, and a nonsymmetrical peak is observed. When nonspecific hydrophobic interactions take place, we can see a strong correlation of compound protein and phospholipid binding with the lipophilicity, which is proportional to the octanol/water partition coefficient.

COMPARING VARIOUS LIPOPHILICITY MEASURES BY THE SOLVATION EQUATION MODEL

The question arises that if the protein binding and the phospholipid binding are governed only by the compound's lipophilicity, why compounds with similar lipophilicity can have different *in vivo* distributions. For example, chlorpheniramine, diazepam, and probenecid have similar $c\log P$ (3.15, 3.17, and 3.37, respectively) and they have very different volumes of distribution (3.4, 1.5, and 0.2 l/kg, respectively). Are we just measuring the same property of the compound in various different ways? How do the octanol/water partition coefficients correlate with the chromatographic measure of lipophilicity? The plot of the chromatographic $\log P$ (CHIlog P) versus the octanol/water $\log P$ is shown in Figure 5.30 as an example. We can observe that the points for compounds without H-bond donor groups form a line, while compounds with H-bond donor groups show higher octanol/water partition ($\log P$).

The major difference between the reversed phase partition and the octanol/water partition is their very different sensitivity to H-bond donor groups on the molecules. The octanol easily accommodates H-bond donor groups based on the H-bond acceptor property of the hydroxyl groups, while the nonpolar alkyl-substituted reversed phase stationary phases will not interact with H-bond donor groups. In order to be able to explain the differences between various lipophilicity measures, we need to have a good understanding of the molecular driving forces behind compound partitioning between the aqueous and nonaqueous phases. We need to know which molecular property drives the various partition processes. For this purpose, the Abraham solvation equation model [60] can provide a quantitative insight. The Abraham solvation equation model starts with the description of the solvation forces/energy from the gas phase to a liquid phase. Figure 5.31 illustrates the basis of the model.

The first step of the solvation is creating a cavity in the solvent in order to fit the solute molecule into it. The cavity formation needs energy as we have to break the interaction between the solvent molecules to create the cavity. This energy is especially large when water is the solvent as the water molecules form a structure via the dipole–dipole interactions and the H-bonding between the water molecules.

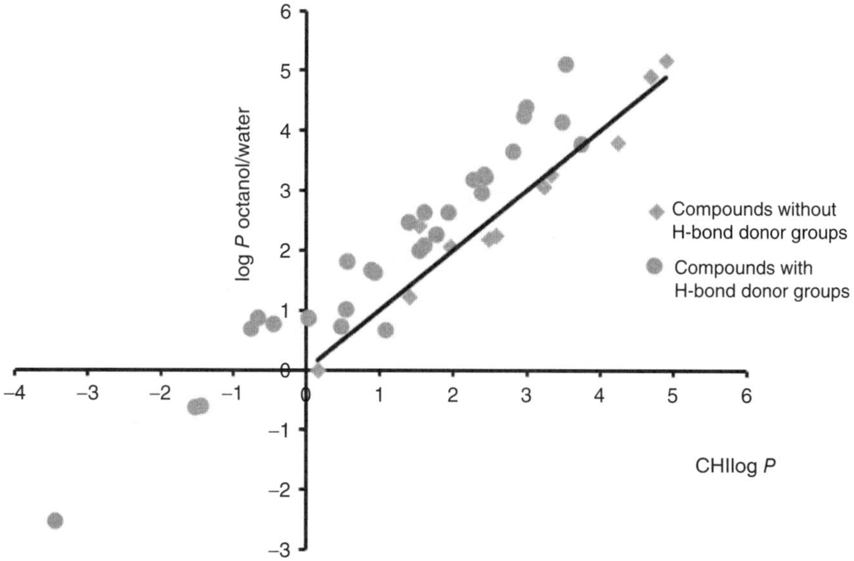

Figure 5.30 The plot of the chromatographic and the octanol/water log *P* for a set of compounds from Reference [34]. Circle: compounds with H-bond donor groups; square: compounds without H-bond donor groups.

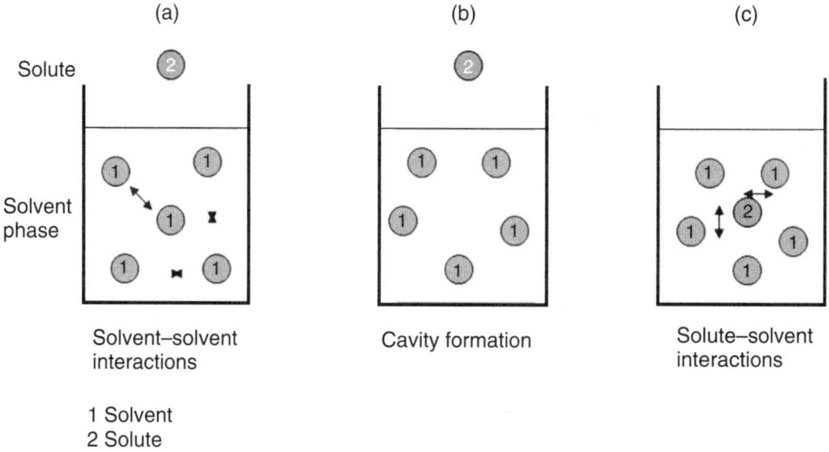

Figure 5.31 The interactions between the solute and the solvent molecules during the solvation process. There are interactions between the solvent molecules, which should be broken for cavity formation (a and b). Energy can be gained when the solute interacts with the solvent molecules (c). (Figure is courtesy of Professor Michael H. Abraham.)

The energy needed depends on the size of the solute, that is, how a large cavity is needed. It also depends on the interactions between the solvent molecules that have to be broken during the cavity formation. Some of the energy invested into the cavity creation can be gained back if the solute and solvent molecules are able to form energetically favorable interactions. These interactions can be H-bond donation, H-bond acceptance, and dipole/dipole interactions. On the basis of the Abraham solvation model, the basic molecular properties that govern the solvation process are the size of the molecules, the H-bond acidity and basicity, and dipolarity and/or polarizability. The theory includes the fifth molecular property called *excess molar refraction*, which accounts for the differences in size caused by the heteroatoms. The excess molar refraction can be estimated from the differences between the McGowan volume and the molar refraction. Both parameters are related to the size of the molecules based on the radius of the electron clouds around the atoms. Figure 5.32 illustrates the difference between the two size parameters. Molecules with heteroatoms have higher molar refraction than their McGowan volume would suggest. Only one heteroatom is an exception, namely, fluorine. This explains the very interesting behavior of perfluorinated hydrocarbons as solvents that form a third immiscible phase when they are mixed with water and hydrocarbon solvents, such as hexane.

Abraham and coworkers at University College London put a huge effort into measuring and compiling molecular descriptors such as H-bond acidity and basicity, dipolarity, and polarizability for several thousand molecules. They found that the partition coefficient in water/octanol, water/cyclohexane, and many other solvent systems can be described by the above-mentioned molecular descriptors by a multivariate regression equation. They have used the logarithmic value of the partition coefficients as dependent variables and the five molecular descriptors as independent variables. The regression analysis provided the coefficients for each descriptor. These regression coefficients are quite different for the different solvent systems and they can be used to describe the solvent properties, as is demonstrated in Figure 5.33.

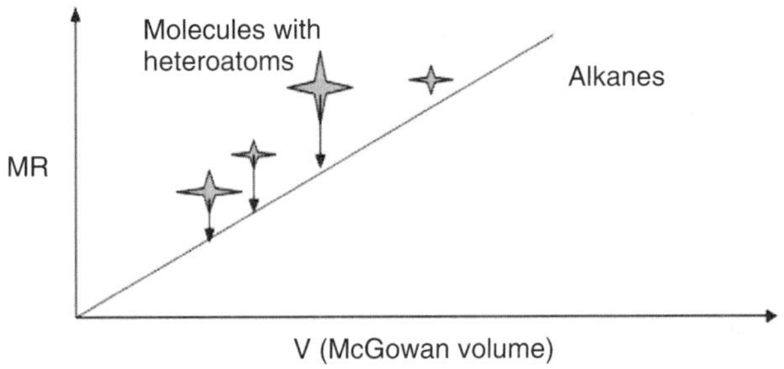

Figure 5.32 The derivation of excess molar refraction. The difference between molar refraction and McGowan volume is called *excess molar refraction*. It is zero for carbon atoms, negative for fluorine, and positive for all other heteroatoms. (Figure is courtesy of Professor Michael H Abraham.)

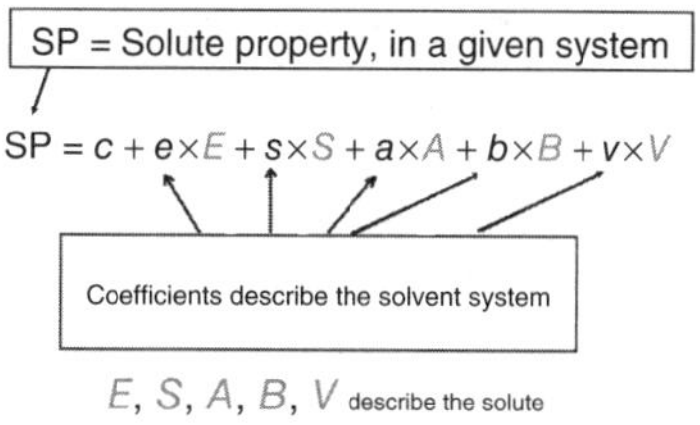

Figure 5.33 The general Abraham solvation equation [60]. The letters *E*, *S*, *A*, *B*, and *V* describe the solute properties (*E*: excess molar refraction, *S*: dipolarity/polarizability, *A*: H-bond acidity, *B*: H-bond basicity, and *V*: McGowan volume), while the coefficients (*e*, *s*, *a*, *b*, *v*) are characteristic of the solvent partition system.

By measuring a linear free-energy-related solute property (SP) such as the logarithmic value of octanol/water partition coefficient (log P) or logarithmic retention factor (log k) values for a set of probe molecules (minimum 25 compounds, preferably 40–60) with known molecular descriptors (E, S, A, B, and V), the solvation equation characteristic for the partitioning system (e, s, a, b, and v) can be constructed using multiple linear regression equation, as shown by Equation 5.17.

$$\log\text{(solute property)} = e \times E + s \times S + a \times A + b \times B + v \times V + \text{constant} \quad (5.17)$$

Such an approach may be used to evaluate and compare various partition and biological distribution systems. Numerous equations have been published [61–67] for the characterization of various reversed phase columns (C-18, perfluorinated hydrocarbons, cyclodextrin, Si–CH_2–CH_2–CN) using acetonitrile and methanol as organic modifier in the mobile phase. The generally good statistical fit of the chromatographic retention data and the molecular descriptors derived from bulk organic solvent/water partition coefficients suggests that they are similar partition processes governed by the same molecular properties. However, it should be noted that the parameters of the solvation equation are often based on the retention data of relatively small, mono- or bifunctional molecules in unionized form. Table 5.5 contains published solvation equations for several solvent partitions and reversed phase chromatographic systems. Solvation equations published for biological partition/distribution processes are also listed in Table 5.5 for comparison [61–71]. The models for biological partition systems are discussed later in more detail among the application examples.

In order to reveal the similarities and differences of various partition systems, the coefficients of each molecular descriptor were divided by the volume coefficient (v). In this way, the absolute values of the coefficients are on the same scale and can be compared. Table 5.6 shows the relative coefficients of the molecular descriptors obtained for various biological, biomimetic, and organic solvent/water partitions. Such comparison of the system coefficients helps in identifying which chromatographic or partition system model is best for various biological distribution processes.

Table 5.5 The parameters of the Abraham solvation equations in various distribution systems[a]

Solute Property	Constant	e	s	a	b	V	R^2
$\log P_{octanol/water}$	0.09	0.56	−1.05	0.03	−3.46	3.81	0.997
$\log P_{alkane/water}$	0.29	0.65	−1.66	−3.52	−4.82	4.28	0.998
$\log k_{C\text{-}18/AcN}$	0.00	0.09	−0.28	−0.47	−1.22	1.21	0.981
$CHI_{C\text{-}18/AcN}$	42.8	5.39	−14.1	−25.4	−70.2	68.4	0.985
$\log k_{IAM}$	−0.37	0.28	−0.08	0.30	−2.65	2.53	0.964
CHI_{IAM}	0.74	8.02	−8.81	6.97	−52.4	50.6	0.973
$\log k_{C\text{-}18/MeOH}$	−0.60	0.26	−0.44	0.19	−2.30	2.38	0.991
$gt_{RCN/MeOH}$	3.93	0.79	−1.05	−0.72	−4.50	5.42	0.924
$gt_{R\,perfluorC8/TFE}$	7.45	−0.12	−0.57	−3.64	−1.89	3.11	0.950
$\log k_{HSA}$	−1.28	0.82	−0.36	0.18	−1.97	1.62	0.910
$\log k_{AGP}$	−1.71	0.51	−0.58	−0.60	−1.52	2.02	0.922
$\log k_{water/skin}$	−5.33	0.00	−0.62	−0.38	−3.34	1.851	0.978
$\log k_{brain/blood}$	0.04	0.51	−0.89	−0.72	−0.67	0.86	0.870
$\%_{(intest.absorption)}$	92	2.9	4.1	−21.7	−21.1	10.6	0.860

[a]References [61–71].

Table 5.6 The relative coefficients of the Abraham solvation equations listed in Table 5.5

System	e/v	s/v	a/v	b/v
$\log P_{octanol/water}$	0.15	−0.28	0.01	−0.91
$\log P_{alkane/water}$	0.15	−0.35	−0.90	−1.12
$\log k_{C\text{-}18/AcN}$	0.07	−0.01	−0.39	−1.01
$CHI_{C\text{-}18/AcN}$	0.09	−0.24	−0.30	−0.98
$\log k_{IAM}$	0.11	−0.03	0.01	−1.05
CHI_{IAM}	0.42	−0.25	0.07	−1.08
$\log k_{C\text{-}18/MeOH}$	0.11	−0.18	0.08	−0.97
$gt_{RCN/MeOH}$	0.15	−0.19	−0.13	−0.83
$gt_{R\,perfluorC8/TFE}$	−0.04	−0.18	−1.17	−0.61
$\log k_{blood/brain}$	0.19	−0.69	−0.72	−1.28
$\log k_{HSA}$	0.02	−0.07	0.16	−1.21
$\log k_{AGP}$	0.46	−0.38	−0.33	−0.85
Water/skin	0.00	−0.33	−0.35	−1.95
$\log k_{blood/brain}$	0.59	−1.03	−0.84	−0.78
$\%_{(intest.absorption)}$	0.27	0.39	−2.05	−1.99

The coefficients were divided by the coefficient of the size parameter, V.

The relative coefficients reveal that the H-bond basicity (H-bond acceptors) parameter is negative in all the listed solvent partition, chromatographic, and biological partition systems. It means that when we have H-bond acceptor groups on the molecules, it restricts partition to the nonaqueous hydrophobic compartment. The dipolarity/polarizability parameter is also negative in most cases, meaning that it increases the hydrophilic character of the compounds (except human intestinal absorption). For the phospholipid partition, serum albumin binding and the C-18 retention with aqueous acetonitrile mobile phase are insensitive to the dipolarity/polarizability parameter, showing close to zero relative coefficients. The biggest variation can be observed for the relative coefficients of H-bond acidity. In most cases, compounds with H-bond donor groups provide a hydrophilic character and discourage compound partitioning out of water. However, there are several systems, where the H-bond acidity coefficient is close to zero; therefore, compounds can partition out from the water regardless of the polar H-bond donor groups. These systems are the octanol/water system, serum albumin binding, and phospholipid membrane partition. In these cases, the nonaqueous hydrophobic phase contains H-bond acceptor groups that help forming H-bonds with the H-bond donor group of the molecules. The explanation for this is that these systems contain water or other H-bond acceptor moieties that accommodate the H-bond donor groups, forming hydrogen bonding. These findings support the idea that the octanol/water partition system is a good model for biologically relevant partitioning of compounds. However, we can see from Table 5.6 that compound distribution between the blood and brain is strongly affected by the H-bond donor groups; the *a*/*v* coefficient has a negative value. This means that molecules with H-bond donor groups prefer the blood to the brain when partitioning between them. The reversed phase chromatographic system and the water/skin partitioning systems also have negative coefficients for the H-bond acidity, but slightly less than the blood–brain system.

The solvation equations published for the various chromatographic systems can be used to select the most orthogonal systems, providing very different selectivity [65]. An example of the possible selectivity difference is shown in Figure 5.34. The figure shows the separation of a mixture of compounds with diverse molecular descriptors using three different chromatographic systems. The three systems used were C-18 reversed phase with acetonitrile gradient, Develosil CN column with methanol gradient, and perfluorooctyl column with tri-fluoroethanol gradient. The structures of the components of the mixtures are shown in Figure 5.35. We can see that on C-18 column with acetonitrile gradient, the retention order is proportional to the lipophilicity order of the compounds. Theophylline is the most polar, followed by caffeine, which has an additional methyl group, then nitro-phenol, the 3,4-dichlorophenol, and finally anisole being the most lipophilic. The CN stationary phase with methanol gradient, however, is more sensitive to the H-bond donor compounds such as phenols, and the theophylline shows relatively longer retention. It is probably due to the adsorbed methanol molecules onto the polar stationary phase surface. Methanol contains OH group, which is an H-bond acceptor. The perfluoro-octyl column with tri-fluoroethanol gradient has an opposite

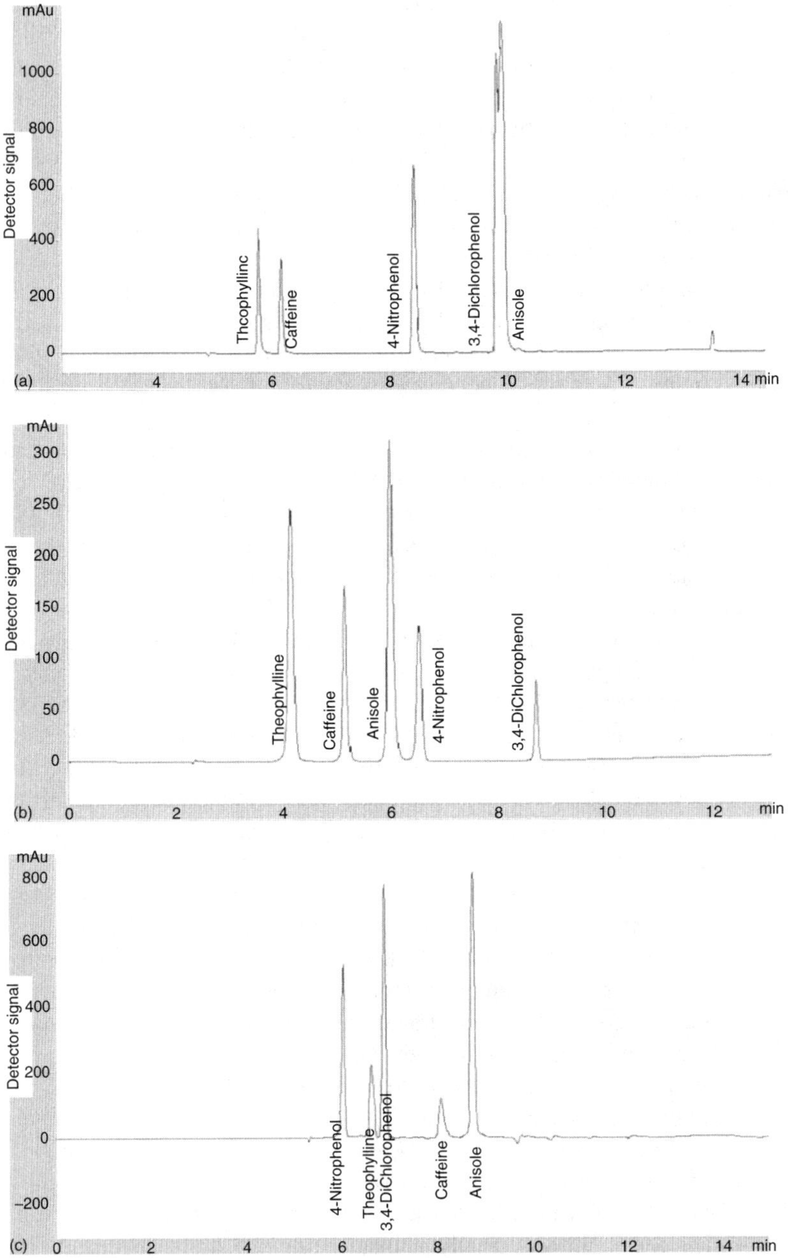

Figure 5.34 Chromatograms of a sample mixture containing theophylline, caffeine, 4-nitrophenol, 3,4-dichlorophenol, and anisole. (a) C-18 Luna column with acetonitrile gradient; (b) Develosil CN column with methanol gradient; and (c) perfluoro-octyl-coated silica with tri-fluoroethanol gradient. (Reproduced with permission from Reference [65].)

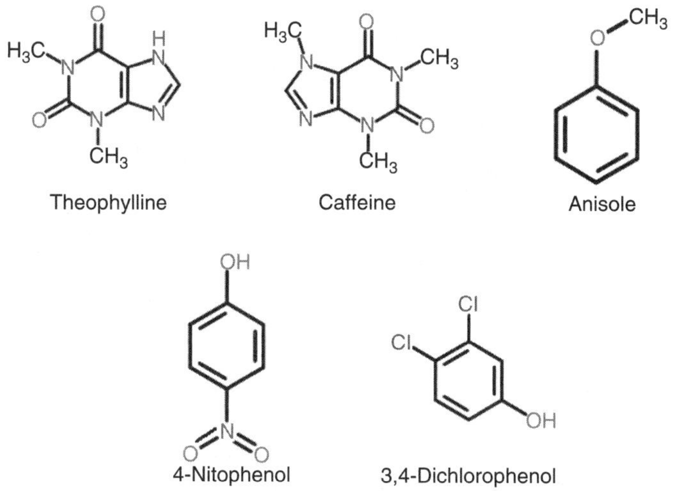

Figure 5.35 The chemical structures of the components of the sample mixture shown in Figure 5.34.

effect toward H-bond donor groups. This OH group only donates the proton and has practically no H-bond acceptor property, unlike methanol and octanol. It is due to the tri-fluoromethyl group that has a strong electron-withdrawing effect on the oxygen in the OH group. The nonpolar fluorinated part of the molecule adsorbs onto the fluorinated stationary phase, positioning the OH group toward the mobile phase. Therefore, compounds with H-bond donor properties elute with shorter retention times. On the basis of these findings, we can vary the selectivity of the chromatographic partition systems, which can be utilized to determine the molecular descriptors of the compounds from three to five retention time measurements. Following on the systematic work of Du et al. [65], we can use the nonlinear mapping procedure to compare the selectivity of various stationary phase/mobile phase systems. In this procedure, we plot the systems in a multidimensional space with the coefficients of the five descriptors being the axes. When a system is similar to the other systems, they are close to each other in the multidimensional space. We can project the positions of the points (representing the HPLC system) in a five-dimensional space to two dimensions, keeping the distances of the points in the two dimensions to the closest of the multidimensional space. Figure 5.36 shows such a nonlinear plot of several investigated HPLC systems [65].

By selecting the four to five most orthogonal systems, we can use the four to five equations to calculate the solute properties of new molecules from the measured data in these orthogonal systems. The following procedure can be applied for the descriptor determination of a compound by HPLC. First, select three to five chromatographic systems with different selectivities. This can be C-18 column with acetonitrile gradient, C-18, and CN columns with methanol gradient and polymer-based and perfluorinated reversed phase columns with tri-fluoroethanol gradient.

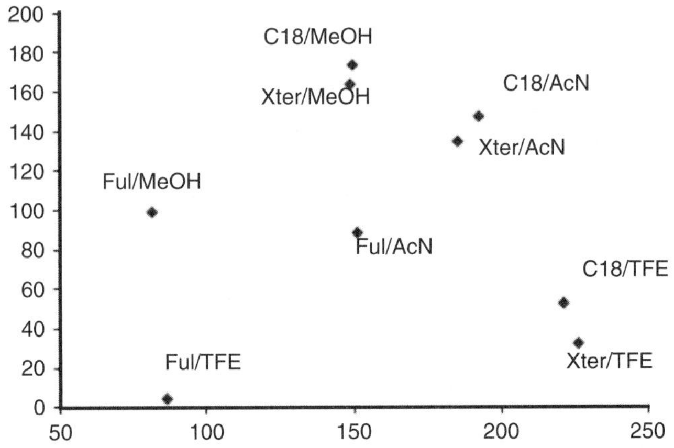

Figure 5.36 The nonlinear map of the five-dimensional space, where each system has coordinates corresponding to the five coefficients of the five molecular descriptors in the Abraham solvation equations. (C18 means C-18 reversed phase stationary phase, Xter means Xterra stationary phase, which is a combination of silica- and polymer-coated C-18 phase, Flu means perfluorinated C-8 stationary phase, MeOH, AcN, and TFE mean methanol, acetonitrile, and tri-fluoroethanol, respectively, the organic phase gradients applied during the gradient retention time measurements).

(Extra safety precautions should be used with the highly acidic and corrosive trifluoroethanol!)

Determine the solvation equations in the five systems using a diverse set of compounds with known molecular descriptors (an example of the set is shown in Table 5.7). Then determine the gradient retention times of the unknown compounds. Use Excel solver to solve the five equations for the five unknown molecular descriptors, as is demonstrated in Figure 5.37.

The McGowan volume and the excess molar refraction can be easily calculated from the molecular structure by various software packages on the Internet (Chemspider, ACD Laboratories, Absolv). The three polar parameters (dipolarity/polarizability, H-bond acidity, and H-bond basicity) can be derived from the

$$\mathrm{SP1} = c1 + e1 \times E + s1 \times S + a1 \times A + b1 \times B + v1 \times V$$
$$\mathrm{SP2} = c2 + e2 \times E + s2 \times S + a2 \times A + b2 \times B + v2 \times V$$
$$\mathrm{SP3} = c3 + e3 \times E + s3 \times S + a3 \times A + b3 \times B + v3 \times V$$
$$\mathrm{SP4} = c4 + e4 \times E + s4 \times S + a4 \times A + b4 \times B + v4 \times V$$
$$\mathrm{SP5} = c5 + e5 \times E + s5 \times S + a5 \times A + b5 \times B + v5 \times V$$

Figure 5.37 Calculation of molecular descriptors using the data of a compound obtained in five partition systems with different selectivities for which the solvation equations are known. Microsoft Excel "solver" add-in can be used to solve the equation systems for *E*, *S*, *A*, *B*, and *V* when all the other parameters are known.

Table 5.7 An example of Abraham training set of compounds with molecular descriptors

Compound	*E*	*S*	*A*	*B*	*V*
n-Octanophenone	0.72	0.95	0	0.5	1.859
n-Heptanophenone	0.72	0.95	0	0.5	1.718
n-Hexanophenone	0.72	0.95	0	0.5	1.578
n-Valerophenone	0.8	0.95	0	0.5	1.437
n-Butyrophenone	0.8	0.95	0	0.51	1.296
n-Propiophenone	0.8	0.95	0	0.51	1.155
Acetophenone	0.82	1.01	0	0.48	1.014
Paracetamol	1.06	1.78	1.09	0.81	1.172
Acetanilide	0.87	1.4	0.5	0.67	1.1133
Theophylline	1.5	1.6	0.54	1.15	1.2223
Dibenzothiophene	1.959	1.31	0	0.2	1.3791
Caffeine	1.5	1.6	0	1.33	1.3632
Indazole	1.18	1.22	0.53	0.35	0.9053
Benzonitrile	0.742	1.11	0	0.33	0.8711
Cyclohexanone	0.403	0.86	0	0.56	0.8611
Chlorobenzene	0.718	0.65	0	0.07	0.8388
Naphthalene	1.34	0.92	0	0.2	1.0854
1,4-Dinitrobenzene	1.13	1.63	0	0.46	1.0648
Hydrocortisone	2.03	3.49	0.71	1.9	2.7976
Cortisone-21-acetate	1.82	3.11	0.21	2.13	3.0521
Pyrene	2.808	1.71	0	0.28	1.5846
Progesterone	1.45	3.29	0	1.14	2.6215
Anisole	0.71	0.75	0	0.29	0.916
Benzamide	0.99	1.5	0.49	0.67	0.9728
Butalbarbital	1.03	1.14	0.47	1.18	1.6557
3,4-Di-Cl-phenol	1.02	1.14	0.85	0.03	1.0199
Phenol	0.805	0.89	0.6	0.3	0.7751
4-Nitrophenol	1.07	1.72	0.82	0.26	0.9493
4-Cl-phenol	0.915	1.08	0.67	0.2	0.8975
4-I-phenol	1.38	1.22	0.68	0.2	1.0333
Resorcinol	0.98	1	1.1	0.58	0.8338
4-CN-phenol	0.94	1.63	0.8	0.29	0.9298
4-Nitrobenzoic acid	0.99	1.07	0.62	0.54	1.1059
Benzoic acid	0.73	0.9	0.59	0.4	0.9317
3-CF3-phenol	0.87	0.72	0.09	0.9691	2.95
4-OH-benzyl alcohol	0.998	1.15	0.88	0.85	0.9747
Salicylic acid	0.89	0.7	0.72	0.41	0.9904
Phenylacetic acid	0.73	0.97	0.6	0.61	1.0726
4-Nitroaniline	1.22	1.91	0.42	0.38	0.9904
Propranolol	1.843	1.5	0.6	1.27	2.148
p-Toluidine	0.923	0.95	0.23	0.52	0.9571
Pyridine	0.631	0.84	0	0.47	0.6753
Aniline	0.955	0.96	0.26	0.5	0.8162
3-Nitroaniline	1.2	1.71	0.4	0.35	0.9904

(continued)

Table 5.7 (*Continued*)

Compound	*E*	*S*	*A*	*B*	*V*
Procaine	1.14	1.67	0.32	1.36	1.9767
Nicotine	0.865	1.44	0	0.9	1.371
Methyl4-hydroxybenzoate	1.37	0.69	0.45	1.1313	2.47
n-Ethyl4-hydroxybenzoate	1.35	0.69	0.45	1.2722	3.04
n-Propyl4-hydroxybenzoate	1.35	0.69	0.45	1.4131	3.57
n-Butyl4-hydroxybenzoate	0.86	1.35	0.69	0.45	1.554
Benzene	0.61	0.52	0	0.14	0.716
Toluene	0.601	0.52	0	0.15	0.857
n-Ethylbenzene	0.613	0.51	0	0.15	0.998
n-Propylbenzene	0.604	0.5	0	0.15	1.139
n-Butylbenzene	0.6	0.51	0	0.15	1.28
n-Hexylbenzene	0.59	0.5	0	0.15	1.562
Nitromethane	0.313	0.95	0.06	0.31	0.424
n-Nitroethane	0.27	0.95	0.02	0.33	0.564
n-Nitropropane	0.242	0.95	0	0.31	0.7055
n-Nitrobuthane	0.227	0.95	0	0.29	0.8464
n-Nitropentane	0.212	0.95	0	0.29	0.9873
n-Nitrohexane	0.203	0.95	0	0.29	1.1282

The coefficients of the solvation equation for a new partition system can be determined by the measurements of properties of this set using the given descriptors and multiple regression analysis.

retention data by solving at least three equations with three unknowns [72,73]. It is important to see that the H-bond acidity (A) and basicity (B) scale provides a more refined scale than using simply the H-bond donor and acceptor counts on the molecule that considers, for example, the H-bond acidity of an NH or OH groups equal. We need to keep in mind that the molecular descriptors refer to the whole molecule, thus adding up the effect of all the H-bond donor and acceptor groups (which may not be additive, when they are in close proximity on the molecule). The molecular descriptors are very unique characteristics and they determine how a molecule will distribute in various systems including the biological partition systems. The only way we can measure these descriptors is to measure the SP in systems of which the equations are known.

Sometimes two partition systems are very similar with the exception of their sensitivity toward only one molecular descriptor, for example, the H-bond acidity (A). In this case, measuring compound property in two systems would be enough to determine that particular molecular descriptor. Du et al. [67] compared the linear solvation equations obtained for isocratic HPLC retention factors (using C-18 stationary phases and acetonitrile as organic modifier in the mobile phase) and for the octanol/water partition coefficients. The major difference between the two partition systems is their sensitivity toward the H-bond acidity of the molecules. While the water-saturated octanol phase easily accommodates the compounds with H-bond donor groups, the chromatographic C-18 stationary phase has no such polar functionality. Therefore, H-bond donor functional groups decrease the chromatographic retention and consequently chromatographic partition coefficients. The correlation between isocratic retention data (with C-18 stationary phases and aqueous acetonitrile mobile phases) and octanol log P can be significantly improved by taking into account a simple H-bond donor count (HBC; $r = 0.95$ improved to $r = 0.96$) or the Abraham H-bond acidity descriptor (A) ($r = 0.95$ improved to $r = 0.98$) in Equation 5.18 and Equation 5.19.

$$\log P_{\text{oct}} = 1.91 \log k + 0.37\text{HBC} + 0.72$$
$$N = 111 \quad r = 0.962 \quad \text{sd} = 0.272 \tag{5.18}$$

$$\log P_{\text{oct}} = 2.07 \log k + 1.09A + 0.52$$
$$N = 111 \quad r = 0.982 \quad \text{sd} = 0.189 \tag{5.19}$$

The gradient chromatographic CHI values obtained for the unionized forms of the molecules (CHIN) have also shown an acceptable correlation with the octanol/water log P values when the HBC or the Abraham H-bond acidity (A) descriptor are included [34]. The correlation coefficient between log P_{oct} and CHIN is only 0.92. Equation 5.20 and Equation 5.21 show the relationships obtained for a training set of 86 known drug molecules.

$$\log P_{\text{oct}} = 0.047\text{CHIN} + 0.36\,(\text{HBC})–1.10$$
$$n = 86 \quad r = 0.943 \quad s = 0.39 \tag{5.20}$$

$$\log P_{\text{oct}} = 0.054\text{CHIN} + 1.32(\text{A})–1.88$$
$$n = 86 \quad r = 0.970 \quad s = 0.29 \tag{5.21}$$

In conclusion, the solvation equation approach is a valuable tool for comparing the similarities and differences between physicochemical partitioning and biological partitioning of the molecules; moreover, it can be used to select the best physicochemical model systems for biological partitioning [74–76]. We have to remember that the above-mentioned solvation equations do not incorporate the effect of ionization of the molecules and only the data of the uncharged molecules should be used for generating these equations for system comparisons. Recent efforts [77–79] that estimate the effect of the ionization on the partitioning of the molecule have been published.

We can conclude from these findings that the physicochemical water/organic solvent partition systems and the biomimetic chromatographic partitioning systems can be used as simple measurements of compound behavior that are able to imitate compound partitioning in biological environment. However, it would be even more useful to use these measured parameters in models that describe the compound's behavior *in vivo*. This is discussed later in Chapter 10.

SUMMARY

Lipophilicity or hydrophobicity of the compound refers to the compound affinity for a lipophilic environment or the compound's tendency to avoid the aqueous environment. Lipophilicity can be measured by the compound partition coefficient between two nonmiscible solvents or two compartments, which are in equilibrium. The partition coefficient is expressed as the equilibrium concentration ratio of the compound in an aqueous and in a lipophilic solvent or compartment. The concentration ratio is constant, regardless of the amount of compound (when the solute concentration is low and no interaction between the solute molecules occurs). The octanol/water partition system has been suggested as the best model for biological partitioning of a compound. The compound concentration in the octanol and in the water has to be determined after equilibrium has been established. The HPLC technique can be used for concentration determination. Special care has to be taken to ensure the equilibration of the solvents before the partition measurements. Constant temperature has to be applied for the equilibration. Separation of the two solvents after equilibration can be facilitated by centrifugation when emulsion occurs because of amphiphilic solutes. The determination of the octanol/water partition coefficients can be performed on the Sirius GLpKa or T3 instruments. Two potentiometric titrations have to be performed, one in water and the other in the presence of octanol. From the shift of the two titration curves, the octanol/water partition coefficient can be calculated.

The lipophilicity of the compounds can be determined from isocratic reversed phase retention times. The retention factor, which is calculated from the retention time and the dead time, is equal to the quotient of the average number of moles of the compounds in the stationary phase and the mobile phase. In order to get the partition coefficient from the retention factor, volumes of the stationary phase and mobile phase have to be included. As this is difficult to measure or estimate, a calibration set of compounds are used for which the lipophilicity is known from other nonchromatographic measurements. Various mobile phase compositions have to be used to elute compounds with very different lipophilicity. Therefore, the logarithmic retention factor is determined for each compound using three to five different organic phase concentrations in the mobile phase. The plot of the logarithmic retention factors as a function of the organic phase concentration in the mobile phase gives a straight line (within a limited range). The intercept of the straight line gives the extrapolated logarithmic retention factor for the 100% aqueous mobile phase, which can be used as the measure of lipophilicity. The extrapolated retention data can be calibrated by compounds with known octanol/water partition coefficients. The reversed phase chromatographic retention is different from the octanol/water partition as different solute–solvent interactions might take place for different molecules. Therefore, numerous approaches have been published, which attempted to make the chromatographic partitioning system more similar to the octanol/water partition system by applying octanol in the mobile phase or dynamically loaded onto the stationary phase. When the extrapolated retention gives different lipophilicity rankings of compounds from retention order at any given mobile phase composition, the slope values of the retention—organic phase concentration straight lines can also be incorporated in the correlation study. The isocratic CHI (φ_0) can be calculated from the quotient of the intercept and the slope of the logarithmic retention factor versus organic phase concentration straight lines. It refers to the organic phase concentration in the mobile phase, by which the compound retention time is exactly double the dead time, meaning a 50%–50% average distribution of the compound between the stationary and mobile phases. The advantage of the φ_0 is that it can be measured and is not an extrapolated value. It has been shown to correlate with octanol/water partition coefficients ($\log P$) when they were obtained for nonionized compounds. The correlation of the $\log P$ and φ_0 makes it possible to estimate the necessary mobile phase composition for HPLC analytical development of the compound.

The isocratic CHI shows linear correlation to the gradient retention time. Converting the gradient retention time to the approximate organic phase concentration in the mobile phase, the CHI can be obtained. The CHI method, as it uses gradient chromatography, is able to cover a wide range of lipophilicities of the molecule without the need to change the chromatographic conditions. The CHI approximates the acetonitrile concentration in the mobile phase at which the compound elutes from the column; thus, it ranges from 0% to 100%. However, it can be converted to the octanol/water scale. Any gradient reversed phase retention time can be converted to CHI when we use a calibration set of mixtures for which we know the isocratically determined φ_0 values. We can use a set of compounds for calibration for which the octanol/water partition/distribution coefficients are known. The gradient retention times show linear relationship with the logarithmic value of the partition coefficients. We have to remember that the chromatographic partition system is different from the octanol/water partition system; so, deviations from the linear relationship can be expected when structurally unrelated compounds are investigated.

When a compound contains ionizable groups, their ionization state depends on the pH of the environment. When a compound is ionized by protonation or dissociation of a proton, it becomes positively or negatively charged. Charged molecules are much more hydrophilic than the uncharged counterparts. Therefore, the lipophilicity, expressed by the partition coefficient, is very different between the charged and uncharged forms of the molecule. As a rule of thumb, the logarithmic partition coefficient is lower by three to four orders of magnitude for the charged species. The pH profile of the lipophilicity expressed by $\log D$ has a sigmoid shape and is described by the Henderson–Hasselbalch equation. Similar to the partition coefficients, the chromatographic lipophilicity also depends on the mobile phase pH. When we use aqueous organic phase as mobile phase, the mobile phase pH changes from the pure aqueous pH; moreover, the organic solvent shifts the acid dissociation constant of the compound as well. Therefore, an extra constant is needed in the Henderson–Hasselbalch equation in gradient reversed phase chromatography, which describes the pH profile of the gradient retention times of an ionizable compound.

The advantage of chromatographic lipophilicity determination is that it is based on dynamic equilibrium distribution of the compound, just like the distribution of the drug molecules *in vivo* between the moving plasma compartment and the "stationary" tissue compartments. This dynamic equilibrium process is developed on the large surface of the stationary phase, which is very similar to the *in vivo* drug partition and distribution processes. The great advantages of the chromatographic measurements are that we can use directly protein or phospholipid stationary phases, which are the main components in the real biological distribution phases. The distribution coefficient of the compounds on biomimetic stationary phases can be determined using isocratic conditions, preferably pure aqueous mobile phase at physiological pH (pH 7.4). The gradient elution mode also can be used, but we have to be careful to use appropriate organic phase (iso-propanol for protein phases up to 30% or acetonitrile for the phospholipid phase up to 70%) in order to retain the natural conformation of the protein and phospholipids.

Reversed phase liquid chromatography provides a valuable tool for high throughput measurements of lipophilicity of the compounds, covering a wide range with very high ranking accuracy. Small amounts of compounds are needed for the determination, and simple and accurate time measurements are used to calculate the lipophilicity. There is no need for concentration determination. Impure compounds, even mixtures, can be measured simultaneously, as separation takes place during the measurements. Both ionizable and neutral compounds can be analyzed. The pH dependence of ionization and consequently, lipophilicity, is very similar in the solvent/solvent partition and the chromatographic distribution systems. Measuring the chromatographic lipophilicity via retention times using mobile phases with different pH values can be used to detect the acid/base character of compounds and even the extent of ionization at physiological pH. However, during chromatographic partition, we measure entropy-driven dynamic partitioning equilibrium. Solvent/solvent partition is more enthalpy driven and we do not know what happens at the interface of the two solvents. Octanol/water partition has been selected as a model system for biological partition processes. However, chromatography can provide the possibility to directly model biological partitioning, especially when using biomimetic stationary phases. In order to compare the chromatographic partition systems with the octanol/water and with other, even *in vivo* partition systems, the Abraham solvation equation approach can be used. On the basis of the solvation theory, compound partitioning in phases can be described by five molecular descriptors, excess molar refraction, dipolarity/polarizability, H-bond acidity and basicity, and size. Knowing the molecular descriptors of a compound, we can estimate its partition coefficients in all the systems for which the parameters of the solvation equations are known. Various chromatographic methods that can be used to measure biomimetic partition of the compounds have been described, that is, compound interactions with proteins and phospholipids. A methodology has been described for the determination of molecular descriptors by chromatographic retention time measurements. The molecular descriptors can be obtained only by measurements of solvent/solvent or chromatographic partition, and they represent a unique scale of polarity, H-bond acidity, and basicity of the compounds. It has been shown that the *in vitro* and *in vivo* partition systems show unique selectivity toward these descriptors and eventually it determines how compounds distribute in various systems.

QUESTIONS FOR REVIEW

Q5.1 What is lipophilicity and how can we measure it?

Q5.2 How can we measure octanol/water partition coefficient?

Q5.3 What is isocratic CHI and how can we determine it?

Q5.4 Why are gradient retention times directly proportional to the logarithmic values of partition coefficients, unlike in isocratic mode, we need to take the logarithmic value of the retention factor?

Q5.5 Approximately how much acetonitrile and methanol do we need to use in isocratic reversed phase chromatography to elute a compound that has a lipophilicity ($\log P$) of 4 to have a retention time approximately double that of the dead time?

Q5.6 Describe the relationship of lipophilicity of ionized compounds as a function of pH.

Q5.7 We have $\log D$ and $\log P$ values of a compound with an ionizable group. Which value must be the highest, $\log P$ or $\log D$?

Q5.8 At which pH range is a basic compound more lipophilic? At which pH range is an acidic compound more lipophilic? Why?

Q5.9 Describe a few biomimetic stationary phases. What HPLC conditions have to be used to mimic the biological partition behavior of a compound?

Q5.10 How can we compare various lipophilicity measures?

Q5.11 What are the similarities and differences between biomimetic HPLC and the octanol/water partition systems?

Q5.12 What are the five molecular descriptors that can be used to describe a compound's partition in various solvent systems?

Q5.13 Describe the principles of the determination of the Abraham molecular descriptors.

REFERENCES

1. Hansch, C., Leo, A. (1964) ρ-σ-π analysis. A method for correlation of biological activity and chemical structure. *Journal of American Chemical Society*, *86*, 1616–1624.
2. Collander, R. (1951) The partition of organic compounds between higher alcohols and water. *Acta Chemica Scandinavica*, *5*, 774–782.
3. Leo, A. J. (1972) Biological Correlations—The Hansch Approach in Advances in Chemistry Ser. No. 114, Ed. Gould, R. F., American Chemical Society, Washington, DC, pp. 51–69.
4. Valkó, K. (2000) Measurements of physical properties for drug design in industry. In: Handbook of Analytical Separations, Vol. 1, Ed. Smith, R., Separation methods in drug synthesis and purification. Ed. Valkó, K., Elsevier, Amsterdam, pp. 564–584.

5. Valkó, K., Snyder, L. R., Glajch, G. L. (1993) Retention in reversed-phase liquid chromatography as a function of mobile-phase composition. *Journal of Chromatography A, 656*, 501–520.
6. Scherrer, R. A., Donovan, S. F. (2009) Automated potentiometric titrations in KCl/water-saturated octanol: Method for quantifying factors influencing ion-pair partitioning. *Analytical Chemistry, 81(7)*, 2768–2778.
7. Horváth, Cs., Melander, W., Molnár, I. (1976) Solvophobic interactions in liquid chromatography with non-polar stationary phases. *Journal of Chromatography, 125*, 129–156.
8. Harnisch, M., Mockel, H. J., Shulze, G. J. (1983) Relationship between log P_{ow}, shake flask values and capacity factors derived from reversed-phase high performance liquid chromatography for n-alkylbenzenes and some OECD reference substances. *Journal of Chromatography, 282*, 315–332.
9. Tomlinson, E. (1975) Chromatographic hydrophobicity parameters in correlation analysis of structure–activity relationships. *Journal of Chromatography, 113*, 1–45.
10. McCall, J. M. (1975) Liquid-liquid partition coefficients by high-pressure liquid chromatography. *Journal of medicinal Chemistry, 18*, 549–552.
11. Braumann, Th., Weber, G., Grimme, L. H. (1983) Quantitative structure-activity relationships for herbicides: Reversed-phase liquid chromatographic retention parameter, log kw versus liquid-liquid partition coefficient as model of the hydrophobicity of phenylureas, s-triazines and phenoxycarbonic acid derivatives. *Journal of Chromatography, 261*, 329–343.
12. Giaginis, C., Tsantili-Kakoulidou, A. (2008) Current state of the art in HPLC Methodology for lipophilicity assessment of basic drugs. (Review) *Journal of Liquid Chromatography & Related Technologies, 31*, 79–96.
13. Brent, D. A., Sabatka, J., Minick, D. J., Henry, D. J. (1983) A simplified high-pressure liquid chromatography method for determining lipophilicity for structure-activity relationships. *Journal of Medicinal Chemistry, 26*, 1014–1020.
14. Minick, D. J., Brent, D. A., Frenz, J. (1986) Modeling octanol-water partition coefficients by reversed-phase liquid chromatography. *Journal of Chromatography A, 461*, 177–191.
15. Mirrlees, M. S., Moulton, S. J., Murphy, C. T., Taylor, P. (1976) Use of high-pressure liquid chromatography for quantitative structure-activity relationship studies of sulphonamides and barbiturates. *Journal of Medicinal Chemistry, 19*, 619–626.
16. Unger, S. H., Cook, J. R., Hollenberg, J. S. (1978) Simple procedure for determining octanol-aqueous partition, distribution and ionization coefficients by reversed-phase high-pressure liquid chromatography. *Journal of Pharmaceutical Sciences, 67*, 1364–1371.
17. Minnick, D. J., Frenz, J. H., Patrick, M. A., Brent, D. A. (1988) A comprehensive method for determining hydrophobicity constants by reversed phase high-performance liquid chromatography. *Journal of Medicinal Chemistry, 31*, 1923–1933.
18. Du, C. M., Valko, K., Bevan, C., Reynolds, D., Abraham, M. H. (2001) Rapid method for estimating octanol-water partition coefficient (log P_{oct}) from isocratic RP-HPLC and a hydrogen bond acidity term (A). *Journal of Liquid Chromatography & Related Technologies, 24 (5)*, 635–649.

19. Kaune, A., Bruggemenn, R., Kettrup, A. (1998) High performance liquid chromatographic measurement of the 1-octanol-water partition coefficient of s-triazine herbicides and some of their degradation products. *Journal of Chromatography A*, *805*, 119–126.

20. Lombardo, F., Shalaeva, M. Y., Tupper, K. A., Gao, F. (2001) $ElogD_{oct}$: A tool for lipophilicity determination in drug discovery. 2. Basic and neutral compounds. *Journal of Medicinal Chemistry*, *44*, 2490–2497.

21. Gocan, S., Cimpan, G., Comer, J. (2005) Lipophilicity measurements by liquid chromatography. *Advances in Chromatography*, Eds Grushka, E., Grinberg, N., *44*, 79–176.

22. Valkó, K. (1984) General approach for the estimation of octanol/water partition coefficient by reversed-phase high performance liquid chromatography. *Journal of Liquid Chromatography*, *7*, 1405–1424.

23. Valkó, K., Friedmann, T., Báti, J., Nagykáldi, A. (1984) Reversed-phase chromatographic system as a model for characterizing the offset rate of action of azidomorphines in guinea-pig ileum. *Journal of Liquid Chromatography*, *7(10)*, 2073–2092.

24. Valkó, K., Slégel, P. (1993) New chromatographic hydrophobicity index (φ_0) based on the slope and the intercept of the log k' versus organic phase concentration plot. *Journal of Chromatography*, *631*, 49–61.

25. Mutton, I. (2000) Fast generic HPLC methods. In: Handbook of Analytical Separations Vol. 1, Ed. Smith, R., Separation methods in drug synthesis and purification. Ed. Valkó, K., Elsevier, Amsterdam, pp. 73–85.

26. Valkó, K., Bevan, C., Reynolds, D. (1997) Chromatographic hydrophobicity index by fast-gradient RP-HPLC: A high throughput alternative to log P/log D. *Analytical Chemistry*, *69*, 2022–2029.

27. Valko, K. (2004) Measurements of lipophilicity and acid/base character using HPLC methods. In: Pharmaceutical Profiling in Drug Discovery for Lead Selection, Eds Borchardt, R., Kerns, E., AAPS, Arlington, VA, pp. 127–182.

28. Young, R. J., Green, D. V. S., Luscombe, C. N., Hill, A. P. (2011) Getting physical in drug discovery II: The impact of chromatographic hydrophobicity measurements and aromaticity. *Drug Discovery Today*, *16(17/18)*, 822–830.

29. Kerns, E. H., Di, L., Petusky, S., Kleintop, T., Huryn, D., McConnell, O., Carter, G. (2003) Pharmaceutical profiling method for lipophilicity and integrity using liquid chromatography-mass spectrometry. *Journal of Chromatography B*, *791*, 381–388.

30. Donovan, S. F., Pescatore, M. C. (2002) Method for measuring the logarithm of octanol-water partition coefficient by using short octadecyl-poly(vinyl alcohol) high performance liquid chromatography columns. *Journal of Chromatography A*, *952*, 47–61.

31. Camurri, G., Zamarella, A. (2001) High-throughput liquid chromatography/mass spectrometry method for the determination of chromatographic hydrophobicity index. *Analytical Chemistry*, *73*, 3716–3722.

32. Du, C. M., Valko, K., Bevan, C., Reynolds, D., Abraham, M. H. (1998) Rapid gradient RP-HPLC method for lipophilicity determination: A salvation equation based comparison with isocratic methods. *Analytical Chemistry*, *70*, 4228–4234.

33. Dolan, J. W., Lommen, D. C., Snyder, L. R. (1998) Drylab © computer simulation for high-performance liquid chromatographic method development II. Gradient elution. *Journal of Chromatography A*, *485*, 91–112.

34. Valkó, K., Du, C. M., Bevan, C., Reynolds, D. P., Abraham, M. H. (2001) Rapid method for the estimation of octanol/water partition coefficient (log P_{oct}) from gradient RP-HPLC retention and a hydrogen bond acidity term ($\Sigma\alpha_2^H$). *Current Medicinal Chemistry*, *8*, 1137–1146.
35. Horváth, Cs., Melander, W., Molnar, I. (1977) Separation of amino acids and peptides on nonpolar stationary phases by HPLC. *Analytical Chemistry*, *142*, 623–640.
36. Canals, I., Valkó, K., Bosch, E., Hill, A. P., Roses, M. (2001) Retention of ionisable compounds on HPLC. 8. Influence of mobile-phase pH change on the chromatographic retention of acids and bases during gradient elution. *Analytical Chemistry*, *73*, 4937–4945.
37. Fuguet, E., Ràfols, C., Bosch, E., Rosès, M. (2007) Determination of the chromatographic hydrophobicity index for ionizable solutes. *Journal of Chromatography A*, *1173*, 110–119.
38. Kaliszan, R., Haber, P., Baczek, T., Siluk, D., Valko, K. (2002) Lipophilicity and pKa estimates from gradient high-performance liquid chromatography. *Journal of Chromatography A*, *965*, 117–127.
39. Wiczling, P., Kawczak, P., Nasal, A., Kaliszan, R. (2006) Simultaneous determination of pKa and lipophilicity by gradient RP-HPLC. *Analytical Chemistry*, *78*, 239–249.
40. Kaliszan, R., Noctor, T. A. G., Wainer, I. W. (1992) Quantitative structure – enantioselective retention relationships for the chromatography of 1,4-benzodiazepines on a human serum albumin based HPLC chiral stationary phase: An approach to the computational prediction of retention and enantioselectivity. *Chromatographia*, *33*, 546–550.
41. Noctor, T. A. G., Diaz-Perez, M. J., Wainer, I. W. (1993) Use of human serum albumin-based stationary phase for high performance liquid chromatography as a tool for rapid determination of drug plasma protein binding. *Journal of Pharmaceutical Sciences*, *82*, 675–676.
42. Tiller, P. R., Mutton, I. M., Lane, S. J., Bevan, C. D. (1995) Immobilized human serum albumin: Liquid chromatography/mass spectrometry as a method of determining drug-protein binding. *Rapid Communications in Mass Spectrometry*, *9*, 261–263.
43. Beaudry, F., Coutu, M., Brown, N. K. (1999) Determination of drug-plasma protein binding using human serum albumin chromatography column and multiple linear regression model. *Biomedical Chromatography*, *13*, 401–406.
44. Valkó, K., Nunhuck, S. B., Bevan, C. D., Abraham, M. H, Reynolds, D. P. (2003) Fast gradient HPLC method to determine compounds binding to human serum albumin. Relationships with octanol/water and immobilized artificial membrane lipophilicity. *Journal of Pharmaceutical Sciences*, *92*, 2236–2248.
45. Bertucci, C., Bartolini, M., Gotti, R., Andrisano, V. (2003) Drug affinity to immobilized target bio-polymers by high-performance liquid chromatography and capillary electrophoresis. *Journal of Chromatography B*, *797*, 111–129.
46. Buchholz, L., Cai, C. H., Cleton, A., Brodfuehrer, J., Cohen, L. (2002) Evaluation of the human serum albumin column as discovery screening tool for plasma protein binding. *European Journal of Pharmaceutical Sciences*, *15*, 209–215.
47. Kratochwil, N., Huber, W., Muller, F., Kansy, M., Gerber, P. R. (2004) Predicting plasma protein binding of drugs – revisited. *Current Opinion in Drug Discovery & Development*, *7*, 507–512.

48. Colmenarejo, G., Alvarez-Pedraglio, A., Lavandera, J-L. (2001) Cheminformatic models to predict binding affinities to human serum albumin. *Journal of Medicinal Chemistry, 44*, 4370–4378.

49. Jewell, R. C., Brouwer, K. L. R., McNamara, P. J. C. (1989) α1-acid glycoprotein high-performance liquid chromatography column (Enantiopac) as a screening tool for protein binding. *Journal of Chromatography, 487*, 257–264.

50. Rapti, A. (2006) Evaluation of α_1-acid glycoprotein(AGP) binding measurements by high performance liquid chromatography & ultrafiltration Methods, *Project Report, Drug Discovery MSc*, School of Pharmacy, University of London, London.

51. Kaliszan, R., Nasal, A., Turowski, M. (1995) Binding site for basic drugs on alpha-1-acid glycoprotein as revealed by chemometric analysis of biochromatographic data. *Biomedical Chromatography, 9*, 211–215.

52. Fournier, T., Medjoubi-N, N., Porquet, D. (2000) Alpha-1-acid glycoprotein. *Biochimica et Biophisica Acta, 1482*, 157–171.

53. Oravcova, J., Bohs, B., Lindner, W. (1996) Drug-protein studies trends in analytical and experimental methodology. *Journal of Chromatography B, 677*, 1–28.

54. Pidgeon, C., Venkataram, U. V. (1989) Immobilized artificial membrane chromatography: Supports composed of membrane lipids. *Analytical Biochemistry, 176*, 36–47.

55. Taillardat-Bertschinger, A., Marca Martinet, C., Carrupt, P.-A., Reist, M., Caron, G., Fruttero, R., Testa, B. (2002) Molecular factors influencing retention on immobilized artificial membranes (IAM) compared to partitioning in liposomes and n-octanol. *Pharmaceutical Research, 19*, 729–737.

56. Taillardat-Bertschinger, P., Carrupt, A., Barbato, F., Testa, B. (2003) Immobilised artificial membrane HPLC in drug research. *Journal of Medicinal Chemistry, 46*, 655–665.

57. Giaginis, C., Tsantili-Kakoulidou, A. (2008) Alternative measures of lipophilicity: From octanol/water partitioning to IAM retention. *Journal of Pharmaceutical Sciences, 97(8)*, 2984–3004.

58. Stewart, B. H., Chung, F. Y., Tait, B., John, C., Chan, O. E. (1998) Hydrophobicity of HIV protease inhibitors by immobilized artificial membrane chromatography: Application and significance to drug transport. *Pharmaceutical Research, 15*, 1401–1406.

59. Valkó, K., Du, C. M., Bevan, C., Reynolds, D., Abraham, M. H. (2000) Rapid-gradient HPLC method for measuring drug interactions with immobilized artificial membrane: Comparison with other lipophilicity measures. *Journal of Pharmaceutical Sciences, 89*, 1085–1096.

60. Abraham, M. H. (1993) Scales of solute hydrogen bonding: Their construction and application to physicochemical and biochemical processes. *Chemical Society Reviews, 22*, 73–83.

61. Abraham, M. H., Cadha, H. S., Leiato, R. A. E., Mitchell, R. C., Lambert, W. J., Kaliszan, R., Nasal, A., Haber, P. (1997) Determination of solute lipophilicity, as logP(octanol) and logP(alkane) using poly(styrene-divinylbenzene) and immobilized artificial membrane stationary phases in reversed-phase high-performance liquid chromatography. *Journal of Chromatography A, 766*, 35–47.

62. Abraham, M. H., Chadha, H. S., Whitin, G. S., Mitchell, R. C. (1994) Hydrogen bonding. 32. An analysis of water-octanol and water-alkane partitioning and the Δlog P parameter of Seiler. *Journal of Pharmaceutical Science, 83*, 1085–1100.

63. Valkó, K., Plass, M., Bevan, C., Reynolds, D., Abraham, M. H. (1998) Relationship between the chromatographic hydrophobicity indices and solute descriptors obtained by using several reversed-phase, diol, nitrile, cyclodextrin and immobilised artificial membrane bonded high-performance liquid chromatography columns. *Journal of Chromatography A*, *797*, 41–55.

64. Nasal, A., Sznitowska, M., Bucinski, A., Kaliszan, R. (1995) Hydrophobicity parameter from high-performance liquid chromatography on an immobilized artificial membrane column and its relationship to bioactivity. *Journal of Chromatography A*, *692*, 83–89.

65. Du, C. M., Valkó, K., Bevan, C., Reynolds, D., Abraham, M. H. (2000) Characterizing the selectivity of stationary phases and organic modifiers in reversed-phase high performance liquid chromatographic systems by a general solvation equation using gradient elution. *Journal of Chromatographic Science*, *38*, 503–511.

66. Abraham, M. H., Roses, M. (1994) Hydrogen bonding. 38. Effect of solute structure and mobile phase composition on reversed-phase high-performance liquid chromatographic capacity factors. *Journal of Physical Organic Chemistry*, *7*, 672–684.

67. Du, C. M., Valko, K., Bevan, C., Reynolds, D., Abraham, M. H. (2001) Rapid method for estimating octanol-water partition coefficient (log P_{oct}) from isocratic RP-HPLC and a hydrogen bond acidity term (A). *Journal of Liquid Chromatography & Related Technologies*, *25*, 635–649.

68. Abraham, M. H., Chadha, H. S., Mitchell, R. C. (1995) The factors that influence skin penetration of solutes. *Journal of Pharmacy and Pharmacology*, *47*, 8–16.

69. Abraham, M. H., Platts, J. A. (2000) The Blood/Brain Barrier and Drug Delivery to CNS, Eds Begley, D. J., Bradbury, M. W., Kreuter, J.. Marcel Dekker, New York, NY, pp. 9–33.

70. Platts, J. A., Abraham, M. H., Zhao, Y. H., Hersey, A., Ijaz, L. Butina, D. (2001) Correlation and prediction of a large blood-brain distribution data set – an LFER study. *European Journal of Medicinal Chemistry*, *36*, 719–730.

71. Zhao, Y. H., Le, J., Abraham, M. H., Hersey, A., Eddershaw, P. J., Luscombe, C. N., Butina, D., Beck, G., Sherborne, B., Cooper, I., Platts, J. A. (2001) Evaluation of human intestinal absorption data and subsequent derivation of a quantitative structure—activity relationship (QSAR) with the Abraham descriptors. *Journal of Pharmaceutical Sciences*, *90*, 749–784.

72. Zissimos, A. M., Abraham, M. H., Barker, M. C., Box, K. J., Tam, K. Y. (2002) Calculation of Abraham descriptors from solvent-water partition coefficients in four different systems; evaluation of different methods of calculation. *Journal of Chemical Society, Perkin Transactions 2*, 470–477.

73. Zissimos A. M., Abraham, M. H., Du, C. M., Valkó, K. Bevan, C., Reynolds, D., Wood, J., Tam, K. Y. (2002) Calculation of Abraham descriptors from experimental data from seven HPLC systems; evaluation of five different methods of calculation. *Journal of Chemical Society, Perkin Transactions 2*, 2001–2010.

74. Roses, M., Bolliet, D., Poole, C. F. (1998) Comparison of solute descriptors for predicting retention of ionic compounds (phenols) in reversed-phase liquid chromatography using the solvation parameter model. *Journal of Chromatography A*, *829*, 29–40.

75. Cimpean, D. M., Poole, C. F. (2002) Systematic search for surrogate chromatographic models of biopartitioning processes. *Analyst, Royal Society of Chemistry*, *127*, 724–729.

76. Hidalgo-Rodriguez, M., Fuguet, E., Rafols, C., Roses, M. (2010) Estimation of biological properties by means of chromatographic systems: Evaluation of the factors that contribute to the variance of biological – chromatographic correlations. *Analytical Chemistry*, *82*, 10236–10245.

77. Abraham, M. H., Acree Jr., W. E. (2010) The transfer of neutral molecules, ions and ionic species from water to wet octanol. *Physical Chemistry Chemical Physics*, *12*, 13182–13188.

78. Abraham, M. H., Acree Jr., W. E. (2010) Equations for the transfer of neutral molecules and ionic species from water to organic phases, *Journal of Organic Chemistry*, *75*, 1006–1015.

79. Abraham, M. H. (2011) The permeation of neutral molecules, ions and ionic species through membranes: Brain permeation as an example. *Journal of Pharmaceutical Sciences*, *100*, 1690–1701.

6

MOLECULAR PHYSICOCHEMICAL PROPERTIES THAT INFLUENCE ABSORPTION AND DISTRIBUTION—SOLUBILITY

DEFINITION OF SOLUBILITY

Solubility is a term that is very common in everyday use. We consider a compound to be soluble when it appears to form a clear solution as we put the solid form of it into a solvent. However, it is more difficult to define solubility from a strictly physicochemical point of view. It is very common to be concerned about low solubility of compounds early in the drug discovery process as it causes problems for preparing solutions for biological testing. For example, low solubility can affect the results of a screening assay, which relies on a known concentration of the molecule in solution under the assay conditions. Low solubility also affects oral absorption when we administer the compound in a solid form. The compound has to be in solution before it can be absorbed through the gastrointestinal tract. We can imagine the detrimental effect of solid particles in the bloodstream. It follows that an ideal drug molecule should be soluble enough to be absorbed and circulate in solution in the bloodstream and reach the site of action. Pharmacopeias usually describe the solubility of a compound by stating how many units of solvent are needed to dissolve a unit weight of the solid compound in a defined solvent. For example, 1 g of caffeine can be dissolved in 60 g of water. The simplest way to measure solubility is to weigh a certain amount of compound and add a known volume of solvents until it forms a clear solution. Some compounds can take longer to dissolve and

Physicochemical and Biomimetic Properties in Drug Discovery: Chromatographic Techniques for Lead Optimization, First Edition. Klara Valko.

need more stirring. Therefore, the definition and precise measurement of solubility are complex. It is even more difficult to define good or bad solubility because it depends on the purpose and condition for which we need to achieve the desired solubility.

The thermodynamic definition of solubility is the concentration of the saturated solution of a substance that is in equilibrium with a defined solid form of the compound. This concentration has to be cited together with the solvent. We also need to define the temperature at which the equilibrium between the solid and dissolved form was achieved. When the solvent is water, we need to state the pH and ionic strength. It is very important to describe the solid form of the compound (crystal or amorphous) as the solubility of different crystal forms of the same compound may differ by orders of magnitude. In order to assess when we have reached equilibrium and obtained a saturated solution, we need to measure the concentration of the compound at various time points. We need to check that the solid form of the compound has not changed during the course of the measurement.

The potency measurement is carried out in aqueous solutions; therefore, in early drug discovery, the aqueous solubility of the compounds is an important parameter. For good oral absorption, we need good aqueous solubility too. The "apparent solubility" of the compound equals the saturated concentration when the dissolved compound is in equilibrium with solid material. It is referred to as S_{app}. When the compound has ionizable groups, we can define "intrinsic solubility" (S_0), which refers to the concentration of the saturated solution of the neutral (unionized) form, which is in equilibrium with the solid material. In order to suppress the ionization, the pH of the solution must be adjusted. When buffer solutions are applied, the measured solubility depends on the type of buffer and the ionic strength as well. Several publications [1–3] describe the definition of various solubility measures and have studied the effect of experimental conditions such as the excess amount of solid, the length of equilibration, the method applied for the separation of the solid from the saturated solution, and the pH at the beginning and at the end of the measurements.

In drug discovery, it may be advantageous to approach solubility from a physiological point of view as well. We have to know the pH, the time course, and the approximate volume of the liquid and other constituents of the gastrointestinal tract, in which the desired amount of drug molecule (dose) should be dissolved in order to facilitate the absorption process. In this respect, we speak about "kinetic solubility," which means the concentration of the compound in solution at a given time interval after adding the solvent to the solid material. We have to mention that even if the dose dissolves in the gastrointestinal tract, absorption of the compound does not necessarily follow as the compound has to permeate through the intestinal wall in order to reach the blood circulation. The FDA in the United States defines the drug as soluble taking into account the physiological requirements for absorption as follows:

> *A drug substance is considered HIGHLY SOLUBLE when the highest dose strength is soluble in <250 ml water over a pH range of 1 to 7.5.*

A drug product is considered to be RAPIDLY DISSOLVING when <85% of the labeled amount of drug substance dissolves within 30 minutes using USP apparatus I or II in a volume of <900 ml buffer solutions.

The FDA definition of high solubility concentrates on the dose and the important physiological conditions because we want to be sure that the required amount of the drug taken is in solution in a volume of equivalent to a glass of water (250 cm^3). Ideally, the drug should remain in solution during transit from the stomach (pH 1) to the small intestine (pH 7.5), where most of the absorption occurs. The FDA introduced the so-called Biopharmaceutical Classification System (BCS), which includes the above-mentioned definition of solubility and also adds a definition of high permeability.

A drug substance is considered HIGHLY PERMEABLE when the extent of absorption in humans is determined to be >90% of an administered dose, based on mass-balance or in comparison to an intravenous reference dose.

When a drug molecule fulfills the requirement for good solubility and good permeability, good absorption and bioavailability may be expected. Therefore, the authorities may waive expensive bioequivalence studies in the clinic, which can save a significant part of the cost of drug development and registration.

A more detailed theoretical basis for the Biopharmaceutical Drug Classification is discussed by Amidon et al. [4]. Figure 6.1 shows the four classes in the BCS.

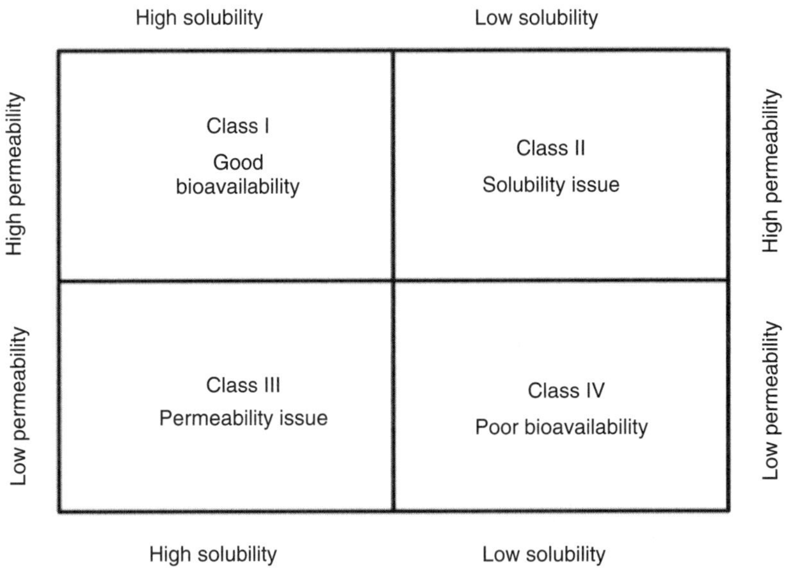

Figure 6.1 The Biopharmaceutical Classification System. (Reproduced with permission from Reference [4].)

Class I contains compounds that have both high solubility and high permeability. Class II contains drugs that have solubility-limited absorption, that is, low solubility but high permeability. We can expect very poor absorption for Class IV drugs, which have not only low solubility but low permeability too. Class III drugs have good solubility, but they have poor permeability.

By the FDA definition, a dose has to be in solution within a wide pH range (pH 1–8) that necessitates the measurement of solubility over a pH range. The pH of the aqueous solution will affect the solubility of all ionizable compounds that contain acidic or basic or both functional groups, which may get protonated or deprotonated depending on the pH. The lowest solubility should be enough to keep the compound in solution.

When we approach solubility from a physiological absorption point of view, we want to know whether the dose of the drug molecule will dissolve *in vivo* or not. In this case, we need to express the solubility in weight per volume unit and relate it to dose per 250 ml, that is, one dose of the drug dissolves or not in a 250 ml glass of water. From a thermodynamic point of view and in early drug discovery stages, potency is expressed in molar concentration units; therefore, we would like to know the solubility expressed in molar term. As it is easy to calculate the molar concentration of the saturated solution when we know the molecular weight of the compounds, one would easily ignore the differences. However, when we want to relate solubility to structure, we would like to know which compounds with which substituents are more soluble and we may not have any idea about the dose of the compounds (we may not even know whether they are active or not). At this stage, it is crucial to know the molar solubility. We can consider one compound as more soluble than another if higher numbers of molecules are in solution. If we express solubility in milligram per milliliter for two molecules that differ in molecular weight, the comparison is misleading. As an example, molecule A has a molecular weight of 200, while the other molecule is 600, and the two molecules have exactly the same solubility in molar term; the bigger molecule shows higher solubility if we express it in milligram per milliliter unit. Moreover, we can draw completely wrong conclusions using different units as measure of solubility, that is, bigger molecules are more soluble, which is not true if we consider the number of molecules in solution. There is a change in the solubility unit used during the drug discovery process; from molar solubility, we change to weight per volume solubility, which we have to be aware of.

At the beginning of the discovery process, we need to know the solubility of the compound in molar term, that is, how many moles (mmol or μmol) are in solution. If a compound has a pIC_{50} *of 6, it means that* 10^{-6} *mol/l (1 μmol) concentration is needed to occupy 50% of the receptor and that concentration should be lower than the saturated concentration. In principle, if we know the amount of the target receptor in each patient, we should give the appropriate customized dose in molar term to the patients. That would be impractical at this stage of the patient care. The ideal situation would be that each patient would take only the minimum amount of drug molecules that would be needed to treat the disease; so, the dosage could be adjusted to the weight of each patient. This type of "customized" drug formulation might be achievable in the future. We shall see later in this book that there is a lot of scope for improvement to adjust the dose or even the type of the drug according to individual patients. We could assess the plasma protein concentration, the volume ratio of the plasma and tissue, and the body fat in each patient and adjust the required minimum dose of the drug.*

MOLECULAR INTERACTIONS WITH WATER

Let us investigate the process of dissolution of a compound from solid in water from a physicochemical point of view. First we need to break the bonds holding the solid crystal together. This process requires energy. Every crystal has crystal lattice energy; this is required to disrupt the crystal packing. For an amorphous material, in general, we need to invest less energy to break the interactions between the molecules in the solid form than in a crystalline form. Once removed from the crystal lattice, the molecules have to create a cavity in the water by disrupting the hydrogen bonds. This dissolution process requires further energy. However, if the solute and solvent molecules can form new interactions (e.g. dipole–dipole and H-bond donor–acceptor), then some of the energy may be regained during the dissolution process. Lipophilic molecules, however, cannot form strong interactions with the polar water molecules; so, energy cannot be retrieved from these interactions, and they have poor solubility in water.

The most popular equation for the estimation of solubility of organic molecules is the Yalkowsky equation [5]. This equation describes the logarithmic value of the water solubility of an organic compound (log S) in relation to its lipophilicity expressed by the octanol/water partition coefficient (log P oct:water) and its melting point (MP).

$$\log S_{\text{solids in water}} = 0.5 - 0.01(\text{MP-25}) - \log P_{\text{oct:water}} \tag{6.1}$$

Note that the MP is set to zero for solutes that are liquid at room temperature; that is why we subtract 25 from the MP: (MP-25) °C. Equation 6.1 relates solubility to lipophilicity and the MP, which is taken as proportional to the energy of the crystal lattice. The octanol/water partition coefficient represents the lipophilicity of the compound, which summarizes the energy of all the interactions between the solute molecule and water. It is also important to note that log S refers to the logarithmic value of the equilibrium solubility of the compound expressed in molar concentration units. We have to be sure that equilibrium has been achieved and other conditions such as particle size and stirring do not unduly influence the final result. From the coefficient of the MP in the equation, we can deduce that an increase of 100 °C in MP will reduce the intrinsic solubility by a factor of 10, that is, a unit lower in log S terms.

As we have seen in Chapter 5, five important molecular descriptors could describe the lipophilicity of the compounds, and similarly the same approach can

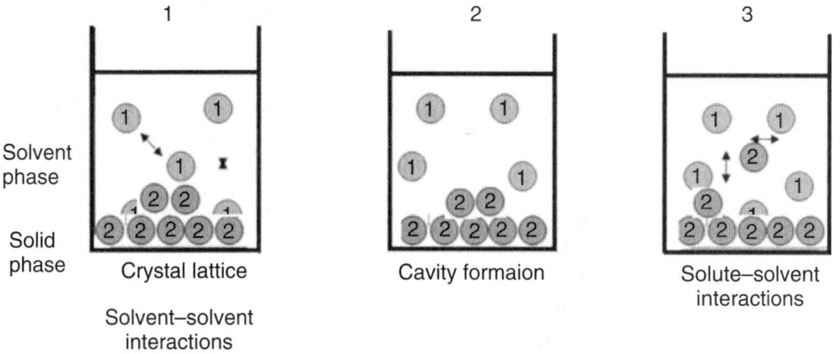

Figure 6.2 The schematic representation of the dissolution process for which the solvation equation model can be applied. (1 represents the solvent molecules and 2 the solutes. 1: the interaction of the solvent and the solute molecules has to be broken; 2: a cavity has to be formed in which the solute molecules can fit; 3: energy is gained by forming interactions between the solute and the solvent molecules.)

be applied to solubility. Abraham and Le [6] applied the solvation equation model to predict the solubility of over 400 diverse compounds for which the molecular descriptors were available. The solvation process is depicted in Figure 6.2.

Abraham and Le [6] constructed a solvation equation containing the five molecular descriptors (E is the excess molar refraction, S is the dipolarity/polarizability, V is the molecular size (McGowan volume), A and B are used for H-bond acidity and basicity, respectively, and MP is the melting point), as is shown by Equation 6.2.

$$\log S_w = 0.58 - 0.58E + 0.98S + 1.23A + 3.39B - 4.08V - 0.01(\text{MP-25})$$
$$n = 411 \quad r^2 = 0.915 \quad \text{sd} = 0.564 \quad F = 724 \qquad (6.2)$$

Note that the +/− signs of the coefficients are the opposite of the signs of the Abraham equation obtained for lipophilicity as there is an *inverse* correlation between lipophilicity and solubility. The coefficients are of similar magnitude to the coefficients in the octanol/water partition, as given in Table 5.5; however, the H-bond acidity term has a positive value and is not zero, meaning that the H-bond acidity enhances the solubility. We have to remember that, while the solvation equation for the octanol/water partition process describes the concentration ratio of the compound in the two solvents, the solubility process can be considered as the partition equilibrium between the solid form and the solute in the water. As the properties of the solid forms vary for different types of molecules, it is very difficult to derive a precise solvation equation for solubility. Taking into account this variability, the solvation equation for solubility was statistically significant for a large number of compounds. Abraham and Le [6] have tried to replace the MP with molecular descriptors by the product term of H-bond acidity and basicity, which resulted in Equation 6.3.

$$\log S_w = 0.52 - 1.00E + 0.77S + 2.168A + 4.24B - 3.36A \times B - 3.99V$$
$$n = 659 \quad r^2 = 0.920 \quad \text{sd} = 0.56 \quad F = 1256 \qquad (6.3)$$

The Abraham solvation equation for describing the logarithmic value of aqueous solubility expressed in molar concentration units as a solute property demonstrates that larger compounds that have both H-bond donor and acceptor groups with low dipolarity and polarizability have low solubility. This can be explained by the fact that molecules that have both H-bond donor and acceptor functionality may form strong intermolecular H-bonding in the solid crystal.

This results in increasing crystal lattice energy and higher MP and consequently a reduction in solubility. Another thermodynamic approach has been published by Perlovich et al. [7], which helps in understanding the molecular solvation mechanism and the crystal lattice energy. Crystal lattice energy can be estimated by measuring the temperature dependence of the vapor pressure of saturated solutions and the temperature dependence of solubility of compounds in organic solvents such as octanol and hexane relative to water. They highlighted the importance of lipophilicity and other polar forces such as the H-bond formation both in the crystal and in water. For the estimation of the effect of H-bond formation on solubility, an "inert" solvent such as *n*-hexane was used.

All of the above-mentioned theoretical considerations for the description of the dissolution process and the equilibrium solubility refer to a pure compound forming a perfect crystalline solid form. In reality, the compounds are never 100% pure and it is commonly observed in pharmaceutical industry practice that solubility of the compound gradually decreases as it moves forward in the drug discovery and development process. The simple explanation for this is that the compound is synthesized in an increasingly pure form. Impurities cause disruptions in the crystals, thus disrupting the forces between the molecules, in turn reducing the MP and crystal lattice energy. The theoretical description of the dissolution process and thermodynamic equilibrium solubility refers to solids irrespective of their particle size. It is a standard practice in development to reduce the particle size of active pharmaceutical ingredients (APIs) by milling and micronization in order to increase the dissolution rate of the drug. The thermodynamic solubility usually refers to large particles. Solubility will increase with decreasing size of solute because of the additional surface energy. This effect is generally small unless particles become very small, typically smaller than 1 μm.

When the particle size gets so small that the surface area relative to the particle volume or mass becomes large, the surface properties will influence the macroscopic thermodynamic properties. The Kelvin equation or the Gibbs–Thompson effect, which relates the surface curvature to chemical potential and surface tension, can enhance the solubility of nanoparticles. So, the recently emerging nanotechnology may help to improve the solubility and as a consequence bioavailability of APIs.

When using nanotechnology [8], that is, reducing the size of the solid particles to nanoscale level by applying surface-active or polymer additives, we can conserve the nanoparticles and make the compound seemingly more soluble. Although there are many pharmaceutical formulation technology approaches that might help overcome the problem of poorly soluble drug molecules, the scope and the aim at early drug discovery lead optimization are to synthesize and progress the potent compounds that have the best physicochemical properties for developability, which includes solubility and oral absorption. To be able to help the drug design and lead optimization process to progress the best compounds, we need to develop solubility measurements that can be carried out on the required number of compounds in a desired time frame and provide results adequate for the particular problem. Typical methods for solubility measurements at early stages of the drug discovery and lead optimization process are described in the following section.

VARIOUS SOLUBILITY MEASUREMENTS THAT CAN BE APPLIED DURING THE DRUG DISCOVERY PROCESS

The solubility of a compound is dependent significantly on the experimental conditions. The measurement of thermodynamic solubility of a pure solid substance may take a week and requires sensitive analytical methods. Even when using a high level of automation, it is practically impossible to characterize thousands of compounds per month. However, if we understand the factors that influence the solubility, and we would like to answer a particular question about the solubility of the compound that is relevant to the drug discovery stage, we can carry out an adequate number of measurements with fit-for-purpose quality of results. Thus, first we need to clarify by the measurements, the question that we would like to have the answer to:

1. Is the compound in solution under the biological screening condition?
2. Will the required dosage dissolve in the gastrointestinal tract?
3. Can formulation enhance the solubility and improve oral absorption of the compound?

The first question arises early in the drug discovery process, and a high throughput precipitative solubility measurement can give a satisfactory answer. It is able to reveal whether the measured biological activity is compromised by low solubility. In this case, the best results can be achieved by applying exactly the same amount of dimethyl sulfoxide (DMSO) and the same buffer (pH and ionic strength) as is used for screening. We have seen the effect of DMSO concentration and the type of buffer on the measured apparent precipitative solubility.

The need for assessing rapidly the solubility of large number of compounds emerged with the introduction of combinatorial libraries and high throughput chemistry followed by high throughput potency screening. As the molecules became larger and more lipophilic in order to increase the potency, the reproducibility of the potency screens decreased. The compounds synthesized by high throughput chemistry are usually dissolved and stored in DMSO (DMSO = $(CH_3)_2SO$). DMSO solutions are typically dispensed using liquid-dispensing robots for various screens. DMSO is considered to be a universally good solvent for both polar and nonpolar compounds. It has a relatively high boiling point (189 °C), which means that it does not easily evaporate when dispensed in microliter (μl) quantities onto 96-well plates. Storing the compounds in liquid form made possible the accurate dispensing of large number of compounds using automated robots.

Another advantage of DMSO is that it is readily miscible with water. In a typical screening procedure, the compounds are dispensed as 0.1–10 mM DMSO stock solutions and are diluted with aqueous buffer solution before adding the appropriate target protein or enzyme. Although the compound might be in solution in the DMSO solvent, when it is diluted by aqueous buffer, the compounds might precipitate out. Typically, a 10 to 100 times dilution is applied, resulting in only 10% to 1% DMSO in the final solution. As the dilution is carried out by robotic systems, the precipitation occurring during the dilution often remains unobserved and unrecorded. Therefore, there is a need to find an automatic procedure that detects the precipitation of compounds during dilution of the DMSO solution with aqueous buffer. Lipinski et al. [9] has published a method that detects the scattering of light due to solid particulates when precipitation occurs. The turbidity measurement has been carried out by instruments for measuring optical characteristics of a solution, and can operate in 96-well and 384-well plate formats. They also have observed that compounds that precipitate out of solution as crystals gave a sudden strong signal. For these compounds, the dilution factor when the signal occurred was more reproducible than for compounds that produced a gentle steadily increasing signal. These latter compounds precipitated out as amorphous solids according to Lipinski et al. [9], and the solubility results were not very reproducible. The turbidity measurement does not give an absolute value of the solubility, which is unlikely to be the same as an equilibrium solubility obtained from a solid; however, it provides a warning signal when the compound precipitates out of solution. Bevan and Lloyd [10] have used the Nephelostar instrument from BMG LabTech to detect precipitation using laser beam. The instrument could detect the laser-light-scattered signals for 96 wells in a plate within a couple of minutes. However, it has also been found that when compounds precipitated out as large crystals, the signal was weaker. Some compounds with surface-active properties caused a change in the background signal because of the meniscus change in the solution in the well as a so-called "lensing" effect. Nevertheless, these technologies are used frequently in screening laboratories, not for providing an accurate solubility measurement but for detecting precipitation during the biological screening processes.

A typical measurement process is to dilute down 5 μl × 10 mM DMSO stock solution with 95 μl phosphate-buffered saline solution. It results in 500 μM total concentration of the compound containing 5% DMSO. The wells can be screened by measuring the light scattering. For these measurements, a simple UV plate reader can also be used. Precipitation should result in a distinct distortion in the compound's UV spectrum. If no precipitation is detected, the compound can be declared soluble as equal to or greater than 500 μM.

CONDITIONS THAT AFFECT SOLUBILITY

We have to remember that the above-mentioned solution contains 5% DMSO, which enhances aqueous solubility. Rubino and Yalkowsky [11] have investigated the solubility enhancement of cosolvents in aqueous systems for poorly soluble compounds (diazepam, benzocaine, and phenytoin). Equation 6.4 describes the solubility enhancement, and Figure 6.3 shows the increase in aqueous solubility by increasing the concentration of DMSO solution.

$$_{\text{total}}S^{\text{cos}} = S + S \times 10^{\sigma\text{ufc}} \tag{6.4}$$

where $_{\text{total}}S^{\text{cos}}$ is the solubility of a neutral compound with the cosolvent, σu is the solubilizing factor of the cosolvent, and fc is the fraction of cosolvent used. The effect of pH on solubility of the ionized compounds is discussed later in more detail; so, the pH effect has been omitted from Equation 6.4.

We can see in Figure 6.3 that 1–5% DMSO does not cause a significant enhancement in aqueous solubility. There are publications dealing with high throughput measurements of solubility, which apply 2% or 5% DMSO solutions [12, 13]. Using Equation 6.4, we can calculate the approximate differences in the measured solubility under these two conditions as follows:

$$S_{\text{increase}} = \frac{1 \times 10^{5\times 0.05}}{1 \times 10^{5\times 0.02}} = 1.4 \tag{6.5}$$

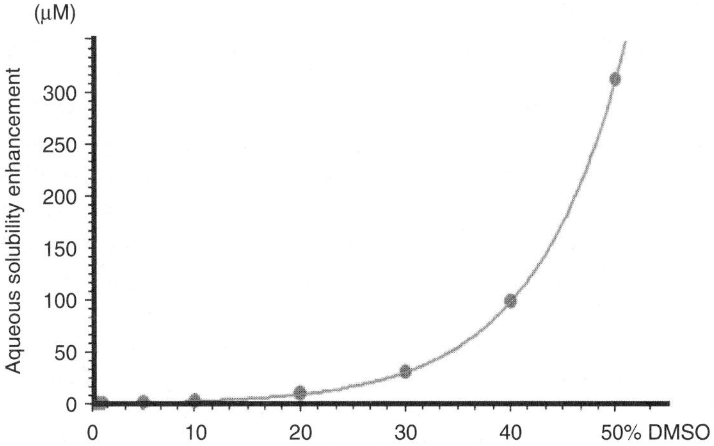

Figure 6.3 Effect of the DMSO concentration on aqueous solubility enhancement. (Figure is courtesy of Chris Bevan.)

It means that we can expect a 1.4 times increase in solubility when the DMSO concentration is raised to 5% from 2%. When the DMSO stock solution concentration is fixed (typically 10 mM), which is the case in most pharmaceutical companies, applying the lower percentage DMSO solution during the solubility measurement also means that the maximum amount of compound we put into the solvent is less (e.g., instead of 500 μM we have only 200 μM). This means that the upper limit of the solubility measurement is lower. As the detection limits of the analytical methods are similar, the solubility range of the measurement will also be smaller. When we use the DMSO precipitative method to assess the solubility of the compound and use nephelometry or turbidity measurements, we get a "yes" or "no" answer to the question: is the compound still in solution under the assay condition? In order to obtain a quantitative measure of solubility, we need to repeat the measurements using a range of dilution of the samples. The highest concentration when the precipitation was detected provides an estimate for solubility. When more sophisticated detection methods that measure the concentration of the compound in solution are used, it is enough to make sure that precipitation has occurred; so, the measured concentration is the result of the equilibrium between the dissolved and solid material.

The upper limit of the solubility measurement depends on the concentration of the compound used. If no precipitation occurs, the compound is not in equilibrium with the solid. In this case, we can only say that the solubility of the compound is equal to or above a given concentration. The range we can quantify the solubility is between the detection limit (approximately 1 μM when UV detection is used) and 200 or 500 μM, depending on the amount of compound (DMSO solution) used.

In early drug discovery, a typical high throughput DMSO precipitative solubility measurement can be carried out as follows. A known volume of DMSO stock solution of the compound of known concentration is dispensed in a 96-well plate. As an example, a method that uses 5% DMSO in buffer and starts off with 5 μl × 10 mM DMSO stock solution is described. We need to dispense the same quantity of the compound into another 96-well plate in the same position. Then, to one of the plates, we add 95 μl buffer solution (typically pH 7.4 phosphate buffer or phosphate buffer saline), while in the other plate, we dispense 95 μl solvent that fully dissolves the compound. Typically, DMSO is used. This plate will be the so-called "standard" plate. The two plates are then set aside to allow the precipitation to occur in the buffer solution in the so-called "sample" plate. The equilibration time may vary from 1 to 24 h. After the equilibration time, the solid particles have to be separated from the solution by applying filtration or centrifugation. The quantitative analysis is carried out on both the plates, and comparisons of the detector signals of the known concentration in the "standard" plate to the detector signal to the unknown concentration in the "sample" plate take place. This measurement starts with DMSO stock solution of the compound; so, it does not involve the dissolution of the solid particles. The measurement of dissolution rate is discussed later in this chapter. The procedure is summarized in Figure 6.4.

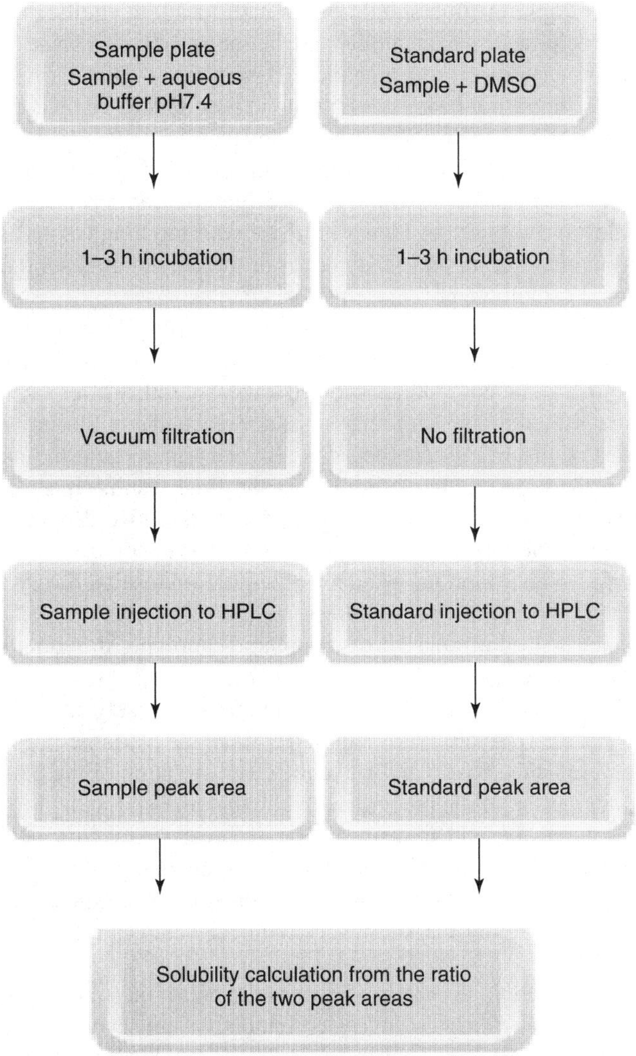

Figure 6.4 Typical high throughput methods for the determination of solubility of drug discovery compounds using 5% DMSO stock solution. Two identical plates contain the same amount (5 μl × 10 mM DMSO) solution of the compounds. The "sample plate" is diluted down with 95 μl pH 7.4 aqueous buffer, and the "standard plate" is diluted down with 95 μl DMSO. The "sample plate" is filtered after the equilibration time, and the same amount of solution is injected to an HPLC from both plates for the relative concentration determination of the compounds in the two plates.

Ideally, equilibration time should be enough to allow equilibrium to develop between the precipitated solid material and the saturated buffer solution of the compound. It can be very compound dependent. If we wait too short a time, then some of the compounds might not have enough time to precipitate. Some compounds tend to form supersaturated solutions, which means that the concentration in the buffer may be temporarily higher than the compound solubility. A little movement of the plate or a tiny dust particle can initiate precipitation and terminate the energetically unstable supersaturation. However, if we wait too long (more than 24/48 h), decomposition of the compound may occur or the precipitated solid material may transform to a less soluble polymorphic form.

Figure 6.5 shows a comparison of the measured solubility using different equilibration times, but otherwise under exactly the same conditions [14]. It can be seen in Figure 6.5 that some compounds showed lower solubility after 3 h than after 1 h, indicating that the precipitation process has not reached equilibrium within 1 h.

The nature of the filtration process is also very important and an example is shown where the type of filter used affects the final results. Inert filters need to be used as adsorption can occur onto the filter (or the 96-well plate wall), which reduces the apparent concentration of the compound in solution. The pore size of the filter may also be important when the precipitation results in very fine particles that can penetrate certain types of filters. Figure 6.6 shows the plot of solubility results obtained with the same set of compounds applying the same equilibration time and buffer, but altering the filter type [14].

The PCTE (polycarbonate track-edge) filter has a precisely controlled pore size, whereas the PTFE (polytetrafluoroethylene) filter has variable pore sizes, but is more resistant to acids, bases, and solvents. Centrifugation of the plates can result in rapid sedimentation of the precipitated solid material and with properly adjusted needle positioning, the HPLC instrument can inject solely from the supernatant liquid. However, this methodology is difficult to apply in high throughput situations as some research compounds may have low density, floating precipitates that will not sediment, and without manual inspection such samples may block the HPLC instrument's injector.

The effect of the ionic strength and type of the buffer on solubility has also been investigated on diverse drug discovery compounds from several programs [14]. It was found that the addition of saline to the phosphate buffer (0.9 w/v NaCl solutions to match the ionic strength of blood, plasma) altered the solubility of several compounds (Figure 6.7). The explanation for this may be that the solubility of protonated bases is different when a phosphate or a chloride salt is formed.

Various analytical techniques can then be applied to quantify the amount of compound in the buffer solution relative to the standard solution. When the plate reader technology is used, the reader will acquire the UV spectrum of the compound in the standard solution and compare the extent of absorbance in the sample solution. The "standard" plate should be prepared by diluting the DMSO stock solution with a solvent that will not precipitate out the compound. It should also be less UV active than the DMSO, which has strong UV absorbance around 230 nm. When comparing UV spectra of the same compound from the standard and from the sample solution,

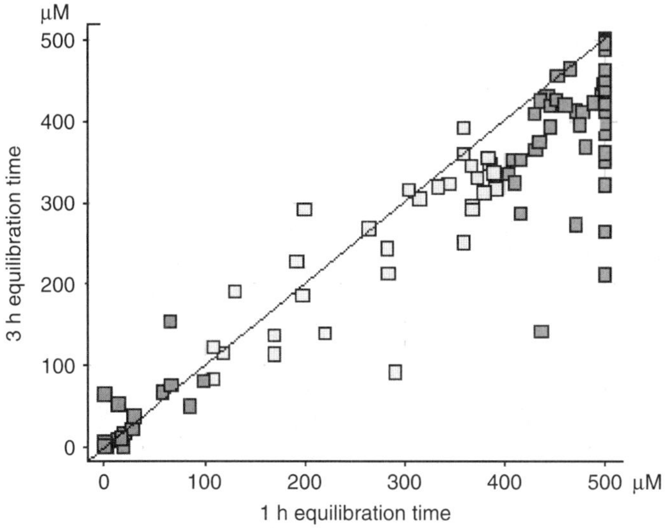

Figure 6.5 Comparison of DMSO precipitative solubility of a diverse set of research compounds obtained after 1 h and 3 h equilibration time. Other conditions (buffer, filter type, etc.) were exactly the same.

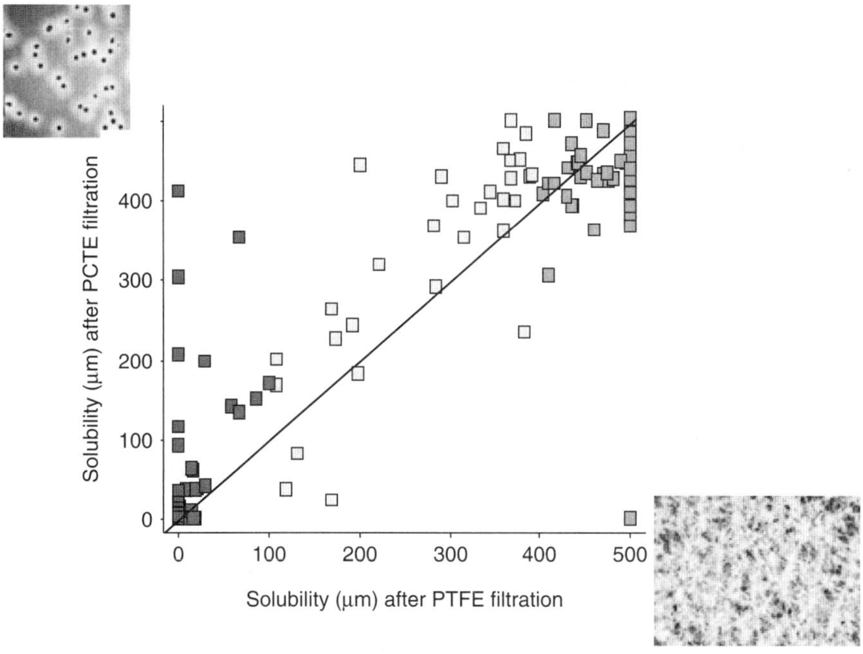

Figure 6.6 Effect of the filter on the DMSO precipitative solubility results using otherwise exactly the same conditions and the same HPLC/UV quantification method. The *Y* axis shows the results obtained using PCTFE (polycarbonate track-edge) filter, while the *X* axis shows the solubility results using PTFE (polytetrafluoroethylene) filter.

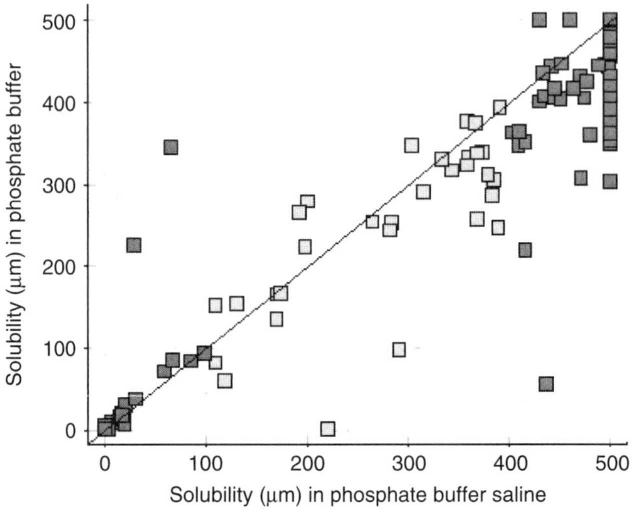

Figure 6.7 The effect of saline (0.9 w/v NaCl) in the buffer solution on solubility of diverse research compounds.

which has different solvents, we have to be aware of the possible solvatochromic effect of the solvent, which may shift the UV maximum or increase or decrease the molar extinction coefficient of the compounds. Avdeef [15] developed a "universal" solvent mixture that dissolves most compounds at the 500 μM level, and has not got a significant solvatochromic effect on the UV spectra.

Applying HPLC/UV quantification has several advantages over the plate reader method. We inject only 5–10 μl solutions from both the "standard" and the "sample" plate; so, the solvent in the standard will elute at the beginning of the chromatogram and it appears only as a solvent peak separated from the retained compound. Thus, UV-active solvents can also be used without influencing the quantification of the compounds. The separation power of the HPLC method also makes it possible to analyze the compounds that are impure because the impurities are separated from the compound of interest. When a generic reversed phase gradient method is applied (Chapter 5), there is no need to develop a specific method for each compound on the plate. By injecting the compound from the standard plate, we can expect to see a peak on the chromatogram with a retention time that is characteristic for the compound. The filtered "sample" is injected after the injection from the "standard" plate. We have to look for a peak with the same retention time. When a peak appears with the same retention time and approximately the same size (peak area), the compound was fully soluble in the buffer. When the peak area is smaller than that of the peak obtained from the "standard" plate, we can calculate the solubility by Equation 6.6.

$$\text{Solubility } (\mu\text{M}) = \frac{\text{peak area (sample)}}{\text{peak area (standard)}} \times \text{conc standard } (\mu\text{M}) \tag{6.6}$$

When there is no peak to be found in the "sample" chromatogram that has the same retention time as the peak in the "standard" sample, the solubility of the compound is less than the detection limit. It is common to observe other peaks in the "sample" chromatogram that are not visible in the standard peak. The explanation for this is that the impurities may be more soluble in the buffer than the compound itself; so, the relative concentration of the impurities may have been increased in the filtered buffer solution. Figure 6.8 shows an example of chromatograms from the "standard" and "sample" solutions monitored at 230 and 254 nm wavelengths, using a 5-min generic gradient reversed phase chromatography with 0% to 100% acetonitrile gradient. As the DMSO has a significant UV absorbance at 230 nm, we can see a large solvent peak at the beginning of the chromatogram. Note that the "standard" chromatogram shows a much bigger solvent peak than the "sample" chromatogram at 230 nm because the "standard" solution contains 100% DMSO and the "sample" solution contains only 5% DMSO (the rest of the solvent is buffer).

The HPLC/UV detection method has another advantage, namely, when we use a reversed phase gradient with acetonitrile and adjust the mobile phase pH to 7.4, we can obtain the CHI of the compound from the retention time at the same time as when the solubility result is obtained. This means that we can obtain the solubility and lipophilicity characteristics of a compound from a single measurement. The limitation of the HPLC/UV quantification is that, in principle, we use only a "one-point" calibration to calculate the concentration of the "sample" solution. We also suppose that the original DMSO stock solution was accurately prepared as a 10 mM solution. In practice, the concentration of the stock solution may deviate from the original value significantly. When the stock solutions go through several freeze–thaw cycles, they can absorb moisture from the air because of the hygroscopic nature of the DMSO. The wet DMSO solution may decrease the compound solubility, resulting in precipitation in the stock solution. When the robot dispenses from the suspension, it either can pick up a crystal, then we have higher concentration of the compound, or just dispense from the moist DMSO solution containing less than 10 mM compound.

There are techniques that can determine absolute concentration of the DMSO stock solution, such as the CLND. It measures the nitrogen content of molecules that can be calibrated by any nitrogen-containing compound. The CLND signal is proportional to the number of nitrogen atoms in a mol quantity of molecules. The compound undergoes pyrolysis at a high temperature (over 1000 °C) and oxidizes the nitrogen to nitric oxide. Nitric oxide reacts with ozone to form electronically excited nitrogen dioxide. The excited nitrogen dioxide emits light in the red and infrared region of the spectrum (600–3200 nm) as it relaxes back to its ground state. The light emitted is directly proportional to the amount of the nitrogen present in the sample. When the detector is calibrated with a known amount of nitrogen-containing compound, the calibration curve obtained can be used to quantify any other compound. For example, various concentrations of caffeine solution can be used to calibrate the CLND instrument signal. The CLND methodology can be used for analyzing a sample directly without a chromatographic separation process. However, without separation, it will measure the total quantity of nitrogen in the

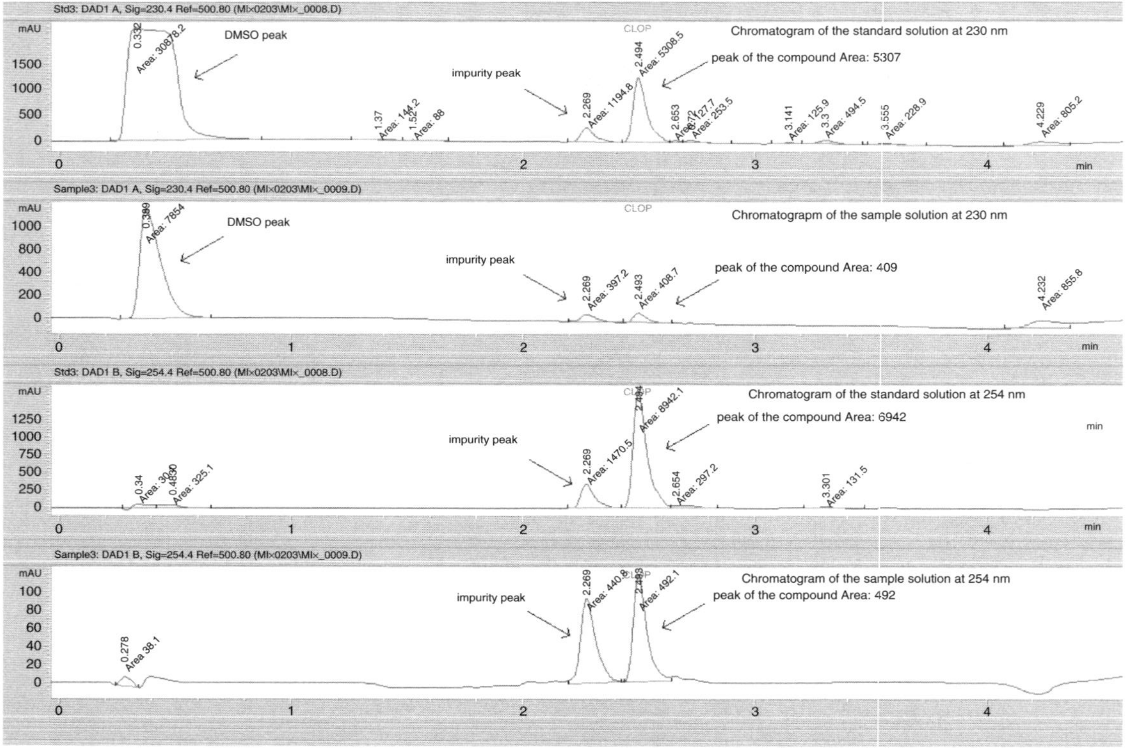

Figure 6.8 Chromatograms obtained from a precipitative solubility measurement from the "standard" and the "sample" plate monitored at 230 nm and at 254 nm UV wavelengths.

sample, which may come from impurities or nitrogen-containing solvent residues, such as acetonitrile. When the CLND is connected to an HPLC system, we have to make sure that the mobile phase does not contain nitrogen (especially do not use ammonium acetate as buffer or acetonitrile as organic modifier). Methanol, ethanol, and propanol can be used as organic solvent additives in the mobile phase. The application of the CLND technology for solubility determination has been discussed by Bhattachar et al. [16] in more detail. When the compound contains adjacent nitrogen atoms, the additivity of the signal decreases as the nitric oxide mole for mole proportional formation from both nitrogen atoms cannot be assured. The CLND method could also be very useful to measure the absolute concentration of a pure standard solution made from solid samples without accurate weighing. The weighing procedure is still carried out manually even in large pharmaceutical companies as the accurate handling of solid samples has not yet been reliably automated. Because of the high sensitivity for nitrogen, even small amounts of ammonia in the air or other nitrogen-containing contamination can distort the results; so, special care and handling should be applied. Fortunately, the nitrogen in the air does not react to give a signal.

In conclusion, the precipitative solubility measurements in early drug discovery are useful for detecting compound precipitation under screening conditions, and may provide an approximate value of solubility, albeit in a relatively narrow range. The amount of DMSO, the equilibration time, the buffer, the filter, and other experimental conditions can alter the results. The data cannot be regarded as precise thermodynamic constant for solubility as it refers only to the solubility at a particular pH, at a particular time, without any description of the precipitated solid form.

Wenlock et al. [17] described an excellent approach to automate the aqueous solubility measurements from solids using the shake-flask equilibrium principle. They used a Tecan Freedom Evo 200 liquid-handling robot equipped with parts for transportation, decapping and recapping, and centrifugation of the sample tubes. The principle of the method is very similar to that described for the solubility determination from DMSO stock solution. The robot prepares several dilutions from a 1 mg solid material with DMSO that serves as standard solutions. The other 1 mg solid compound is dissolved in phosphate buffer; they used longer equilibration time, a minimum of 18 h. Then, the tubes are centrifuged and several dilutions are prepared by the robot automatically from the supernatant. Each dilution is analyzed by HPLC-UV with mass spectrometric identification of the compound. The solubility range by this method is up to 1.33 mg/ml, which is equivalent to 1–2000 μM depending on the compound's molecular weight.

Another semiautomated procedure has been described by Alsenz et al. [18]. They added the solid material in hexane, which is a nonpolar solvent. Then they used a liquid-dispensing robotic system to dispense the hexane solutions equally to a 96-well plate. They used the hexane solution dispensing instead of weighing precise amounts of compounds for the solubility measurements. After evaporating the hexane, they obtained the known amount of solid in each well. Then the buffer solutions were added to the solid material. After appropriate equilibration time, they filtered the undissolved solid particles out from the solution and determined the concentration of the compound that remained in solution in the buffer.

The question may arise as to what are the expected differences between the solubility values obtained from solid material and obtained from DMSO stock solution, when the pH, equilibration time, and other significant conditions are constant. We have seen in Figure 6.3 that the presence of a small quantity of DMSO causes a slight increase in solubility. There could be a significant difference in the solid material's crystallinity when freshly precipitated from solution or is prepared as solid material. The DMSO precipitates are more likely to be amorphous. This also makes it more likely to measure higher solubility for the same compound when it is measured from DMSO stock solutions than from the solid. Figure 6.9 shows the plot of solubility data of around 160 compounds obtained from the solid and from DMSO stock solutions using the phosphate buffers of pH 7.4 [14]. The compound set contained a diverse set of known drugs and research compounds. For the precipitative solubility measurements, 10 mM DMSO stock solution was used and diluted down with phosphate buffer saline (pH 7.4) to obtain 5% DMSO in the final solution. As expected, the precipitative solubility was equal to or higher than the solubility obtained from the solid for the same compounds using exactly the same filters, buffers, and analytical method (HPLC-UV) for the concentration determination.

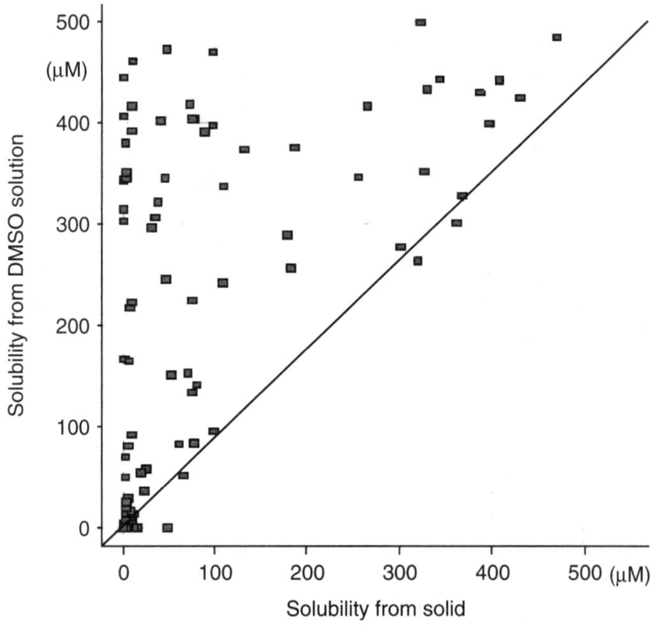

Figure 6.9 The plot of solubility data of a diverse set of 160 compounds [14] obtained from solid and from 5% DMSO solution in pH 7.4 phosphate buffers. The same batch of compounds, the same analytical technique, and instrument have been used for both measurements.

A proper thermodynamic aqueous intrinsic solubility (S_0) measurement should be carried out with a buffer where the pH is adjusted to ensure that the compound is completely unionized. Sufficient amounts of well-characterized solid material should be added to the buffer. Then, a sufficient amount of time is given to progress to equilibrium. The concentration of the saturated solution can be checked every 3–6 h and when it is constant, it suggests that the equilibrium has been established. The appropriate amount of clear supernatant solution (preferably to avoid any application of filtration) can be diluted and subjected to some form of quantitative analysis (using UV spectrophotometer or HPLC/UV/MS). The pH of the saturated solution should be checked at the end of the equilibrium process to ensure that the compound is still in its fully unionized state. The proper characterization of the solid form in equilibrium with the saturated solution should be carried out in order to ensure that the solubility value relates to a well-defined solid form of the compounds. The temperature should be controlled for the whole equilibration process. Applying smaller particle size solid material and stirring may accelerate the equilibrium process, but should not affect the results [3].

Variables such as temperature, equilibration time, type of buffer, and phase separation techniques may alter the solubility results, while the excess amount of solid material should not cause any deviations. Baka et al. [3] suggest a minimum 24-h equilibration time that consists of 6 h stirring followed by 18 h of sedimentation. They claim that when these conditions are strictly controlled, the experimental error of the solubility measurements should be less than ±4%.

It is much more difficult to design an experiment for solubility measurements to answer the second question, namely, whether the required amount of drug will dissolve in the gastrointestinal tract within a certain time frame. We usually do not know the pharmacologically active dose at an early stage of the drug discovery process, although it is usual that the program team has a target dose and the target potency before they start the lead optimization. As an example, let us say we require 100 mg of the compound to be dissolved in the gastrointestinal tract. The next thing we need to consider is the volume of liquid we want to dissolve the required amount of compound in. When we take the definition of solubility based on the BCS, the dose should be dissolved in 250 ml water (i.e., a glass of water). The pH change in the gastrointestinal system has to be taken into account as well, as it changes from acidic (pH 1) in the stomach to basic (pH 8) in the intestine. Solubility of ionizable compounds will vary depending on the pH. Therefore, the pH dependence of solubility should also be measured or at least understood.

To answer the question whether we can enhance the solubility at a later stage by formulation, we need to understand the causes of poor solubility. When poor solubility is due to the lipophilicity of the compound, there is a hope that the compound will dissolve in biorelevant media as it can partition into phospholipid and taurocholate micelles having a reasonably large lipophilic environment. When poor solubility is due to the high crystal lattice energy, only expensive formulation studies may help. Micronization, including polymer coating of the drastically reduced size of the crystals, may improve the dissolution rate and solubility just enough to get the formulated drug absorbed. The effect of solubility on absorption is discussed in Chapter 9. The required solubility for good oral absorption also depends on the permeability of the compound. The interrelation of solubility and permeability and the absorption potential of the compound are also discussed in more detail in Chapter 9.

SOLUBILITY–pH PROFILE

As we have described in Chapter 5, the Henderson–Hasselbalch equation can be used to calculate the percentage of ionized form of a compound at a particular pH. Similar to the lipophilicity pH profile, we can describe the pH dependence of solubility by the solubility of the neutral form and the solubility of the ionized form, as is shown by Equation 6.7.

$$\log S_{\mathrm{pH}} = \log S_{\mathrm{intrinsic}} + \log S_{\mathrm{ionized}}(1 + 10^{\mathrm{ch}(\mathrm{pH}-\mathrm{p}K_{\mathrm{a}})}) \tag{6.7}$$

where S_{pH} is the molar solubility of the compound at a particular pH, $S_{\mathrm{intrinsic}}$ and S_{ionized} are the molar solubilities of the compound in neutral and fully ionized forms, ch is the charge (+1 for acids and −1 for bases), pH is the negative logarithm of the proton concentration, and is the acid dissociation constant of the compound (the pH where 50% ionized and 50% nonionized forms are present). Note that while ionization decreases the lipophilicity of compounds, it increases the aqueous solubility. Figure 6.10 shows a typical solubility–pH profile of a weakly basic compound.

It is important to keep in mind that the solubility of the fully ionized species depends on the type of buffer, especially on the type of counterion in the buffer. It means that we can approximate the solubility increase by the pH up to a certain

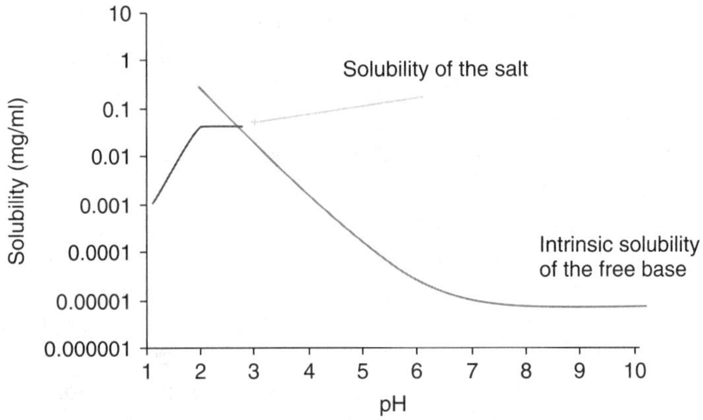

Figure 6.10 Typical solubility pH profiles of a weakly basic compound.

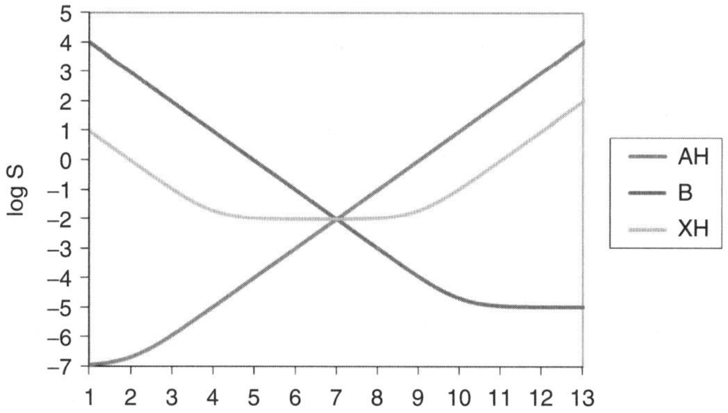

Figure 6.11 The solubility pH profiles of basic, acidic, and amphoteric compounds. AH = weak acid, $pK_a = 2.0$; B = weak base, $pK_a = 10.0$; XH = ampholyte, pK_a = 9.0 and 4.0. All the three compounds have 10^{-2} mol solubility at pH 7.

point, but it is very difficult to calculate the solubility of the fully ionized form of the compound. It is possible that, for example, a phosphate salt of a base has very low solubility, while the HCl salt will be very soluble. That would alter the asymptote of the curve at the lower pH range. Figure 6.11 illustrates the importance of the solubility–pH profile and it shows the solubility values of hypothetical acidic, basic, and amphoteric compounds, which have 10^{-2} mol solubility at pH 7. However, by changing the pH, they will have quite a different solubility. Acidic compounds may precipitate in the stomach, while the basic compounds may precipitate in the small intestine because of the significant decrease in solubility (ionization).

In the BCS, adequate solubility is defined as the lowest solubility of the compound in the pH range 1–8. The reason for this is that we want to avoid any precipitation to occur *in vivo*. In this respect, the ampholyte in Figure 6.11 has the most useful solubility profile as the solubility is never lower than at pH 7. The solubility–pH profile for a compound can be obtained by measuring the solubility at various pH values between 1 and 8 or by measuring the intrinsic solubility and the acid dissociation constant (pK_a) in order to estimate the lowest solubility value within the pH range 1–8.

The pH dependence of solubility of ionized compounds has been utilized by the recently introduced "chasing equilibrium method" by Sirius Analytical, United Kingdom [19]. This method has an advantage in that we do not have to wait for the equilibrium as we force the precipitation by changing the pH; so, the ionizable compound turns to a less soluble neutral form. A known amount of solid compound should be accurately measured and a known quantity of acid (HCl) for basic compounds or base (KOH) for acidic compounds is added until the solid compound is fully dissolved because of salt formation. Then, we start titrating the acidic or alkaline solution with a known amount and strength of base or acid until precipitation is detected because of the loss of ionization of the compound. The precipitation is detected using a spectroscopic dip probe. From the amount of the alkaline and acid solutions, we can calculate the exact pH when precipitation occurs. The kinetic solubility of the compound is derived from the pH at the point of precipitation. In order to calculate the equilibrium solubility of the compound, we titrate back the sample with an acid or base, until the compound starts to dissolve. Then, additional aliquots of titrant are added to produce additional solid, after which the pH change is measured. The direction of the pH gradient on precipitation gives an indication of the saturation level of the solution. An unionized base precipitating from a supersaturated solution, for example, produces a negative pH gradient, contrary to an unionized base dissolving from a subsaturated solution and producing a positive pH gradient. These changes of the sign of the pH gradient can be used to determine the exact pH at which the solid is in equilibrium with the saturated solution. The concentration of the saturated solution, that is, the intrinsic solubility is calculated from the pH at equilibrium by applying the principles of mass and charge balance [20]. The advantage of this method is that we do not need to wait 18–48 h to reach the equilibrium between the solid and the saturated solution. We do not need to separate the solid material and use an analytical method to determine the concentration of the saturated solution. As we know exactly the amount of solid we started the experiment with and exactly how many moles of acid or base are used to dissolve the compound and precipitate it out again, we can calculate how many moles are in solution at the time of precipitation. Another interesting observation has been made using the "chasing equilibrium method": some compounds form supersaturated solutions very easily ("chasers"), while other compounds instantly precipitate out without supersaturation when the equilibrium has been achieved ("nonchasers") [21]. Box et al. [21] had observed that around 5% of acids and 25% of bases are "nonchasers." Typical "nonchasers" are chlorpromazine, imipramine, and nortriptyline. They have relatively large rigid aromatic systems with an attached flexible sidechain containing hydrophilic basic nitrogen. The chaser's kinetic solubility is much higher (it can be four times higher) than its equilibrium solubility. These types of experiments highlight the important differences between the kinetic and equilibrium solubilities and reveal that the kinetic solubility is always higher or equal to the equilibrium solubility. The difference shows the ability of the compound to form supersaturated solutions. On the basis of the requirement for good solubility by the BCS, it was enough to assess the solubility of the compounds in water in a pH range (pH1–pH7.5). However, we also know that there may be several solubilizing agents present in the gastrointestinal tract.

SOLUBILITY AND DISSOLUTION IN BIORELEVANT MEDIA

The methods for dissolution testing and the solubility of an API from formulated medicine are described in several pharmacopoeias and FDA guidelines. The aspects and aim of solubility and dissolution measurements of early drug discovery compounds are slightly different. At early stages of the lead optimization process, we would like to apply measurements that help to reveal some basic structure–property relationships that can be used to help modify the structure. However, it is important even at this stage to increase the probability that the compound will succeed during the late stage development process. For this reason, the aim is to understand the fundamental relationships between structure and behavior of the compound in biorelevant media, too. For example, the American Pharmacopeia (USP) describes devices that create similar hydrodynamic and kinetic test conditions as *in vivo* to measure the solubility and dissolution rate of formulated drugs. Here, we would like to concentrate on the composition of biorelevant fluids and how each component may affect the solubility of the molecules based on the physicochemical properties. The gastric fluid is about 300 ml with pH 1–3 and surface tension lower than water. It also contains pepsin, which is an enzyme that degrades proteins into peptides under acidic conditions. The simulated gastric fluid usually contains sodium and potassium chloride, adjusted to pH = 1 with hydrochloric acid. To reduce the surface tension, surfactant can be added (e.g., 0.1% Triton 100). At this point, the simulated gastric fluid can significantly alter the compound's thermodynamic solubility in water by the pH that has been discussed earlier. The ionic strength and the presence of sodium, potassium, and chloride ions might affect the solubility of the compound if it forms salt with these ions, which has also been discussed earlier. When we consider the composition of intestinal fluid besides a variety of enzymes that might cause metabolic transformations, the most important components are the micelles that are formed mostly from phosphatidyl choline and taurocholate. Figure 6.12 shows the structure and shape of such micelles. They have a discus shape. The edges are formed by the negatively charged taurocholate molecules, while the center is formed by a phosphatidyl choline bilayer.

The intestinal fluid is about 500 ml with increasing pH from 3 to 8 and containing various concentrations of the micelles. In the fasted state, the pH is higher than in the fed state. The micelle concentration also increases in the fed state. Several biorelevant media have been suggested by Dressman and coworkers [22–25],

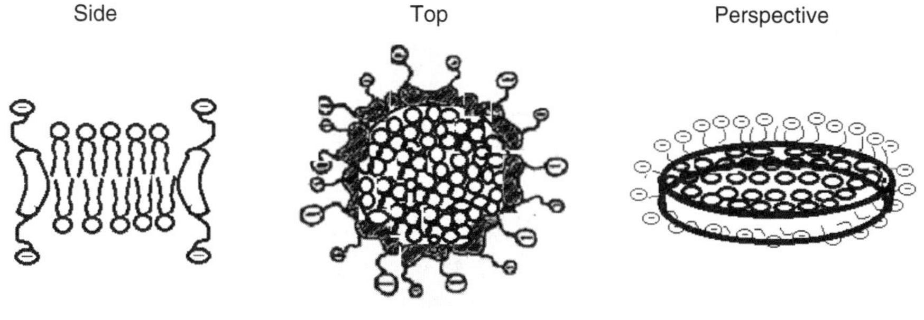

Figure 6.12 Shape and structure of the phosphatidyl choline–taurocholate micelles found in the intestine.

which are also adopted by the US Pharmacopeia. Typical compositions of simulated intestinal fluid fasted or fed states are as follows:

COMPOSITION OF FASTED STATE SIMULATED INTESTINAL FLUID (FaSSIF)

Sodium taurocholate 3mM
Lecithin 0.75 mM
NaOH (pellets) 0.174 g
$NaH_2PO_4{\cdot}H_2O$ 1.977 g
NaCl 3.093 g
Purified water qs 500 ml
Media has a pH of 6.50 and an osmolality of about 270 mOsmol/kg.

PREPARATION OF FaSSIF SOLUTION

Dissolve 1.74 g of NaOH (pellets), 19.77 g of $NaH_2PO_4 \cdot H_2O$, or 17.19 g of anhydrous NaH_2PO_4, and 30.93 g of NaCl in 5 l of purified water. Adjust the pH to exactly 6.5 using 1N NaOH or 1N HCl. This will be the buffer stock solution. Dissolve 3.3 g of sodium taurocholate in 500 ml of previously prepared buffer stock solution. Add 11.8 ml of a solution containing 100 mg/ml lecithin in methylene chloride, forming an emulsion. The methylene chloride is evaporated under vacuum at about 40 °C. Draw a vacuum for 15 min at 250 mbar, followed by 15 min at 100 mbar. This results in a clear, micellar solution having no perceptible odor of methylene chloride. After cooling to room temperature, adjust the volume to 2 l with the previously prepared solution.

COMPOSITION OF FED STATE SIMULATED INTESTINAL FLUID (FeSSIF)

Sodium taurocholate 15 mM
Lecithin 3.75 mM
NaOH (pellets) 4.04 g
Glacial acetic acid 8.65 g
NaCl 11.874 g
Purified water 1000 ml
Media has a pH of 5.00 and an osmolality of about 670 mOsmol/kg.

PREPARATION OF FeSSIF SOLUTION

Dissolve 20.2 g of NaOH (pellets), 43.25 g of glacial acetic acid, and 59.37 g of NaCl in 5 l of purified water. Adjust the pH to exactly 5.0 using 1N NaOH or 1N HCl. This will be the buffer stock solution for fed state simulated intestinal fluid (FeSSIF) media.

Dissolve 16.5 g of sodium taurocholate in 500 ml of blank FeSSIF. Add 59.08 ml of a solution containing 100 mg/ml lecithin in methylene chloride, forming an emulsion. The methylene chloride is eliminated under vacuum at about 40 °C. Draw a vacuum for 15 min at 250 mbar, followed by 15 min at 100 mbar. This results in a clear to slightly hazy, micellar solution having no perceptible odor of methylene chloride. After cooling to room temperature, adjust the volume to 2 l with blank FeSSIF.

It is important to mention that these biorelevant solutions are not stable; they expire within 3 months of the preparation and should be kept at a cool temperature. We can see from the composition of the two solutions that the FeSSIF solution not only differs in pH but also contains five times higher concentration of micelles. When we use methylene chloride to dissolve the lecithin (which contains mostly phosphatidyl choline) we can ensure a reasonable homogenous size distribution of the micelles. The emulsion formation also helps the taurocholate molecules to surround and stabilize the micelles. The easiest and most reproducible way to prepare these biorelevant solutions is using SIF powder, which is available at Biorelevant.com.

Dressman and coworkers [23, 24] have reported that the solubility of some lipophilic and poorly soluble drugs is enhanced in these simulated intestinal solutions. The explanation for that is that lipophilic and often positively charged molecules can partition into the micelles, which causes the solubility enhancement. Glomme et al. [26] set up Equation 6.8 for the estimation of solubility enhancement of lipophilic drugs as a function of the micelle concentration and lipophilicity of the compound expressed by the octanol/water partition coefficient (log P).

$$S_{\mathrm{m}} = c_{\mathrm{m}} \times \frac{S_{\mathrm{o}}}{c_{\mathrm{H_2O}}} \times 10^{(a \log P + b)} + S_{\mathrm{o}} \tag{6.8}$$

Equation 6.8 expresses the solubility of a compound in a simulated intestinal fluid solution (S_{m}) that contains sodium taurocholate (NaTC) and lecithin (Lec) in a 4 : 1 proportion present in c concentration (c_{m}), S_{o} is the solubility of the compound in water (intrinsic solubility of the unionized form). Log P is the logarithmic value of the octanol/water partition coefficient and a and b are constants. The $c_{\mathrm{H_2O}}$ stands for the concentration of water in mole per liter, which is 55.56. By rearranging the equation and taking the logarithm on both sides, Equation 6.9 can be obtained, which expresses the logarithm of the solubility enhancement (log $S_{\mathrm{enhancement}}$) in a linear relation to the lipophilicity expressed as log P and the logarithm of the micelle concentration.

$$\log S_{\mathrm{enhancement}} = \log\left(\frac{S_{\mathrm{m}} - S_{\mathrm{o}}}{S_{\mathrm{o}}}\right) = a \log P + b + \log\left(\frac{c_{\mathrm{m}}}{c_{\mathrm{w}}}\right) \tag{6.9}$$

The equation suggests that the solubility enhancement will increase with increasing lipophilicity of the compounds and with the increasing concentration of the micelles in the solution. However, this equation is valid only for neutral, uncharged compounds at the pH of the micelle solution as the authors have not investigated charged compounds.

Valkó [27] presented the results of an experiment in an early drug discovery setting, where the solubility enhancement was determined from solubility measurements in phosphate buffer solution at pH 7.4 (served as S_o) and in FeSSIF solution at pH 6.5 containing 15 mM micelle solution. It has been found that the solubility enhancement correlated better with the CHI obtained on IAM stationary phase (see Chapter 5 for more details) than with the octanol/water partition coefficients of the compounds. The compounds are related in structure. The observation of the solubility enhancement is not possible for compounds that are highly soluble in the buffer solution, reaching the upper limit of the solubility measurements (500 μM). These compounds appear as outliers, as is demonstrated in Figure 6.13.

It has been found that very insoluble compounds that partition to phospholipid (IAM), strongly giving CHI IAM value above 40, have very significant solubility enhancement. This finding can be considered as further evidence that the enhancement of solubility is due to the partitioning of compounds into the phospholipid micelles.

The solubility enhancement of the micelles depends on the ability of the compounds to partition into the micelles, which depends on the lipophilicity or more exactly their partitioning ability into phospholipids. Higher micelle concentration produces a greater enhancement of solubility. However, very little is known about the effect of the size and composition of the micelles, and about the absorption process of the compound partitioned into the micelles. Matsuoka et al. [28] have investigated in more detail the structure of the micelles using dynamic light

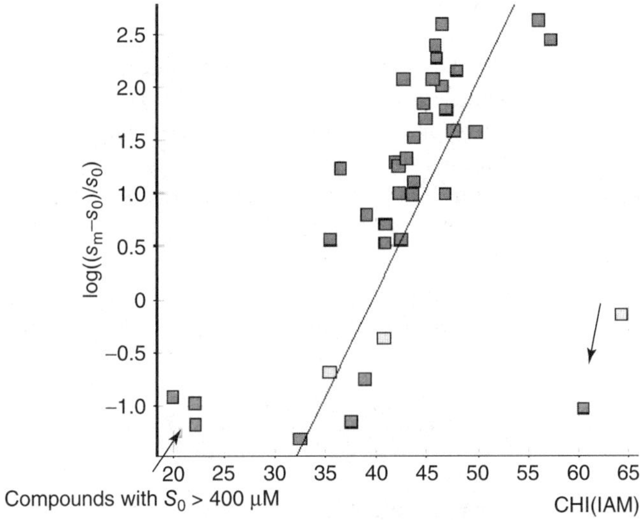

Figure 6.13 The correlation of solubility enhancement for a set of structurally related compounds with the phospholipid binding (expressed as CHI(IAM)).

scattering and turbidimetric methods. They have found aggregates and polydispersity in size of phosphatidyl choline and bile salts (sodium taurocholate and sodium glycol-deoxy-cholate) micelles.

On the basis of the solubility measured in biorelevant media such as the fasted state simulated intestinal fluid (FaSSIF), the BCS [4] has been further refined by Butler and Dressman [29]. It is called *Developability Classification System (DCS)*. The Class II drugs that show low solubility but good permeability are further divided into Class IIa and Class IIb groups, as is illustrated in Figure 6.14. The drugs that belong to Class II/a have slow dissolution rate, which limit good absorption. The absorption of these compounds can be enhanced by relatively simple developability methods, such as reducing the particle size and increasing the surface area and wettability. The Class II/b drugs are more difficult to develop; some solubility enhancement formulations should be used in order to increase the *in vivo* absorption of these drugs.

It is also important to consider, even at early stages of the drug discovery process, finding a candidate molecule that will be easier and cheaper to develop. It requires very different chemical modifications to improve the solubility of the compounds that are dissolution rate limited. These are the so-called "greasy" compounds. Introducing weakly basic nitrogen and reducing the hydrophobicity by introducing polar, hydrophilic groups will solve the solubility issue.

However, compounds that belong to Class II/b are much more difficult to develop; they might not be too lipophilic and still have low solubility because of high crystal lattice energy. They are difficult to dissolve not only in aqueous solvent but in organic solvents too, and they do not partition readily to the taurocholate–lecithin micelles, too. In these cases, the chemists might be able to change the shape of the molecules, reducing the number of aromatic rings that tend to produce flat structure causing π–π stacking in the crystal. We shall discuss

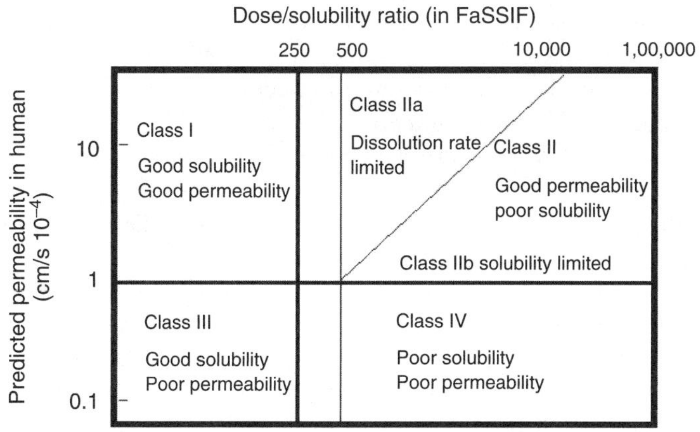

Figure 6.14 Developability Classification Systems. (Reproduced with permission from Reference [29].)

this in more detail in Chapter 12 when we go through the applications of the physicochemical properties in drug discovery.

There are other simulated body fluids that can be used to assess the solubility apart from gastrointestinal absorption purposes. For example, for inhaled administration, it is good to know whether the compound dissolves at the lung's surface, which is covered by mucus, a highly glycosylated protein. Marques et al. [30] describes several biorelevant fluids and thus the composition of fluids that simulate the environment in the lungs. The simulated lung fluids (SLFs) usually contain potassium and sodium chloride close to the isotonic molar concentration. Some SLFs contain calcium, acetate, and citrate ions as well. These ions can influence the compound solubility, especially when it is in ionized form, and the solubility is decreased in the presence of the above-mentioned counterions in the SLF. It is common to apply surfactants in SLF as the lung surface contains surfactants that help maintain the large surface area for the effective exchange of gases. Typical surfactants to use are di-palmitoyl phosphatidyl choline and sodium dodecylsulfate. The pH value of the lung surface can be variable; the typical pH range is between 4.5 and 7.4.

We have seen that solubility is measurement dependent and can be expressed by various terms, such as intrinsic solubility, kinetic solubility, and apparent solubility, and it depends on experimentally variable conditions such as temperature, filtration, sedimentation time, ionic strength, and pH of the solution, which cause significant variations in the data. An alternative approach could be the determination of intrinsic dissolution rate (IDR) of compounds. The dissolution rate is probably more important for evaluating the *in vivo* absorption of the compounds because it considers the appropriate time frame when a sufficient amount of drug can be dissolved. The dissolution rate is important as the drug molecules stays in the gastrointestinal tract for only a limited time, normally 3–5 h. We need to have an approximate idea of whether the discovery compound shows low absorption because of low dissolution rate during the first animal experiments. Therefore, a reasonably fast measurement of dissolution rate that requires a small amount of compounds has a value even in early drug discovery. If we can standardize the conditions for testing the intrinsic dissolution of a compound, we may be able to predict the *in vivo* absorption more accurately.

The IDR can be defined as the increase of concentration of a pure molecule per unit time, where the surface area, temperature, agitation, the pH, and ionic strength of the media are controlled and kept constant [31]. The IDR of a compound can be determined using specific equipment containing solid material that is in contact with the solvent. The contact surface area is determined by the size of the well containing the compressed solid material. The dissolution rate may be described by the Noyes–Whitney equation (from 1897) as a function of the surface area of the solid (A), concentration of the compound in the solvent (c), concentration of the compound in the diffusion layer surrounding the solid particles (c_s), the diffusion coefficient (D), and the thickness of the diffusion layer (L), as is shown by Equation 6.10.

$$\frac{dW}{dt} = \frac{DA(c_s - c)}{L} \tag{6.10}$$

We can express the dissolution flow (j) by Equation 6.11.

$$j = \frac{Vdc}{dt} \times \frac{1}{A} \tag{6.11}$$

where V is the volume of the dissolution medium, c is the concentration of the compound in the medium, A is the surface area of the sample, and t is the time. Although the Noyes–Whitney equation is widely used, sometimes it cannot describe the experimentally determined shape of the dissolution curve. Therefore, a more general equation is suggested to be used [32], as is described by Equation 6.12.

$$M = 1 - \exp(-\alpha t^{\beta}) \tag{6.12}$$

M is the accumulated fraction of the compound in solution at time point t, α is a scaling factor, and β is 1 for a typical exponential dissolution curve shown in Figure 6.15. When $\beta > 1$, it describes an S-shaped curve, and when $\beta < 1$, it means there is a faster initial dissolution rate. The first step during a dissolution process involves the detachment of the molecules from the solid surface and a solvation process by the solvent molecules. It causes a quick saturation of the solution close to the solid interface. The next step during dissolution is the migration of the solvated molecules away from the solid surface, which is controlled by diffusion. Most of the dissolution process is controlled by the second step, the diffusion. The saturation process at the liquid–solid interface is considered to be fast. The diffusion of the

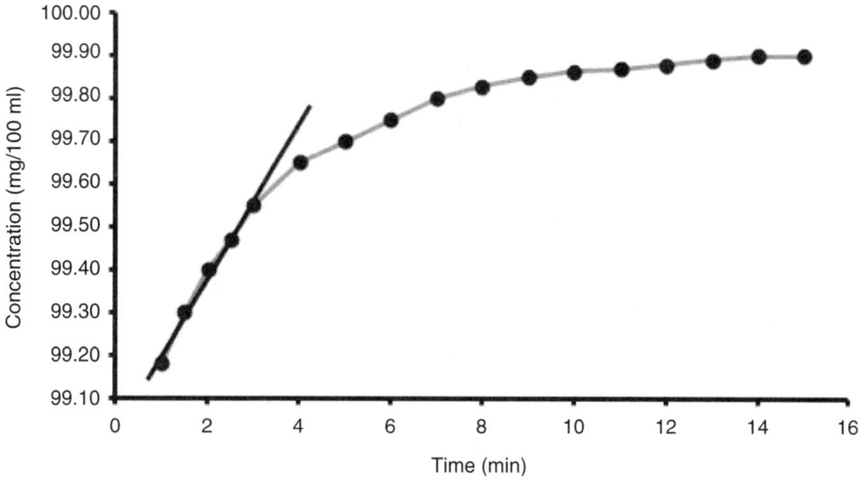

Figure 6.15 Typical dissolution–time profile. The dissolution rate is determined by the slope of the straight line fitted to the steeper initial points.

molecules across this saturated layer is the rate-limiting step. The diffusion process can be influenced by stirring, temperature, and the size of the solid–liquid interface. The slope of the straight line fitted to the initial points of the curve can be considered as the IDR of the compound.

Pharmacopoeias describe two types of apparatus for the measurement of dissolution, a fixed disk system and a rotating disk system, known as *Wood's apparatus* [33]. The principle of such measurements is to compress the solid form of the compound into a usually stainless steel, or strong inert plastic holder, which has a known surface area. The solid surface is in contact with the solvent. Then we apply the solution media on top of the compressed solid material. In a fixed disk system, the media should be stirred, while in the rotating disk system, the solid surface is rotated. Then we need a sampling device and an analytical technique to determine the concentration of the compound in solution as a function of time. Several apparatus are described in the literature [34–37]. The μDISS Profiler (pION Instruments, Billerica, MA, United States) measures the intrinsic solubility and dissolution rate of small amounts of discovery compounds in solid form, which is described by Avdeef and Tsinman [38]. Pettersson et al. [39] have constructed a new and improved miniaturized rotating disk equipment for dissolution rate studies. They connected the rotating disk and the solvent to a HPLC system. They used the HPLC to monitor the compound concentration as a function of time. The dissolution measurements were carried out using various media (buffer and simulated intestinal fluid). They also investigated the relationship between the dissolution rate and solubility [40, 41].

There are compounds with water-repellent solid surface properties that are difficult to wet. It is nearly impossible to measure the solubility or dissolution rate for such materials as the solid particles may just float in the aqueous buffer solution, seemingly remaining completely dry. In these cases, formulation with surface-active compounds could be the only help for wetting the solid surface and accelerating the dissolution process. Persson's [42] PhD thesis from Uppsala University describes a thorough study of the correlation between the dissolution rate and solubility using a newly designed and constructed apparatus with a low volume flow cell using about 5 mg of solid drug and a gold-plated stirrer. The thesis is available on the Internet and provides an excellent review of testing dissolution rate of compounds from all the BCS classes. A reasonably good correlation was found between the logarithmic values of the dissolution rate and apparent solubility of 13 drug molecules; therefore, the dissolution rate measurement can be a good indication for not only the time dependence of the solubility but for the apparent solubility as well.

Another interesting relationship has been found between the solubility and frequent nonselective binding of compounds, which produces hits in a variety of biological screening assays [43]. Certain type of compounds form aggregates in aqueous solutions and produce false-positive results in biological assays with a variety of target proteins. Mashimo et al. [43] found that these low solubility compounds appear to be soluble as no solid precipitates can be observed. They have reported that over 90% of the compounds have molar solubility less than 10^{-5} and molecular weight larger than 400 and less than four heteroatoms form aggregates. The aggregation includes micelle colloid formation that causes nonspecific interactions with several proteins. NMR techniques or sensitive light scattering techniques can detect these aggregates. We need further studies of the phenomena as the traditional methods for solubility measurements, especially those involving filtration, might not be able to detect these colloid particles.

Murdande et al. [44] have published an excellent critical review of the challenges in solubility measurements in drug discovery settings described in this chapter. They have put special emphasis on the aqueous solubility of crystalline and amorphous drugs and their possible interconversions during the measurement and the drug development process.

SUMMARY

Low solubility of a potent compound causes various problems during the drug discovery and development process. Sparingly soluble compounds may give false results in biological assays and hinder the absorption of the drug from the gastrointestinal tract. Solubility is usually expressed as the highest concentration of the compound that can be dissolved in a solvent. From a drug discovery aspect, the aqueous solubility is a critical property. As the solubility is a condition-dependent property, there are numerous ways to measure and define solubility.

Definitions

The *thermodynamic equilibrium solubility* refers to the concentration of the saturated solution, which is in equilibrium with a well-defined solid form at a given temperature. The *intrinsic solubility* refers to the thermodynamic solubility of the neutral form of the molecules.

Apparent solubility refers to the concentration of a saturated solution also at a given pH without an accurate characterization of the solid material and without ensuring that the saturation equilibrium has been achieved.

Kinetic solubility refers to an aqueous concentration of the compound in a given time frame after adding the aqueous solvent to a solid form of material or to a dissolved form usually in a DMSO solvent. When a DMSO stock solution is diluted down by an aqueous buffer and we measure the concentration that remains in the solution after precipitation, we speak about *precipitative kinetic solubility*.

The aqueous solubility of the compound depends on its ability to create a cavity in water and form new energetically favorable interactions with the water molecule. In order to leave the crystal form, the solvent molecules have to break the crystal lattice energy too. The Yalkovsky equation describes the aqueous solubility using two important parameters, namely, the lipophilicity of the compound and the MP of the solid form. The more lipophilic (or hydrophobic) is the molecule, the less soluble it is in water. The higher the melting of the solid form is, the lower the aqueous solubility. The lipophilicity is described by the logarithm of octanol/water partition coefficient. One log unit increase in lipophilicity decreases the logarithm of molar solubility by one unit. A 100 °C increase in MP reduces the solubility by one log molar unit. While the lipophilicity of the compound characterizes the energetic aspects of compound's interaction with water, the MP represents the magnitude of the crystal lattice energy.

The intrinsic solubility can also be described by the five important molecular descriptors, size, polarizability/dipolarity, H-bond acidity and basicity, and the excess molar refraction, just like lipophilicity. In order to evaluate the effect of the crystal lattice energy on solubility and additional term, namely, the product term of H-bond acidity and basicity has to be added to the general solvation equation. Both the Yalkowsky equation and the solvation equation can provide a way of estimating the aqueous solubility of the neutral (unionized) form of the compounds. In general, lipophilic compounds are less soluble in water. Compounds that can form strong crystal lattice energy via H-bonding or π–π stacking because of flat aromatic structure have low solubility.

In terms of biological screening and *in vivo* absorption, the apparent and kinetic solubilities are more relevant than the thermodynamic solubility. We would like to know whether the compound is in solution under the biological assay conditions, or the required dose is soluble and able to absorb from the gastrointestinal tract during its transport. Therefore, it is important to understand how various conditions affect the solubility of the compound. The pH has a significant effect on solubility only for ionizable drug molecules that contain acidic, basic, or both groups. When ionization increases, the aqueous solubility increases too, proportionally to the percentage of ionization, which is described by the Henderson–Hasselbalch equation. Unlike the case for lipophilicity, the solubility of the ionized form cannot be calculated accurately, as it depends on the type of compound and the type of the counterion and as different salt forms of the same compound may have different solubilities.

The FDA defines solubility as the lowest solubility in the pH range of the gastrointestinal tract (pH 1–8), which is equivalent to or a higher concentration than the dose amount in 250 ml. The definition reflects that a single dose will dissolve in a glass of water even when the pH is minimizing the aqueous solubility. The FDA definition of solubility is used in the BCS that groups compounds into four classes depending on the high or poor solubility and permeability. A typical solvent used for storing and dispensing known quantities of drug discovery compounds, is DMSO, which is able to dissolve polar and nonpolar compounds and is totally miscible with water. It has strong solubilizing properties. The effect of organic solvent concentration on solubility can be described by the concentration and solubilizing power of the solvent. On an average, DMSO increases the solubility of compounds by approximately 1.5 times when its concentration is increased from 2% to 5%. The intestinal fluids contain micelles that are formed from phospholipids and bile acids. The simulated intestinal fluid contains phosphatidyl choline and taurocholate micelles in 1 : 4 proportions. The FeSSIF contains a five times higher concentration of micelles than the FaSSIF. The micelles are able to enhance the solubility of lipophilic compounds that partition into the micelles. The solubility enhancement is proportional to the lipophilicity (more exactly the phospholipid partition) of the compound and the concentration of the micelles. The pH value of the simulated intestinal fluids ranges from 5.5 to 6.5.

Typical solubility measurements at early drug discovery stages are performed using DMSO solutions by diluting down the compounds with a pH 7.4 buffer. After an appropriate equilibration time, filtration or centrifugation is applied to separate the solution from the solid. The concentration of the compound is determined by a suitable analytical method, such as UV, HPLC/UV, or CLND in comparison with a standard solution containing a known amount of dissolved compound. The classical determination of solubility from solid material is the shake-flask method that allows the solid particles to equilibrate with the solvent, after which a concentration determination of the saturated solution is carried out. In order to obtain a true thermodynamic solubility, the temperature, the pH, and the equilibration time should be well defined. The pH should be adjusted to ensure that the compound is in an unionized state. An accelerated way to measure thermodynamic solubility is the so-called "chasing equilibrium" method. It is applicable for ionizable compounds only. The principle of the method is that a known amount of compound is dissolved in a known amount of acidic or basic buffer depending on the compound acid/base character. Then, slow titration takes place and the pH is measured when the first precipitate appears in the solution. After applying acidic or basic titrants again, the compound is redissolved and precipitated again, until the accurate pH is determined for the saturated solutions. The concentration of the saturated solution is calculated from the amount of acid or base added to the mixture to create the saturated solutions. Some compounds called *chasers* easily form supersaturated solutions; this means the actual concentration of the compound is higher than the saturated solution concentration.

The dissolution rate is the amount of compound dissolved in a unit of time, starting from the solid form of the compound. It depends on the solubility of the compound, the surface area between the solid and the solvent, the thickness of the diffusion layer around the solid that contains the saturated concentration of the compound, and, of course, the diffusion coefficient of the compound. Thus, the surface area between the solid and solvent should be kept constant; the speed of stirring (or rotating the disk containing the solid material) should be controlled as it affects the thickness of the diffusion layer. The Noyes–Whitney equation is the most often used to describe the dissolution rate. However, in practice, the measured curve can be different from a concave saturation equation. The importance of the dissolution rate measurements lies in the estimation of solubility in a given time frame that is imitated by the time of the gastrointestinal transit time of the drug molecules. The required amount of active compound (dose) should be dissolved during the intestinal transit, as it is the prerequisite for oral absorption. The permeability of the compound through the gut wall is another prerequisite for absorption that is discussed in Chapter 7.

QUESTIONS FOR REVIEW

Q6.1 Define thermodynamic equilibrium, apparent, kinetic, and precipitative kinetic solubilities.

Q6.2 How does the FDA define good solubility for the BCS?

Q6.3 Why should solubility be expressed in molar concentration at lead optimization stage?

Q6.4 There are two compounds A and B with molecular weight 250 and 500, respectively. Both compounds showed aqueous solubility 10 μg/ml. Which compound is more soluble in terms of the number of molecules dissolved?

Q6.5 Which properties of molecules have an effect on their solubility?

Q6.6 Why is the MP of the solid material used in the Yalkowsky equation?

Q6.7 How does the pH affect the solubility of ionizable compounds?

Q6.8 How does the presence of 5% DMSO in the aqueous buffer affect the solubility?

Q6.9 Explain the enhanced solubility in simulated intestinal fluids.

Q6.10 Is solubility enhanced in simulated intestinal fluids for every compound?

Q6.11 Describe a simple method for measuring DMSO precipitative and thermodynamic solubilities.

Q6.12 Which experimental conditions should be taken into consideration for the measurement of thermodynamic solubility?

Q6.13 What is IDR?

Q6.14 What experimental conditions should be kept constant to determine the IDR of a compound?

Q6.15 Why do we need to measure the rate of dissolution as well as solubility?

REFERENCES

1. Sugano, K., Okazaki, A., Sugimoto, S., Tavornvipas, S., Omura, A., Mano, T. (2007) Solubility and dissolution profile assessment in drug discovery (Review). *Drug Metabolism and Pharmacokinetics*, *22(4)*, 225–254.
2. Sinko, P. J. (2006) Martin's Physical Pharmacy and Pharmaceutical Sciences. (5th Edition) Baltimore, Lippincott Williams & Wilkins.
3. Baka, E., Comer, J. E. A., Takács-Novák, K. (2008) Study of equilibrium solubility measurement by saturation shake-flask method using hydrochlorothiazide as model compound. *Journal of Pharmaceutical and Biomedical Analysis*, *46*, 335–341.
4. Amidon, G. L., Lennernas, H., Shah, V. P., Crison, J. R. (1995) A theoretical basis for a Biopharmaceutic Drug Classification: The correlation of *in vitro* drug product dissolution and *in vivo* bioavailability. *Pharmaceutical Research*, *12(3)*, 413–419.
5. Yalkowsky, S. H., Valvani, S. C. (1980) Solubility and partitioning I: Solubility of nonelectrolytes in water. *Journal of Pharmaceutical Sciences*, *69*, 912–922.
6. Abraham, M. H., Le, J. (1999) The correlation and prediction of the solubility of compounds in water using an amended solvation energy relationship. *Journal of Pharmaceutical Sciences*, *88*, 868–880.

7. Perlovich, G., Volkova, T. V., Bauer-Brandl, A. (2006) Towards an understanding of the molecular mechanism of solvation of drug molecules: A thermodynamic approach by crystal lattice energy, sublimation, and solubility. Exemplified by paracetamol, acetanilide, and phenacetin. *Journal of Pharmaceutical Sciences*, *95*, 2158–2169.
8. Bhatt, D. A., Pethe, A. M. (2010) Nanotechnology: A promising drug delivery for poorly water soluble drugs. *Journal of Pharmacy Research*, *3*, 1748–1751.
9. Lipinski, C. A., Lombardo, F., Dominy, B. W., Feeney, P. J. (1997) Experimental and computational approaches to estimate solubility and permeability in drug discovery and development settings. *Advanced Drug Delivery Reviews*, *23*, 3–25.
10. Bevan, C. D., Lloyd, R. S. (2000) A high-throughput screening method for the determination of aqueous drug solubility using laser nephelometry in microtiter plates. *Analytical Chemistry*, *72*, 1781–1787
11. Rubino, J. T., Yalkowsky, S. H. (1987) Co-solvency and co-solvent polarity. *Pharmaceutical Research*, *4(3)*, 220–230.
12. Pan, L., Ho, Q., Tsutsui, K., Takahashi, L. (2001) Comparison of chromatographic and spectroscopic methods used to rank compounds for aqueous solubility. *Journal of Pharmaceutical Sciences*, *90*, 521–529.
13. Colclough, N., Hunter, A., Kenny, P. W., Kittlety, R. S., Lobedan, L., Tam, K. Y., Timms, M. A. (2008) High throughput solubility determination with application to selection of compounds for fragment screening. *Bioorganic & Medicinal Chemistry*, *16*, 6611–6616.
14. Kalayani, M. (2005) Drug Discovery MSc Project Work at School of Pharmacy, University of London.
15. Avdeef, A. (2001) Physico-chemical profiling (solubility, permeability and charge state). *Current Topics in Medicinal Chemistry*, *1*, 277–351.
16. Bhattachar, S. N., Wesley, J. A., Seadeek, C. (2006) Evaluation of the chemiluminescent nitrogen detector for solubility determinations to support drug discovery. *Journal of Pharmaceutical and Biomedical Analysis*, *41*, 152–157.
17. Wenlock, M. C., Austin, R. P., Potter, T., Barton, P. (2011) A highly automated assay for determining the aqueous equilibrium solubility solubility of drug discovery compounds. *Journal of the Association of Laboratory Automation*, *16*, 276–284.
18. Alsenz, J., Meister, E., Haenel, E. (2007) Development of a Partially Automated Solubility Screening (PASS) Assay for early drug development. *Journal of Pharmaceutical Sciences*, *96*, 1748–1762.
19. Stuart, M., Box, K. (2005) Chasing equilibrium: Measuring the intrinsic solubility of weak acids and bases. *Analytical Chemistry*, *77*, 983–990.
20. Yalkowski, S. H., Banerjee, S. (1992) Aqueous solubility: Methods of Estimation for Organic Compounds. Marcel Dekker, New York, pp. 149–154.
21. Box, K. J., Völgyi, G., Baka, E., Stuart, M., Takács-Novák, K., Comer, J. E. A. (2006) Equilibrium versus kinetic solubility measurements of aqueous solubility and ability of compounds to supersaturate in solution–validation study. *Journal of Pharmaceutica Sciences*, *95*, 1298–1307.
22. Dressman, J. B., Reppas, C. (2000) *In vitro–in vivo* correlations for lipophilic, poorly soluble drugs. *B. T. Gattefosse*, *93*, 91–100.
23. Nicolaides, E., Symillides, M., Dressman, J. B., Reppas, C. (2001) Biorelevant dissolution testing to predict the plasma profile of lipophilic drugs after oral administration. *Pharmaceutical Research*, *18(3)*, 380–388.

24. Horter, D., Dressman, J. B. (2001) Influence of physicochemical properties on dissolution of drugs in the gastrointestinal tract. *Advances in Drug Delivery Reviews*, *47*, 75–87.
25. Marques, M. (2004) Dissolution media simulating fasted and fed state. *Dissolution Technologies*, *11(5)*, 16–16.
26. Glomme, A., März, J., Dressman, J. B. (2006) Predicting the intestinal solubility of poorly soluble drugs. In: Pharmacokinetic Profiling in Drug Research: Biological, Physicochemical, and Computational Strategies, Eds Testa, B. S. D. Kramer, H. Wunderli-Allenspach, Wiley, Weinheim.
27. Valkó, K. (2005) *Role of high throughput solubility measurements and calculations in early drug discovery*, ACD Laboratories User Meeting, UK, 2005. (www.acdlabs.com/company/events, 2005). Accessed 2013 Aug 19.
28. Matsuoka, K., Maeda, M., Moroi, Y. (2004) Characteristics of conjugate bile salt–phosphatidylcholine- cholesterol–water system. *Colloids and Surfaces B: Biointerfaces*, *33*, 101–109.
29. Butler, J. M., Dressman, J. B. (2010) Developability Classification System: Applications of biopharmaceutics concepts to formulation development. *Journal of Pharmaceutical Sciences*, *99*, 4940–4954.
30. Marques, M. R., Loebenberg, R., Almukanzi, M. (2011) Simulated biological fluids with possible application in dissolution testing. *Dissolution Technologies*, *8*, 15–28.
31. Issa, M. G., Ferraz, H. G. (2011) Intrinsic dissolution as a tool for evaluating drug solubility in accordance with the Biopharmaceutical Classification System. *Dissolution Technologies*, *18(8)*, 6–13.
32. Granero, G. E., Ramachandran, C., Amidon, G. L. (2008) Gastrointestinal dissolution and absorption of drugs. In: Drug Bioavailability–Estimation of Solubility, Permeability, Absorption and Bioavailability, Eds Van de Waterbeemd, H. Lennernas, H., Artursson, P., Methods and Principles in Medicinal Chemistry. Wiley-VCH, Weinheim, Vol. *18*, Chapter 8, pp. 191–214.
33. Wood, J. H., Syarto, J. E., Letterman, H. (1965) Improved holder for intrinsic dissolution rate studies. *Journal of Pharmaceutical Sciences*, *54*, 1068–1070.
34. Josefson, M., Johansson, E., Torstensson, A. (1988) Optical fiber spectrometry in turbid solutions by multivariate calibration applied to tablet dissolution testing. *Analytical Chemistry*, *60*, 2666–2671.
35. *U.S. Pharmacopeia National Formulary* (2006); *29*, 2923–2924 (Chapter 1087).
36. Sathe, P., Lionberger, R. A., Lee, S. L., Yu, L. X. (2008). Regulatory aspects of dissolution for low solubility drug products. Water-Insoluble Drug Formulation, Chapter 6. in "Water insoluble drug formulation" Ed. Liu, R., CRC Press Boca Raton, USA, pp. 101–112.
37. Peltonen, L., Liljeroth, P., Heikkila, T., Kontturi, K., Hirvonen, J. (2004) A novel channel flow method in determination of solubility properties and dissolution profiles of Theophylline tablets. *Journal of Drug Delivery Science and Technology*, *14*, 389–394.
38. Avdeef, A., Tsinman, O. (2008) Miniaturized rotating disk intrinsic dissolution rate measurement: Effects of buffer capacity in comparisons to traditional Wood's apparatus. *Pharmaceutical Research*, *25*, 2613–2627.
39. Persson, A. M., Baumann, K., Sundelof, L.-O., Lindberg, W., Sokolowski, A., Pettersson, C. (2008) Design and characterization of a new miniaturized rotating disk

equipment for *in vitro* dissolution rate studies. *Journal of Pharmaceutical Sciences*, *97*, 3344–3355.

40. Persson, A. M., Sokolowski, A., Pettersson, C. (2009) Correlation of *in vitro* dissolution rate and apparent solubility in buffered media using a miniaturized rotating disk equipment. Part I. Comparison with a traditional USP rotating disk apparatus. *Drug Discoveries & Therapeutics*, *3*, 104–113.
41. Persson, A. M., Pettersson, C., Sokolowski A. (2009) Correlation of *in vitro* dissolution rate and apparent solubility in buffered media using a miniaturized rotating disk equipment. Part II. Comparing different buffer media. *Drug Discoveries & Therapeutics*, *3*, 114–122.
42. Persson, A. (2010) Evaluation of a miniaturized rotating disk apparatus for *in vitro* dissolution rate measurements in aqueous media. Correlation of *in vitro* dissolution rate with apparent solubility. *Digital comprehensive summary of Uppsala dissertations from the Faculty of Pharmacy 118*, Uppsala Universitet.
43. Mashimo, T., Fukunishi, Y. Orita, M., Katayama, N., Fujita, S., Nakamura, H. (2010) Quantitative, analysis of aggregation-solubility relationship by in-silico solubility prediction. *International Journal of High Throughput Screening*, *1*, 99–107.
44. Murdande, S. B., Pikal, M. J., Shanker, R. M., Bogner, R. H. (2011) Aqueous solubility of crystalline and amorphous drugs: Challenges in measurements. (Review) *Pharmaceutical Development and Technology*, *16*, 187–200.

7

MOLECULAR PHYSICOCHEMICAL PROPERTIES THAT INFLUENCE ABSORPTION AND DISTRIBUTION—PERMEABILITY

BIOLOGICAL MEMBRANES

Biological membranes cover every cell. They protect the inner environment against the changes occurring in the outside world. Biological membranes are very flexible and represent a physical state between liquid and solid. The membrane is formed from phospholipid bilayer. In 1972, Singer and Nicolson described the unique structure of biological membranes as a "fluid mosaic structure." The polar head groups of phospholipids cover the outer sides of the membrane, while the inner part consists of completely nonpolar hydrocarbons. The polar head group is often covered with carbohydrates and proteins and hydrated with structured water molecules. The inner part of the membrane contains saturated or unsaturated hydrocarbon chains. The hydrophobic forces between the fatty acid chains hold the membrane firmly, but at the same time they make it very flexible too. There are many "holes" in the membrane that are usually covered with proteins. While polar water molecules cannot go through the inner hydrocarbon part of the phospholipid bilayer, they can go through the pores of the membrane together with other small polar molecules. The surface characteristics of the membrane depend on the nature of the polar head groups. There are several books and reviews that discuss the structure and physicochemical properties of cell membranes [1–6].

Physicochemical and Biomimetic Properties in Drug Discovery: Chromatographic Techniques for Lead Optimization, First Edition. Klara Valko.

As the membranes form a selective barrier to exogenous molecules, including drugs, it is important to understand the fundamental physicochemical principles of the passive membrane transport processes. The measurement of a compound's ability to pass through biological membranes is essential for successful drug discovery. Although the importance of active membrane transport processes is well known, the active transport processes are difficult to explain only by the physicochemical properties of compounds; therefore, it is not discussed here in detail.

A schematic picture of a cell membrane is shown in Figure 7.1. Numerous receptors and drug targets are within the cell; so, the potent compound when administered *in vivo* need to reach the target protein in sufficient concentration at a given time. Even if the compound is soluble and, based on its lipophilicity, can reach the target tissue, it still has to go through the cell membranes to reach the site of action before it is excreted from the body.

The most frequent route of administration of the drug is via oral absorption. The cellular membranes in the gastrointestinal tract can prevent the drug molecules from being absorbed. In the event of successful absorption, the drug molecules get into the bloodstream. The blood's natural role is to deliver important nutrients to every cell in the body. Therefore, the blood vessels are easily permeable even for larger molecules in the capillaries. However, there are several other barriers in the body that protect certain important organs and tissues from potentially harmful compounds. The blood–brain barrier is very unique in this sense of protecting the brain. It has small tight junctions that prevent large hydrophilic molecules from going through. The hydrophobic molecules also can be pumped out by active efflux process by active transporters, such as PgP. Other organs such as the kidneys and the lungs have very different permeability properties for different types of compounds.

Several cultured epithelial cell lines of human or animal origin can be used as *in vitro* models for cell permeability and absorption studies. The Caco-2 human

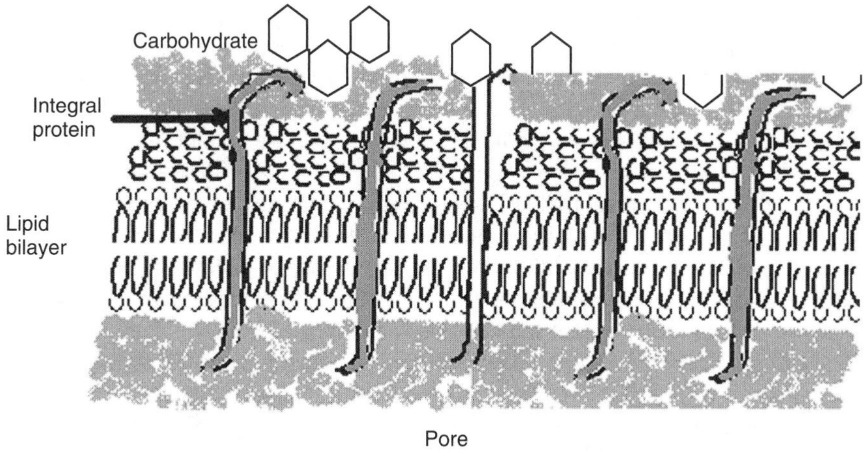

Figure 7.1 The schematic structure of biological membranes.

colonic cell line is the most widely used for this purpose. Caco-2 cells adopt a polarized morphology and express many intestinal transport proteins and other enzymes characteristic of epithelial cells of the small intestine, when cultured under proper conditions. They also form tight junctions with each other. The cell monolayers can be grown on polycarbonate membrane filters separating two chambers filled with culture medium. The cell growth may take 3 weeks. The cells adopt polarized epithelial morphology under these culturing conditions with distinct apical and basolateral sides. Electrical resistance across each cell monolayer is measured to confirm proper barrier function without large holes. The apical and basolateral wells of the assay plate are filled with buffer. Then the test compound is added to the apical wells. Samples are taken from the basolateral wells at various times over a couple of hours. At the end of the experiment, the apical well is sampled in order to measure recovery of the test compound. The parameter that is obtained is the apparent permeability (P_{app}). It can be calculated from the basolateral concentration versus time curve divided by the concentration of compound in the apical dosing chamber and the total area of the Caco-2 cell monolayer. Yee [7] described sigmoid relationships between the Caco-2 cell permeability and human intestinal absorption. It is because a rate is compared to a fraction absorbed unit. Such plots are useful for the discrimination between poor and good absorption, but the estimation of absorption in the high slope section of the curve is less accurate. The cause of the sigmoid relationship can be explained by the fact that permeability is usually expressed as a rate constant, that is, the speed of the compound by which it goes through the cell and expressed as nm/s, while the human intestinal absorption is expressed as a percentage or fraction. It is determined from the administered dose and the amount of compound that eventually appears in the systemic circulation before it has a chance to be eliminated by the first-pass metabolism in the liver. The exact value of P_{app} is influenced by the conditions under which the assay is conducted. Therefore, a validation set of compounds with established human fractional absorption is recommended to be measured under the same conditions. Typical compounds used for the assay validation are atenolol that has around 50% and propranolol that has around 90% human intestinal absorption. In order to accelerate the throughput of such *in vitro* cell permeability assays, Irvine et al. [8] investigated the applicability of the MDCK (Madin-Darby canine kidney) cell line, which can form monolayers on a filter plate in a couple of days. Both cell lines are widely used for *in vitro* measurements of permeability. We shall discuss later in Chapter 9 that absorption depends both on solubility and permeability of the compound, so it is not realistic to expect that a cell permeability assay on its own can predict absorption.

The permeability of a compound through the phospholipid bilayer is also important for delivering the potent molecule to the site of action, which is often inside the cell. It is often observed that the potency of the compound is lost in a cellular assay. This is the so-called "drop off" in potency. It can be very different for various compounds, resulting in no correlation between the cellular potency and enzyme potency. The initially derived structure–potency relationships often break down when the cellular potency is considered. In these cases, the poor cell permeability of the compound is often responsible for the poor correlation between the enzyme and cellular potency. Not just poor permeability but several other factors can cause potency drop off, such as low solubility or nonspecific binding to proteins and lipids inside the cell that reduces the free concentration of the active compound near the target enzyme. For the study of cellular assay potency drop off, we need to have a better understanding of how compounds permeate through the cell membrane. Therefore, it is important to establish structure–permeability relationships to aid the structure design process. The cell permeability assays measure the complex behavior of the compounds that may involve active transport processes; therefore, it may be difficult to establish structure–permeability relationships. In order to exclude the effect of active transporters, it is common to measure the permeability of a compound not only from the apical to basolateral direction but also the opposite direction (basolateral to apical). It is also possible to add inhibitors of certain transporters to the permeability assay, such as PgP inhibitor. From the difference between the measured permeability without the inhibitor and in the presence of the inhibitor, we can reveal the effect of active transport. In order to study and understand the influence of physicochemical properties on permeability without any active biological processes, several artificial membrane permeability (AMP) assays have been developed.

ARTIFICIAL MEMBRANES

Avdeef [9] has described the historical developments of artificial membranes from the work of Hooke and Newton in the seventeenth century through to Langmuir's work in the early twentieth century up to the present "fluid mosaic" model of phospholipid bilayers by Singer and Nicolson [10,11]. The black lipid membrane, which was one of the first studied artificial membranes by Mueller et al. [12], was formed by phosphatidylcholine in the capillaries of a Teflon or polyethylene filter. When the proper bilayer was formed, the membrane turned the filter optically black and hence the name black lipid membrane. The first parallel artificial membrane permeability assay (PAMPA) for the characterization of drug discovery compounds in a high throughput way was described by Kansy and colleagues [13] in 1998. Since then, several versions of PAMPA have been described, studied, and developed. A commercially available highly automated assay setup is offered by pION Inc. (http://www.pion-inc.com/). Almost every major pharmaceutical company has its own in-house AMP assay in order to fulfill the high demand of medicinal chemists. However, the interlaboratory correlations of the permeability assays are very poor because of the significant effects of the assay conditions on the measured values [14]. Many debates and approaches have been published in the literature about the measurements and use of the PAMPA data in drug discovery. The major differences among the artificial membrane assays are in the composition and thickness of the membrane, the filter material, the pH and composition of the donor and acceptor compartments, and the stirring/shaking during the measurements. We have to investigate the effect of all these parameters in order to have a better understanding of the measured permeability data and make appropriate decisions about the discovery compounds.

The first and most important question is the composition and preparation of the artificial membranes. The phospholipids are glycerin derivatives. Two of the hydroxyl groups form esters with fatty acids of various lengths with varying saturation, while the third hydroxyl group forms substituted phosphate ester. The phosphate group is negatively charged. When the substituent is choline or ethanolamine, there is an additional positive charge on the molecule as well. Various amino acids can also form esters with the phosphate group, such as serine, but sugar molecules such as inositol can also be attached, as is shown in Figure 7.2. These molecules have a very strong amphiphilic nature, which means that one part of the molecule is polar (containing charges) while the other part of the molecule consists of hydrophobic nonpolar alkyl chains. It is difficult to dissolve these molecules in either water or nonpolar solvents as they prefer to be at the interface. When the phospholipid molecules are unable to organize themselves at an interface, they can self-associate, forming micelles, liposomes, or bilayers, as is depicted in Figure 7.3.

The most frequently used artificial membrane component is phosphatidylcholine, which is the major component of lecithin, most often isolated from egg yolk. To measure the permeability of compounds through the bilayer, these

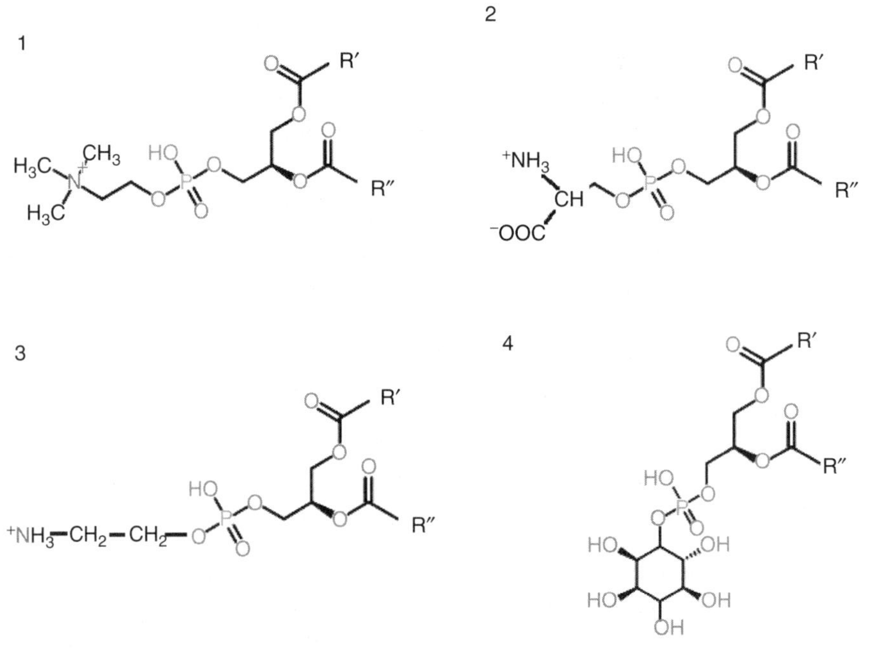

Figure 7.2 Phospholipid molecules. (1: phosphatidylcholine; 2: phosphatidylserine; 3: phosphatidylethanolamine; and 4: phosphatidylinositol).

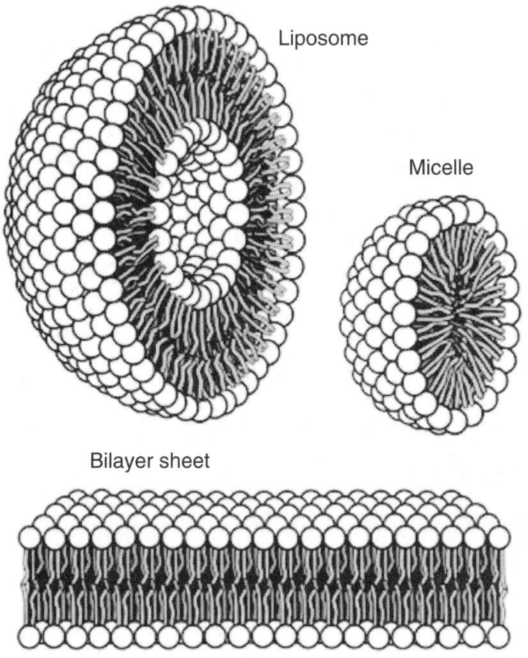

Figure 7.3 Various arrangements of amphiphilic molecules, such as micelles, liposomes, and bilayer.

amphiphilic molecules have to be spread evenly on a filter that separates the two compartments containing aqueous buffers. We shall investigate the basic physicochemical principles of permeability before the detailed description of various types of PAMPA assays and the discussion of the effect of experimental conditions.

PHYSICOCHEMICAL PRINCIPLES OF PERMEABILITY

Permeability is expressed as the rate by which compounds go through membranes. The basic principle of permeability is based on Fick's law of diffusion. It describes the spread of the molecules from the high concentration to the lower concentration side via diffusion. The major driving force is the concentration difference. The rate of transport (dQ/dt) through a phospholipid bilayer can be described by Equation 7.1.

$$\frac{dQ}{dt} = \frac{DAK}{h}(c_a - c_b) \tag{7.1}$$

where D is the diffusion coefficient of the compound, K is the membrane partition coefficient, A is the surface area of the membrane, h is the membrane thickness, and c_a and c_b are the concentrations of the compound in the donor (apical) and

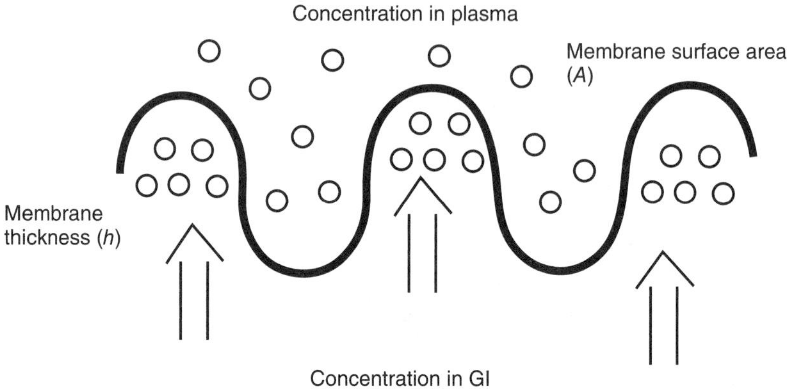

Figure 7.4 The major factors that influence the flux of a compound through biological membrane.

acceptor (basolateral) sides of the membrane. Equation 7.1 shows the factors that influence the flux. Figure 7.4 illustrates the process schematically. Equation 7.1 shows that the concentration difference, the diffusion coefficient (D), the membrane partition coefficient (K), and the membrane surface area increase the rate of transport, while the membrane thickness reduces it. The diffusion coefficient of the compound can be described by the Stokes–Einstein equation applied for membranes and valid for small spherical molecules. The diffusion coefficient is smaller for bigger compounds. Temperature increases the diffusion, while the viscosity of the solvent (in biological permeability, it is the viscosity of the water with possible dissolved proteins and other small molecules) decreases it. The rate of transport is bigger when the membrane surface (A) is bigger. The flux is the normalized rate of transport per surface area unit. The membrane partition coefficient (K) is an important factor as well because the compound should be able to partition into the lipid membrane from the aqueous phase to be able to go through. However, when the compound is too lipophilic, it might stay in the membrane, which means that the equation is true only up to a maximum K. At high K values, the rate of the transport decreases.

The apparent permeability (P_{app}) of the molecules can be derived from the rate of the concentration increase in the acceptor compartment as a function of time according to Equation 7.2.

$$\frac{dC_{\mathrm{a}}}{dt} = \frac{P_{\mathrm{app}}A}{V_{\mathrm{a}}}(c_{\mathrm{d}} - c_{\mathrm{a}}) \tag{7.2}$$

where A is the membrane surface area, V_{a} is the volume of the acceptor compartment, and the $(c_{\mathrm{d}} - c_{\mathrm{a}})$ is the concentration difference between the donor and the acceptor compartments. From the differential Equation 7.2, we can obtain Equation 7.3, which contains the parameters that we can measure and use to obtain the apparent permeability (P_{app}). The apparent permeability is expressed as a unit of the surface; therefore, it is independent of the applied surface area of the membrane.

$$P_{\text{app}} = -\frac{V_{\text{d}} \times V_{\text{a}}}{(V_{\text{d}} + V_{a}) \times A \times t} \times \ln(1 - R) \tag{7.3}$$

where V_{d} is the volume of the donor compartment and R is the concentration ratio (UV absorbance or peak area ratio in the acceptor compartment relative to the donor compartment). When the volumes are expressed in cm^3, the membrane surface area is expressed in cm^2, and the time in second, the unit of the P_{app} is expressed as cm/s. As P_{app} is usually a small number (10^{-6} to 10^{-4}) for drug discovery compounds, it is often expressed in nm/s. When R reaches 1, it means that the concentration of the compound is the same on both sides of the membrane and the permeability drops to zero. It is common to measure the permeability of the compounds at a given time point; so, compounds can be ranked according to the concentration ratio (R) they achieve during a given time interval. It has been observed that some compounds retain in the membrane; so, the sum of the amount of compounds in the donor and acceptor compartments after the experiment is less than that at the beginning. In this case, the effective permeability (P_{eff}) is less than the apparent permeability (P_{app}). Therefore, it is important to check the mass balance after the permeability experiment. Membrane retention is a typical phenomenon during cell permeability measurements, and it is often observed during the AMP measurements as well.

As the concentration difference is one of the major driving forces for permeability, it is important to note that when the diffusion process is slow, the concentration near the other side of the membrane can be temporarily high, thus reducing the rate of transport. It has been observed during the permeability measurements that shaking or stirring the acceptor compartment significantly influenced the rate of transport. It is due to the unstirred water layer (UWL), which is the aqueous layer around the polar head groups on the membrane surface. In this layer, the solute concentration is controlled purely by diffusion [15]. While Kansy et al. [13] have not applied shaking or stirring during the PAMPA, Avdeef studied in great detail the effect of stirring [16]. It has been observed that stirring the aqueous compartments reduces the thickness of the unstirred water later. Owing to the slower movement of the molecules in the UWL, the apparent permeability of some molecules measured lower than its actual membrane permeability. In order to calculate the compound's permeability both in the membrane and in the UWL, we can describe the effective permeability as the sum of the reciprocal values of the permeability in each layer, as is shown by Equation 7.4.

$$\frac{1}{P_{\text{eff}}} = \frac{1}{P_0} + \frac{1}{P_{\text{uwl}}} \tag{7.4}$$

where P_0 is the intrinsic permeability, which refers to the permeability of a neutral form of a molecule without its permeability through the UWL. P_{uwl} is the permeability of the compound through the UWL. For ionizable molecules, we need to add the reciprocal value of the permeability of the ionized species. Equation 7.3 and Equation 7.4 consider the permeability of one molecular species, assuming that the molecules are in the same form on both sides of the membrane. However, most of the drug molecules have ionizable groups and they may be ionized at physiological pH values. Ionized polar molecules have great difficulty in going through inner hydrophobic region of the membrane. The pH partition hypothesis [17] describes the permeability of charged molecules via temporarily losing the charge. This means that basic compounds lose the protons, and acidic compounds gain a proton when passing through. When the pH is different in the donor and acceptor sides, a so-called "sink condition" occurs. A typical "sink condition" for basic compounds happens when the pH in the gut is higher than that in the bloodstream. The bases are less ionized in the gut at pH 8, for example. The percentage of the nonionized form is higher, which drives the compound through the membrane. The pH on the other side of the membrane is lower (pH 7.4), which means that when the neutral form of the base reached the other side, it gains a proton and a positive charge; so, the concentration of the uncharged form does not increase as fast as it decreases on the other side. This creates a "sink," which means an extra driving force for permeability. The same happens with acids in the stomach, for example, where the pH is lower than in the bloodstream. The unionized acid is high on one side but it does not increase on the other side because of the pH difference; the unionized molecules get ionized. When we want to imitate these conditions during the AMP measurements applying "sink condition" due to pH differences, the apparent permeability of the compound is increased. In order to take into account the effect of the "sink condition," Gutknecht and Tosteson [18,19] suggested a modification for Equation 7.4 shown by Equation 7.5 for a weak base and Equation 7.6 for a weak acid.

$$\frac{1}{P_{\text{eff}}} = \frac{[B_{\text{T}}]}{[B]P_0} + \frac{1}{P_{\text{uwl}}} \tag{7.5}$$

$$\frac{1}{P_{\text{eff}}} = \frac{[A_{\text{T}}]}{[A]P_0} + \frac{1}{P_{\text{uwl}}} \tag{7.6}$$

where B_{T} and A_{T} stand for the total concentrations of base and acid, while B and A stand for only the neutral form concentration of the base and acid, respectively. There may be other forces *in vivo* that results in a so-called "double-sink condition," which is due to plasma protein binding of compounds. When the molecules that permeated through the membrane bind to plasma proteins, it reduces the concentration on the other side. It means that the decrease in concentration in the donor compartment is not equivalent to the increase in the concentration in the acceptor compartment. As the free concentration difference on the two sides of the membrane drives the permeability, the concentration difference does not decrease over time. The acceptor compartment behaves like a sink; the molecules that permeated through the membrane bind to proteins and are not present as free species.

These conditions can be imitated in the AMP measurements too. By applying different pH values in the donor and acceptor compartments and using serum albumin in the acceptor compartment only, the "double-sink conditions" can be set up and studied.

EXPERIMENTAL METHODS TO MEASURE ARTIFICIAL MEMBRANE PERMEABILITY

The *in vivo* permeability or absorption studies are carried out by comparing the plasma concentration of a compound after oral and intravenous administration. FDA considers the compound to be highly permeable when the extent of absorption in humans is determined to be greater than 90% of an administered dose based on the mass balance or in comparison to an intravenous reference dose. This criterion is fulfilled when the permeability of a drug through the intestinal wall is above 10^{-4} cm/s. The extent of absorption can be studied in humans using mass balance pharmacokinetic studies or absolute bioavailability studies. It is also possible to carry out *in vivo* intestinal perfusion studies in humans. It is obvious that these *in vivo* experiments are labor and resource intensive; so, they cannot be carried out at early stages of the lead optimization process on large number of molecules. That is why FDA introduced the BCS [20], which makes it possible to waive *in vivo* bioequivalence studies in humans for compounds that have good solubility and good permeability (Class I). Class I compounds are completely soluble; so, the rate of absorption depends only on the permeability.

In order to model these conditions, several AMP methods have been developed and published. The principles of a general methodology are described in this chapter, which may help set up and validate an "in-house" permeability measurement procedure. Figure 7.5 shows a typical experimental setup for a high throughput measurement of AMP. We can use 24-well to 96-well plate formats. Therefore, the assay is called *PAMPA*. We need special plates containing an upper layer with a filter bottom part that fits into a lower layer of wells. The compound is placed in the

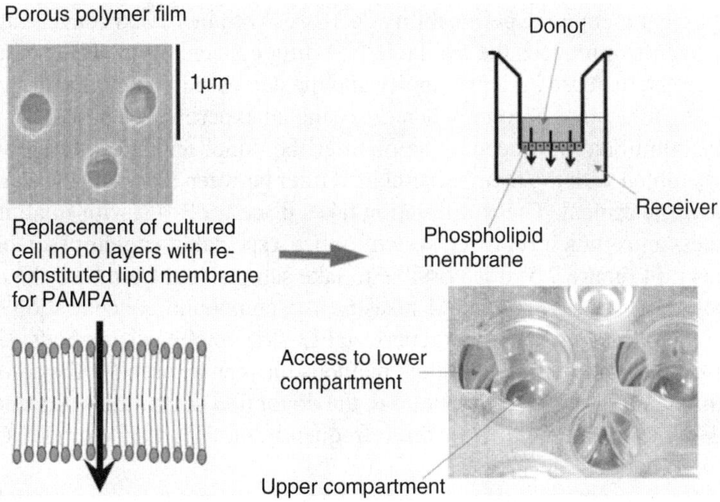

Figure 7.5 The experimental setup for high throughput artificial membrane permeability measurements. (*See insert for color representation of this figure.*)

donor compartment. After allowing the compound to permeate through the artificial membrane, the concentration in the acceptor compartment is determined. There are special wells that make it possible to sample from both the top and the bottom compartments without dismantling the plates. It is advisable to use a liquid-handling robot to form the membrane and add accurate volumes of buffer in the donor and acceptor compartments. The phospholipid membrane component in PAMPA has to be dissolved in an organic solvent with a low boiling point that will evaporate quickly after it has been spread onto the filter. Some methods apply the phospholipid, typically egg lecithin, in dodecane solution which stays on the filter. Five or so microliters of the lipid solution is applied to a 125-μm thick polyvinylidene fluoride (PVDF) microfilter with 0.45 μm pores by a liquid-handling robot. After a few seconds, the membrane covers the filter evenly. The membrane has to be wetted by aqueous buffer from both sides. Then, the sample has to be introduced into the donor compartment, which is typically the lower part of the assembly. The donor compartment typically contains 200 μl buffer solutions, which contains 10 to 100 μM of the test compound. When the sample is added in a DMSO stock solution, it might precipitate out when diluting down with the buffer. This has to be avoided, and the sample has to be dissolved in the buffer. Precipitation causes low permeability results that have often been observed for low solubility compounds. Some experimental protocols (e.g., from pION Inc.) suggest filtering the diluted sample solution. The acceptor compartment contains another 200 μl buffer solution when the experiment is carried out in a 96-well assay format. Other protocols may suggest adding cyclodextrin or other solubilizing agents to the donor and acceptor compartments. However, the solubilizing cyclodextrin that forms an inclusion complex with many molecules containing an aromatic ring will influence the measured permeability [21]. It has been found that the cyclodextrin also influences the thickness of the UWL; it decreases with increased cyclodextrin concentration. However, the apparent membrane permeability decreases with increased concentration of the solubilizer as it reduces the available free drug concentration. There is a trade-off between the increase in the solubility and the decrease in the permeability. So, this should be taken into account when assessing an experimental protocol.

The equilibration procedure starts after the donor and acceptor compartments are assembled with the membrane-coated filter between them and the sample in the donor compartment. The equilibration takes place for 2–4 h with small membrane thickness and efficiently stirred system. When a special filter plate assembly is used, as shown in Figure 7.5, it is possible to take samples out from both the donor and the acceptor compartments and measure the compound concentration at several time intervals using sensitive generic HPLC/MS analytical methods. Of course, there are several types of analytical methods that can be used for the concentration determination of the test compound in the donor and acceptor compartments, such as UV-plate reader method, or more frequently the HPLC-UV or HPLC-UV-MS method is applied.

At the end of the permeability experiment, the compound concentration is determined in both the donor and acceptor compartments. It is always advisable to check the mass balance, that is, the sum of the amounts of the test compound in both the donor and the acceptor compartments compared with the initial concentration of the compound. When it is less than 100%, it suggests that the test compound has accumulated in the membrane. The other important quality control test is the homogeneity of the artificial membrane surface. It may happen that the membrane lipid fails to cover the filter surface properly. It may happen that there remain some holes or an air bubble sticks between the membrane surface and the buffer. These affect the measured permeability significantly and provide erroneous high or low permeability. The best way to carry out the quality control in each well is by adding a small amount of a standard compound for which the permeability is known and which does not interfere with the test compound.

When the sample is applied to the buffer as DMSO stock solution and HPLC-UV detection is used for the quantification, the proportion of the DMSO peaks in the donor and acceptor compartments may indicate such errors. The DMSO peak should be significantly smaller in the acceptor compartment than in the donor compartment even after a 3-h incubation.

The calculation of the apparent permeability is carried out using the volumes of the donor and acceptor compartments (V_d and V_a), the surface area of the membrane (A), and the peak area (or absorbance) ratio of the test compounds (R). Equation 7.7 is used to calculate P_{app}, using the time point (t) when the analysis has been carried out.

$$P_{app} = -\frac{V_d \times V_a}{(V_d + V_a)A \times t} \times \ln(1 - R) \tag{7.7}$$

The typical plot of the apparent permeability (P_{app}) as a function of the peak area ratio obtained after 3 h (7200 s) of equilibration using the same volumes in the donor and acceptor compartments ($V_d = V_a$) and 0.33 cm^2 artificial membrane surface is shown in Figure 7.6.

The typical permeability of drug discovery compounds obtained by the GSK "in-house" AMP experiment ranges from 3 to 500 nm/s. This means that the peak area ratio is a small number, typically less than 0.2. Therefore, it is important to have a sensitive analytical method to detect the small concentration of the test compound in the acceptor compartment. Permeability values less than 50 nm/s are considered as low, while values between 50 and 300 nm/s are considered as medium. Permeability values measured above 300 nm/s are categorized as high.

The *in vitro* cell permeability measurements can be carried out in a similar manner. However, a bigger surface area is used for growing the cells (MDCK or Caco-2) on the filter plate. The cell monolayer has to be washed by the buffers to remove the components of the cell culture liquid; the membrane integrity has to be checked before the assembly of the donor and acceptor compartments. It is essential to put the compound into the apical side of the cell to measure the permeability through A to B direction as the permeability of a compound can be very different from the B to A direction because of active transporters present in the cells. The cell culturing procedure may take 21 days for the Caco-2 cells and a couple of days for the MDCK cells. The cellular permeability assay is usually carried out at 37 °C in order to maintain normal cell functions. The concentration measurements of the

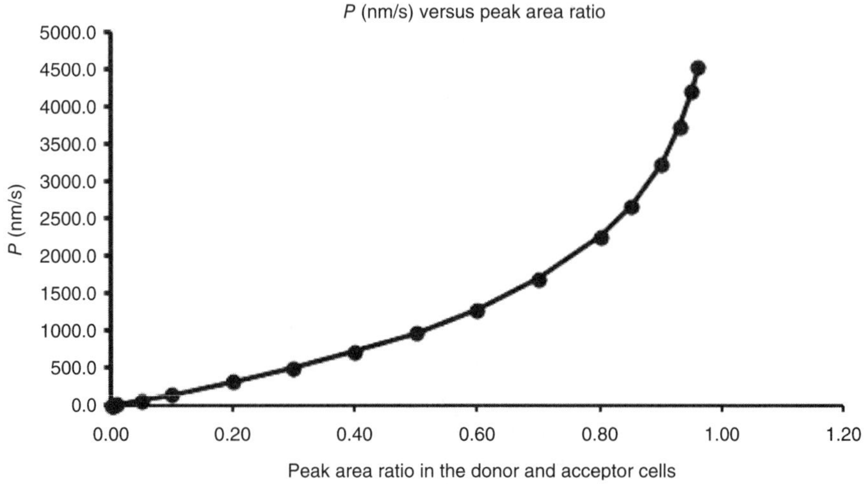

Figure 7.6 The apparent permeability (P_{app}) as a function of peak area ratio of the test compound obtained after 3 h of equilibration time using membrane surface area of 0.33 cm^2 and equal volumes of the donor and acceptor compartments.

compounds are carried out after a 3-h equilibration time. The initial test compound concentration ranges typically from 100 to 200 μM.

Balimane et al. [22] have compared typical values of artificial membrane and Caco-2 cell permeability of known drug molecules together with the human intestinal absorption expressed in percentage. The values are shown in Table 7.1. The authors have given the average and the standard deviation of the measurements obtained from a minimum of three parallel determinations.

A plot of the artificial membrane and Caco-2 cell permeability is shown in Figure 7.7. We can see that compounds showing low permeability in the PAMPA assay show low cellular permeability with the exception of two compounds, ketoprofen and naproxen. Both are acidic compounds that are generally considered as low permeable compounds. However, as they bind strongly to albumin and are substrate to several types of monocarboxylate transporters (MCTs), they may be able to go through cellular barriers [23]. The low, medium, and high category permeability values show reasonably good agreement. As the human intestinal absorption process is even more complex than the cellular permeability, we cannot expect very good correlation between the two either. Figure 7.8 shows a plot of cellular permeability and percentage of human intestinal absorption.

One of the major differences between the cell permeability assay results and the human intestinal absorption from a physicochemical point of view is the unit of measure. One is a rate and the other is a percentage, which would result in a sigmoid curve. The other aspect is the possible pH difference between the donor and acceptor sides in the intestine. The pH can range from 2 to 8 in the gastrointestinal tract, while the cell permeability assay uses one fixed pH, typically pH 7.4 on both

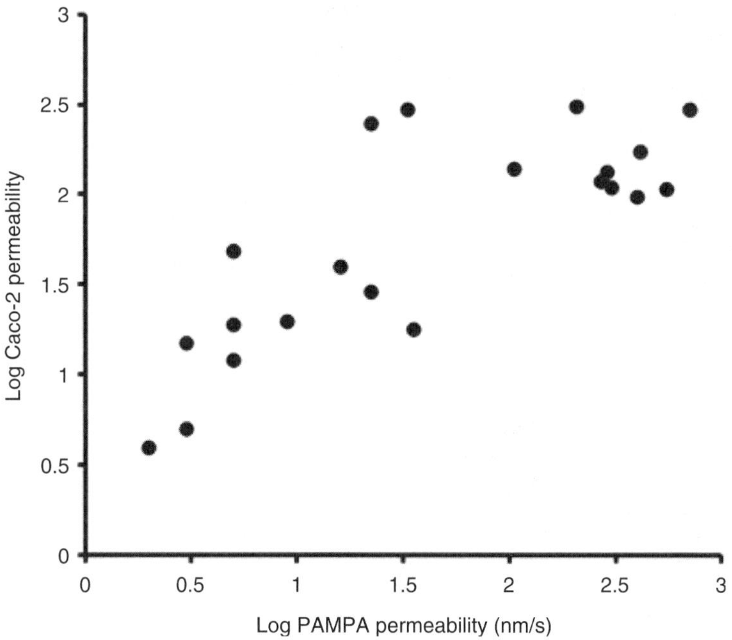

Figure 7.7 A plot of PAMPA and Caco-2 cell permeability of known drugs listed in Table 7.1. (Reproduced with permission from Reference [22].)

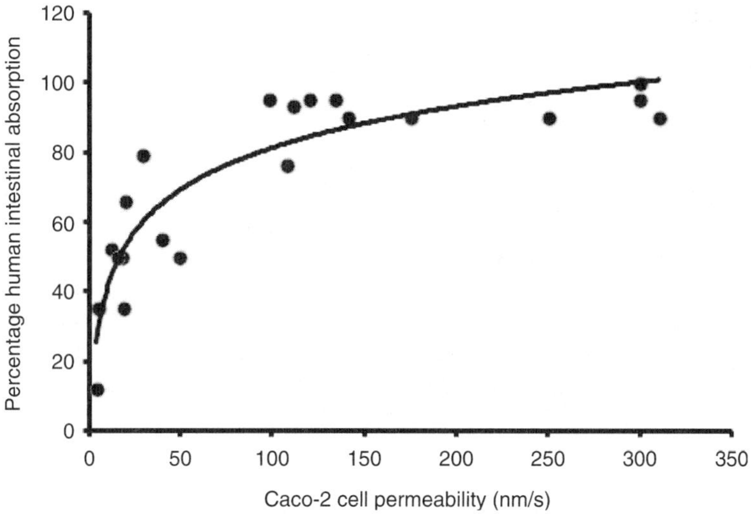

Figure 7.8 The plot of Caco-2 cell permeability and percentage of human intestinal absorption for the known drugs listed in Table 7.1. The trend line shows the logarithmic function. (Reproduced from the data in Reference [22].)

Table 7.1 Artificial membrane and Caco-2 cellular permeability of selected known drugs with their human intestinal absorption[a]

Drug	PAMPA Permeability (nm/s)	Caco-2 Cell Permeability (nm/s)	% Fraction Absorbed in Humans
Acebutolol	16 ± 11	40 ± 4	55
Alprenolol	299 ± 68	111 ± 30	93
Amiloride	5 ± 3	49 ± 8	50
Cimetidine	22 ± 3	29 ± 12	79
Desipramine	700 ± 170	300 ± 21	95
Dexamethasone	287 ± 11	134 ± 13	95
Etoposide	35 ± 4	18 ± 10	50
Furosemide	9 ± 1	20 ± 4	66
Hydralazine	105 ± 4	141 ± 16	90
Ketoconazole	542 ± 37	108 ± 18	76
Ketoprofen	22 ± 6	250 ± 26	90
Metformin	5 ± 2	12 ± 3	52
Metoprolol	266 ± 16	120 ± 10	95
Naproxen	33 ± 3	300 ± 41	100
Norfloxacin	5 ± 1	19 ± 3	35
Phenytoin	204 ± 9	310 ± 15	90
Propranolol	411 ± 110	175 ± 26	90
Sulfasalazine	2 ± 1	4 ± 1	12
Sulpiride	3 ± 1	5 ± 2	35
Terbutaline	3 ± 1	15 ± 6	50
Verapamil	399 ± 112	98 ± 19	95

[a]Reference [22].

sides of the cells. It may result in a so-called "double-sink" condition in the intestine. One sink is due to the disappearance of the neutral form of the molecule on the other side of the membrane, and the other sink condition is due to the presence of plasma proteins on the other side of the intestinal wall that might bind the compound, reducing its free concentration. Both effects result in higher permeability than it is expected from the Caco-2 measurement.

In conclusion, the measured permeability values of compounds can differ significantly depending on the experimental conditions and do not necessarily show good correlation with intestinal absorption. The various models for human intestinal absorption are discussed in more detail in Chapter 9. Cellular permeability can be very different from the AMP as the cells contain transporters; the compound has to go through the cell membrane twice and can bind to cell components, causing significant decrease in mass balance. The cell membrane contains pores at the junctions and transporter enzymes too, which affect the intrinsic permeability of the compounds. The AMP results obtained by different experimental protocols are dependent highly on the conditions. The composition of the membrane, the applied membrane thickness, the pH applied in the donor and acceptor compartments, the shaking or agitation of the assembly, and the

Table 7.2 Comparison of various parallel artificial membrane permeability assays (PAMPA)

Lipid Model	Composition	Filter Type	Solvent	Agitation
Egg lecithin Kansy et al. [13]	10% egg lecithin	PVDF	*n*-Dodecane	None
Dioleoylphos-phatidyl-choline (DOPC) Avdeef et al. [24]	2% DOPC	PVDF	*n*-Dodecane	Shaking
Double-sink methods Different pH values in donor and acceptor and albumin Avdeef [9]	20% phospholipid mixture	PVDF	*n*-Dodecane	Individual well stirring
n-Hexadecane Wohnsland and Faller [25]	100% *n*-Hexadecane	Polycarbonate	*n*-Hexadecane	Shaking
Phospholipid mixtures Sugano et al. [26]	Phosphatydil-choline, -serine, -ethanolamine, cholesterol	PVDF	1,4-Octadiene	None
Lecithin Zhu et al. [27]	Lecithin (1%)	Hydrophilic filter	*n*-Dodecane	shaking

temperature all have significant effects on the measured values. Variations of the measured values are often observed because of the type of filter material used. Even the material of the plate assembly may influence the results because of the compound's nonspecific adsorption to the filter and plate. Table 7.2 summarizes the most well-known published experimental approaches for the determination of AMP.

Special compositions of artificial membranes have been developed to model other membranes, such as blood–brain permeability by Di et al. [28], containing more negatively charged phospholipids. To model the skin permeability, 70% silicon oil and 30% iso-propyl myristate have been suggested as a model membrane by Ottaviani et al. [29]. Faller [30] reviewed the various aspects of the published AMP methods and found that the logarithmic values of intrinsic permeability that refer to the permeability of the neutral form of the molecule show very good correlation to the logarithmic values of alkane/water partition coefficients. The explanation for this is that the inner core of any biological membrane bilayer is basically a hydrocarbon chain, where the compound has to partition into. Permeable compounds should be able to go through this inner hydrophobic layer from an aqueous compartment; therefore, permeable compounds have a relatively high alkane/water partition coefficient.

RELATIONSHIPS BETWEEN PERMEABILITY, LIPOPHILICITY, AND SOLUBILITY

It has already been shown that the lipophilicity or hydrophobicity of the compound determines, or at least significantly influences, other physicochemical properties, such as solubility. Membrane permeability is also influenced by the compound's lipophilicity and hydrophobicity. We have seen that the biomembrane consists of a structured assembly of amphiphilic phospholipids with the hydrophilic polar head groups pointing toward the two aqueous sides of the membrane and the hydrophobic alkyl chains form the inner bilayer core. When a compound permeates through it, it has to shed its solvate shell and has to lose its charge. Thus, it is essential that the compound has a certain degree of lipophilicity. However, if the compound is too lipophilic, it would be energetically more favorable to stay in the membrane rather than coming out to the polar aqueous layer on the other side. Therefore, the lipophilicity of the compound measured by the partition coefficient should show a relationship with the membrane permeability. However, as has been discussed in more detail in Chapter 5, the octanol/water partition coefficients are probably not the best model to describe membrane permeability. Octanol contains a lot of water and polar hydroxyl groups, which is able to form hydrogen bonding (both acceptor and donor). This polar functionality is missing from the inner core of the membrane. As has been demonstrated by Wohnsland and Faller [25], the *n*-dodecane partition coefficients are much better predictors of membrane permeability than the octanol/water partition coefficients. The Abraham solvation equation model described in Chapter 5 may provide a better insight on the molecular descriptors that influence the octanol/water and alkane/water partition. As given in Table 5.5, the major difference between the octanol/water partition and the alkane/water partition is in their sensitivity toward H-bond donor groups. The alkane/water partition has a large negative coefficient for the H-bond acidity, which means that H-bond donor compounds do not prefer to partition into alkane from water. The size parameter also has a large negative coefficient. The solvation equations for cell permeability are shown by Equation 7.8 and Equation 7.9 for Caco-2 and MDCK cell permeability, respectively. These equations were obtained using the literature cell permeability data of known drugs and their calculated molecular descriptors.

$$\log P_{\text{app}}\,(\text{Caco-2}) = -4.4 - 0.20E + 0.26S - 1.27A - 0.24B + 0.09V$$
$$n = 53 \quad r^2 = 0.77 \quad \text{sd} = 0.38 \qquad (7.8)$$

$$\log P_{app}\,(\mathrm{MDCK}) = -4.31 + 0.10E + 0.19S - 1.73A - 0.79B - 0.17V$$
$$n = 30 \quad r^2 = 0.81 \quad \mathrm{sd} = 0.48 \qquad (7.9)$$

where n is the number of compounds, r^2 is the multiple regression coefficient, sd is the standard error of the estimate, E is the excess molar refraction, S is the dipolarity/polarizability, A is the H-bond acidity, B is the H-bond basicity, and V is the McGowan size descriptor. The cell permeability equation differs from the alkane/water equation especially in the V coefficient, which is much smaller value in the permeability equations. Interestingly, the dipolarity/polarizability has a positive coefficient, while H-bond acidity is negative. The equations suggest that H-bond donor and acceptor groups reduce the permeability, while polarizability slightly favors the permeability. The size has no or little influence, which might be misleading. It is possible that the permeability of small lipophilic molecules increases with size, but when the size increases above a certain point, the permeability decreases That would result in an overall low coefficient of the size, but it does not mean that permeability does not depend on it. The solvation equation obtained for cell permeability shows great similarity to the equation obtained for diffusion coefficients in water, which has been presented by Draper [31]. It suggests that during the cell permeability measurements, the speed of the molecules is dominated by the diffusion through the UWL. The permeability through artificial membrane, especially when agitation is applied, is controlled by the membrane partition of the compounds. The ionization, which is pH dependent, has a unique effect on permeability. While ionization, in general, reduces lipophilicity and increases solubility, it decreases the permeability. The positive and negative charges have very different effects on permeability. The negative charge definitely decreases the compound's ability to go through cell membranes or artificial membranes. The positive charge is not so detrimental, especially considering phospholipid membrane permeability. The proton is removed from the charged basic compounds; so, they become neutral when passing through the nonpolar inner layer and they gain back the proton (and as a consequence the charge) at the other end. As the concentration of the neutral form on the other side of the membrane disappears continuously because of regaining the proton and charge, it provides a "sink" condition that also helps permeability. Just like for lipophilicity and solubility, we can use the term *intrinsic permeability* for the uncharged species of the molecules. It can be determined using the pH on both sides of the membrane, which keeps the compound in an unionized form. This can be very different from the physiological pH values; so, it can only be measured using artificial membrane, preferably the hexadecane membrane. Figure 7.9 shows a typical example of permeability–pH profile for an acidic drug molecule, diclofenac [25]. The effective permeability ($\log P_e$) is that we measure at a certain pH. The effective permeability is very low at high pH values and it increases as the pH drops because the acidic diclofenac molecules lose the negative charge. This part of the $\log P_e$ versus pH profile is a straight line with a slope of −1. The apparent pK_a of diclofenac is when the slope of the straight line is −0.5, where 50% of the molecules are in ionized form and 50% of the molecules are in

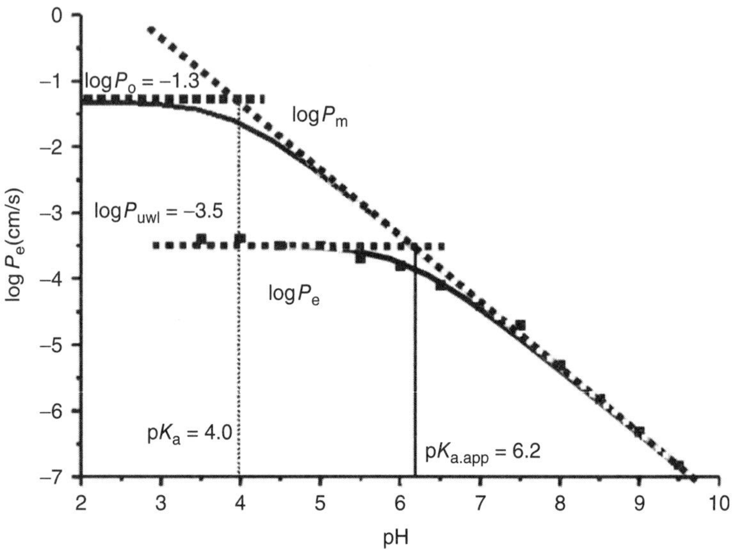

Figure 7.9 The plot of the effective permeability (log P_e) and the pH profile of the weak acid diclofenac. It reaches a maximum (log P_{max} = −3.5 cm/s), where the diffusion in the unstirred water layer limits the permeability. The dotted line shows the theoretical curve, which gives the intrinsic permeability, log P_o as −1.3 cm/s. log P_m refers to the membrane permeability of diclofenac. (Reproduced with permission from Reference [25].)

neutral form. As we see in Figure 7.9, it has been observed at pH 6.2, which has a much higher pK_a than the pK_a of diclofenac in water, which is only 4. The theoretical curve shown by the dotted line suggests that the pK_a of diclofenac is from 4 to 6.2. The fact that the diclofenac reaches the maximum permeability at pH 6.2 indicates that the effective permeability is dominated by the diffusion through the UWL between pH 6.2 and 4.

The pH profile of permeability shows that the actual process of compounds going through membrane layer is quite complex and it can be divided into several physicochemical processes. The intrinsic permeability (log P_0) refers to the speed of the compound when going through membrane in its neutral form. The *intrinsic permeability* does not include any sink condition nor the slowing effect of the diffusion in the UWL that covers biological membranes. The *apparent permeability* is that we actually measure under a defined experimental condition. The *effective permeability* is the maximum apparent permeability that we can observe for a neutral compound that also includes the effect of the diffusion in the UWL. The intrinsic permeability is the most useful parameter to obtain structure–permeability relationships, which actually can guide the compound design process. The acid dissociation constant of the compounds is also important as it determines the proportion of ionized and neutral molecules that are present at physiological pH. The apparent permeability provides information about the compound's behavior under physiological conditions as long as the measurements can appropriately imitate those conditions. When comparing permeability with lipophilicity, the permeability usually refers to the membrane permeability and excludes the permeability through the UWL. The effective membrane permeability should be compared with log *P*, the lipophilicity of the neutral form of the molecules, while the apparent permeability at a given pH should be compared with the log *D*, the lipophilicity of the compound at that given pH. It is important to remember that while lipophilicity is a dimensionless measure (the equilibrium concentration ratio of the compound in an organic and aqueous solvent system), permeability is a rate. Several authors [6,13,32] have reported a parabolic relationship between apparent permeability and lipophilicity expressed by octanol/water partition coefficients. It means that permeability is low for hydrophilic and very hydrophobic compounds too. Compounds that have log *D* values around 2 and 3 are highly permeable. Compounds with very high lipophilicity show low permeability as they partition into the membrane, but it is energetically more favorable for them to stay in the lipophilic membrane environments than coming out to the aqueous, polar environment on the other side. The reason for the decreasing permeability of highly lipophilic compounds is due to membrane retention. Huque et al. [33] published a solvation equation model for intrinsic membrane permeability (log Perm_0) based on 2% phosphatidylcholine in dodecane (DOPC) PAMPA model.

Table 7.3 shows the calculated Abraham descriptors and measured intrinsic membrane permeability of 35 compounds. Equation 7.10 shows the obtained solvation equation for 43 compounds [33].

$$\log \mathrm{Perm}_0 = -4.03 + 1.06E - 0.95S - 2.26A - 2.84B + 2.58V$$

$$n = 43 \quad r^2 = 0.78 \quad s = 0.92 \qquad (7.10)$$

where n is the number of compounds, r^2 the multiple regression coefficient, sd the standard error of the estimate, E the excess molar refraction, S the dipolarity/polarizability, A the H-bond acidity, B the H-bond basicity, and V is the McGowan size descriptor. The equation is similar to the cell permeability Equations 7.8 and 7.9, in the sense that it shows that compounds with H-bond acidity and basicity are less permeable. However, the V size parameter has a significant positive value for the AMP using dodecane and 2% phosphatidylcholine. It is because the intrinsic membrane permeability does not include the reduction of permeability because of the diffusion in the UWL that would decrease permeability with the increasing molecular size.

Comparing the solvation equation for intrinsic permeability and octanol/water partition coefficient, the marked difference is in the constant for the H-bond acidity (A). Compounds with H-bond donor groups are less permeable than would be expected from the octanol/water lipophilicity. While the octanol/water partition system was a very good model for compounds binding to albumin and IAM (see Chapter 5 for more details), the actual membrane permeability is closer to the reversed phase (C-18) chromatographic lipophilicity. Both C-18 partition with acetonitrile gradient, and membrane permeability drops when

Table 7.3 The intrinsic membrane permeability, calculated octanol/water partition coefficients and calculated molecular descriptors of 35 model compounds[a]

Compound	$\log P_0$ profile	$\log P_{oct}$	E	S	A	B	V
2,4-Dichloro-phenoxyacetic acid	−4.38	3.51	1.21	1.18	0.77	0.60	1.38
2-Naphthoic acid	−3.40	3.44	1.51	1.32	0.59	0.54	1.30
3,5-Dimethoxyphenol	−4.12	2.24	0.88	1.11	0.53	0.71	1.17
4-Hydroxybenzoic acid	−5.61	2.32	0.93	0.90	0.81	0.56	0.99
4-Nitrophenol	−5.25	2.63	1.07	1.65	0.94	0.22	0.95
4-Phenylbutylamine	−2.13	3.19	0.81	0.87	0.10	0.72	1.38
4-Propoxybenzoic acid	−3.55	3.93	0.88	1.19	0.59	0.46	1.41
Aniline	−3.36	1.71	0.96	1.00	0.11	0.51	0.82
Aspirin	−5.57	1.94	0.78	0.80	0.49	1.00	1.29
Carazolol	−3.99	3.19	2.08	2.02	0.68	1.68	2.38
Carbendazim	−4.82	2.54	1.64	1.82	0.77	0.74	1.36
Chlorpromazine	0.94	6.24	2.44	1.37	0.06	1.08	2.41
Chlorthalidone	−6.32	0.57	2.65	3.17	1.25	1.96	2.17
Desipramine	−0.83	4.90	1.99	1.38	0.07	1.23	2.26
Diclofenac	−2.58	5.12	1.84	1.58	0.90	0.83	2.03
Diltiazem	−2.67	4.05	2.34	2.65	0.00	2.11	3.14
Diphenhydramine	−0.94	4.89	1.39	1.43	0.00	1.04	2.19
Eserine	−4.62	1.98	1.58	1.87	0.27	1.73	2.14
Fluoxetine	0.07	5.05	1.24	1.33	0.08	1.06	2.24
Flurbiprofen	−2.55	5.14	1.30	1.38	0.59	0.59	1.84
Furosemide	−4.94	2.82	2.05	2.14	1.35	1.45	2.10
Haloperidol	−2.15	5.07	1.90	1.39	0.40	1.76	2.80
Ibuprofen	−2.46	4.64	0.73	0.97	0.60	0.70	1.78
Indomethacin	−3.54	5.05	2.24	2.85	0.40	1.08	2.53
Lidocaine	−2.51	3.10	1.23	1.43	0.01	1.39	2.06
Metoprolol	−4.11	2.64	1.00	1.69	0.31	1.63	2.26
Nifuroxime	−5.08	1.61	1.07	1.09	0.34	0.70	0.97
Penbutolol	0.30	5.94	1.31	1.06	0.27	1.20	2.52
p-F-deprenyl	−0.73	3.94	0.83	1.05	0.08	0.84	1.73
Phenylacetic acid	−5.03	0.64	2.46	2.28	0.49	0.96	1.07
Propranolol	−2.12	3.77	1.14	1.57	0.42	1.24	2.15
Tetracaine	−1.98	3.12	2.47	1.07	0.66	1.92	2.26
Theophylline	−6.77	−0.77	1.50	2.74	0.96	1.24	1.22
Tramadol	−1.49	4.57	1.12	0.69	0.40	1.37	2.23
Warfarin	−2.90	3.91	1.13	1.32	0.38	1.45	2.31

[a]Reference [33].

H-bond acidity of the compound increases. In conclusion, lipophilicity plays an important role in membrane permeability as compounds have to be lipophilic enough to partition from the aqueous environment of one side of the membrane into the nonpolar inner core of biological membranes. However, when the compound is too lipophilic, it is retained in the membrane and the rate of transport is decreased. An optimum lipophilicity is required for good permeability. Membrane partitioning decreases with the presence of H-bond donor and acceptor groups in the molecules, which is not the case for octanol/water partition. While hydrophobicity increases with increasing size, apparent permeability reaches an optimum; after that the further increase in molecular size will reduce the apparent permeability due to the slow diffusion of the molecules in the UWL. While the presence of charge increases solubility and decreases lipophilicity, it also decreases the permeability as charged molecules cannot go through the membrane. However, positively charged compounds may lose the proton and consequently their charge temporarily while passing through the membrane and they gain back the proton and the charge on the other side according to the pH partition hypothesis. This provides a sink condition that increases the permeability for positively charged compounds. In general, the permeability of negatively charged compounds is low at pH 7.4, but there is published evidence that even negatively charged compounds are able to go through membranes. We have to keep in mind that permeability is a rate parameter, while lipophilicity is dimensionless (partition coefficient); therefore, permeability units will effect direct correlation with lipophilicity.

CHROMATOGRAPHY AS A POTENTIAL TOOL FOR MEASURING THE RATE OF PERMEATION

Chromatography plays a very important role in the permeability measurements. It is mostly used for measuring the compound concentration in the donor and acceptor compartments after a defined period of time. When the compound has been administered to the AMP assay in DMSO solution, the DMSO peak can be detected during the reversed phase chromatographic measurements (it is usually the first peak in the chromatogram). The DMSO peak should be always much smaller (approximately one-third) in the acceptor compartment than in the donor compartment. This can be used as the quality control of the artificial membrane, showing that it covered the filter properly. The HPLC method for concentration determination also provides a quality control of the sample, namely, no decomposition or degradation occurred during the assay when the compound appears as a single peak. Typical generic reversed phase HPLC methods with fast acetonitrile gradient can be used, as described in Chapter 4. The obtained peak area ratio (R) is used for the calculation of permeability, as is shown in Equation 7.3. Although the chromatography is not essential, it has several advantages. A UV plate reader that reads the absorbance ratio provides a much faster method, but it does not give information about the permeability of each component in the sample, for example, about the permeability of DMSO. An internal standard compound for further quality control also cannot be used without the chromatographic separation during the quantification. The generic gradient reversed phase method, especially when using small columns and fast flow rate, can reduce the analysis time to less than 2 min per compound. For the cell permeability assay, the compound quantification by HPLC is very useful as the cells might excrete UV-active components in the receiver compartment from which the peak of interest has to be separated. It is very common to use HPLC/MS quantification to ensure the precise concentration determination of the test compound.

We discussed in Chapter 4 that the chromatographic retention provides information about compound's distribution between the mobile and stationary phases; however, this does not provide information about the rate of compound's partitioning into the stationary phase. Contrast to this principle, several authors tried to apply the chromatographic retention times obtained on IAM to correlate with permeability and absorption [34–36]. Genty et al. [34] described a convincing relationship between passive rat intestinal absorption and the logarithmic retention factor obtained on using IAM chromatography and the size of the investigated compounds. The correlation coefficient has been reported to be above 0.9 for 12 diverse set of compounds. The explanation for the good correlation between phospholipid partitions minus the size of the molecules may be that the IAM retention explains the membrane partition of the compounds, which drives the permeability, while the size parameter explains the effect of the diffusion in the UWL. Although the two parameters do not measure permeability in the true physicochemical sense, they represent the two most important factors that affect the rate of passive permeation through the rat intestinal wall. It is also not surprising that the IAM partition is a better model for biological membrane partition than octanol/water lipophilicity or C-18 retention, although the authors have not carried out a comparative study. Kotecha et al. [35] has published an interesting approach using IAM chromatography for modeling human oral absorption. They have determined the IAM retention of 28 known drug molecules with known human oral absorption data using a pH range from 4.5 to 7.4. They have found that the highest IAM retention factor ($\log k_{\mathrm{IAM}}$) at any of the pH values showed the best sigmoid relationship with the percentage of oral absorption. They argued that the investigated pH range can be found in the gastrointestinal tract, and the drug molecules absorb at that pH where they have the least charge present and exert the highest possible membrane partition. This is how they justified using IAM partition data obtained at different pH values for the different compounds. IAM chromatography has been extensively studied for modeling the blood–brain barrier permeability of compounds too. The early study of Salminen et al. [37] compared the IAM retention factors and octanol/water partition coefficients as descriptors to estimate the blood–brain penetration of 23 known drug molecules. They concluded that the octanol/water partition coefficients do not show good correlation with brain penetration and nor does the IAM retention factors on their own. However, including the size and ionization parameters, the estimation of the brain penetration has been successful regardless of which lipophilicity parameter they used. Since then, several authors [38,39] have tried to use the IAM retention data that describe a compound's partitioning into phospholipids to model brain penetration. The various models using IAM chromatography have been reviewed [40,41]. Obviously, the membrane partition and other physicochemical properties such as ionization and size can explain only the passive permeation. Therefore, it is common to have outliers, and the coefficients vary greatly on the set of compounds that is used to derive the models. Lazaro et al. [42]

have adopted a more systematic approach using the Abraham solvation equation model to compare various partition/distribution and permeation models with the chromatographic partition. They have derived a similarity factor from the coefficients of the five molecular descriptors. They have found that molecular properties such as H-bond acidity and dipolarity/polarizability have a different effect on the blood–brain distribution and the permeability. The IAM partition is a closer model for the blood–brain permeability than the C-18 partition, while the skin permeation can be modeled better with C-18 partition behavior of the compounds.

However, these chromatographic approaches measure compound distribution behavior that may show correlation with their speed of going through biological membranes. The true measure of the kinetic aspects of the partitioning is the chromatographic peak-broadening process, as it has been discussed in more detail in Chapter 4. The peak broadening is a consequence of three different effects, namely, the eddy diffusion, longitudinal diffusion, and the kinetic of the mass transfer between the mobile and stationary phases. The eddy diffusion depends only on the column packing, and the longitudinal diffusion depends on the diffusion coefficient of the compound in the liquid mobile phase and the mobile phase velocity. The third force on the peak broadening is directly related to the speed of the molecules entering the stationary phase and reemerging from the stationary phase, and it is inversely related to the mobile phase velocity. The relationship between the peak broadening and mobile phase flow rate is described by the van Deempter equation, which is discussed in Chapter 4. To date, only one publication has shown the possibilities of using the van Deempter equation in chromatography to measure the rate of permeation by measuring the mass transfer speed [43]. Meng et al. [43] used the van Deempter curve to derive the relative permeability of compounds. As is shown in Figure 7.10, they have plotted the theoretical plate height (H) as a function of the linear velocity of the mobile phase (u). The theoretical plate height can be obtained from the length of the column (L) divided by the theoretical plate number (N) observed from the peak width (t_w) and the peak retention time (t_R) of the compounds (Equation 4.15). The authors have used C-18-derivatized polystyrene–divinylbenzene polymer-based columns and meperidine, alfentanil, fentanyl, and sufentanil opioid drug molecules as model compounds. As is shown in Figure 7.10, they have obtained straight lines with different slopes of the compounds but with the same intercept values. They have derived the diffusion coefficients of the compounds and found good correlation with their potency ratio after intravenous administration that reflects the dynamics of the binding of the drugs to the opioid receptors in the brain. Figure 7.11 shows a plot of the derived diffusion coefficients from the chromatographic $H - u$ plots and the potency ratio of the compounds observed after intravenous administration. The paper demonstrates very well the potential of chromatography to determine diffusion coefficients and mass transfer speed from one phase to another. This is an equivalent parameter to permeability as it describes the rate of partition to the stationary phase. The authors have compared the obtained diffusion coefficients with the octanol/water lipophilicity that showed reasonable correlation.

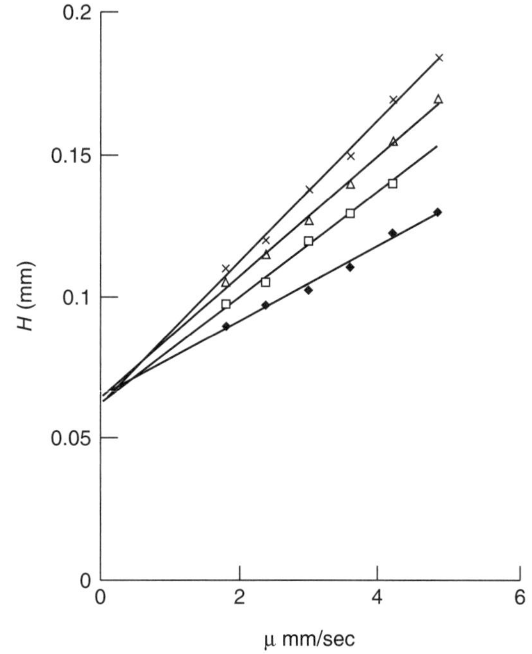

Figure 7.10 The plot of the theoretical plate height (H) as a function of mobile phase velocity (u) for four opioid drugs. (Reproduced with permission from Reference [43].)

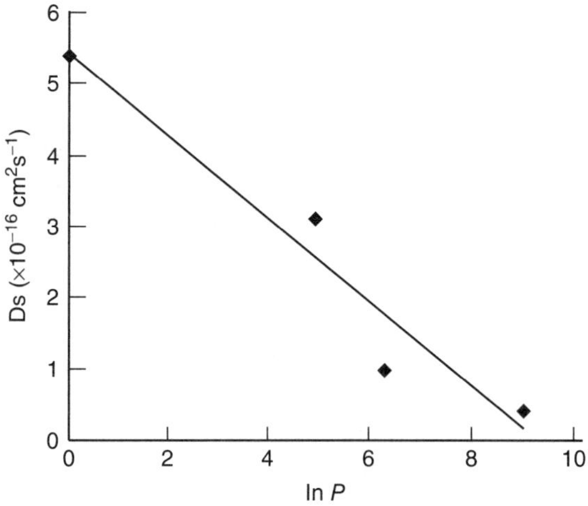

Figure 7.11 The diffusion coefficient (Ds) of the opioid drugs as a function of their relative potency (ln P) measured after intravenous administration. (Reproduced with permission from Reference [43].)

The potential of using the peak width for the determination of permeation rate, or more exactly relative permeation rate of the molecules, is evident from pioneering paper of Meng et al. [43]. *There is a possibility of developing even a more pragmatic approach to measure and rank compound's rate of partitioning by relative peak width measurements. We have frequently observed that compounds with distinct chemical structures from certain program teams gave much wider chromatographic peaks than would be expected from the retention times. As the subsequent peaks showed narrow peaks, again the wide peak most probably indicated slow partitioning process. The wide peaks on IAM column most often have been observed for compounds that have delocalized positive charge and long flat structure. Most of the HPLC integration software can accurately measure peak width and retention time, and the calculation of the theoretical plate number of each peak could be easily automated. The plate numbers (N) could be calibrated by compounds with slow, medium, and high permeabilities. In this way, chromatography would provide a fast detection of low, medium, and high permeability compounds, especially using the IAM stationary phase. The drawback of such a method would be the expected small differences between the N numbers for each compound; thus, it would require precise calibration and accurate peak integration. In principle, when the mass transfer process is quick, the theoretical plate number (N) should be the same for every peak. The potential reproducible differences could be used for measuring at least the permeability category of the compounds. It is worth mentioning that the authors used polymer-based HPLC columns that have much slower mass transfer process than the silica-based stationary phases. Perhaps column manufacturers can develop special stationary phases with slow mass transfer. It is also possible to develop generic chromatographic conditions that are falling into the dominant mass transfer speed region of the H–u curve. As usual, the chromatographic technique would not be able to provide absolute measurements of permeability. It has to be calibrated with a set of compounds for which the permeability or the rate of phase transfer is determined by another absolute method. Chromatography in this respect may provide an easily automated high throughput tool to measure the rate of transfer of compounds from one phase to another. However, it still needs careful experimental design and validation of the procedure by an enthusiastic scientist.*

SUMMARY

Biological membranes protect cells from the environment; on the other hand, they need to let through important compounds in and out of the cells. Cell membranes consist of amphiphilic molecules that form bilayer with the polar head groups facing toward the aqueous environment both inside and outside of the cells. The inner core of the bilayer contains aliphatic side chains with a few double bonds, which only provide a completely hydrophobic environment. The hydrophobic forces between the alkyl chains provide the force that keeps the membrane molecules together. The most common components of the membrane are the phospholipids, which are the fatty acid esters of glycerol having only one of the hydroxyl groups, forming esters with small polar acids, such as amino acids (serine) or phosphatidylcholine, or sugar molecules. Phosphatidylcholine is one of the major components of biological membranes containing both a positive and a negative charge in the polar head group. Cell membranes contain tight junctions and ion channels as well, which are filled with various types of protein molecules that may serve as targets for various drugs. When the drug target is inside the cell, it is important that the drug molecule is able to penetrate through the cell membrane. Permeability barriers can distort the normal physicochemical equilibrium process and may cause delays for the drug to reach the target. The intestinal absorption process is also controlled by the compound's permeability through the intestinal wall. Drug molecules have to be able to get into the blood circulation to reach the target. The two main reasons for the need of permeability measurements at early stages of drug discovery are the detection of possible problems of potent molecules with the intestinal absorption and cellular permeability. The permeability of the compounds is expressed as a rate constant, usually centimeter or nanometer per second. It is determined from the concentration of the compound in both sides of the membrane after a certain amount of time when the compound is introduced to one side of the membrane. The concentration determination can be carried out using a chromatographic technique that provides information about the quality and the quantity of the compound. Besides the time and the concentration ratio, we need to know the membrane surface area and the volumes of the acceptor and donor compartments to be able to calculate permeability. The major driving force for the compounds to go through the membrane is the concentration difference between the two sides. The rate of permeation also depends on the surface area of the membrane and the thickness of the membrane. Larger surface area increases the rate of permeation, while membrane thickness is inversely related to the rate of permeation. The stirring or agitation in the donor and acceptor compartments increases the permeability of lipophilic molecules because of the reduction in the thickness of the UWL that covers the polar head groups on both sides of the membrane. While the membrane partition coefficient controls the speed of the molecules inside the membrane, the diffusion controls the compound's movement in the UWL. The compound's concentration can be temporarily high in the UWL that reduces the rate of transport. It is assumed that polar and charged compounds cannot go through the hydrophobic inner layer of the membrane unless they temporarily lose their charge and shed their hydrate shell. The shape and size of the compounds are also important in the permeation process. Various *in vitro* cell permeability measurements are used for modeling intestinal absorption. The cells can be grown as monolayers on a filter plate. The human carcinoma cell line (Caco-2) and the canine kidney cell line (MDCK) cells are used most frequently. The Caco-2 cells contain similar active transporters as human intestinal cells, but they grow very slowly; it usually takes several weeks. The MDCK cell line grows much quicker in only a couple of days, but it contains less active transporters. Using transporter inhibitors, it is possible to detect and measure active transport processes. It is also common to measure cell permeability in two directions, apical to basolateral and basolateral to apical. As the transporters work only in one direction, the difference obtained in the measured permeability values indicates active transport process.

The physicochemical principles of permeability can be best studied using artificial membranes. Several AMP assays have been published in the literature. The artificial membranes most commonly used are phospholipids (such as egg lecithin, phosphatidylcholine dissolved in dodecane) and mixtures of these components. During the experimental study of the AMP, the effect of membrane partition coefficient, the diffusion coefficients, the pH of the donor and acceptor compartments, and the effect of agitation have been studied. Besides the *effective permeability* that is determined under a given condition, the *intrinsic permeability* is defined as the permeability of the unionized compound through the membrane, excluding the effect of the diffusion in the UWL and corrected for ionization. The *effective permeability* includes both the membrane permeability and the permeability through the UWL, which is diffusion controlled. The sum of the reciprocal values of each permeability components provides the reciprocal value of the effective permeability. There are two types of "sink conditions" that may accelerate the compound's crossing of the membranes. The first "sink condition" may be caused by the pH difference between the two sides of the membrane. Compounds go through the membrane in their neutral, unionized state. If the pH on the other side of the membrane is such that it causes the compound's ionization (e.g., acids get ionized at pH higher than their pK_a values, or bases get charged if the pH is lower than their pK_a values), it immediately reduces the concentration of the unionized form. It means that in spite of the compound passing through the membrane, the concentration of the neutral form does not increase on the other side, which would slow down the permeation. The acceptor compartment behaves like a sink. The second type of "sink conditions" can occur when the acceptor compartment contains proteins, for example, plasma proteins, that bind the compound that permeated through the membrane, thus immediately reducing its concentration. During the intestinal absorption, both types of sink conditions are likely to occur, which is called *double-sink condition.* During the AMP measurements, the "sink condition" and the thickness of the UWL can be controlled by agitation, and thus it can be adjusted to imitate physiological conditions. However, from a drug design point of view, it is important to understand which molecular property controls which type of permeability to be able to modify the structure accordingly in order to achieve the desired physiological behavior. The Abraham solvation equation can provide a better understanding of the molecular properties that influence compound's permeability. On the basis of the coefficients of the molecular descriptors in the Abraham solvation equation, the cell permeability is slower for compounds with H-bond acidity and basicity and dipolarity/polarizability. When a complex cell permeability process was studied, it was found that polar properties, such as H-bond acidity and basicity as well as dipolarity/polarizability reduce the cell permeability. Interestingly, the coefficient for the size of the compounds was small, almost negligible. This may be because an optimum size is necessary for good permeability. Both small and polar or large and lipophilic compounds have reduced the permeability. When the intrinsic membrane permeability was studied by the solvation equation approach, the size parameter was found to be significant. Larger molecules show higher intrinsic membrane permeability. In this respect, the solvation equation was found to be similar to an alkane/water partition equation.

In conclusion, the compound's permeability through cell or artificial membrane is a complex procedure that depends on the conditions, such as the composition of the membrane, membrane surface area, agitation, volumes of the donor and acceptor compartments, and pH of the compartments and other constituents that may bind the compound. Therefore, it is important to know the physiological conditions, where the permeability is the rate-limiting factor for the drug molecule to reach the pharmacological target.

QUESTIONS FOR REVIEW

Q7.1 What is the major constituent of biological membranes?

Q7.2 What are the two major reasons for permeability measurements in drug discovery?

Q7.3 Which are the major factors that influence the rate of permeation?

Q7.4 Explain the experimental setup for *in vitro* permeability measurements.

Q7.5 How is HPLC used for the permeability measurements?

Q7.6 How can we calculate the apparent permeability from effective permeability and the permeability through the UWL?

Q7.7 How does the UWL affect permeability?

Q7.8 What is "sink condition" and "double-sink condition?"

Q7.9 How can we determine the intrinsic permeability?

Q7.10 Which of the molecular descriptors reduce permeability?

Q7.11 How does lipophilicity influence permeability?

Q7.12 How does solubility influence permeability?

REFERENCES

1. Avdeef, A. (2003) Absorption and Drug Development, Solubility, Permeability and Charge State. Wiley-Interscience, Hoboken, NJ, pp. 116–246.
2. Yeagle, P. L., Ed., (2011) Structure of Biological Membranes. CRC Press, Boca Raton.
3. Sten-Knudsen, O. (2002) Biological Membranes: Theory of Transport, Potentials and Electric Impulses. Cambridge University Press, Cambridge, UK.
4. Camenisch, G., Folkers, G., Van de Waterbeemd, H. (1996) Review of theoretical passive drug absorption models: Historical background, recent developments and limitations. *Pharmaceutica Acta Helvetiae*, *71*, 309–327.
5. Andersen, O. S., Koeppe, R. E. (2007) Bilayer thickness and membrane protein function: An energetic perspective. *Annual Review of Biophysics and Biomolecular Structure*, *36*, 107–130.
6. Mälkiä, A., Murtomäki, L., Urtti, A., Kontturi, K. (2004) Drug permeation in biomembranes *in vitro* and *in silico* prediction and influence of physicochemical properties. *European Journal of Pharmaceutical Sciences*, *23*, 13–47.

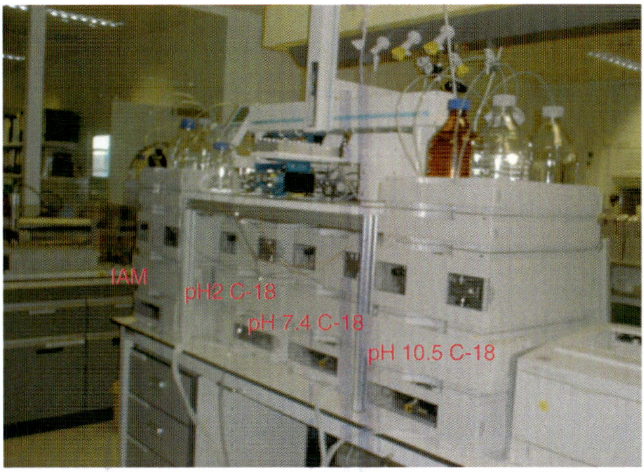

Figure 5.20 An example of a four-way Agilent HPLC system that has a CTC autosampler injecting the same sample into four columns; one computer controls the four HPLC gradient pumps and collects data from four detectors.

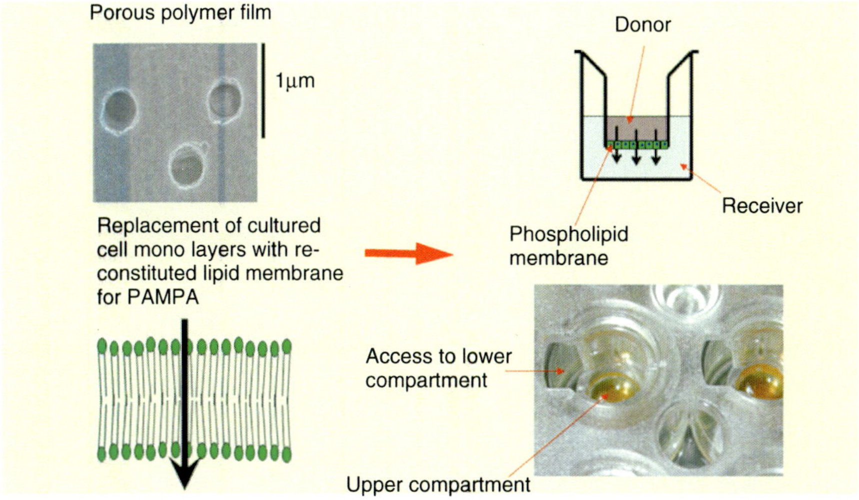

Figure 7.5 The experimental setup for high throughput artificial membrane permeability measurements.

Physicochemical and Biomimetic Properties in Drug Discovery: Chromatographic Techniques for Lead Optimization, First Edition. Klara Valko.

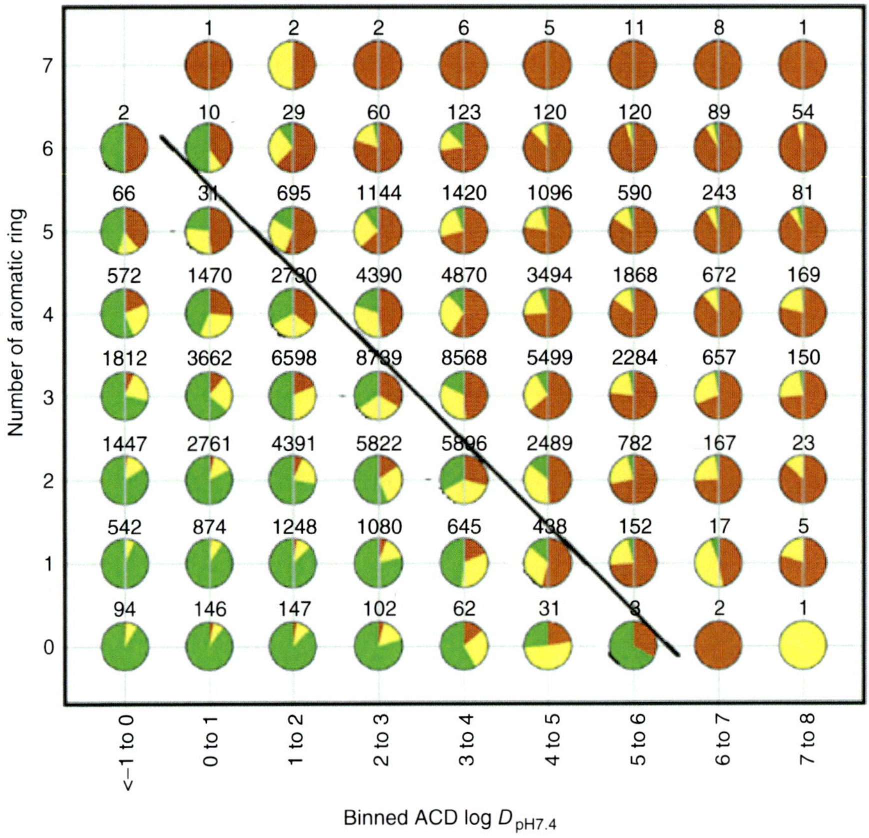

Figure 12.15 The distribution of low (red), medium (yellow), and high (green) solubility compounds with various number of aromatic rings and in various bins of calculated ACD log $D_{pH7.4}$. Compounds with the same lipophilicity bin are less soluble when they had more aromatic rings. The line shows the value when SFI is around 6, where approximately 50% of the compounds are insoluble. The numbers above the circle shows the number of compounds falling into that particular bin. (Adopted from Reference [40] with permission.)

7. Yee, S. (1997) *In vitro* permeability across CaCo-2 cells (colonic) can predict *in vivo* (small intestinal) absorption in man–Fact or myth. *Pharmaceutical Research, 14(6)*, 763–766.

8. Irvine, J. D., Takahashi, L., Llockhart, K., Cheong, J., Tolan, J. W., Selick, H. E., Grove, J. R. (1999) MDCK (Madin-Darby canine kidney) cells: A tool for membrane permeability screening. *Journal of Pharmaceutical Sciences*, *88*, 28–33.

9. Avdeef, A. (2003) Absorption and Drug Development, Solubility, Permeability and Charge State. Wiley-Interscience, Hoboken, NJ, USA, 118–124.

10. Singer, S. J., Nicolson, G. L. (1972) The fluid mosaic model of the structure of cell membranes. *Science*, *175*, 720–731.

11. Singer, S. J. (1974) The molecular organization of membranes. *Annual Reviews of Biochemistry*, *43*, 805–834.

12. Mueller, P., Rudin, D. O., Tien, H. T., Westcott, W. C. (1962) Reconstruction of cell membrane structure *in vitro* and its transformation into an excitable system. *Nature*, *194*, 979–980.

13. Kansy, M., Senner, F., Gubernator, K. (1998) Physicochemical high throughput screening: Parallel artificial membrane permeability assay in the description of passive absorption processes. *Journal of Medicinal Chemistry*, *41*, 1007–1010.

14. Volpe, D. A. (2008) Variability in CaCo-2 and MDCK cell-based intestinal permeability assays. *Journal of Pharmaceutical Sciences*, *97*, 712–725.

15. Pohl, P., Saparov, S. M., Antonenko, Y. N. (1998) The size of the unstirred water layer as a function of the solute diffusion coefficient. *Biophysics Journal*, *75*, 1403–1409.

16. Avdeef, A. (2003) Unstirred water layer effect (Transport across barriers in series and in parallel). In: Absorption and Drug Development. Wiley-Interscience, Hoboken, NJ, pp. 199–210.

17. Shanker, L. S., Tocco, D. J., Brodie, B. B., Hogben, C. A. M. (1958) Absorption of drugs from the rat small intestine. *Journal of American Chemical Society*, *123*, 81–88.

18. Gutknecht, J., Tosteson, D. C. (1973) Diffusion of weak acids across lipid bilayer membranes: effects of chemical reactions in the unstirred layer. *Science*, *182(4118)*, 1258–1261.

19. Gutknecht, J., Walter, A. (1981) Histamine, theophilline and tryptamine transport through lipid bilayer membranes. *Biochimica et Biophysica Acta*, *649*, 149–154.

20. Amidon, G. L., Lennernas, H., Shah, V. P., Crison, J. R. (1995) A theoretical basis for a Biopharmaceutic Drug Classification: The correlation of *in vitro* drug product dissolution and *in vivo* bioavailability. *Pharmaceutical Research*, *12*, 413–419.

21. Dahan, A., Miller, J. M., Hoffman, A., Amidon, G. E., Amidon, G. L. (2010) The solubility–permeability interplay in using cyclodextrins as pharmaceutical solubilizers: Mechanistic modeling and application to progesterone. *Journal of Pharmaceutical Sciences*, *99*, 2739–2749.

22. Balimane, P. V., Han, Y.-H., Chong, S. (2006) Current industrial practices of assessing permeability and P-glycoprotein interaction. *The AAPS Journal*, *8*, Article 1 E1-E13 (http://www.aapsj.org).

23. Halestrap, A. P., Meredith, D. (2004) The SLC16 gene family–from monocarboxylate transporters (MCTs) to aromatic amino acid transporters and beyond. *European Journal of Physiology*, *447*, 619–628.

24. Bermejo, M., Avdeef, A., Ruiz, A., Nalda, R., Ruell, J. A., Tsinman, O., Gonzalez, I., Fernandez, C., Sanchez, G., Garrigues, T. M., Merino, V. (2004) PAMPA–a drug absorption *in vitro* model: 7. Comparing rat in situ, CaCo-2, and PAMA permeability of fluoroquinolones. *European Journal of Pharmaceutical Sciences*, *21*, 429–441.

25. Wohnsland, F., Faller, B. (2001) High-throughput permeability pH profile and high-throughput alkane/water logP with artificial membranes. *Journal of Medicinal Chemistry*, *44*, 923–930.

26. Sugano, K., Hamada, H., Machida, M., Ushio, H. (2001) High throughput prediction of oral absorption: Improvement of the composition of the lipid solution used in parallel artificial membrane. *Journal of Biomolecular Screening*, *6*, 189.

27. Zhu, C., Jiang, L. Chen, T. M., Hwanq, K. K. (2002) A comparative study of artificial membrane permeability assay for high throughput profiling of drug absorption potential. *European Journal of Medicinal Chemistry*, *37*, 399–407.

28. Di, L., Kerns, E. H., Fan, K., McConnell, O. J., Carter, G. T. (2003) High throughput permeability assay to predict oral absorption. *European Journal of Medicinal Chemistry*, *38*, 223–232.

29. Ottaviani, G., Martel, S. Carrupt, P.-A. (2006) Parallel artificial membrane permeability assay: A new membrane for the fast prediction of passive human skin permeability. *Journal of Medicinal Chemistry*, *49*, 3948–3954.

30. Faller, B. (2008) Artificial membrane assays to assess permeability. *Current Drug Metabolism*, *9*, 886–892.

31. Draper, E. (2007) Prediction of diffusion using the Abraham descriptors. Presentation at Physchem Forum (http://physchem.org.uk/symp03/symp03_ed.pdf). Accessed 2013 Aug 13.

32. Faller, B., Grimm, H. P., Loeuillet-Ritzler, F., Arnold, S., Briand, X. (2005) High-throughput lipophilicity measurements with immobilized artificial membranes. *Journal of Medicinal Chemistry*, *48*, 2571–2576.

33. Huque, F. T., Box, K., Platts, J. A., Comer, J. (2004) Permeability through DOPC/dodecane membranes: measurement and LFER modelling. *European Journal of Pharmaceutical Sciences*, *23(3)*, 223–232.

34. Genty, M., Gonzales, G., Clere, C., Desangle-Gouty, V., Legendre, J.-Y. (2001) Determination of the passive absorption through the rat intestine using chromatographic indices and molar volume. *European Journal of Pharmaceutical Sciences*, *12*, 223–229.

35. Kotecha, J., Shah, S., Rathod, I., Subbaiah, G. (2007) Relationship between immobilized artificial membrane chromatographic retention and human oral absorption of structurally diverse drugs. *International Journal of Pharmaceutics*, *333*, 127–135.

36. Chan, E. C. Y., Tan, W. L., Ho, P. C., Fang, L. J. (2005) Modeling Caco-2 permeability of drugs using immobilized artificial membrane chromatography and physicochemical descriptors. *Journal of Chromatography A*, *1072*, 159–168.

37. Salminen, T., Pulli, A., Taskinen, J. (1997) Relationship between immobilised artificial membrane chromatographic retention and brain penetration of structurally diverse drugs. *Journal of Pharmaceutical and Biomedical Analysis*, *15*, 469–477.

38. Yoon, C. H., Kim, S. J., Soo, B., Choon, K., Yoo, S. D. (2006) Rapid screening of blood-brain barrier penetration using the immobilized artificial membrane phosphatidylcholine column chromatography. *Journal of Biomolecular Screening*, *11*, 13–20.

39. Ducarme, A., Neuwels, M., Goldstein, S., Massingham, R. (1998) IAM retention and blood brain barrier penetration. *European Journal of Medicinal Chemistry*, *33*, 215–223.
40. Stewart, B. H., Chan, H. (1998) Use of immobilized artificial membrane chromatography for drug transport applications. *Journal of Pharmaceutical Sciences*, *87*, 1471–1478.
41. Giaginis, C., Tsantili-Kakoulidou, A. (2008) Alternative measures of lipophilicity: From octanol-water partitioning to IAM retention. *Journal of Pharmaceutical Sciences*, *97*, 2984–3004.
42. Lazaro, E., Rafols, C., Abraham, M. H., Roses, M. (2006) Chromatographic estimation of drug disposition properties by means of immobilized artificial membranes (IAM) and C18 columns. *Journal of Medicinal Chemistry*, *49*, 4861–4870.
43. Meng, Q. C., Johansson, J. S., Eckenhoff, R. G. (2002) Chromatographic approach for determining the relative membrane permeability of drugs. *Journal of Chromatography B*, *774*, 89–95.

8

MOLECULAR PHYSICOCHEMICAL PROPERTIES THAT INFLUENCE ABSORPTION AND DISTRIBUTION—ACID DISSOCIATION CONSTANT—pK_a

DEFINITION OF pK_a

The presence of charge on the molecules dramatically influences many of their physicochemical properties, such as lipophilicity, solubility, and permeability. The presence of charge depends on the acid dissociation constant of the ionizable groups and the pH of the solution/environment. The pH is defined as the negative logarithm of the proton or, more precisely, the hydronium ion concentration in aqueous solutions. The product of the concentrations of hydronium and hydroxide ions in water is constant (10^{-14}); thus, the pH normally ranges from 1 to 14. The acid dissociation constant, or pK_a, is defined as the pH where an ionizable group is 50% in ionized form. In other words, the acid dissociation constant, K_a, is the equilibrium constant for the reaction in which a weak acid is in equilibrium with its conjugate base in aqueous solution. For example, for acetic acid, the following equilibrium takes place:

$$CH_3COOH + H_2O \rightleftharpoons CH_3COO^- + H_3O^+ \qquad (8.1)$$

$$K_a = \frac{[CH_3COO^-][H_3O^+]}{[CH_3COOH]} \qquad (8.2)$$

Physicochemical and Biomimetic Properties in Drug Discovery: Chromatographic Techniques for Lead Optimization, First Edition. Klara Valko.

When the acetate ion concentration is equal to the acetic acid concentration, K_a equals the H_3O^+ concentration. The negative logarithm of the H_3O^+ concentration is the pH. The smaller the value of pK_a, the stronger is the acid. For basic compounds, Equation 8.3 and Equation 8.4 can be used.

$$BH^+ + H_2O = B + H_3O^+ \tag{8.3}$$

$$K_a = \frac{[B][H_3O^+]}{[BH^+]} \tag{8.4}$$

Again, the negative logarithm of K_a equals the pH of the aqueous environment, where 50% of the basic group is in a protonated charged form, while 50% is in a neutral, unionized form. The percentage of the ionized molecules depends on the proton concentration (pH) and can be calculated at any pH using the Henderson–Hasselbalch equation [1]. Equation 8.5 describes the relationship between the percentage of ionized molecules and pH for a given pK_a.

$$\%\ \text{ionized form} = \frac{100}{1 + 10^{\text{charge}(\text{pH}-\text{p}K_a)}} \tag{8.5}$$

In Equation 8.5, the charge is −1 for acid and 1 for basic compounds, and it is assumed that only one charge is present on the molecule. Figure 8.1 shows a plot of the percentage of ionized species at a given pH for an acid with pK_a = 8, while Figure 8.2 shows the same plot for a base with pK_a = 8. It can be seen that an acid with pK_a = 8 is a weak acid, and only around 20% of the molecules have negative charge at physiological pH, whereas a base with pK_a = 8 can be considered as relatively strong because 80% of the molecules have a positive charge at pH 7.4. The physiological pH is taken as the pH of human plasma under healthy conditions, which is pH 7.4. A wide pH range can be found in the gastrointestinal tract. For example, the stomach is very acidic, where the pH can be as low as 1.2, and in the small intestine, the pH can reach above 7.

It is important to realize the difference between the H-bond acidity and basicity and the actual acid/base dissociation constants. While the H-bond acidity measures how strongly the proton is donated to take part in hydrogen bond interaction with an H-bond acceptor compound, dissociation of the proton and formation of a charge do not take place in hydrogen bonding. The energy of an H-bond is around 8 kcal/mol, while a charge–charge interaction energy is orders of magnitude higher. An ionized molecule has very strong affinity toward water, which has a high dielectric constant ($\epsilon = 73$), while nonpolar solvents have dielectric constant values close to zero. The presence of charge and ionization has a dramatic effect on the physicochemical properties of the compounds, too. For example, lipophilicity can drop by three orders of magnitude for one charge, while the aqueous solubility can increase by three or more orders of magnitude. Therefore, the charge state of the molecule is very important when we assess other physicochemical and biological properties. Usually, the intrinsic values of lipophilicity, solubility, and permeability refer to the uncharged species. The charge–charge interactions can be very influential in

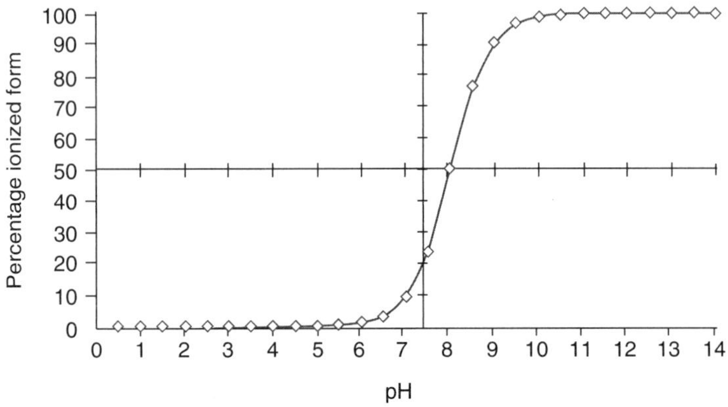

Figure 8.1 The pH dependence of the percentage of ionized monoprotic acid with pK_a = 8.

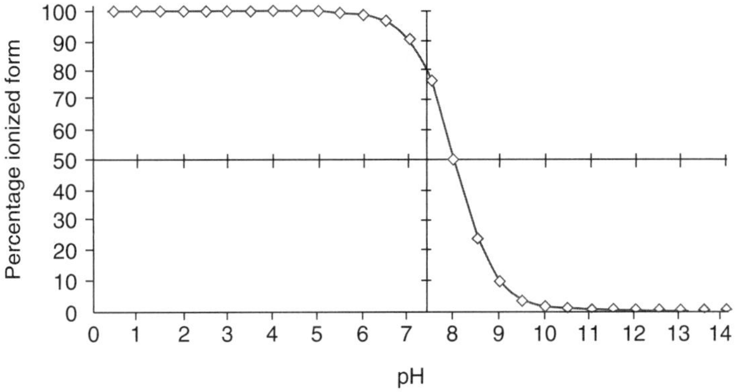

Figure 8.2 The pH dependence of a monoprotic base with pK_a = 8.

the drug–receptor binding as well. Several amino acids contain positive or negative charges at physiological pH values that may form the binding site of the target protein. Therefore, the knowledge of the pK_a values of ionizable groups can be important for the estimation of the drug–target interaction. It is also important whether the charge is negative or positive as it hugely influences the drug distribution between the plasma and the tissue compartments in the body as is discussed later in more detail. Changing the acid/base character and the percentage ionized form of the drug molecule provides an excellent tool for medicinal chemists to tune the compound's distribution in the body. Before the experimental descriptions of the measurement of pK_a values, it is important to emphasize that the definition of pK_a refers to an aqueous environment. When the water is mixed with organic solvents or the compound is near the membrane or proteins, its pK_a shifts depending on the H-bond donor, acceptor property of the solvent, and its dielectric constant,

Table 8.1 pK_a values in methanol/water mixtures for simple basic and acidic compounds[a]

Compound	0% MeOH	20% MeOH	40% MeOH	60% MeOH	80% MeOH
Aniline	4.62	4.46	4.32	4.17	4.07
Methyl aniline	4.85	4.70	4.48	4.18	3.9
Dimethyl aniline	5.15	4.96	4.73	4.31	3.82
Acetic acid	4.76	5.01	5.33	5.81	6.5
Benzoic acid	4.20	4.51	4.97	5.54	6.29

[a]Reference [2]

dipolarity/polarizability. The Born equation describes the changes in the pK_a values (ΔpK_a) of the compounds due to the electrostatic effects of the solvents [2], as is shown by Equation 8.6.

$$\Delta pK_a = \frac{121.6\,n}{r} \times \left(\frac{1}{\epsilon} - 0.0218\right) \tag{8.6}$$

In Equation 8.6, n is the number of charged species in the ionization process, r is the common radius of all the ions, and ϵ is the dielectric constant of the solvent. Typically, when the water is mixed with organic solvents or proteins and membranes are in the aqueous compartment, the dielectric constant, H-bond acidity, and basicity of the surrounding are decreased relative to the net water that results in the weakening of both acids and bases. The pK_a of an acid increases and the pK_a of a base decreases, which means that ionization process is suppressed. Table 8.1 shows examples of the pK_a shift of simple basic and acidic compounds in increasing methanol concentration [2].

Another factor that affects the pK_a value is the ionic strength of the solution. The effect of ionic strength on the acid dissociation constant can be calculated using the Debye–Hückel equation [3]. The concentration of the compounds in ionic solutions cannot be considered as ideal even with low ionic strength. The concentration should be multiplied with the so-called activity coefficient (γ) that describes more precisely the activity of various species of molecules in solution, and the activity values should be used instead of concentration when describing chemical equilibrium constants. Thus, the pK_a values of the compounds differ in solutions of varying ionic strength, where the activity constant depends on the concentration of the other ions in solution. The activity coefficient takes into account the interaction energy of ions in solutions. Typically, both acids and bases get stronger at higher ionic strength. To be able to compare the pK_a values of various molecules, it is extrapolated to zero ionic strength. Table 8.2 and Table 8.3 contain pK_a values of the most common acidic and basic groups.

Typical acidic groups found in organic molecules are the carboxyl group, phenolic hydroxyl group, enolic hydroxyl group, and sulfonamide groups, as shown in Figure 8.3. Tetrazole, also shown in Figure 8.3, is an often used carboxyl group isostere containing one acidic proton on the nitrogen that has a very similar pK_a

Table 8.2 The pK_a values of simple acids

Substance	pK_a
Acetic acid	4.76
Chloroacetic acid	2.85
Trichloroacetic acid	0.70
Phenol	9.89
4-Chlorophenol	9.18
2,4-Dinitrophenol	3.96
Tetrazole	4.86

Table 8.3 The pK_a values of simple basic compounds

Substance	pK_a
Pyridine	5.25
2,3-Dimethyl-pyridine	6.57
Piperidine	11.12
Piperazine	9.83; 5.7
Cylohexylamine	10.66
Ethylamine	10.70
Aniline	4.63

value to the carboxyl group. Figure 8.4 shows the typical basic functional groups. The primary, secondary, and tertiary amines are the most common basic groups. Secondary amines, in general, are stronger bases with higher pK_a values than the tertiary amines, while the primer amines are weaker bases than the tertiary amines. When the nitrogen is in an aromatic ring, such as in pyridine, its basicity is weakened. Acids get stronger when a strong electron-withdrawing group is nearby, while bases get weaker when electron-withdrawing substituent is close. It is difficult to estimate the pK_a or the acid/base character of complex organic molecules even for experienced physicochemists. Most of the *in silico* pK_a estimation programs are based on the assessment of the Hammett equations [4], set up using measured pK_a values in the presence and absence of substituent that can be compiled in extensive databases. ACD Laboratories, Inc and Chemaxon Ltd offer *in silico* pK_a calculation software. Chemaxon's pK_a calculator is available free in the ChemSpider, a free chemical database (chemspider.com).

The Hammett equation describes the effect of a substituent on a chemical equilibrium constant, which, in the case of the pK_a calculation, is the acid dissociation constant. Hammett in the 1950s studied the effect of meta and para substituents on the dissociation constant of benzoic acid and observed that the same substituent has a similar effect on other equilibrium or reaction rate constants. The linear free energy relationship described by the Hammett equation is shown by Equation 8.7.

$$\log \frac{K}{K_0} = \sigma\rho \qquad (8.7)$$

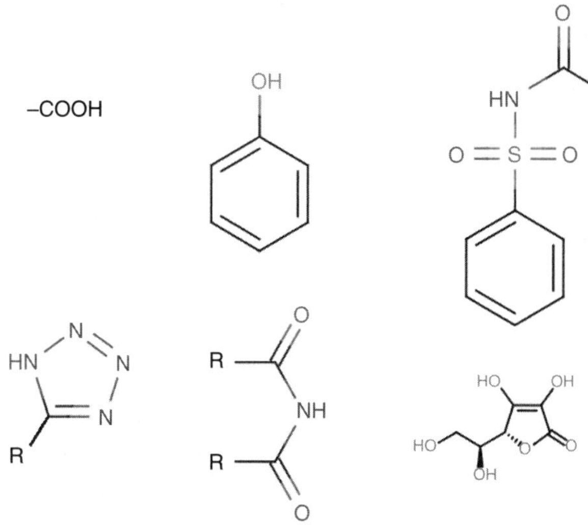

Figure 8.3 Structures of typical acidic functional groups (carboxyl, phenol, sulfonamide, tetrazole, imide, and enolic hydroxyl groups in ascorbic acid).

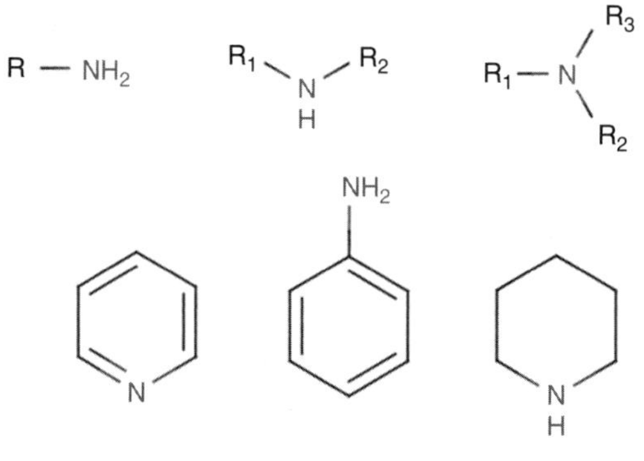

Figure 8.4 Structures of typical basic groups (aliphatic primary amine, secondary amine, tertiary amine, pyridine, aniline, and piperidine).

In the equation, K is the equilibrium constant of a given reaction with an R substituent, while K_0 is the equilibrium constant of the same reaction when the R substituent is hydrogen. The substituent constant, the Hammett σ, is characteristic of the substituent, while ρ is the characteristic constant for the reaction. Typical σ constants for several substituents on an aromatic ring are shown in Table 8.4.

The Hammett σ constants are widely used not only for estimating the acidity or basicity of compounds but also in quantitative structure–activity relationship (QSAR) studies.

Table 8.4 Hammett σ constants for some substituents on a phenyl ring in meta and para positions

Substituent	σ-para	σ-meta
Amine	−0.66	−0.16
Methoxy	−0.27	0.12
Ethoxy	−0.25	0.02
Dimethylamino	−0.21	−0.21
Methyl	−0.17	−0.07
Hydrogen (no substitution)	0	0
Fluoro	0.06	0.34
Chloro	0.23	0.37
Bromo	0.23	0.39
Iodo	0.28	0.35
Nitro	0.78	0.71
Cyano	0.66	0.56

In conclusion, the definition of acid dissociation constant, pK_a, is showing the pH at which 50% of an ionizable group is charged. However, drug molecules often contain more than one ionizable group. When a molecule contains both acidic and basic functional groups, it can be zwitterionic or amphoteric. Zwitterions contain charge at any pH, while amphoteric compounds may be uncharged at a certain pH range. Ordinary ampholytes are neutral at the pH equivalent to their isoelectric point.

METHODS FOR DETERMINING pK_a

Potentiometric Titration for the Determination of pK_a

The presence of charge on a molecule changes its characteristics dramatically. Therefore, the measurements of any properties that are dependent on the charge in a pH range can provide an experimental approach for the determination of pK_a. Lipophilicity, solubility, UV absorbance, NMR spectra, or electrophoretic mobility are all dependent on the charge state and they can be measured as a function of pH to obtain the pK_a. The most commonly used method, however, is the potentiometric titration, which is an acid/base titration using potentiometric endpoints. The titration is monitored with a precision pH electrode. The theoretical pH may be calculated at any point of the titration from the known amount of acid or base added to the solution. Any perturbation for this is caused by the ionization changes within the test molecule. It may be difficult to measure pK_a values when the compound has more than one ionizable group. Avdeef [5] has described in more detail the approaches one can take in these instances. By subtracting the titration curve in the presence of the substances from the curve obtained in the absence of the substance and rotating 90°, the so-called Bjerrum plot [6] can be obtained. The *y*-coordinate of the plot shows the number of protons that are bound to the test molecules as a function of pH. Sirius Analytical Inc (United Kingdom) has developed an instrument that contains the automatic titrator, the precision pH electrode, and software for the calculations and plots to derive the pK_a of the test compounds. The titration vessel is thermostated and covered with nitrogen or argon to insulate from the carbon dioxide present in the air as it can dissolve in the water and alter the pH. The pH electrode should be carefully standardized and calibrated before the measurements. The requirement of the test compound is 50–500 μM in 2–20 ml of water. It is common to use 0.1–0.2 M sodium or potassium chloride in order to keep the ionic strength constant, which makes the measurements more precise.

We can carry out the potentiometric titration without a dedicated instrument. A simpler way to evaluate the results of a potentiometric titration, especially when only one ionizable group is present on the molecule, is using an Excel spreadsheet. The commercial software developed by Gans et al. [7], which is called *HYPERQUAD 2008* (Protonic Software Inc.), uses the experimental data to fit the titration curve and calculates the pK_a.

We can also plot the volumes of the base added gradually during the titration of an acid as a function of the difference between the volumes of the base added (V_B) and the volumes of the base at the equivalence point (V_E) as a function of pH, as is described in Equation 8.8.

$$[H^+] \times V_B = K_a(V_E - V_B) \tag{8.8}$$

Plotting the left-hand side of the equation as a function of the difference of the added volume of the base and the added volume of the base at the equivalence point gives a straight line, which is called the *Gran's plot* [9]. The slope of the straight line provides the pK_a values [9]. Figure 8.5 shows a Gran's plot for a hypothetical compound with pK_a = 5. The pH and pK_a values are used in Equation 8.8.

A potential problem during the potentiometric determination of pK_a values may arise from low solubility of the test compound when it is unionized. In this case, the compound is usually soluble when we start the titration and it is in its fully ionized form.

Precipitation may be observed during the titration because of the decrease in ionization. The precipitation distorts the potentiometric curve and makes the calculation of the free proton concentration impossible. In order to avoid this problem, application of a cosolvent to enhance the solubility can be a solution. If solubility is an issue, the potentiometric titration can be carried out in the presence of an organic solvent such as methanol, ethanol, or DMSO, dioxane, and acetonitrile.

However, the presence of cosolvent alters the dielectric constant of the mixture that also alters the compound pK_a value according to the Born equation (Equation 8.6). The pK_a determination can be carried out using several concentrations of the cosolvent and based on the Yasuda–Shedlovsky procedure [8,10–13]; it is possible to extrapolate the pK_a value to 100% water.

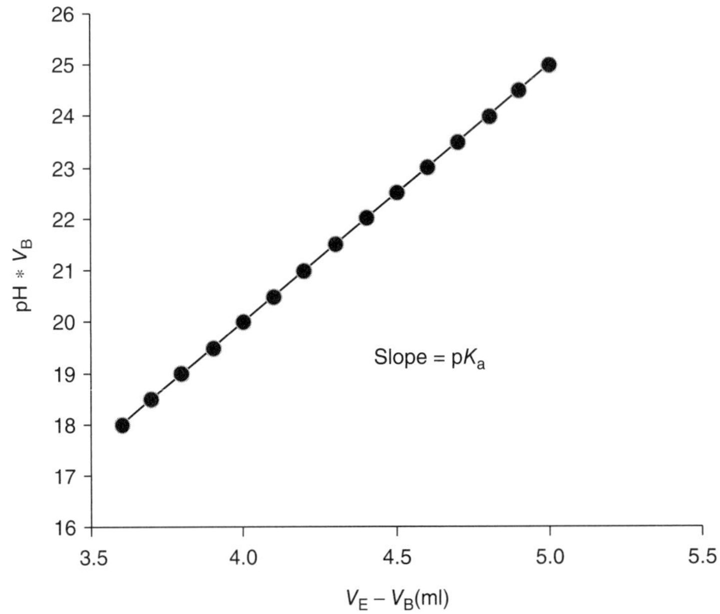

Figure 8.5 The Gran's plot of a hypothetical basic compound with pK_a = 5. The volume of the titration acid at equivalence point was 8 ml (V_E). The X-axis is the difference between the volume of the base (V_B) and V_E at each point of the titration, while the Y-axis is the pH × V_B. The slope of the obtained straight line gives the pK_a value of the base.

The pK_a measured in a range of water–solvent mixtures (pK_{a_s}) usually gives a straight line when plotting the concentration of the organic solvent and the obtained pK_a values. The slope of the straight line depends on the dielectric constant (ϵ) of the mixture, and the intercept provides the extrapolated pK_a for the pure water, as is shown by Equation 8.9.

$$\mathrm{p}K_{a_s} + \log[\mathrm{H_2O}] = \frac{a}{\epsilon} + b \tag{8.9}$$

Takács-Novák et al. [13] have demonstrated that the extrapolated pK_a values of 25 molecules using 15–65 wt% methanol were in agreement with the pK_a values obtained in pure water by spectroscopic methods. The solubility was a problem with the potentiometric method that normally requires a higher concentration than the spectroscopic method. A general rule of thumb is that both acids and bases become weaker in the presence of organic phase because of the decrease in the dielectric constant of the organic solvent–water mixtures. When there are multiple pK_a values in the molecule, especially when both acidic and basic pK_a values are present, the slope of the Yasuda–Shedlovsky plot helps identify which is the acidic pK_a and which is the basic pK_a. Acidic groups have positive values, while basic groups have negative slope values. However, there are exceptions to this rule. It may happen that some of the compounds with sulfonamide acidic functional group show a negative slope.

It is important to note that the effect of the dielectric constant of the solvent on the pK_a values is often forgotten when the ionization of a compound near the biological membranes is considered. For example, the dielectric constant of pure methanol is around 32. This value has been associated with the dielectric constant near the surface of phospholipid bilayers. The dielectric constant of acetonitrile is 36, very similar to methanol and the phospholipid bilayer surface. So, the pK_a values referring to water may shift near the membranes, and the extent of this shift may well be modeled by the pK_a shift in a methanol-rich environment. The extent of the pK_a shift near the membranes may vary for different functional groups. It also means that, for example, a pK_a order of discovery compounds obtained in pure water might be different from the rank order near the biological membranes. Therefore, the investigation of the Yasuda–Shedlovsky plot can be very useful to assess the ionization behavior of compounds in physiologically relevant conditions.

There is another question when using pH measurements in organic solvent and water mixtures. What is the meaning of the pH in an organic solvent–water mixture? The pH has been defined in water, having meaningful values between 1 and 14. When we mix the water with a significant amount of organic solvents, the proton concentration, and consequently the pH, is changing purely because of the "dilution" of the water and its hydronium concentration. How does this affect the constant product term of the proton and hydroxyl ion concentration? How do we calibrate the pH electrode? There are three possible procedures, which result in three different measured pH values (${}^{w}_{w}$pH, ${}^{s}_{w}$pH, and ${}^{s}_{s}$pH). The pH electrode can be calibrated in pure water (subscript w) and in the solvent/water mixture (subscript s), and the pH of the solution can be measured before (superscript w) and after (superscript s) mixing the buffer with the organic solvent. Ràfols et al. [14] have studied and compared the pH values obtained by the three methods. Canals et al. [15] and Espinosa et al. [16] studied the pH change by mixing the buffer with methanol and acetonitrile, respectively. The ${}^{s}_{w}$pH and the ${}^{s}_{s}$pH scale provides a closer measure of the pH in the organic solvent/water mixture, and they suggested the use of ${}^{s}_{w}$pH scale. Owing to its simplicity and the fact that it provides results closer to the true pH, we need to determine the pH after mixing the buffer with the organic solvent using the pH electrode that is calibrated in pure water. It is also important to keep a constant temperature during the measurements as the temperature has an effect on the equilibrium constant. Most of the pK_a measurements and the reported pK_a values refer to 25 °C.

In conclusion, the potentiometric determination of pK_a values is regarded as the "gold standard" for pK_a determination in water. It is based on an acid/base titration in the presence and the absence of the test substance. The 90° rotation of the differential titration plot, the Bjerrum plot, indicates how many protons have been associated or dissociated from the molecule at a particular pH. It provides a tool for the determination of multiple pK_a values on the same molecule. The limitation of the potentiometric titration is that it requires a relatively high concentration of test substance (20–500 μM) in a reasonable large volume, 2–20 ml. Potential drug molecules are often insoluble when they are in unionized forms; thus, precipitation might occur during titration. It is possible to use a cosolvent such as methanol, acetonitrile and DMSO solution. Völgyi et al. [17] suggested a special universal mixture (MDM, from MeOH, 1,4-Dioxan and MeCN) containing an equal amount of methanol, dioxane, and acetonitrile that dissolves a wide range of drug discovery compounds.

When a cosolvent is used, the pK_a values have to be determined using various concentrations of the organic solvents. Plotting the obtained pK_a values as a function of organic solvent concentration (Yasuda–Shedlovsky plot) usually results in a straight line for which the intercept provides the pK_a value in pure water. Acidic pK_a values usually increase with the increasing concentration of the cosolvent, while basic pK_a values decrease, which means both the acidity and the basicity of the molecules are weakened in cosolvent mixtures. The slope value depends on the functional group and the dielectric constant of the cosolvent mixture. The pK_a shift from pure water can be important near the biological membranes and protein solutions that have significant biological relevance. The suggested method for the pH determination in organic solvent–water mixtures is to calibrate the pH electrode with aqueous buffer and measure the pH in the buffer–cosolvent mixture.

SPECTROPHOTOMETRIC DETERMINATION OF pK_a

As mentioned previously, we can determine pK_a by measuring the pH dependence of any property of the molecules that changes because of the ionization. When the ionizable group is close to a UV chromophore, the presence or absence of a proton on the ionizable group will shift the UV absorbance of the compound depending on the pH of the solution. Thus, monitoring of the UV spectrum at various pH values can be used for the measurement of pK_a. When 50% of the compound is in ionized from, the UV spectrum shows an inflection point. The first derivative of the pH profile of the UV spectra produces a maximum at the original inflection point. The pH where the maximum occurs is the pK_a of the given functional group that caused the spectral change. Figure 8.6 shows a schematic illustration of the spectral change during pH titration of a molecule with multiple ionizable groups. The accurate determination and detection of pK_a is easy when using the spectrophotometric method; however, it does not provide information about the group that is ionized, nor can it reveal whether the molecule lost or gained proton (charge) during the titration process. Therefore, an expert has to look up the structure and assign the pK_a values to the appropriate functional group. The expert has to assign the ionization center whether it is basic or acidic. This is often not very simple and *in silico* pK_a calculation software (for example, ACD Labs, Inc or Chemaxon Ltd.) can provide help in the assignment of the pK_a values. The *in silico* calculation uses various fragment methods and the Hammett equation to assess the electron-withdrawing or electron-donating effects of the neighboring substituents. The major advantage of the spectroscopic method is that it is very sensitive, and a low concentration

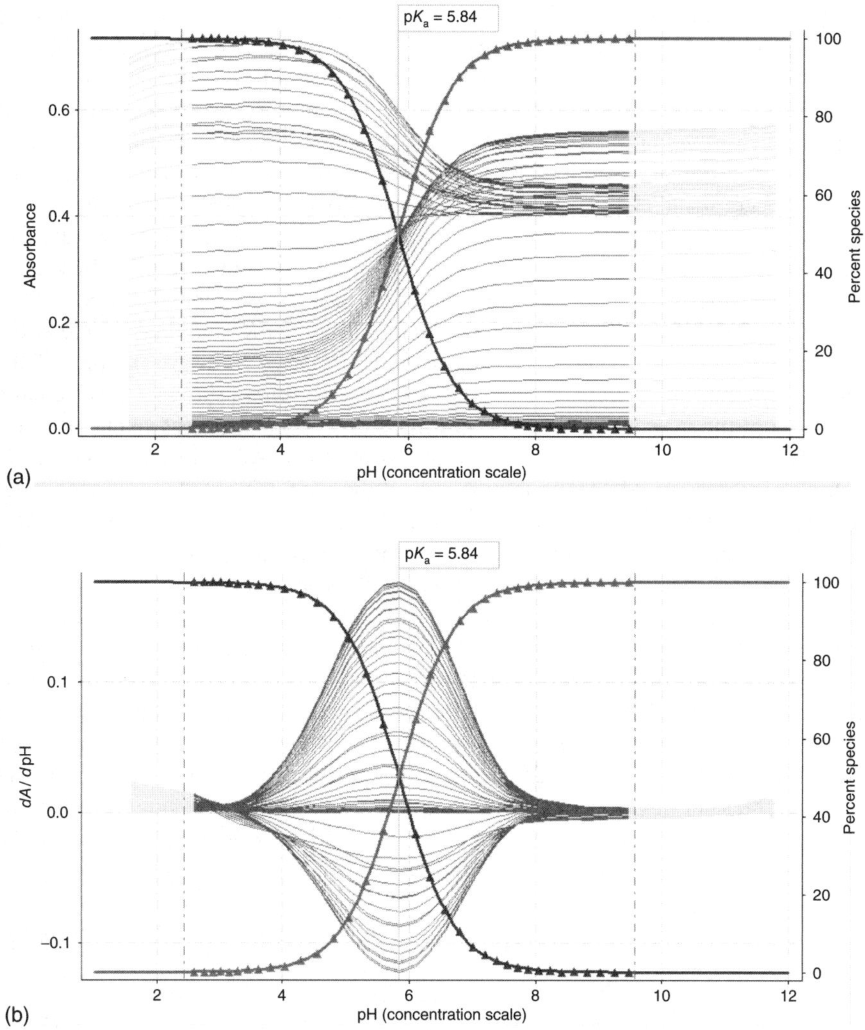

Figure 8.6 (a, b) The pH dependence of the UV spectra of a molecule with pK_a = 5.84 and its first derivative plot (absorbance change/pH change). (Figure is courtesy of Iain Reid obtained from Sirius T3 pK_a measurement.)

of sample is needed. This might prevent the precipitation at pH values where the molecule has the least amount of charge.

Sirius Analytical Inc, United Kingdom, has developed an automated instrument (Sirius T3) that is able to perform both the potentiometric and spectroscopic pK_a determination from small amounts of samples using the so-called dip-probe UV sensor and the automated titrator. The instrument provides all the software tools that are needed for the calculation of the measured pK_a. Sirius Analytical Ltd. has also developed a high throughput instrument for spectroscopic pK_a determination [18]. The principle is to use a linear pH gradient by continuously changing the volumes of the acid and base buffers using an HPLC gradient controller and introducing the sample mixture into the pH gradient. The whole mixture then goes into a diode array detector commonly used in HPLC instrumentation. The spectrum is continuously monitored, and the software evaluates the inflection points using a timescale of the pH gradient. The pH change is so fast that pH electrodes cannot be used for *in situ* pH measurements as they need time to settle. Therefore, the buffers are formulated in such a way that they produce a linear pH gradient with time. Instead of the pH measurements, samples with known pK_a values are applied for calibration and conversion of the timescale to a pH scale. From the pK_a and time profile of the calibration set of compounds, the timescale can be used to derive the pK_a of the test compound. The method is called *pH gradient titration* [19,20]. The schematic principle of the instrument is shown in Figure 8.7. The instrument can be used for 10 – 50 μl of 10 mM DMSO sample solutions in 96-well plate formats. For poorly soluble samples, the MDM (equal volumes of dioxane, methanol, and acetonitrile) universal solvent [17] can be added to the buffer mixture. The instrument can produce a pH gradient in 120 s, and it is usually repeated from pH 2 to pH 12 and pH 12 to pH 2. Practically, the pK_a of a test compound can be measured in 5 min. The very same principle has been used by Chiang et al. [21]. They used a complete HPLC system with two pumps that mix the acidic and basic pH buffers; the autosampler injected the sample solution, and without a column, the solvent mixture has been introduced to an HPLC diode array detector. The absorbance has been monitored at various wavelengths as a function of pH. The acidic buffer of pH 1.9 contained acetate (0.025 M), borate (0.025 M), and phosphate (0.025–0.03 M), and NaCl to increase the ionic strength. The alkaline buffer used was 0.05 M sodium hydroxide with NaCl to adjust the ionic strength. They used Equation 8.10 to calculate the pK_a of an acid and Equation 8.11 for a base.

$$\mathrm{p}K_\mathrm{a} = \mathrm{pH} + \log\left(\left|\frac{A_\mathrm{ion} - A_\mathrm{obs}}{A_\mathrm{obs} - A_\mathrm{m}}\right|\right) \tag{8.10}$$

$$\mathrm{p}K_\mathrm{a} = \mathrm{pH} + \log\left(\left|\frac{A_\mathrm{obs} - A_\mathrm{m}}{A_\mathrm{ion} - A_\mathrm{obs}}\right|\right) \tag{8.11}$$

In Equation 8.10 and Equation 8.11, A_{ion} stands for the absorbance of the ionized species, A_m for absorbance of the unionized species, and A_{obs} for the absorbance observed at a given pH.

The pK_a values obtained by spectroscopic methods have to be assigned by an expert to a particular functional group, which sometimes is more time consuming than the measurement.

In conclusion, the UV spectroscopic method provides an advantage over the potentiometric method, in that it requires smaller amount of sample; therefore, solubility is less likely to disturb the measurement. In general, the method is faster than a potentiometric titration that requires proper equilibration of the pH electrode. However, it can be used only when the compounds have UV absorbance and

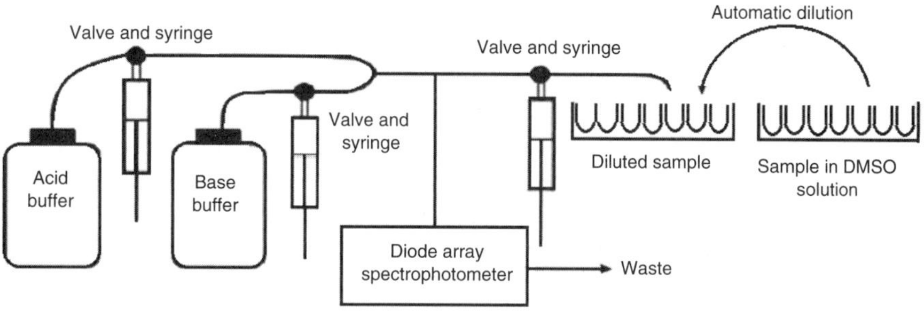

Figure 8.7 The schematic principle of the Sirius SGA high throughput pK_a instrument. (Figure is courtesy of K. Box, Sirius Analytical, United Kingdom.)

the UV spectrum changes because of the ionization state of the molecule. When the ionization occurs far from the UV chromophore, the UV spectrum does not alter by altering the pH, and the method is not suitable for pK_a determination.

Similarly, other spectroscopic methods, such as NMR, can be used for pK_a determination. During proton NMR spectroscopy, it is possible to measure the concentration of the protonated molecules from the chemical shifts at several pH values and obtain titration curves that can be used for the evaluation of pK_a. An example of using NMR spectroscopy has been published by Szakács et al. [22] for monitoring the pH titration of a new signal transduction inhibitor drug (Gleevec) for the determination of site-specific basicity. Grycová et al. [23] reported a detailed NMR method for the determination of antimalarial indoloquinoline compounds. The NMR spectroscopic method is even more useful for the determination of pK_a values of peptides and proteins, where multiple pK_a values are present on a single molecule. The NMR spectroscopy provides information about the site of the protonation as well, and an example is demonstrated by Song et al. [24]. They determined 31 pK_a values out of 32 in turkey ovomucoid third domain. Croke et al. [25] determined 26 pK_a values in α-synuclein, all below 7 by NMR, which would not be possible using any other method. The work has clearly demonstrated the shift in pK_a values near a sodium dodecyl sulfate micelle, too.

DETERMINATION OF pK_a BY CAPILLARY ELECTROPHORESIS

Capillary electrophoresis (CE) is another separation method that is often used for analytical purposes. The separation is based on the differential migration of charged molecules. The mobility of the charged analyte is induced by electric forces by the application of high voltage through the capillary, which is filled with electrolyte. As the capillary is usually made of quartz that is UV transparent, UV detection can be used. The migration time depends on the charge and the hydrodynamic radius of the compound. The migration time, which is very similar to the retention time in HPLC, can be related to the electroosmotic flow, which is basically the velocity of the ions in the capillary filled with aqueous solution. The migration time of the compound depends on the presence of charge and it can be measured at various pH values. A titration curve can be constructed by plotting the migration time of the compound as a function of pH. Then, using the traditional methods of analyzing the titration curve, we can read the pH value, where the compound is 50% ionized. This pH provides the pK_a value. The CE technique is mostly used for protein separation. Cleveland and Gluck [26–28] have investigated the applicability of the methodology for pK_a determination of small molecules. Although the CE technique is not widely used for the analysis of drug discovery compounds, it has great potential. The instrumentation is well developed, and several manufacturers offer computer-controlled sophisticated CE instruments (Applied Biosystem Inc., Bio-Rad, Hewlett-Packard, Wynsep, MegaBace from GE Healthcare Life Sciences, etc.). The CE technique does not require expensive columns like HPLC; capillaries can be cut to the desired length and are available in several diameters. The sample requirement for the measurements is very small; 20–100 μM material can be easily detected. The solvent consumption is minimal, and only small percentages of organic solvent are applied, which is environment friendly. It can be automated; moreover, multiple parallel measurements can be easily performed. It has the advantage that the pK_a measurement is based on migration time measurements at several pH values, so there is no need for concentration determination of the test compound. However, it is difficult to measure the actual pH in the capillary; the composition of various buffers to adjust the pH in a wide range has to be varied, which introduces variability in the electroosmotic flow. Cleveland et al. [26] introduced equations to handle potential discontinuities in solute mobility between different buffer solutions that may cause changes in the shape of the solute and the viscosity of the buffer, thus affecting the migration time. Gluck et al. [27] suggested applying an internal standard compound with known pK_a, which makes it possible to avoid the accurate pH determination when the test compound mobility is measured. As CE is a separation technique, several compounds can be analyzed in one mixture in one analysis. Gluck et al. [27] have also compared various calculation approaches and suggested that the nonlinear fit of the mobility versus pH curve provided the least biased results. Gluck et al. [28] further proposed the application of dynamically coated fused silica capillaries with positively charged polymer to make the electroosmotic flow stable and less dependent on pH. They evaluated the procedure for monoprotic acids, and also found that diprotic acids with pK_a difference less than 1 could not be accurately determined. Pharmaceutical companies have become interested in the methodology as the previously described methods for pK_a determination were difficult to automate and perform on large amounts of samples. Ishihama et al. [28–30] from Tsukuba Research Laboratories, Eisai Co, Japan, have published their significant efforts to improve the method and reduce the analysis time. First, they introduced miniaturization and checked the applicability of the method for compounds with multiple pK_a values. Later they applied pressure to accelerate the experiment, and diode array UV detector has been attached to the capillaries to detect the spectral changes of the compound as a function of pH. In this way, they obtained the mobility change of the compound as well as the UV spectrum change caused by the pH variations. The run time has become less than 1 min. They validated their methodology using 77 drug molecules with 82 pK_a values. Soon AstraZeneca [31,32] and Pfizer [33] set up fast pK_a measurements. Because of the low solubility of discovery compounds that would cause capillary blockage and would stop the automated measurements, Buckenmaier et al. [31] investigated the applicability of acetonitrile up to 60% for the analysis of basic compounds. Wan et al. [32] applied the pressure-assisted CE with mass spectrometric detection. They applied volatile buffers in the pH range 2.5–10.5, which increased the sensitivity of the MS detection and has made it possible to pool samples. With the described methodology, they could analyze 50 compounds in 150 min, and pK_a values of poorly soluble and non-UV active molecules could be determined. Shalaeva et al. [33] in Pfizer chose to use multiplexed capillaries connected with a detector equipped with multiple UV cells. In this way, they measured the electrophoretic mobility of one test compound at 24 different pH values simultaneously. They found very good interlaboratory and instrument-to-instrument reproducibility and could measure compounds with low solubility in methanol/aqueous buffer mixtures. Poole et al. [34] have published a detailed review of both the theory and experimental approaches of CE for the application of pK_a determination, which is strongly recommended for further reading.

These approaches have sorted out the experimental difficulties and have made it possible to measure pK_a for large numbers of early drug discovery compounds using the CE technique. The difficulties come after the pK_a measurements, when an expert has to assign the obtained values to the appropriate functional groups in each molecule. The task can be transferred to the chemists as they have a good understanding for the rational of introducing ionizable group to the molecule. The reasons for that can be to increase potency or increase solubility and reduce lipophilicity in terms of log D. Chemists also have a good knowledge of the electronic effects of neighboring substituents to the ionizable group too and they can assess their effect on pK_a.

CHROMATOGRAPHIC APPROACHES FOR THE DETERMINATION OF pK_a

As we have already discussed in Chapters 5 and 6, both lipophilicity and solubility depend on the pH and the ionization state of the molecules. The pH dependence of lipophilicity follows the Henderson–Hasselbalch equation and provides a sigmoidal curve with the pH. The pH at which the inflection point of the curve is gives the pK_a value. The reversed phase chromatographic retention factor is directly proportional to the compound's lipophilicity. As the dissociation is a much quicker process than the compound partitioning into the mobile phase, the partially charged compound's retention is the average lipophilicity of the ionized and unionized molecules present at a particular mobile phase pH. Horváth et al. [35,36] were the first to describe the relationship between the retention factor and the mobile phase pH in 1976–1977. Since then, the mobile phase pH has been used to modify the reversed phase retention and to develop and optimize separation methods

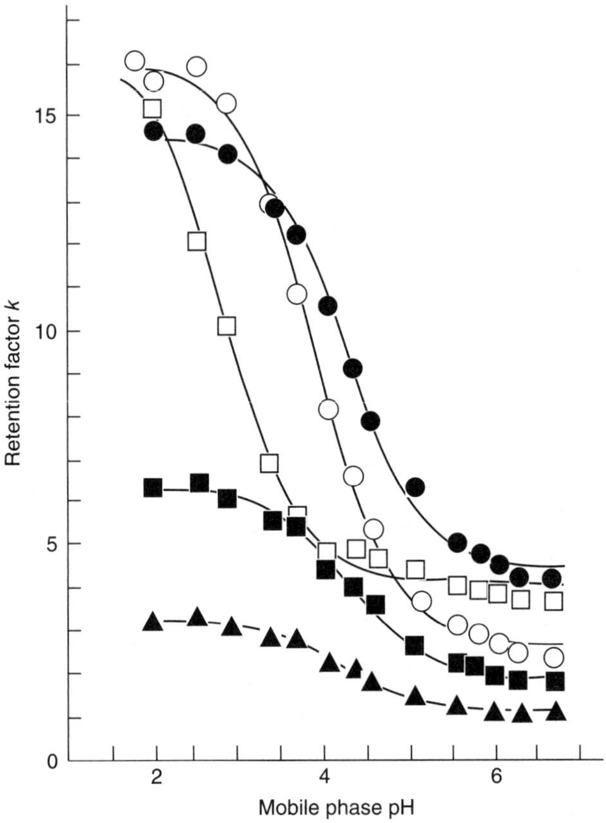

Figure 8.8 The pH dependence of reversed phase retention factors for clinically important catechols. (Adopted with permission from Reference [35].)

[37]. Figure 8.8 shows the pH dependence of retention of clinically important catecholamines [35].

Horváth et al. [36] presented the first mathematical description of the pH dependence of reversed phase retention of partially ionizable compounds. Equation 8.12 shows the relationship between pH and retention factor (k) using the dissociation constant (K_a) and retention factors of the fully unionized (k_u) and fully ionized (k_i) species for monoprotic acids.

$$k = \frac{k_u + k_i \frac{[H^+]}{K_a}}{1 + \frac{[H^+]}{K_a}} \tag{8.12}$$

As stated earlier, the dissociation constant refers to the compound pK_a in the mobile phase. In the past, the limitation for investigating a large pH range was the

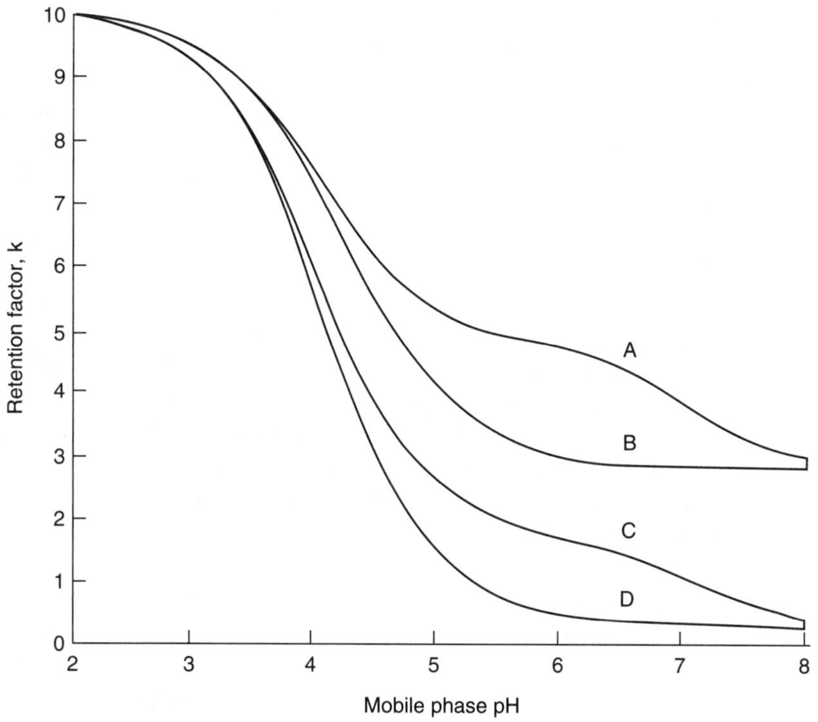

Figure 8.9 The pH dependence of reversed phase retention of several diprotic acids. (Reproduced from Reference [36] with permission.)

pH stability of the silica-based reversed phase stationary phase. The silica gel dissolved and the phase quickly decomposed when a higher than pH 8 mobile phase was used. The aim of the work of Horváth and his colleagues was to describe and understand the theoretical principles of the reversed phase retention; therefore, the method was not suggested for pK_a determination of compounds. Figure 8.9, which has been reproduced from Reference [36], shows the pH dependence of the retention of diprotic acids; it shows a clear resemblance to a titration curve, with two inflection points that would have been suitable to derive pK_a values. They used polar compounds in the experiments; therefore, there was no need to apply an organic modifier in the mobile phase that would have shifted the pH and pK_a values. Thus, the experimental data have provided a clear proof of the theory of retention of ionizable compounds in a reversed phase chromatographic system. It is important to remember that if we plot the logarithmic retention factor as a function of pH, the pK_a is not the pH value at the inflection point of the curve. It is the pH when the logarithmic retention factor is log 2 (0.301) less than the logarithmic retention factor of the unionized form of the molecule (see Figure 5.15 and Figure 5.16).

It took more than 20 years for scientists to start using the theory of reversed phase retention for actual determination of dissociation constants. First, the stationary phase development had to be improved so that a high pH mobile phase could be used. Second, the idea of measuring the HPLC retention factor of the same compound using various mobile phase pH values seemed far too tedious in comparison with a potentiometric titration. However, the emergence of computer-controlled accurate gradient pumps provided a simple tool to develop the pH gradient automatically. The third problem was the need for application of an organic modifier in the mobile phase to achieve reasonably short retention times. This causes shifts in pH and pK_a values, which again complicates the applicability of HPLC retention for pK_a determination.

Kaliszan et al. [38] have reported a very appealing solution for the simultaneous determination of lipophilicity and apparent pK_a using a few gradient reversed phase chromatographic measurements. The first few gradient runs should be carried out at the mobile phase pH where the compound is completely unionized, that is, high pH for basic compounds and low pH for acidic compounds using acetonitrile or methanol gradient. Preferably, a slow and a faster organic phase gradient should be applied. From the obtained gradient retention times, it is possible to calculate a theoretical retention factor for the compound when no organic modifier is present in the mobile phase ($\log k_w$) using an equation published by Snyder et al. [39]. The next step is to choose a suitable organic phase concentration that provides a reasonably long retention time of the unionized form of the molecule. Using the selected organic phase concentration, we then have to perform a pH gradient run starting from the pH, where the compound is unionized and changing the pH toward the complete ionization. The apparent pK_a value of the compound can be estimated from the retention time measured during the pH gradient. Wiczling et al. [40] have applied the method for 90 diverse drug molecules with acidic and basic functionalities. The obtained chromatographic pK_a values showed good correlation with the literature pK_a values with the correlation coefficient of above 0.89 and the root mean square error 0.65. In general, the pK_a values can be determined up to the second digit precision (this is the precision of a good pH meter). The error (0.65) of the method is still acceptable precision for drug discovery purposes. We have to remember that the authors have not suggested any corrections to the pK_a because of the presence of varying amount of organic solvents. The method can be considered as medium/low throughput as after the first two organic phase gradient runs (that can be the same for a wide range of compounds), the pH gradient run has to be adjusted for each compound. The pH range and the organic modifier concentration have to be selected on the basis of the expected pK_a and lipophilicity of the compound. The determination of the pK_a is more precise when the pH gradient is small and covers the expected pK_a of the compound. Depending on the lipophilicity of the uncharged molecule, the organic phase concentration has to be varied in order to get a well-measurable, not too long and not too short a retention time. The calculation of the results also has to be automated with extensive programming. Nevertheless, the three gradient retention times also provide a lipophilicity value of the compounds together with the dissociation constant. The determination of the pK_a values in the presence of multiple ionizable groups is not discussed.

Another problem with the application of pK_a values in drug discovery is that it is not a whole molecular property as it refers only to one particular functional group. It cannot be used on its own in any models of absorption or distribution. We have to derive the fraction-ionized form of the molecules at physiological pH values from the pK_a values to be able to use it as a physicochemical descriptor in models. For a drug discovery point of view, the important properties of the compound such as solubility and lipophilicity and distribution coefficient depend on the presence of a positive or a negative charge; so, we need to measure somehow the manifestation of the acid dissociation constant in properties of molecules. For such purposes, a simple generic gradient retention time, or its calibrated form, the CHI obtained at several pH values, would be sufficient. (see section "Lipophilicity Measurements by Reversed Phase Chromatography with Gradient Elution" in Chapter 5). It has been tested [41] that CHI values of acids and bases depend on the starting mobile phase pH. The CHI versus pH profiles resemble a titration curve, as shown in Figure 8.10. It has to be remembered that the CHI values are proportional to the logarithmic partition coefficients of the compounds in the mobile and stationary phases, and the increasing CHI value means increasing acetonitrile concentration in the mobile phase. It means that the more lipophilic a compound is, the higher the acetonitrile concentration that is needed for the elution. At higher acetonitrile concentration, the mobile phase pH is lower than the starting mobile phase pH. The pK_a also shifts higher for acids and lower for bases at high concentration of organic solvent. These conditions contribute to the distortion of the theoretical titration curves constructed from gradient retention time versus mobile phase pH. In order to get the CHI versus pH profile for a large number of compounds, the gradient retention time measurements have to be repeated at least 10 times using 10 different starting mobile phase pH values.

Thereafter, we need to make corrections to the pH change caused by the increasing acetonitrile concentration in the mobile phase. The CHI value of the compound indicates the volume percentage of acetonitrile when the compound elutes. Then the pK_a and pH shifts due to the presence of organic solvent have to be estimated. Instead of this, we have taken a pragmatic approach to measure the gradient retention times using only three different mobile phases with acidic, neutral, and basic pH values.

The acidic mobile phase was 0.01 M phosphoric acid or 0.01 M formic acid if volatile mobile phase has to be used because of the MS detection. This provided a starting mobile phase pH of approximately 2.6.

To assess the compound's lipophilicity at the blood pH (pH 7.4), a carefully adjusted 50 mM ammonium acetate buffer has been used, while for high mobile phase pH, the ammonium acetate buffer pH has been increased to pH 10.5 by adding significant amount of concentrated ammonia solution [42]. The buffers are compatible with both UV and MS detections.

It has already been discussed in Chapter 5 and demonstrated in Figure 5.17 that the mobile phase pH changes with increasing acetonitrile concentration during the short (2–3 min) gradient run. Fortunately, the neutral pH remains close to neutral, the acidic pH increases from pH 2.6 up to pH 5, while the basic pH decreases

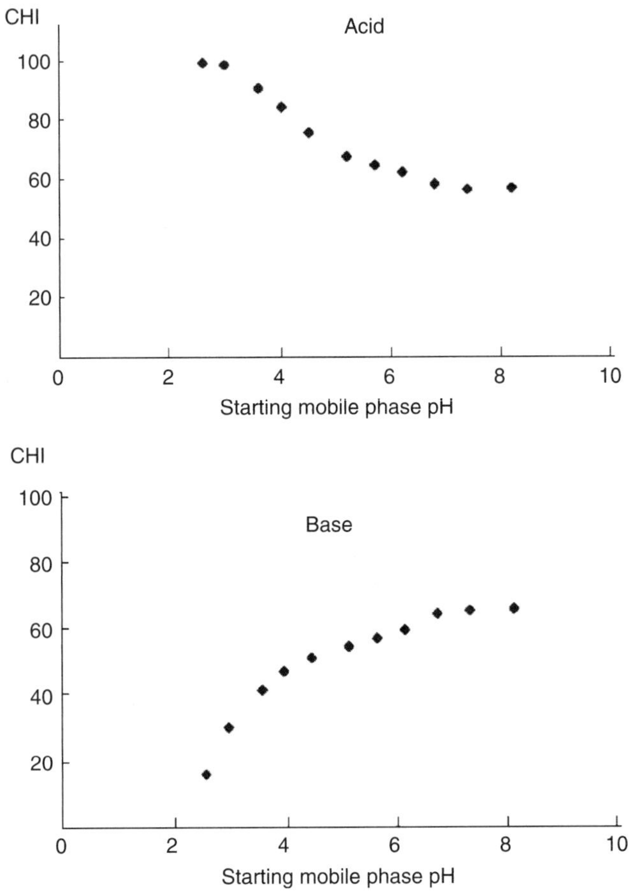

Figure 8.10 The pH dependence of the Chromatographic Hydrophobicity Index of an acidic and a basic compounds. (Reproduced with permission from Reference [41].)

to 8.3 when the acetonitrile composition reaches 90%. Interestingly, the pK_a shift for acids and bases makes them weaker; so, practically, at higher pH values, due to the higher concentration of the acetonitrile, the acid pK_a shifts to higher, while the base pK_a shifts lower. It means that we can assume that, for example, a basic compound with a pK_a 9, then at the starting mobile phase pH 10.5, is practically unionized, and it stays so when the acetonitrile concentration increases, regardless of whether the pH drops to 8.5. From the CHI measurements that are carried out at three pH values (using parallel HPLC systems as is described in Reference [43]), we cannot derive the pK_a values of the compounds, but we can group them confidently into five different groups, such as acidic, basic, neutral, weak acid, and weak base. These groups are automatically calculated from the differences at the three pH CHI values. Supposing that the accuracy of the CHI determination is within 5 CHI units, a larger difference than 5 obtained in any of the three pH values refers to the change in ionization of the molecule. It means that when the CHI values are the same (more precisely, they are within 5 units) at all the three pH values, then the compound is said to be neutral. If the CHI values obtained at pH 7.4 and 10.5 are the same but lower at the acidic pH, then the compound is a weak base and it is not ionized at physiological pH. If the acidic pH CHI is higher, then the compound is an acid and it has significant negative charge at physiological pH values. If the pH 2 and pH 7.4 CHI values are the same and the high pH CHI value of the compound is larger, then the compound is a base and it has a significant amount of positive charge at physiological pH. When the higher pH CHI is the lowest, then the compound is a weak acid, gets negatively charged only at high pH, and neutral at physiological pH. With a simple algorithm that compares the three pH CHI values, the discovery compounds can be grouped into classes and they can be identified as neutral or charge bearing at physiological pH. From the direction of the change in the CHI values, it is easy to identify whether the compound has positive or negative charges without looking up the structure. A typical example of the three pH CHI values in the five groups of the molecules is shown in Figure 5.19. We shall see in the following chapters how helpful this information can be in color coding the compounds based on their acid/base character. From the three gradient retention time measurements, we can derive the lipophilicity at physiological pH values, the lipophilicity of the neutral form of the molecule, and the charge state at physiological pH values, as well as an approximation of the percentage of the ionized form. The automatic grouping of the compounds, however, might be misleading. For example, zwitterionic compounds may have the same lipophilicity at all the three pH values, and in this respect they behave like neutral compounds. Their lipophilicity is practically independent of the pH in a wide range. Very strongly basic compounds (pK_a above 11 or quaternary ammonium salts) also look like neutral compounds as they are fully charged at all the investigated pH values. Very strong acids (pK_a less than 3) may be fully ionized too at all the three pH values and wrongly grouped as neutral. Besides these extreme conditions, the automatic assignment of the acid/base character based on the chromatographic lipophilicity measurements at the three pH values is quite robust. Obviously, the same is true when the octanol/water lipophilicity or the electrophoretic mobility of the compound is measured at these pH values. Another advantage of the determination of acid/base character via the chromatographic lipophilicity is that solubility is not a problem during the HPLC analysis, as the nanomolar amount of injected material is diluted down with the mobile phase. Table 8.5 shows the measured CHI values of some drug molecules published by Hollósy et al. [44]. The pK_a values have been calculated using the ChemSpider free database (chemspider.com) with the Chemaxon pK_a calculator plug-in.

The pK_a calculator is a useful tool for the assignment of the measured pK_a values to the functional groups. Figure 8.11 shows a few examples of the calculated pK_a values for known drug molecules.

In conclusion, the fast chromatographic lipophilicity determination of compounds provides valuable information about the acid/base character of drug discovery compounds, their ionization state at physiological pH, and the nature of the charge. The deltaCHI values, that is, the difference in lipophilicity obtained at

Table 8.5 The measured CHI values at acidic, neutral, and basic pH values for known drug molecules and their calculated pK_a values

Name	CHI pH 2.6	CHI pH 7.4	CHI pH 105	Acid/Base Group	Calculated pK_a Acidic	Calculated pK_a Basic
Acetaminophen	17.4	16.8	−30.6	Weak acid	9.46	
Acetanilide	42	41.2	41.8	Neutral		
Acetazolamide	39.9	33.3	21	Weak acid	10.35	
Alclofenac	72.9	39.4	36.2	Acidic	3.71	
Alosetron HCl	29	45.3	49	Weak base		3
Amantadine	65.9	60.1	61.2	Neutral		
Amiloride	12.5	18.6	15.7	Neutral		
Amoxapine	36.4	61.5	89.6	Basic		8.83
Amoxicillin	40.9	18.1	8.4	Acidic	3.23	
Ampicillin	33.6	31.5	27	Neutral		
Apomorphine HCl	22.9	49.5	78.6	Basic	9.3	7.8
Aspirin	56.7	19.7	14.9	Acidic	3.41	
Bamethan	14.6	24	42.4	Basic		9.96
Betamethasone	59.7	57.6	59.7	Neutral		
Bromazepam	44.1	54.2	56.3	Weak base		2.69
Bromocriptin	49.1	91.6	93	Weak base		6.4
Budesonide	77.2	75	76.6	Neutral		
Bumetanide	76	49.1	45.2	Acidic	4.69	
Cefazoline	39.3	22	22.8	Acidic	3.03	
Cefixime	30	13.8	10.3	Acidic	3.26	
Cephalexin	21.6	23.8	23.8	Neutral		
Chlorpheniramine	28.1	55.3	98.6	Basic		9.47; 3.5
Chlorpropamide	68.2	34.9	33	Acidic	4.33	
Chlorprothixene	111.2	112.5	113	Neutral		
Cinoxacin	46.7	15.6	12.5	Acidic	4.93	
Clonazepam	66.5	66.2	65.1	Neutral		
Clonidine	15.1	34.3	53.2	Basic		8.16
Cloxacillin	73.4	45.7	43.5	Acidic	3.75	
Colchicine	46.8	44.8	47.5	Neutral		
Cytarabine	−45.2	−21.7	−27.1	Neutral		
Diazepam	70.9	80.6	82.3	Neutral		
Diazoxide	49.2	47.7	27.8	Neutral		
Diclofenac	89.3	52.1	49.1	Acidic	4.15	
Diflunisal	91.8	52.9	47.6	Acidic	2.69	
Diltiazem	40.7	71.7	87.2	Basic		8.18
Diphenhydramine	40	59.9	98	Basic		8.87
Diprophylline	19.7	18.2	18.1	Neutral		
Dipyridamole	38.2	68.9	72.1	Neutral		
Doxepine	41.2	64.3	108.2	Basic		9
Droperidol	36.3	68	76.4	Basic		6.75
Encainide	33	48.2	93.6	Basic		9.48

(continued)

Table 8.5 (*Continued*)

Name	CHI pH 2.6	CHI pH 7.4	CHI pH 105	Acid/Base Group	Calculated pK_a Acidic	Calculated pK_a Basic
Ethinyl estradiol	76.9	76.3	77.4	Neutral		
Famotidine	10.7	23.5	25.1	Neutral		
Felbamate	64.7	63.6	65.8	Neutral		
Felodipine	98.7	99.4	99.5	Neutral		
Fenclofenac	90.4	55.5	47.7	Acidic	3.55	
Fenoprofen	83.4	48.7	42.7	Acidic	3.96	
Finasteride	75.9	73.2	75.9	Neutral		
Flucloxacillin	76.3	48.6	45.4	Acidic	3.75	
Flumazenil	49	48.5	50.3	Neutral		
Flunarizine	55.2	121.2	125.8	Weak base		7.6
Flunitrazepam	71	71.1	72.3	Neutral		
Fluoxetine	48.1	67.5	107.3	Basic		9.8
Flurazepam	38.2	69.1	93.4	Basic		8.71
Flurbiprofen	85.5	47.5	44.8	Acidic	4.42	
Furosemide	32.4	−21.7	−31.7	Acidic	4.42	
Gemfibrozil	51.6	79.7	49.3	Neutral		
Glipizide	67.9	46	38.9	Acidic	5.9	
Griseofulvin	71.4	70.9	72.5	Neutral		
Hexobarbital	63.5	63.3	30.7	Neutral		
Hydralazine	36.9	31.9	33.2	Neutral		
Hydrochlorothiazide	32	30.8	−30.9	Neutral		
Hydrocortisone	52.7	50.5	52.4	Neutral		
Imipramine	46.1	69.1	118.6	Basic		9.5
Indomethacin	88.6	54	48.5	Acidic	4.5	
Indoramin	35.2	52.2	71.1	Basic		7.7
Isradipine	89	89.8	90.4	Neutral		
Ketoconazole	40.9	78.8	81.3	Basic		6.51
Ketoprofen	74.3	41.2	38.1	Acidic	4.29	
Labetalol	34.5	46.3	42.2	Neutral		
Levamisol	13.3	44.5	61	Basic		6.98
Levonorgestrel	85.6	85	85.8	Neutral		
Lignocaine	22.6	73.9	86.7	Basic		7.75
Lorazepam	64.6	63.9	65.4	Neutral		
Maprotiline	45.2	61.1	115.3	Basic		10.54
Mebendazole	48.2	61	59.8	Neutral		
Methyl-prednisolone	57.5	56.5	58.2	Neutral		
Metronidazole	8.5	21	22.4	Neutral		
Mianserin	67.1	65.9	68.5	Neutral		
Minoxidil	23.1	33.7	35.9	Neutral		
N-Dealkyl-flurazepam	66.8	68.1	70.6	Neutral		
Nabumetone	87.9	88.7	89.4	Neutral		
Nadolol	19.4	25.9	45.2	Basic		9.76

Table 8.5 (*Continued*)

Name	CHI pH 2.6	CHI pH 7.4	CHI pH 105	Acid/Base Group	Calculated pK_a Acidic	Calculated pK_a Basic
Naproxen	75.2	40.2	36.1	Acidic	4.15	
Neostigmine	5.6	20	37	Basic		
Nicardipine	46.3	101.1	103.9	Weak base		5.42; 8.18
Nifedipine	76.1	77.3	78.7	Neutral		
Nisoldipine	95.8	98	98.7	Neutral		
Nitrazepam	59.6	64.2	63.9	Neutral		
Nitrendipine	87.9	88.4	89	Neutral		
Nitrofurantoin	39.9	33.3	21	Weak acid	9.23	
Nizatidine	5.6	29.9	33.6	Weak base		6.83
Nordazepam	55.5	70.3	72.9	Weak base		2.85
Nortriptyline	44	62.9	118.3	Basic		9.7
Ondansetron	31.7	51.8	60.3	Basic		7.34
Oxacillin	69.3	42.4	40.4	Acidic	3.75	
Oxazepam	62.3	62	63.7	Neutral		
Papaverine	32.6	65.8	68	Weak base		6.4
Pentobarbital	62	59.3	31.3	Weak acid	8.14	
Pentoxifylline	39.5	37.8	39.2	Neutral		
Perphenazine	38.8	76	93.7	Basic		7.94
Phenytoin	61	60.6	37.3	Weak acid	7.5	
Pindolol	20.9	31.5	56.7	Basic		9.7
Piroxicam	19.8	33.6	35	Weak base		3.79
Prazosin	30.1	48.7	52.9	Basic		6.5
Prednisolone	51.9	49.5	51.5	Neutral		
Prednisone	52.2	51.4	53.2	Neutral		
Primidone	37.7	36.1	36.9	Neutral		
Probenecid	79.8	47.4	44	Acidic	3.53	
Procainamide	−34.8	16.6	41.7	Basic		9.32
Procyclidine	91.3	90.9	89.3	Neutral		
Propranolol	37	50.3	83	Basic		9.53
Propylthiouracil	30.1	28	8.2	Weak acid	8.09	
Proxyphylline	26.5	24.9	25.4	Neutral		
Quinidine	14.6	46.4	71.2	Basic		9.05; 4.02
Quinine	16.3	46.7	69.4	Basic		9.05; 4.02
Ranitidine	8.9	23.7	41.7	Basic		8.08
Sulfachlor-pyridazine	45.7	23.2	17.5	Acidic	6.6	
Sulfumetine	39	28.2	−26.1	Acidic	7.06	
Sulfamethoxypyridazine	36	33.8	14.4	Weak acid	6.84	
Sulfametopyrazine	43.4	22.3	−22.7	Acidic	5.91	
Sulfapyridine	23.5	29.9	15.4	Amphoteric	6.24	2.63
Sulfisoxazole	51.7	19.5	14.8	Acidic	7.3	
Sulfadimethoxine	56.1	38.4	18.9	Acidic	6.91	
Sulfadimidine	31.7	35.6	19.1	Weak acid	6.99	

(*continued*)

Table 8.5 (*Continued*)

Name	CHI pH 2.6	CHI pH 7.4	CHI pH 105	Acid/Base Group	Calculated pK_a Acidic	Calculated pK_a Basic
Sulpiride	9.3	23.2	48.2	Basic	10.24	8.39
Tamoxifen	58.6	107.9	137.2	Basic		8.76
Temazepam	70.9	71	73	Neutral		
Terbutaline	8.8	15.8	18.7	Basic		9.76
Theobromine	14.6	13.1	7	Weak acid	9.28	
Tinidazole	33.3	35.6	35.8	Neutral		
Tolbutamide	71.8	46	35.3	Acidic	5.16	
Tolfenamic acid	100.7	61.1	52.8	Acidic	3.88	
Tolmetin	71.7	41.5	38.9	Acidic	3.96	
Tramadol	27	41.2	86.4	Basic		9.41
Trazodone	34.8	70.9	75.6	Weak base		7.09
Trimethoprim	21.1	37.1	40.4	Basic		7.16
Viloxazine	27.8	42.2	55.4	Basic		8.19
Vinblastine	33.8	82.6	93.7	Basic		8.06; 8.74
Vincristine	31.9	74.1	84.5	Basic		7,88; 8.79
Warfarin	81.5	45.7	34.1	Acidic	5.08	
Zidovudine	31.6	30.1	22.5	Weak acid	9.96	

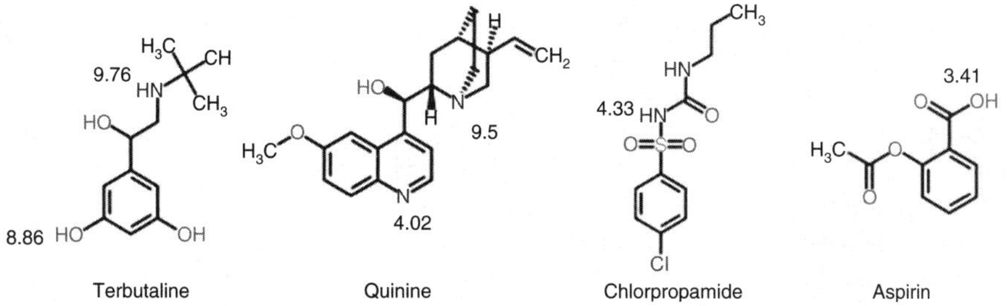

Figure 8.11 Calculated pK_a values for a few known drug molecules using Chemaxon calculator plug-in of Chemspider (Free chemical database at chemspider.com.)

two different pH values, can be used to estimate the percentage of ionized species, and with a closely related series of compounds, it may correlate with the measured pK_a values.

SUMMARY

The negative logarithm of the acid dissociation constant (pK_a) provides information about the pH, where a given functional group is in 50% ionized form. It can be determined by measuring any properties of the compounds that depend on the ionization state. Typical acidic groups in drug molecules are the carboxyl group, the phenolic OH group, and amide groups that have electron-withdrawing substituents in close proximity. Typical basic groups are the primary, secondary, and tertiary amine groups. Electron-withdrawing substituents or aromaticity significantly reduces the basicity of the nitrogen. The Hammett (σ) constants can be used to estimate the inductive and resonance effects of the substituents near the ionizable group, and the Hammett equation can be applied to estimate the pK_a values, too. The "gold standard" method for pK_a determination is the potentiometric titration. From the acid/base titration curve that is obtained in the presence and absence of the test molecule, the pK_a values can be derived. The titration curve is rotated at 90° in order to construct the Bjerrum plot, which shows the number of protons associated with the molecules as a function of pH. Alternatively, the Gran's plot can be constructed from the volume of the titrating solution, the measured pH, and the volume of the titrating solution at equivalence point. The slope of the plot provides the acid dissociation constant. The determined pK_a value depends on the ionic strength and the temperature. The Debye–Hückel theory describes the effect of ionic strength. The potentiometric method requires a reasonably high concentration (amount) of compounds. Therefore, the pK_a values of poorly soluble compounds, especially at pH values where they lose ionization, cannot be measured using aqueous solutions. The addition of an organic solvent is needed to increase the solubility. The pK_a is affected by the dielectric constant of the solvent; therefore, the addition of the organic solvent causes a shift in pK_a values. As the dielectric constants of the organic solvent–water mixtures are smaller than the dielectric constant of pure water, the acidic pK_a values shift toward the larger values, while the basic pK_a shifts to the lower values. In both cases, the result is the weakened acidity and basicity of the compound in the presence of organic solvents. The Yasuda–Shedlowsky plot, the plot of the pK_a value as a function of the organic cosolvent concentration, provides a tool to estimate the pK_a values in pure water from the intercept. It is important to acknowledge that such pK_a shifts of the compound may occur under physiological conditions near biological membranes or concentrated protein solutions such as the plasma. The other popular way to determine the pK_a values of compounds is to measure the UV spectrum change as a function of pH. The method requires lower concentration of compounds; thus, solubility is less of an issue. However, the spectrum change does not reveal whether the ionizable group is basic or acidic. Furthermore, the ionizable group has to be close to the UV chromophore to be able to detect any changes in the spectrum. The method can be easily automated, and the pK_a values of a large number of compounds can be determined in a short period of time. It needs an expert to assign the obtained pK_a values to the appropriate functional groups on the molecule. CE is a very useful and high throughput technique for the measurements of acid dissociation constants. The electrophoretic mobility of the compound has to be measured at a range of pH values from which a titration curve can be constructed and evaluated in the traditional way. The pH dependence of lipophilicity provides the basis for the fast HPLC determination of the acid/base character of compounds. From a few gradient retention time measurements, it is possible to obtain not only the compound's lipophilicity at physiological pH but also the lipophilicity of the neutral form and the pK_a. Alternatively, the compound's ionization state can be assessed at physiological pH values from the measurements of CHIs at three different pH values (acidic, neutral, and basic pH values). The Henderson–Hasselbalch equation can be used to calculate the fraction of the ionized form of the molecules at any pH.

QUESTIONS FOR REVIEW

Q8.1 What is the definition of pK_a?

Q8.2 What types of techniques can be used for the pK_a determination?

Q8.3 How does the measured pK_a depend on the ionic strength of the solution?

Q8.4 How does the organic cosolvent influence the acid dissociation constants?

Q8.5 What are the limitations of the potentiometric pK_a determination?

Q8.6 What are the advantages and disadvantages of the spectrophotometric determination of pK_a?

Q8.7 How can we increase the speed of the spectroscopic pK_a determination?

Q8.8 What are the principles of the pK_a determination by CE?

Q8.9 How have big pharma companies increased the throughput and reduced the limitations of the pK_a determination by CE?

Q8.10 How does the reversed phase retention factor depend on the mobile phase pH?

Q8.11 How can we use the reversed phase gradient chromatography for the determination of pK_a and the acid/base character of compounds?

Q8.12 What are the limitations of the chromatographic determination of the acid/base character?

Q8.13 Can we use the pH dependence of solubility for estimating pK_a?

Q8.14 How can we calculate the fraction of ionized form at any pH from the pK_a values?

REFERENCES

1. Po, H. N., Senozan, N. M. (2001) Henderson–Hasselbalch equation: Its history and limitations. *Journal of Chemical Education*, *78*, 1499–1503.
2. Bacarella, A. L., Grunwald, E., Marshall, H. P., Purlee, E. L. (1955) The potentiometric measurements of acid dissociation constants and pH in the system methanol–water. *Journal of Organic Chemistry*, *20*, 747–762.
3. Debye, P., Hückel, E. (1923) Zur Theorie der Elektrolyte. I. Gefrierpinktserniedrigung und verwandte Erscheinungen (Theory of electrolytes. I. Lowering of freezing point and related phenomena). *Physikalische Zeitschrift*, *24*, 185–206.
4. Hammett, L. P. (1935) Some relations between reaction rates and equilibrium constants. *Chemistry Reviews*, *17*, 125–136.
5. Avdeef, A. (2003) Absorption and drug development. Solubility, permeability and charge state. Wiley-Interscience, Hoboken, NJ, USA, pp 22–28.

6. Avdeef, A., Bucher, J. J. (1978) Accurate measurements of the concentration of hydrogen ions with a glass electrode: Calibrations using the Prideaux and other universal buffer solutions and a computer-controlled automatic titrator. *Analytical Chemistry*, *50*, 2137–2142.
7. Gans, P., Sabatini, A., Vacca, A. (1996) Investigation of equilibria in solution. Determination of equilibrium constants with the HYPERQUAD suite program. *Talanta*, *43*, 1739–1753.
8. Shedlovsky, T., Kay, R. L. (1956) The ionization constant of acetic acid in water-methanol mixtures at 25° from conductance measurements. *Journal of Physical Chemistry*, *60*, 151–155.
9. Gran, G. (1952) Determination of the equivalence point in potentiometric titrations – Part II. *Analyst*, *77*, 661–671.
10. Diamond, D., Hanratty, V. C. A. (1997). Case studies in physical chemistry. In: Spreadsheet Applications in Chemistry using Microsoft Excel, Chapter 5. Wiley, New York, pp. 130–145.
11. Shedlovsky, T. (1962) The behaviour of carboxylic acids in mixed solvents. In: Electrolytes, Ed. Pesce, B., Pergamon Press, New York, NY, pp. 146–151.
12. Avdeef, A., Comer, J. E. A., Thomson, S. J. (1993) pH-metric logP. 3. Glass electrode calibration in methanol-water, applied to pK_a determination of water-insoluble substances. *Analytical Chemistry*, *65*, 42–49.
13. Takács-Novák, K. Box, K. J., Avdeef, A. (1997) Potentiometric pK_a determination of water-insoluble compounds: Validation study in methanol/water mixture. *International Journal of Pharmaceutics*, *151*, 235–248.
14. Ràfols, C., Rosès, M., Bosch, E. (1997) Dissociation constants of several non-steroidal anti-inflammatory drugs in isopropyl alcohol/water mixtures. *Analytica Chimica Acta*, *350*, 249–255.
15. Canals, I. I., Portal, J. A., Bosch, E., Rosès, M. (2000) Retention of ionizable compounds on HPLC. 4. Mobile-phase pH measurements in methanol/water. *Analytical Chemistry*, *72*, 1802–1809.
16. Espinosa, S., Bosch, E., Rosès, M. (2002) Retention of ionizable compounds in high-performance liquid chromatography: IX. Modelling retention in reversed phase liquid chromatography as a function of pH and solvent composition with acetonitrile-water mobile phases. *Journal of Chromatography*, *964*, 55–66.
17. Völgyi, G., Ruiz, R., Box, K., Comer, J., Bosch, E., Takács-Novák, K. (2007) Potentiometric and spectrophotometric pK_a determination of water-insoluble compounds: Validation study in a new co-solvent system. *Analytica Chimica Acta*, *583*, 418–428.
18. Comer, J., Box, K. (2003) High-throughput measurement of drug pK_a values for ADME screening. *Journal of the Association of Laboratory Automation*, *8(1)*, 55–59.
19. Hill, A., Bevan, C., Reynolds, D. (1999) Analytical method and apparatus. *International Patent Application* No WO99/13328.
20. Box, K., Bevan, C., Comer, J., Hill, A., Reynolds, D. (2003) High-throughput measurements of pK_a values in a mixed-buffer linear pH gradient system. *Analytical Chemistry*, *75*, 883–892.
21. Chiang, P.-C., Su, C.-C., Block, K. M., Pretzer, D. K. (2005) Medium throughput pK_a determination of drugs and chemicals using an HPLC equipped with a DAD as a flow injection apparatus. *Journal of Liquid Chromatography*, *28*, 3035–3043.

22. Szakács, Z., Béni, Sz., Varga, Z., Örfi, L., Kéri, Gy., Noszál, B. (2005) Acid-base profiling of Imatinib (Gleevec) and its fragments. *Journal of Medicinal Chemistry, 48*, 249–255.

23. Grycová, L., Dommisse, R., Pieters, L. and Marek, R. (2009) NMR determination of pK_a values of indoloquinoline alkaloids. *Magnetic Resonance in Chemistry, 47*, 977–981.

24. Song, J., Laskowski, M., Oasim, M. A. Markley, J. L. (2003) NMR determination of pK_a values for Asp, Glu, His, and Lys mutants at each variable contiguous enzyme-inhibitor contact position of the turkey ovomucoid third domain. *Biochemistry, 42*, 2847–2856.

25. Croke, R. L., Patil, S. M., Quevreaux, J. Kendall, D. A., Alexandrescu, A. T. (2011) NMR determination of pK_a values in α-synuclein. *Protein Science, 20*, 256–269.

26. Cleveland, J. A., Benkö, M. H., Gluck, S. J., Walbroehl, Y. M. (1993) Automated pK_a determination at low solute concentrations by capillary electrophoresis. *Journal of Chromatography A, 652*, 301–308.

27. Gluck, S. J., Cleveland Jr., J. A. (1994) Investigation of experimental approaches to the determination of pK_a values by capillary electrophoresis. *Journal of Chromatography A, 680*, 49–56.

28. Gluck, S. J., Steele, K. P. Benkö, M. H. (1996) Determination of acidity constants of monoprotic and diprotic acids by capillary electrophoresis. *Journal of Chromatography A, 745*, 117–125.

29. Ishihama, Y., Oda, Y., Asakawa, N. (1994) Microscale determination of dissociation constants of multivalent pharmaceuticals by capillary electrophoresis. *Journal of Pharmaceutical Sciences, 83*, 1500–1507.

30. Ishihama, Y., Nakaruma, M., Miwa, T., Kajima, T., Asakawa, N. (2002) A rapid method for pK_a determination of drugs using pressure-assisted capillary electrophoresis with photodiode array detection in drug discovery. *Journal of Pharmaceutical Sciences, 91*, 933–942.

31. Buckenmaier, S. M. C., McCalley, D. V., Euerby, M. R. (2003) Determination of pK_a values of organic bases in aqueous acetonitrile solutions using capillary electrophoresis. *Journal of Chromatography A, 1004*, 71–79.

32. Wan, H., Holmén, A. G., Wang, Y., Lindberg, W., Englund, M., Någård, M. B., Thompson, R. A. (2003) High-throughput screening of pK_a values of pharmaceuticals by pressure-assisted capillary electrophoresis and mass spectrometry. *Rapid Communications in Mass Spectrometry, 17*, 2639–2648.

33. Shalaeva, M., Kenseth, J., Lombardo, F., Bastin, A. (2008) Measurement of dissociation constant (pK_a values) of organic compounds by multiplexed capillary electrophoresis using aqueous and cosolvent buffers. *Journal of Chromatographic Sciences, 97*, 2581–2606.

34. Poole, S. K., Patel, S., Dehring, K., Workman, H., Poole, C. F. (2004) Determination of acid dissociation constants by capillary electrophoresis. *Journal of Chromatography A, 1037*, 445–454.

35. Molnár, I., Horváth Cs. (1976) Reverse-phase chromatography of polar biological substances: Separation of catechol compounds by high-performance liquid chromatography. *Clinical Chemistry, 22*, 1497–1502.

36. Horváth, Cs., Melander, W., Molnár, I. (1977) Liquid chromatography of ionogenic substances with nonpolar stationary phases. *Analytical Chemistry, 69*, 142–154.

37. Hanai, T., Huber, J. (1984) Optimization of retention time of aromatic acids in liquid chromatography from logP and predicted pK_a values. *Journal of High Resolution Chromatography*, *7*, 524–528.

38. Kaliszan, R., Haber, P., Baczek, T., Siluk, D., Valkó, K. (2002) Lipophilicity and pK_a estimates from gradient high-performance liquid chromatography. *Journal of Chromatography A*, *965*, 117–127.

39. Snyder, L. R., Dolan, J. W. (1996) Initial experiments in high-performance liquid chromatographic method development I. Use of a gradient run. *Journal of Chromatography A*, *721*, 3–14.

40. Wiczling, P., Kawczak, P., Nasal, A., Kaliszan, R. (2006) Simultaneous determination of pK_a and lipophilicity by gradient RP HPLC. *Analytical Chemistry*, *78*, 239–249.

41. Valkó, K., Bevan, C., Reynolds, D. (1997) Chromatographic Hydrophobicity Index by fast-gradient RP-HPLC: A high-throughput alternative to logP/log D. *Analytical Chemistry*, *69*, 2022–2029.

42. Valkó, K., Du, C. M., Bevan, C., Reynolds, D. P., Abraham, M. H. (2001) Rapid method for the estimation of octanol/water partition coefficient (logP) from gradient RP-HPLC retention and a hydrogen bond acidity term ($\Sigma\alpha_2$H). *Current Medicinal Chemistry*, *8*, 1137–1146.

43. Valkó, K. (2004) Measurements of lipophilicity and acid/base character using HPLC methods. In: Pharmaceutical Profiling in Drug Discovery for Lead Selection, Eds Borchardt, R., Kerns, E., AAPS Arlington, VA, pp. 127–182.

44. Hollósy, F., Valkó, K., Hersey, A., Nunhuck, S., Kéri, Gy., Bevan, C. (2006) Estimation of volume of distribution in humans from high throughput HPLC-based measurements of human serum albumin and immobilized artificial membrane partitioning. *Journal of Medicinal Chemistry*, *49*, 6958–6971.

9

MODELS WITH MEASURED PHYSICOCHEMICAL AND BIOMIMETIC CHROMATOGRAPHIC DESCRIPTORS—ABSORPTION

Most drugs are taken orally. These drugs go through the gastrointestinal (GI) tract in about 12–48 h, traveling through a pH range of 1.2–9, starting from 600–800 ml of enzymatic digestive juice in the stomach up to 1500 ml of alkaline pancreatic juice in the small intestine and 400–800 ml of hepatic bile in a day. During this transit, the drugs may make contact with total surface of about 200 m^2 of the intestine including the microvilli [1]. When the drug molecules have managed to dissolve and pass through the intestinal wall, there is an estimated 150,000 km of blood capillaries, where the drug molecules diffuse into and are carried away. However, this blood does not go straight into general circulation, it has to pass through the liver via the hepatic portal vein. The liver ensures that the blood pH and composition remain in a normal homeostatic state. It also transforms absorbed compounds that are potentially harmful. Most of the metabolic transformations are carried out by the liver CYP450 enzyme isoform families. During the transformation, the compound usually becomes less lipophilic, more hydrophilic, and secreted out from the body via the kidney or back to the intestine. In this chapter, we investigate certain steps of the complex absorption process that can be related to the physicochemical properties of the compounds.

Measurements of oral absorption and bioavailability in humans are difficult and cannot be applied for large number of compounds. As we discussed in Chapter 3, absorption and bioavailability are not the same. Bioavailability is the result of

Physicochemical and Biomimetic Properties in Drug Discovery: Chromatographic Techniques for Lead Optimization, First Edition. Klara Valko.

absorption and first-pass metabolism in the liver. Also, permeability and absorption are not interchangeable; intestinal permeability is only one part of the absorption process.

In the early 1990s, the emerging high throughput chemistry in pharmaceutical companies led to the production of large numbers of potent inhibitors of target enzymes. However, these compounds often could not be developed for drugs because of poor absorption or more precisely poor bioavailability. The need to understand the relationships between physicochemical properties and absorption emerged. These efforts resulted in numerous models that have been applied to the drug design process. Nowadays, we have got a better understanding of the requirements in terms of general physicochemical properties of compounds for good absorption. In this chapter, we shall discuss the most well-known approaches and models, such as the Lipinski rule of five, the BCS based on solubility and permeability, the role of lipophilicity and size, the estimation of the absorption potential (AP), and maximum absorbable dose (MAD). The application of the Abraham solvation equation model discussed in more detail in Chapter 5 will also be presented for modeling absorption. To be able to develop meaningful models that can guide the medicinal chemistry efforts, the absorption process has to be divided into simple steps, where the physicochemical properties have the greatest influence. When measured, physicochemical properties are applied in modeling complex biological processes, and it is always worthwhile to compare them with the models that use only calculated descriptors. Measurements are always more expensive and time consuming than calculations; therefore, we need to justify the invested resources by understanding the gains.

LIPINSKI RULE OF FIVE

In 1997, Lipinski and coworkers [2] in Pfizer described experimental and computational approaches to estimate absorption in drug discovery. They observed that, in general, drug discovery compounds had become bigger and more lipophilic. It has already been realized that solubility and permeability (both properties are related to lipophilicity) play an important role in oral absorption. The drug has to be released from the formulation and then be dissolved before it can permeate through the GI cells. Passive diffusion is considered to be another significant transport mechanism for the majority of drugs. Diffusion depends mostly on the size of the molecules. In Lipinski's seminal work, it had already been recognized that solubility, permeability, and diffusion are the important physicochemical properties that can be related to absorption. Camenish et al. [3] also reviewed the role of physicochemical properties in membrane permeability and absorption in 1996. They concentrated mostly on the lipophilicity and solubility as underlying molecular properties.

However, measurements of these properties on a large number of compounds are resource intensive, and they needed better understanding of the property space of orally absorbed drugs. Lipinski et al. [2] have decided to investigate the compounds in the World Drug Index containing over 50,000 drug molecules and extracted those that were known to be orally absorbed, excluding peptides, polymers, and drugs containing unusual functional groups. They analyzed the calculated properties, such as $c \log P$, molecular weight, and the number of H-bond donor and acceptor groups that might be related to solubility and permeability. Reliable *in silico* models for the calculation of solubility and permeability were not available at that time. (We are still in need of more reliable *in silico* solubility and permeability predictions from the chemical structure.)

Lipinski and colleagues carried out the first computational analysis of calculated physicochemical properties on such a large number of compounds. They found that around 90% of the orally administered drug molecules possess a limited range of calculated properties. On the basis of the findings, they formulated the "rule of five," stating that poor absorption or permeation is more likely when two of the following conditions met.

1. There are more than five H-bond donor groups (expressed as the sum of OH and NH groups) in the molecule.
2. The molecular weight is over 500 Da.
3. The $c \log P$ is over 5.
4. There are more than 10 H-bond acceptor groups in the molecule (expressed as the number of nitrogen and oxygen).
5. Compounds that are substrates for biological transporters are exceptions.

As the rules are based on the distribution of calculated properties of several thousands of orally available drugs, it is inevitable that there are some drugs that lie outside the parameter cutoff rules. They noticed that drugs belonging to certain therapeutic areas were often outliers, such as antibiotics, vitamins, and cardiac glycosides. Some of them were bigger in size and hydrophilic, containing more H-bond donor and acceptor groups. The original paper published by Lipinski et al. [2] has been cited in over 5000 scientific papers, books, and it has had a huge impact on medicinal chemistry ever since. Now the calculations for the application of the "rule of five" are available for free on the Internet (ChemSpider free chemical database and Chemaxon Calculations) for over 26 million compounds. Several other drug-like and lead-like filters have been published since then [4–7]. For example, the Veber [8] filter based on the analysis of 1100 drugs suggests that drug molecules, in general, have less than 12 rotatable bonds and the polar surface area is less than 140 Å. One of the reoccurring calculated properties is the lipophilicity, characteristic of the molecules expressed by the logarithmic value of the octanol/water partition coefficient, $c \log P$. As we have discussed in Chapter 5, the lipophilicity is pH dependent for ionizable compounds. Solubility is also pH dependent; so, probably the calculated $\log D$ would be more appropriate to describe a compound's lipophilicity and solubility at physiological pH values. However, to be able to calculate the $\log D$, the calculation of acid dissociation constant pK_a is also required. As it has been discussed in Chapter 8, the pK_a calculations are not as precise as the $\log P$ calculations. ACD Labs developed software for pK_a calculations, and the

Chemaxon pK_a calculator is available for free for the 26 million compounds in the Chemspider database. Bhal et al. [9] have suggested the modification of the "rule of five." They have demonstrated that using c log D instead of the c log P is more relevant for physiological conditions and reduces the number of potentially good molecules to be filtered out. These computational filters are useful for screening large compound libraries and they are increasing the probability of success. On the other hand, many drug molecules might be exceptions from general rules. Therefore, it is always advisable to carry out measurements for selected representative compounds from each series of interest.

Lipinski and coworkers have also realized that there is a need for high throughput experimental methods for solubility measurements. At early stages of the drug discovery process, there is no need for the precise thermodynamic equilibrium solubility measurements used for formulated drugs. Also, large quantities of crystalline compounds are not available in pure form at this stage. Therefore, it would not be feasible to carry out the traditional physicochemical property measurements on a large number of compounds. Instead they have developed high throughput turbidimetric solubility measurements [2] that have been followed by similar approaches by other pharmaceutical companies (see Chapter 6 for more detail). Artificial membrane permeability assays have also emerged. They enabled the measurement of permeability of a large number of compounds. The high throughput solubility and permeability data have made it possible to relate the properties to the chemical structure, and the structure-based design has been replaced by property-based design as Van de Waterbeemd et al. [10] explained in their "perspective" publication in 2001.

It is interesting to note that the "rule of five" does not mention solubility or permeability. The rules mention only the underlying physicochemical properties that can be easily calculated and related to solubility and permeability. In 1932, two botanists, Collander and Bärlund [11] recognized lipophilicity and size as two important properties for membrane uptake. Sha'afi et al. [12] studied the permeability of red cell membranes in 1971 and observed the importance of the hydrogen bonding ability of molecules in reducing permeability. The "rule of five," in principle, summarizes the effects of lipophilicity, size, and hydrogen-bonding ability of the molecules on solubility and permeability. Solubility is inversely related to lipophilicity, which means that very lipophilic molecules are likely to have poor solubility. For good permeability, the molecules have to have a certain level of lipophilicity. Achieving good solubility and permeability at the same time requires a fine balance between the hydrophilic and lipophilic natures of the molecules. When the molecule has too many H-bond donor and acceptor groups, they lose their ability to go through the hydrophobic inner part of the membrane. The molecular size limits the fast diffusion of the compounds in the unstirred water layer near the membrane. The size also affects the solubility as large molecules require higher energy to create cavity in the structured water molecules that are connected via hydrogen bonding. The "rule of five" is a very good example of the multiple optimization tasks that medicinal chemists have to perform in order to find a good drug molecule (it is simple as it contains cutoff values only for certain properties).

When the properties are within these cutoff values, the molecule is more likely to fulfill the requirements for good solubility and permeability, which are essential properties for good absorption.

ABSORPTION MODELS WITH LIPOPHILICITY AND SIZE

One of the first systematic applications of lipophilicity and size for optimizing oral absorption in drug discovery was described by Hill et al. [13] in 1994. Scientists at Wellcome Research laboratories have been working on antimigraine molecules that act via 5-HT_{1D} receptor. They found several potent molecules but with very poor oral bioavailability. The plot of the octanol/water distribution coefficient (log D at pH 7.4) as a function of cMR of orally available drugs and the research compounds revealed that the new research compounds occupied a different property space, as shown in Figure 9.1. They were too large and too hydrophilic

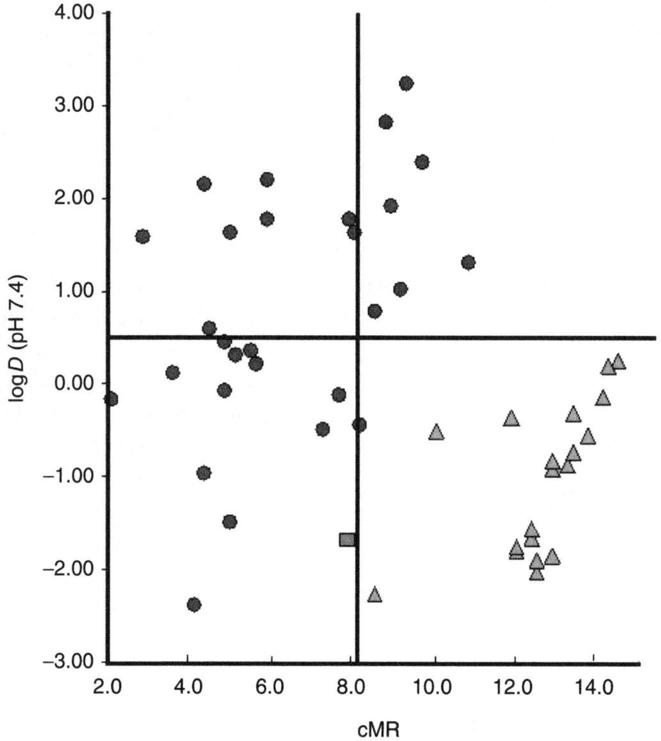

Figure 9.1 The plot of octanol/water log D (pH 7.4) as a function of calculated molecular refractivity (cMR) of known drugs and antimigraine 5-HT_{1D} antagonist research compounds Drugs: circle; research compounds: triangle; and zolmitriptan: square. (Figure is courtesy of Alan Hill, GSK based on the data from Reference [13].)

in comparison to other orally bioavailable drugs. On the basis of the lipophilicity and size, zolmitriptan has been designed as an orally available antimigraine drug by adjusting the physicochemical properties to the right property space. The application of the distribution coefficient at pH 7.4 is significant as it takes into account the lipophilicity of the charged molecule. The molecular size is very rarely measured; usually it is expressed by the molecular weight of the compounds. The molar refraction is proportional to the size of the molecules, which can be derived from the refractive index of the aqueous solutions. It takes into account the radius of all the atoms instead of the weight. Although there is a good correlation between the two size parameters (r^2 is above 0.9), in QSAR studies, it is more common to use the cMR.

The size descriptor may have two roles in relation to absorption. Firstly, large molecules may have difficulties in permeability, but they also have reduced solubility. When a large molecule contains several polar functionalities that are able to form hydrogen bonding, energy can be gained back. However, the presence of several polar groups reduces the permeability as the molecule becomes too hydrophilic to go through the inner hydrophobic part of the membrane. Therefore, it is not surprising that large and hydrophilic molecules, in general, have poor absorption (lower right quadrant of the log D vs cMR plot). On the basis of the data, when a compound has a cMR value larger than 8, it should have log D value higher than 0.5 to be able to be absorbed. Zolmitriptan is hydrophilic, but the size is reduced significantly; see Figure 9.2. The question remains whether we should use log D or log P as the lipophilicity parameter.

We should consider that a compound with good absorption has to have two essential physicochemical properties, solubility and permeability. Solubility depends on ionization to a great extent and is inversely related to lipophilicity. For this reason, the lipophilicity at a pH range of 5–8 should be more relevant than the lipophilicity of the neutral form. The permeability of ionized compounds is affected by the type of charge too. While positively charged compounds may go through the membrane as long as the neutral form is lipophilic, negatively charged compounds will have reduced permeability. As an approximation, it seems that the log D is a more appropriate lipophilicity parameter in relation to intestinal absorption. Log D is equal to log P for unionized compounds, but it can be 3–4 unit less for fully ionized compounds. When there is more than one ionizable group

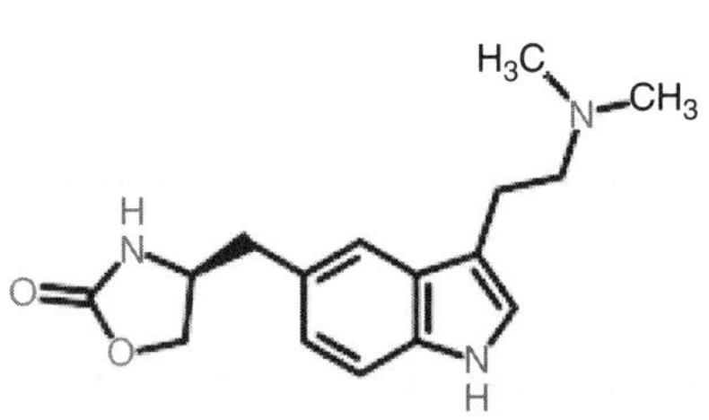

Figure 9.2 The structure of zolmitriptan. ($c \log D = -1.47$, molecular weight = 287.4 Da, and cMR = 8.)

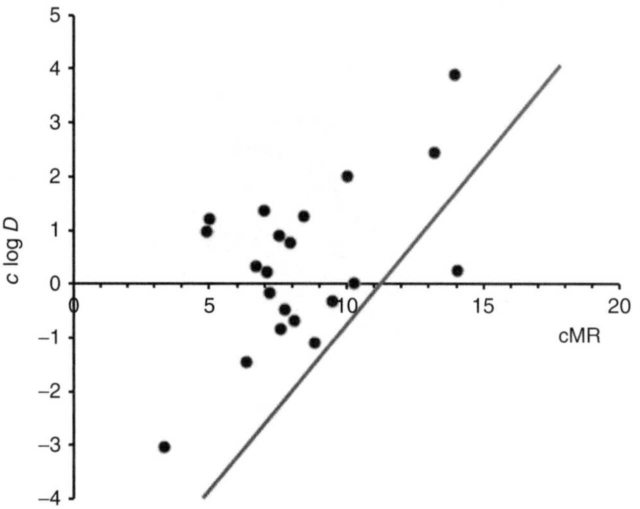

Figure 9.3 The lipophilicity (*c* log *D*) against the molecular size (cMR) plot for orally available known drug molecules (listed in Table 7.1). Compounds that have greater cMR values than 10 should have log *D* above 0 to be orally absorbed. (The line represents the discrimination line between orally absorbed and not absorbed drugs based on the model shown in Figure 9.4.)

present on the molecule, the difference between log *P* and log log *D* can be even greater. Reducing the log *D* without reducing the size of the molecule, however, may lead to poor absorption according to the log *D*–cMR model. It suggests that it is advisable to avoid multiple charges on the molecules. Figure 9.3 shows the plot of *c* log *D* (pH 7.4) versus cMR for several orally available drug molecules given in Table 7.1.

This observation has been developed further [14]. Rob Young from GSK presented the plot and the discrimination line at the Physchem Forum in 2008 [15], which are used as *in silico* estimation of the probability of absorption of research compounds. The model has been extended using lipophilicity and size data of over 400 known drug molecules. The plot shown in Figure 9.4 contains the data of research compounds as well. The discrimination line between the absorbed and not absorbed compounds has been fitted to get the best possible discrimination of the 400 drug molecules. Compounds that are positioned below the line are expected to be poorly absorbed. The model is very useful for identifying classes of compounds or series of compounds with different scaffolds that are more likely to have absorption problem and helps the program teams focusing on optimizing series that fall into a better property space.

The *in silico* prediction may fail for compounds in new chemical space as the log *D* calculation becomes unreliable.

Chromatographic methods can also be used for lipophilicity (log *D*) measurements of newly synthesized research compounds. Borthwick et al. [16] published a good example of using the measured chromatographic lipophilicity and cMR model

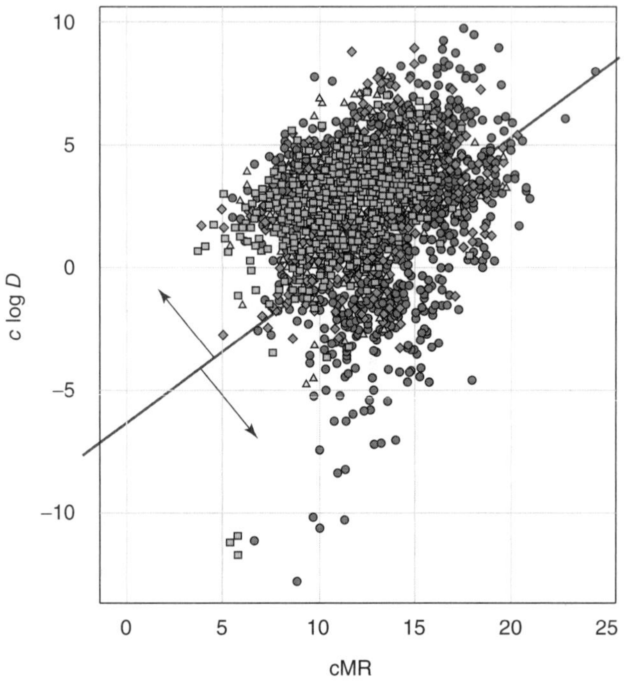

Figure 9.4 The plot of calculated distribution coefficient ($c\log D$) and the size parameter (cMR) for known drugs and research compounds. The line shows the boundaries between good and bad absorptions. The further away the compound is from the line, the higher is the probability of the absorption classification. The lighter squares represent orally absorbed known drug molecules. (Figure is courtesy of Rob Young and Anne Hersey based on References [14, 15].)

for designing orally bioavailable oxytocin antagonist 2,5-diketopiperazine derivatives (Figure 9.5).

They have used the model for estimating percentage oral absorption using the CHIs obtained using methanol gradient reversed phase chromatographic retention time data and the cMR size parameter [17]. The compounds were generally too big in comparison to their lipophilicity. Some compounds' cMR values reached above 18. By reducing the size of the molecule and still maintaining the oxytocin antagonist potency, a molecule called *Retosiban* was designed, which underwent clinical trials for the treatment of preterm labor. The estimation of human intestinal absorption (EstHIA%) is shown by Equation 9.1, which was obtained from the analysis of 52 known drugs with measured human oral absorption (%) [17].

$$\text{EstHIA\%} = 1.31\ \text{CHI(MeOH)} - 10.93\text{cMR} + 88.6$$
$$N = 52 \qquad r = 0.81 \qquad s = 19.7 \qquad F = 15 \tag{9.1}$$

where N represents the number of compounds, r is the multiple correlation coefficient, s is the standard error of the estimate, and F is the Fisher-test statistic.

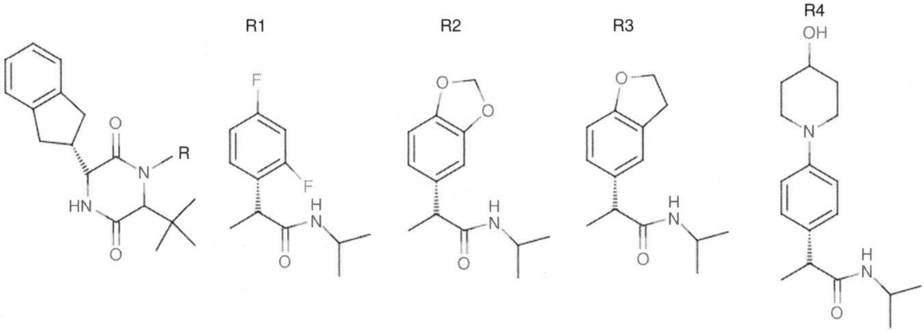

	R1	R2	R3	R4
CHIlog *D*	3.4	3.3	3.4	2.4
cMR	13.51	14.54	14.85	16.61
estHIA	60%	48%	46%	19%

Figure 9.5 The structure of selected oxytocin antagonist 2,5-diketopiperazines with their measured log *D* (derived from CHI(MeOH)) and cMR size descriptor. By reducing the size, the estimated human intestinal absorption (estHIA) has increased significantly. (Reproduced with permission from Reference [16].)

The model is crude as the error of the estimate is ±20%, which means that the oral absorption should be estimated above 60% in order to expect reasonable result for an orally administered drug. These two examples show that the plot of the lipophilicity and size of the molecules can be used to explore the two-dimensional property spaces for orally bioavailable compounds. The most critical quadrant of the plot is where lipophilicity is low and the size of the molecules is large. The plot can be used in a more predictive way than the "rule of five." However, it does not provide a strict cutoff value for the lipophilicity of large molecules, which may fall outside the drug-like property space. It predicts, rather, the permeability part of the absorption process and it does not take into account solubility and dissolution rate.

BIOPHARMACEUTICS CLASSIFICATION SYSTEM (BCS)

The "Lipinski rule of five" and the lipophilicity–size models for estimating the probability of good absorption for a newly synthesized research compound show the realization that one single property is not adequate to model the absorption process. The properties are not completely independent from each other; for example, larger size, in general, shows a trend with greater lipophilicity of the molecule, or more lipophilic compounds tend to be less soluble. However, although hydrophilic compounds may be more soluble, they show poor permeability. This makes the optimization process very difficult. Sometimes we need to find functional groups that break such intercorrelations between the properties in order to get the compound into the right property space. For good absorption, we need good solubility and good permeability.

Amidon et al. [18] in their seminal paper have suggested grouping drugs and potential drugs into four different classes. Class I drugs are highly soluble and highly permeable, Class II drugs have good permeability but poor solubility, Class III drugs have poor permeability but good solubility, and Class IV drugs have both poor permeability and solubility (see Figure 6.1). The FDA defined the criteria and definition of good solubility and permeability more precisely and proposed a waiver of *in vivo* bioequivalence and bioavailability studies for Class I compounds. They defined a drug substance highly soluble when the highest dose strength is soluble in less than 250 ml water over a pH range of 1–7.5. The dissolution rate is considered acceptable when more than 85% of the dose dissolves within 30 min using the USP-defined dissolution testing apparatus in a volume of less than 900 ml buffer solutions. For permeability, a drug substance is considered highly permeable when the extent of absorption in humans is determined to be greater than 90% of an administered dose based on mass balance or in comparison to an intravenous reference dose. Amidon et al. [18] also compared the correlation between *in vitro* and *in vivo* expectation of absorption. For Class I compound, the expected *in vitro–in vivo* correlation is good when the dissolution rate is faster than the gastric emptying rate. For Class III and IV compounds, obtaining *in vitro* and *in vivo* correlations is usually very difficult because of the poor membrane permeability that controls the absorption rate process. The major advantage of grouping compounds according to the BCS is that we are able to identify potential issues that cause poor absorption. In order to improve the permeability, lipophilicity should be increased, but conversely, in order to improve solubility, lipophilicity should be decreased. Identifying the causes of poor absorption helps the medicinal chemists to modify the structure and properties of lead compounds. The aim is to design Class I compound as it potentially saves expenses during the development as formulation development is simpler and the registration agencies might waive the *in vivo* bioequivalence and bioavailability studies. Lindenberg et al. [19] has published the list of drug molecules and their BCS class that had reliable solubility and intestinal permeability data available. Table 9.1 shows the selected drugs with reliable BCS classification and the dose, calculated lipophilicity ($c \log P$ and $c \log D$), molecular weight, and cMR values. The oral bioavailability has been taken from Veber et al. [8], which is an average of the literature values they used in their absorption models.

The BCS concept can be applied at later stages of development, particularly when considering biowaivers. At the lead optimization stage, the question is to understand which property of the compound limits the absorption. The thermodynamic equilibrium solubility at multiple pH values and permeability data are not available at lead optimization stage. Also, the clinical dose is not known for discovery compounds; so, it is not possible to categorize the solubility to high or low for the BCS class. Butler and Dressman [20] modified the BCS concept to be able to apply it at early stages. For permeability, they suggest using a wider range of models, including *in silico* methods to predict human permeability. They also suggest categorizing the compounds based on a dose estimate and a measured solubility in a single biorelevant medium such as in FaSSIF. They subdivided Class II

Table 9.1 The Biopharmaceutical Classification System categories and calculated lipophilicity (ACD Labs) and size parameters (cMR and molecular weight) of selected drugs taken from the literature [18] and their oral bioavailability [8]

Drug	Solubility	Permeability	Dose	BCS Class	$c \log P$	$\log D$	MW	cMR	Oral bioavailability
Amiloride (HCl)	High	High	5	I	1.22	1.22	229.6	4.99	50
Chloroquine phosphate	High	High	100	I	4.41	1.59	319.9	9.74	80
Cyclophosphamide	High	High	25	I	0.73	0.73	261.1	5.81	74
Diazepam	High	High	2, 5	I	2.8	2.8	284.7	8.09	100
Digoxine	High	High	0.625, 0.25	I	1.29	1.29	789.9	19.64	70
Doxycycline HCl	High	High	100	I	−0.6	−4.86	444.4		93
Fluconazole	High	High	50	I	0.45	0.45	306	7.61	90
Levodopa	High	High	100, 10, 250	I[a]	−1.15	3.67	197.2	4.93	41
Metronidazole	High	High	200–500	I	−0.14	−0.14	171.2	4.1	99
Phenobarbital	High	High	15–100	I	0.53	0.31	232.3	5.92	100
Prednisolone	High	High	5	I	1.64	1.64	360.5	9.55	82
Propranolol HCl	High	High	20, 40	I[a]	2.75	0.79	259.4	7.9	26
Stavudine	High	High	15, 20, 30, 40	I	−0.65	−0.65	224.2	5.39	82
Theophylline	High	High	100, 200, 300	I	−0.02	-0.05	180.2	4.32	96
Zidovudine	High	High	300	I[a]	0.05	0.05	267.2		63
Carbamazepine	Low	High	100, 200	II	1.9	1.9	236.3	6.97	70
Dapsone	Low	High	50, 100	II	0.99	0.99	248.3	6.75	93
Ibuprofen	Low	High	200, 400	II	3.5	0.58	206.3	6.08	80
Nifedipine	Low	High	10	II[a]	3.58	3.58	346.3	8.79	50
Nitrofurantoin	Low	High	100	II	−0.4	−0.63	238.2	5.31	87
Phenytoin	Low	High	25, 50, 100	II	1.42	1.38	252.3	6.96	90
Sulfamethoxazole	Low	High	100, 400	II	0.66	−0.54	253	6.25	100

(continued)

Table 9.1 (*Continued*)

Drug	Solubility	Permeability	Dose	BCS Class	$c \log P$	$\log D$	MW	cMR	Oral bioavailability
Trimethoprim	Low	High	100, 200	II	0.59	0.47	290	8.03	63
Valproic acid Na	Low	High	200, 500	II	2.58	0.02	144.2	4.06	100
Abacavir (sulfate)	High	Low	300	III	1.16	1.16	286.3	7.58	83
Acetylsalicylic acid	High	Low	100–500	III[a,b]	1.39	−1.68	180.2	4.45	68
Acyclovir	High	Low	200	III	−0.62	−0.62	225	5.24	23
Allopurinol	High	Low	100	III	−1.33	−4.29	136.1	3.47	53
Atenolol	High	Low	50, 100	III	0.33	−1.76	266.3	7.43	56
Captopril	High	Low	25	III	1.99	−1.46	217.3	5.44	65
Chloramphenicol	High	Low	250	III	1.1	1.1	323.1	7.26	82
Cimetidine	High	Low	200	III	0.56	0.24	252.3	7.07	60
Sodium cloxacillin	High	Low	500	III	2.83	−0.9	435.9	10.62	43
Codeine phosphate	High	Low	30	III	1.2	0.27	299.4	8.29	50
Hydralazine HCl	High	Low	25, 50	III[a]	1.02	0.99	160.2	4.88	16
Hydrochlorothiazide	High	Low	25	III	−0.02	−0.04	297.7	6.27	71
Metformine HCl	High	Low	500	III#	−1.02	−3.01	129.2	3.34	52
Methyldopa	High	Low	250	III[a,c]	−0.74	−3.26	211.2	5.39	25
Paracetamol (acetaminophen)	High	Low	100–500	III[a]	0.48	0.47	151.2	4.24	88
Pyridostigmine Br	High	Low	60	III	−2.84	−2.84	181.2		14
Furosemide	Low	Low	40	IV[a]	2.3	−0.82	330.8	7.58	71
Indinavir sulfate	Low	Low	200, 300, 400	IV[a]	3.44	3.43	613.8	17.61	65
Saquinavir	Low	Low	200	IV	5.08	5.05	670.9	18.9	13

[a]First-pass effect.
[b]Degradation in the GI tract.
[c]Active transport.

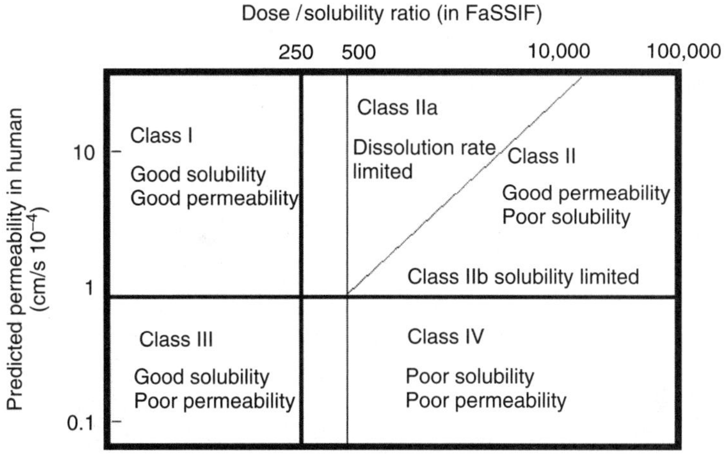

Figure 9.6 Developability Classification System that discriminates the low solubility, high permeability Class (Class II) between compounds with low dissolution rate (Class IIa) and low solubility (Class IIb). (Reproduced with permission from Reference [20].)

(good permeability and poor solubility) depending on the cause of low solubility (Figure 9.6).

Compounds in Class IIa have dissolution-rate-limited absorption unless factors such as particle size are carefully controlled. These compounds are usually lipophilic. However, with appropriate particle size control, the dissolution rate may be enhanced to achieve complete or close-to-complete oral absorption. When the drug falls into the solubility-limited absorption (Class IIb or IV), it means that there is not enough fluid in the GI tract in the fasted state to dissolve the administered dose even if very small particle sizes are used. These compounds need solubilization for oral absorption to be complete, and therefore they tend to have additional challenges in drug development. Dose is also a factor in both BCS and DCS, with higher doses leading to greater challenges for drug solubilization. The FaSSIF solubility models neutral and weak acidic compounds very well; however, it may underestimate the solubility of weak bases. The gastric fluid pH is much lower, which makes it possible to dissolve weakly basic compounds. However, it is very difficult to predict this effect as the gastric pH and residence times are more variable than the intestinal residence time. It is also affected by age, food, and disease state. For example, the bioavailability of ketoconazole is highly dependent on the gastric pH [21]. Weakly basic drugs dissolved in the gastric fluid may precipitate when they reach the higher pH of the small intestine. The precipitation should be avoided as it has serious physiological consequences in the patients.

It is interesting to see how the BCS and the lipophilicity–size (log *D*–cMR) models overlap. The plot of the data in Table 9.1 can be seen in Figure 9.7. It can be seen that the lipophilicity–size plot can separate the poor permeability compounds (Class III) reasonably well.

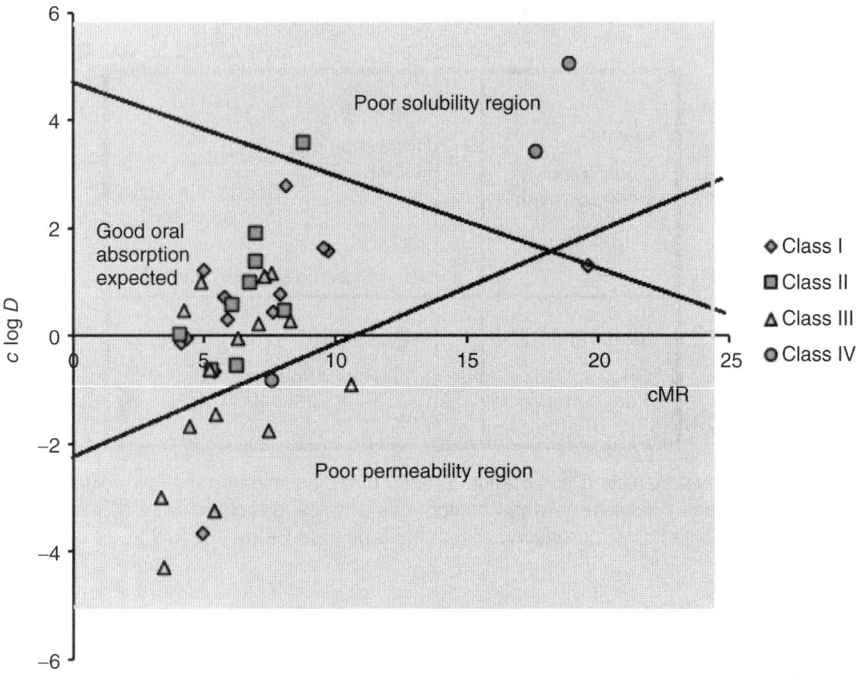

Figure 9.7 The lipophilicity–size plot for the drugs listed in Table 9.1 and their BCS class. An additional line is needed to separate the compounds with poor solubility because of high lipophilicity.

We can expect poor solubility for compound with high lipophilicity. However, most of the Class II drugs that have good permeability but poor solubility fall into the good oral absorption area of the plot. We can see that lipophilicity and size of the molecules are not enough to distinguish between good and poor solubilities. We need an additional line in order to separate the compounds with poor absorption because of too high lipophilicity that reduces the solubility. Nevertheless, the plot shown in Figure 9.7 is very useful for checking the physicochemical property space of new drug discovery compounds and identifying the possible causes of the poor absorption. The plot shows the position of known drugs, including those that have got to market in spite of the disadvantageous BCS category (Class IV for example). The property space of discovery compounds is usually much larger than the space shown in the figure. Therefore, the plot helps in selecting the compounds that fall into the small property space that marks good absorption. The "Golden Triangle" proposed by Johnson et al. [22] is basically the 90 °C rotated form of Figure 9.7. The X-axis is the lipophilicity, and the Y-axis is the size parameter (molecular weight). The triangle identifies the space for compounds with good oral exposure due to good permeability and metabolic stability.

ABSORPTION POTENTIAL—MAXIMUM ABSORBABLE DOSE

The correlation between *in vivo* absorption and physicochemical properties of the drugs is impeded by a wide range of factors including the drug-related properties and the physiological conditions. The drug-related properties include solubility, dissolution rate, particle size, crystal form of the solid, ionization properties, etc. The physiological variables include the motility of the GI tract, the volume, and composition of the intestinal fluids, the pH, the permeability of the intestinal wall, blood flow, etc. Therefore, it is very difficult to include all of these variables into a single model to estimate the absorption of new drug discovery compounds. Dressman et al. [23] have tried to include as many variables as possible into one single dimensionless number that they called *AP*. Excluding the effect of the first-pass metabolism and metabolism in the gut, the fraction of a dose absorbed (F_{abs}) is a function of the permeability of the gut wall (P_w), the aqueous permeability of the drug (P_{aq}), the intrinsic aqueous solubility of the neutral form of the drug (S_0), the dose (X), the unionized fraction of the drug at pH 6.5 (F_{non}), and the volume of the luminal content (V_L), as shown in Equation 9.2.

$$F_{abs} = f(P_w, P_{aq}, S_0, X, F_{non}, V_L) \tag{9.2}$$

Making a few further assumptions, such as having only the neutral form of the drug being permeable and the average pH of the intestinal lumen where the majority of absorption takes place is pH 6.5, the variables of interest can be arranged in a dimensionless number, called *AP*, as described by Equation 9.3.

$$\mathrm{AP} = \log\left(P \times F_{non} \times \frac{S_0 \times V_L}{X}\right) \tag{9.3}$$

The authors have shown data for a few known drugs for which the AP showed a sigmoidal relationship with the fraction absorbed. In principle, the AP is proportional to the product term of intrinsic solubility, permeability, and the fraction of nonionized form of the molecule at pH 6.5. It means that if any of the product terms are close to zero, that is, either the solubility or the permeability is poor, the absorption will be poor. The V_L parameter can be considered constant using the average volume of the intestinal fluid in the lumen. For a given series of drug discovery compounds, we can aim at a given dose (X); so, it can also be considered constant. Therefore, the product term of solubility, permeability, and fraction of nonionized form can be used for assessing and ranking drug discovery compounds. This model is based on mechanistic description of the pure physicochemical processes involved in the absorption and cannot be considered as a statistical model. It does not take into account any first-pass metabolism or active transport processes.

Macheras and Symillides [24] have revealed a qualitative relationship between the AP and the fraction of dose absorbed (FA) by passive mechanism, as is shown by Equation 9.4.

$$\mathrm{FA} = \frac{(10^{\mathrm{AP}})^2}{(10^{\mathrm{AP}})^2 + F_{\mathrm{non}}(1 - F_{\mathrm{non}})} \tag{9.4}$$

The AP term was calculated from the logarithmic value of the product term of octanol/water partition coefficient instead of permeability, intrinsic solubility, fraction of the nonionized form of the molecule at pH 6.5, and the luminal volume divided by the dose. They used 1000 as octanol/water partition coefficient when the partition coefficient was above 1000 and they used 1 for the intrinsic solubility times luminal volume divided by the dose when it was less than 1. Sanghvi et al. [25] further simplified the equation for AP using the lipophilicity and solubility of the ionized form of the molecule, thus avoiding the use of the fraction of nonionized form. They substituted the luminal volume with 0.25 l and came up with a modified absorption potential (MAP) (Equation 9.5).

$$\mathrm{MAP} = \log\left(\frac{K_{\mathrm{ow}}S_{\mathrm{w}}}{4D}\right) \tag{9.5}$$

The K_{ow} is the octanol/water partition coefficient, S_{w} is the intrinsic solubility of the drug molecule in water (mg/ml), and D is the dose in milligram.

Ni et al. [26] have demonstrated that when the solubility and the partition coefficient were measured at the same pH, the product term remains more or less constant. The explanation for this is that ionization affects solubility in an opposite way to lipophilicity. Ionization decreases the lipophilicity, while it increases the solubility, as shown in Figure 9.8, as an example for an acidic drug.

The compensating effect of ionization on AP provides the opportunity to use the high throughput solubility and lipophilicity or permeability measurements at a given pH. There is usually a debate on as though which pH is the best to determine the aqueous solubility of the compounds and the lipophilicity. The recommendation is that we need to ensure that we use the same pH for both measurements, for example, pH 6.5 or pH 7.4; this way the AP calculation should give approximately the same result. Ni et al. [26] showed the correlation between fraction absorbed and the MAP for the data of 26 diverse drug molecules listed in Table 9.2. The trend between the fraction absorbed and the logarithmic value of MAP is shown in Figure 9.9.

At the lead optimization stage, the dose is not known; however, the project teams have a target dose when they optimize the potency of the compounds that can be included in the estimation of the required solubility. To be able to calculate the MAD that has been introduced by Johnson and Swindell [27], we have to introduce the GI transit time and intestinal surface area available for the compound to be absorbed. The absorbable dose also can be expressed from Equation 9.5, which gives an estimate based on measured solubility and lipophilicity. The AP equations clearly show that there should be a balance between dose, solubility, and permeability. Lower solubility and/or permeability are acceptable for lower dose compounds, while the required solubility is much higher for higher dose drugs. Lipinski [28] has

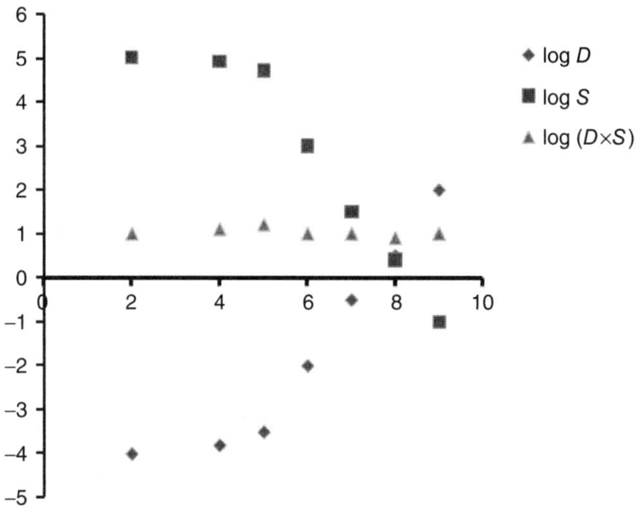

Figure 9.8 The independence of the product term of lipophilicity (K_{ow}) and solubility (S_w) from the pH. An example of a basic drug molecule. Low pH solubility is high and lipophilicity is low because of ionization. The product term remains the same at all pH values. (Adopted with permission from Reference [26].)

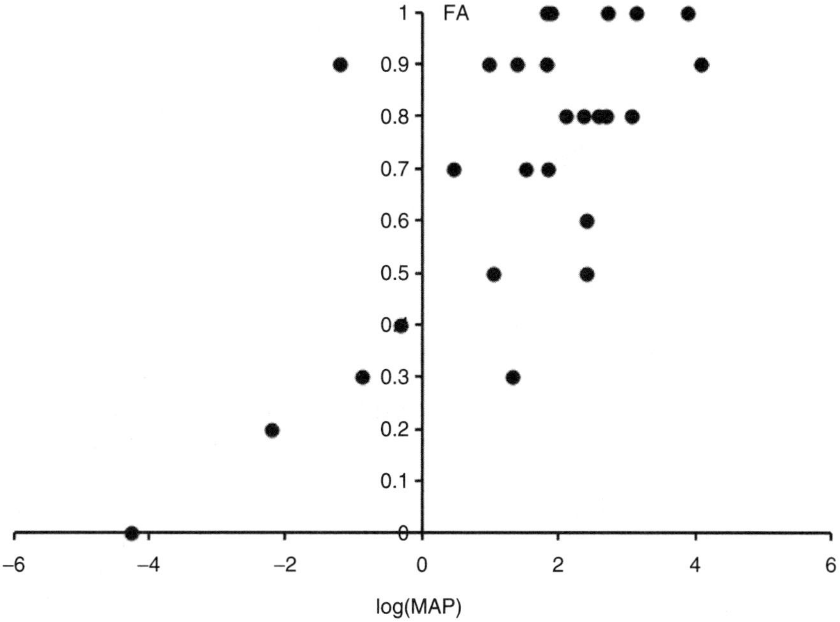

Figure 9.9 The plot of the logarithmic value of MAP calculated from measured solubility, lipophilicity, dose, and the fraction absorbed dose (FA) for 26 drugs listed in Table 9.2.

Table 9.2 The fraction absorbed (FA), dose, log *P*, and log S_w for 26 drug molecules with the logarithm of modified absorption potential (MAP) using Equation 9.5 from Reference [24] with permission

Drug	FA	Dose (mg)	log *P*	log S_w	MAP
Acetylsalicylic acid	0.9	500	1.02	−1.59	1.38
Acyclovir	0.2	200	−2.52	−2.14	−2.21
Allopurinol	0.9	300	−0.88	−2.38	−1.21
Amiloride	0.5	5	−0.69	−0.97	2.4
Atenolol	0.6	50	−0.11	−0.61	2.4
Chlorothiazide	0.3	250	−0.31	−3.05	−0.89
Diclofenac	1	50	4.32	−4.37	3.12
Digoxin	0.9	0.25	2.27	−4.08	4.08
Famotidine	0.5	40	0.26	−2.53	1.05
Fluoxetine	0.8	30	4.57	−5.4	2.58
Furosemide	0.7	40	1.87	−3.66	1.53
Griseofulvin	0.4	250	1.75	−4.61	−0.31
Hydrochlorothiazide	0.7	25	−0.4	−2.62	0.46
Ibuprofen	0.8	200	3.68	−3.99	2.1
Miconazole	0.3	250	5.81	−7.1	1.33
Moxonidine	1	0.3	1.42	−2.85	3.87
Nizatidine	1	300	−0.2	−0.36	1.88
Olanzapine	0.8	10	4.02	−5.22	2.69
Paromomycin	0	250	−6.52	−0.53	−4.26
Phenytoin	0.9	100	2.08	−3.9	0.98
Prednisolone	1	20	1.38	−3.21	1.83
Propranolol	1	80	2.75	−2.96	2.7
Terbinafine	0.8	250	5.96	−6.06	2.36
Xipamide	0.7	20	1.89	−3.7	1.84
Zidovudine	0.9	100	0.04	−1.03	1.83
Zopiclone	0.8	8	1.17	−2.2	3.05

published guidance for the required solubility and permeability for low, medium, and high dose drugs, which is demonstrated in Figure 9.10.

When the dose is low, for example, 0.1 mg/kg (it means around 5–10 mg active ingredients in a tablet or capsule), the required solubility of a medium permeability compound is 5 μg/ml, while for a high dose compound (typically 10 mg/kg) with average permeability, the required solubility is over 2 mg/ml. Therefore, it is advantageous to have an estimate of the dose of the potential candidate molecule to be able to assess its physicochemical properties in terms of estimated oral absorption.

When considering the solubility in terms of AP, it is expressed as milligram per milliliter because it has to be related to the dose, which is expressed in milligram per kilogram body weight. Therefore, it is important to convert the molar solubility. However, when we want to relate the solubility back to structural changes, the molar units should be used. In Chapters 10 and 11, we shall discuss how to estimate the

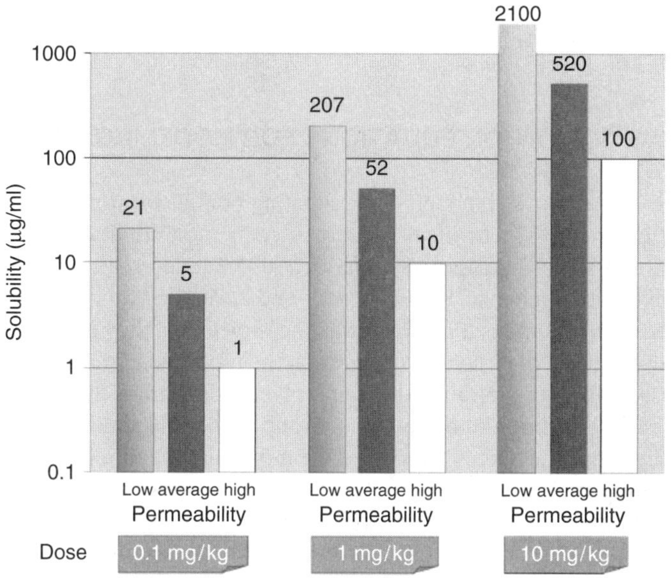

Figure 9.10 The balance between dose, solubility, and permeability. (Adopted with permission from Reference [28].)

necessary dose that results in the necessary receptor occupancy at the site of action. Even in this case, we need to use the molar concentrations.

Sun et al. [29] have provided an excellent guideline using *in vitro* solubility and Caco-2 permeability data to predict the *in vivo* absorption using the data of 24 drug molecules. They differentiated between *in vitro* and *in vivo* permeabilities of drugs with carrier-mediated or passive absorption.

In conclusion, *in vitro* high throughput solubility, lipophilicity, and permeability measurements can be used to estimate the AP of discovery compounds. The lipophilicity–size model involves only one measurement of gradient chromatographic lipophilicity, and the calculated molar refraction can be used to assess the permeability property space of the molecules. However, it does not contain a cap for high lipophilicity that can result in very low solubility.

In order to include the effect of poor solubility on absorption, the BCS can be applied, which is extremely useful in identifying the property that should be modified to improve the absorption. Owing to the opposite effect of ionization on solubility and lipophilicity, one measurement of each is enough to estimate a compound's AP as long as they are measured at the same pH. The permeability can be replaced by lipophilicity measurement for this purpose. In order to give advice to medicinal chemists for the required optimum solubility and permeability, we need to have an idea of the dose; this is usually unknown at early stages. It is discussed in Chapters 10 and 11 that the required concentration at the site of action can be estimated from *in vitro* biomimetic measurements and the *in vitro* potency. On the basis this information, the required efficacious dose can be estimated for compounds with low, medium, and high bioavailabilities at earlier stages of the drug discovery process.

ABRAHAM SOLVATION EQUATIONS FOR MODELING ABSORPTION

The previously discussed models for estimating oral absorption of compounds are based on basic physicochemical properties such as solubility, lipophilicity, permeability, and size. The Lipinski "rule of five" included structural descriptors as well such as the number of H-bond donor and acceptor groups in the molecules. When medicinal chemists try to optimize the molecules, they need more information about the effect of structural changes on the physicochemical properties of the molecules. In this respect, the Abraham solvation equation models provide general guides. The addition of a functional group to the molecule changes more than just one property; it changes the size of the molecule, the dipolarity/polarizability, H-bond acidity and basicity as well. It is important to know how these changes influence lipophilicity, solubility, permeability, and the chance of good oral absorption of the compound. Zhao et al. [30] have published an extensive study on human intestinal absorption using the Abraham solvation equation model. The first difficulties they had to overcome were to find reliable human intestinal absorption data in the literature.

The solvation equation approach has already been used to describe several physicochemical processes such as lipophilicity, solubility, and permeability as discussed in Chapters 5–7, respectively. It can be applied successfully only for thermodynamic equilibrium or rate processes. We cannot expect it to work on biological active transport processes or metabolism. Therefore, it was an important task to collect literature human absorption data that excluded the effect of active transport or metabolism. They evaluated the data of 241 drug molecules, taking into account bioavailability, excluding the metabolism and elimination processes when comparing the drug concentration in plasma after oral and intravenous administrations. Absorption data obtained using a special formulation technique has been examined, and the formulation effect has been excluded from absorption of the drug molecules. This is important because we aim to understand the molecular features that affect absorption at lead optimization stage. Many drug molecules show dose-dependent absorption because of saturation of metabolizing enzymes or active transporters, which should have also been excluded from the study. As a result of the thorough expert judgment of the original absorption data, Zhao et al. [30] selected the data that was marked as "good" for the model building. They used *in silico* calculation for obtaining the molecular descriptors for the drugs. They, then, carefully selected a relatively small training set of drugs for the construction of the solvation equation. Most of the known drugs have high percentage of human intestinal absorption (otherwise they would not pass the drug registration procedure); therefore, the authors paid special attention to the even distribution of poor, medium, and high absorption compounds in the training set. The original solvation equation model refers to any solute property that is linear free energy related. However, the percentage or fraction absorbed data are not, strictly speaking, linear free energy related. The authors still have chosen to model the percentage description of absorption as they have found no significant difference between the models when absorption is converted to a linear scale. It is probably because of the relatively large "noise" in the data. Equation 9.6 shows the obtained solvation equation for the test set of drug molecules.

$$\%\text{Abs} = 90 + 2.11E + 1.70S - 20.7A - 22.3B + 15V$$
$$n = 38 \quad r^2 = 0.83 \quad s = 16 \qquad (9.6)$$

where E is the excess molar refraction, S is the dipolarity/polarizability, A is the H-bond acidity, B is the H-bond basicity, and V is the size parameter of the molecules. The number of compounds is n, r^2 is the multiple regression coefficients, and s is the standard error of the estimate. The coefficients of the descriptors show the impact of the descriptor on the overall value of percentage of absorbed dose (%Abs). The equation reveals that if all the molecular descriptors were zero, we would have still 90% absorption based on the constant. However, the introduction of H-bond donor or acceptor group would decrease the percentage absorbed drug by 20%–22%. The descriptor A is approximately 1 for an average H-bond donor group, while B is approximately 1 for an average H-bond acceptor group. The H-bond acidity and basicity scale can be considered as continuous numbers as it takes into account the electronic effects of neighboring substituents on the H-bond acidity and basicity and calculates an overall property of the molecules when more than one H-bond donor or acceptor groups are present. The question can be asked as how it is possible to have up to 5 H-bond donor groups and up to 10 H-bond acceptor groups on the molecules with potentially good absorption based on the Lipinski "rule of five." It would give a negative value for the estimated absorption in the solvation equation (Equation 9.6). It is possible because the size parameter has a positive coefficient. For every unit increase of McGowan volume term, the absorption increases by 15%. So, large molecules may have relatively large number of H-bond donor and acceptor groups; the absorption will be still reasonably good. It is worth comparing the equation obtained for human intestinal absorption with the equations obtained for cellular and intrinsic permeability measures. The equation suggests very similar molecular properties for good absorption, as shown by Equation 9.7, Equation 9.8, and Equation 9.9.

$$\log P_{\text{app}}(\text{Caco2}) = -4.4 - 0.20E + 0.26S - 1.27A - 0.24B + 0.09V$$
$$n = 53 \quad r^2 = 0.77 \quad s = 0.38 \qquad (9.7)$$
$$\log P_{\text{app}}(\text{MDCK}) = -4.31 + 0.10E + 0.19S - 1.73A - 0.79B - 0.17V$$
$$n = 30 \quad r^2 = 0.81 \quad s = 0.48 \qquad (9.8)$$
$$\log \text{Perm}_{\text{o}} = -4.03 + 1.06E - 0.95S - 2.26A - 2.84B + 2.58V$$
$$n = 43 \quad r^2 = 0.78 \quad s = 0.92 \qquad (9.9)$$

The polar terms for H-bonding always have large negative coefficients, indicating the negative effect on both permeability and absorption. The solvation equation obtained for solubility (see Chapter 6) contains even the product term of H-bond acidity and basicity with negative coefficient. This suggests that the presence of both H-bond donor and acceptor group reduces the chance for good solubility and as a consequence good absorption. The excess molar refraction (E) and the dipolarity/polarizability parameters (S) have small coefficients and they both have positive signs for absorption and MDCK cell permeability. Zhao et al. [31] collected literature data for intestinal absorption in rats as well and constructed the Abraham equation. Rat is the most often used animal species for *in vivo* studies in drug discovery. The obtained data are supposed to be indicative of the behavior of compounds in human. Therefore, it is important to see whether the same molecular properties influence the rat absorption. The solvation equation obtained for the training set of molecules (58) is shown by Equation 9.10.

$$\%\text{Abs(rat)} = 96 + 6.35E + 2.47S - 20.7A - 15.9B + 6.33V$$
$$n = 58 \quad r^2 = 0.77 \quad s = 10 \tag{9.10}$$

There are small differences in the coefficients. The coefficient for the excess molar refraction (E) is slightly higher, while the coefficients for the H-bond basicity (B) and size (V) are smaller in rat than in human absorption. The E and S parameters and even the size (V) parameters are not highly significant in the equations; so, practically the H-bond donor/acceptor properties of the molecules drive the absorption.

Strictly speaking, the Abraham models make use of *in silico* calculated molecular descriptors as independent variables; these descriptors can be measured, as discussed in Chapter 5. The main advantage of the models is that it always uses the same molecular property description; so, it is possible to compare models without the necessity of using the same set of compounds to check correlations between various dependent variables. It is also not a "black box" model, which means that medicinal chemists can get information about the effect and the magnitude of changing certain molecular properties on absorption.

There are other *in silico* approaches to predict intestinal absorption using molecular descriptors. Klopman et al. [32], for example, have published a model that uses 37 structural descriptors (fragments) and the H-bond donor and acceptor counts for the estimation of absorption using the data of 417 known drug molecules. Instead of using elementary fragments obtained from the chemical structure, they used molecular weight and calculated lipophilicity ($c \log P$) as well. Interestingly, they found that calculated lipophilicity and solubility were not sufficiently significant to be included in their model. The most significant descriptor was the H-bond donor count, more precisely, the number of OH and NH_2 groups; they have been found to reduce the absorption significantly. They identified other molecular fragments that are not advantageous for good oral absorption; these include carbonyl group connected to tetrahydropyrrol ring, quaternary nitrogen, and aliphatic parts of condensed ring; SO_2 group connected to aromatic ring, sulfides, and amides; and dialcohols connected to aromatic rings. The Klopman fragments turned to be very useful for developing *in silico* calculation of the Abraham descriptors too, and they have been used successfully in other ADME property predictions.

Some *in silico* models try to model the more complex bioavailability that includes potential clearance and metabolism of the molecules too. Such *in silico* models are able to pick up structural features of the molecules, which may be the target of metabolizing enzymes, thus reducing the exposure of the compounds not necessarily because of low absorption but because of high clearance. One of the most interesting findings has been published by Martin [33] when developing an *in silico* bioavailability score. She has found that different structural features determine oral bioavailability when the compound has different charge states. The bioavailability of negatively charged molecules is above 85% if the polar surface area of the molecule is less than 75 Å. When the polar surface area of a negatively charged drug falls between 75 and 150 Å, the bioavailability drops to an average 56%, while anions with larger polar surface area have poor absorption (average 11%). On the other hand, neutral molecules, zwitterions, and positively charged compounds have higher than 10% absorption when they pass the "rule of five." While the Abraham equation does not distinguish between charged molecules, because the descriptors relate to the neutral form of the molecules, the "bioavailability score" suggests a different model for negatively charged molecules. This finding is in agreement with the observation regarding cell permeability. Negatively charged molecules do not partition into the positively charged surface of phosphatidylcholine membrane, while positively charged compounds may be able to permeate through according to the pH partition hypothesis [34]. Dressman et al. [35] reviewed the computational approaches for the simulation of oral absorption, including the time course of dissolution and permeability as well. Several computer simulation programs are available, which are based on physiological as well as physicochemical considerations, such as Gastroplus from Simulation Plus Inc. (http://www.simulations-plus.com/Products.aspx?grpID=3&cID=16&pID=11). The accuracy of the simulation can be improved significantly when the appropriate measured physicochemical properties serve as input, for example, the solubility in biorelevant media and dissolution rate of a dosage formulation enhance the accuracy of the predicted absorption.

EFFECT OF ACTIVE TRANSPORT AND METABOLIZING ENZYMES ON ORAL ABSORPTION AND BIOAVAILABILITY

The scope of this book is to describe the physicochemical properties and their measurements and applications that influence the *in vivo* behavior of drug discovery compounds. Although it is essential to understand the expected behavior of compounds based solely on their physicochemical properties, it is also necessary to mention at least some of the important biological processes that may change the behavior of compounds *in vivo*. When the physicochemical principles of the absorption process are investigated, it is inevitable to have an insight into the active transport and metabolic transformation processes. These processes are governed by specific structural features of the molecules, but the overall physicochemical properties of the molecules also play an important role. Thus, several organic acid transporters (OATs) may help in the absorption of negatively charged acids. Hepatic uptake transporters help to extract drugs with a molecular weight of over 400 from the blood and there are hepatic efflux transporters that help to excrete high molecular weight ionizable compounds into the bile. When the compound undergoes metabolic transformation in the liver, it also reduces the oral exposure. The CYP isoforms are responsible for most of the biotransformations in the liver [36]. Thus, CYP1A2 (12% of total hepatic CYP) targets compounds with an average log *P* of 2, which contain poly(hetero)aromatic amines and amides. The CYP3A4 is another major component of the CYP isoforms, and it takes 30% of the total hepatic CYP, but it also can be found in the gut. CYP3A4 is responsible for metabolizing large and lipophilic compounds (up to log *P* 7) that are uncharged at physiological pH. Amides and amines are mostly metabolized by CYP 2C19, while basic positively charged compounds are transformed by CYP 2D6. There is an interesting association between some CYP enzymes with the important transmembrane pump protein, PgP expressed by the MDR1 gene. PgP can also be found in the blood–brain barrier, protecting the brain from large lipophilic compounds. PgP works in concert with CYP3A4 to eliminate potentially harmful xenobiotics. If PgP is not able to efflux all the potential harmful exogenous compounds, then CYP3A4 transforms them. Interestingly, these processes mostly target large and lipophilic compounds that have already been considered to be poorly absorbed in several absorption models.

SUMMARY

Bioavailability of a drug molecule is a result of complex physicochemical and biological processes. It depends significantly on the dose and the formulation of the drug molecules, the individual transit time, GI fluid volume and composition, and quantity and quality of the metabolizing enzymes in human. The human intestinal absorption can also be divided into processes that are governed by physicochemical properties of the molecules and the physiological conditions. At early drug discovery stage, we can only control the physicochemical properties of new molecules. On the basis of the analysis of large number of orally available known drug molecules, it has been found that the physicochemical property space is limited for drugs that are orally available with the exception of some natural compounds that are most probably absorbed by active biological transport processes. The lipophilicity of orally available drug molecules does not exceed log *P* 5, the molecular weight is less than 500 Da, and they contain less than 5 H-bond donor and less than 10 H-bond acceptor groups according to the Lipinski "rule of five." These physicochemical properties affect the solubility and permeability of the compounds, which are the two major attributes for good oral absorption. The amount of drug molecules defined by the dose has to be dissolved in the GI tract to be able to permeate through the GI wall. However, even if the dose is fully dissolved, it is not necessarily able to permeate well. The requirements for good permeability are somewhat in contradiction to the requirement of good solubility. Polar compounds are generally soluble in water, but they are not able to go through the hydrophobic inner core of membrane bilayers. Therefore, an optimum lipophilicity of the compound has to be found, which provides reasonably good solubility and permeability of the compound at the same time. The size of the molecules affects both the solubility and permeability. Small polar compounds may have good solubility and may be absorbed via the paracellular route through the tight junctions. Larger molecules require higher lipophilicity for good permeability, which is described in the log *D*–cMR permeability model. This model, however, does not include the upper limit of lipophilicity, which would reduce the absorption because of loss of solubility. When the dose is large, the solubility has to be higher too. Low dose compounds may need lower solubility. It is important to consider the permeability and the solubility of the compounds relative to each other. Higher solubility compounds may be well absorbed even if their permeability is not too great, and lower solubility compounds with higher permeability may show good absorption.

The AP of the compound is based on the product term of solubility and permeability (or lipophilicity). The product term shows, mathematically, the interplay between these two physicochemical properties of the compounds. If any of the terms are too low, the product term (i.e., the AP) will be low. As both solubility and lipophilicity are dependent on the charge in an opposite way, the product term will be independent of the charge as long as both solubility and permeability (lipophilicity) are measured at the same pH. The presence of charge decreases the lipophilicity but increases the solubility; so, the product term is practically not affected. This finding helps the development of fast assays of solubility and permeability of large number of compounds in the molecular design stage because there is no need to measure the pH dependence of these properties. However, it has also been observed that there is a difference between the negatively charged and the positively charged compounds. The presence of negative charge greatly reduces the chances of good permeability, while the positive charge is not so detrimental for good permeability, although both types of charge increase the solubility.

The Abraham solvation equations that have been constructed for intestinal absorption of human and rat as well as solubility, lipophilicity, and permeability provide an invaluable help to medicinal chemists to design potent compounds with good oral absorption. The addition or removal of a functional group changes the five molecular descriptors of the compound that is used in the model (excess molar refraction, dipolarity/polarizability, H-bond acidity and basicity, and McGowan volume). From the coefficients of the descriptors, it is possible to estimate the overall change of the expected absorption, solubility, and permeability of the molecules. From the equation, it can be concluded that addition of polar substituents, especially H-bond donor and acceptor groups, dramatically reduces the estimated oral absorption; this is mainly due to the reduced permeability of polar compounds. The coexistence of H-bond donor and acceptor groups in a molecule may reduce the solubility. The size parameter has a smaller influence on the overall oral absorption. The size normally helps a compound's partitioning out of the aqueous phase as it requires large amount of energy to create large cavity in water by destroying the H-bonding and dipole–dipole structure of the water.

In order to optimize the absorption of drug discovery compounds, all these molecular properties and their effects on solubility and permeability have to be considered. Measured values of solubility and permeability (lipophilicity) greatly improve the estimation of *in vivo* oral absorption of the compounds. However, it is important not to underestimate the role of active transporters in the gut (PgP) and the metabolism both in the gut (CYP 3A4) and in the liver (various P450 isoforms) as they may reduce the exposure and bioavailability of the compounds in spite of optimum physicochemical properties.

QUESTIONS FOR REVIEW

Q9.1 How can the Lipinski "rule of five" be related to solubility and permeability?

Q9.2 How can we model the absorption using the lipophilicity and the size of the molecules as descriptors?

Q9.3 What are the limitations of the log *D*–cMR model of absorption?

Q9.4 What are the criteria of good absorption based on the definition of Class I compounds by the BCS?

Q9.5 How can the MAD be calculated?

Q9.6 How the pH affects the solubility and permeability and ultimately the AP?

Q9.7 Which molecular descriptors have the most significant effect on oral absorption?

Q9.8 What are the limitations of the Abraham solvation equation models for absorption?

Q9.9 Which measured physicochemical properties can be used to estimate human intestinal absorption of drug discovery compounds and how?

Q9.10 What are the limitations of using only physicochemical properties in modeling absorption?

REFERENCES

1. Kay, C. (2011) Introduction. In: ADMET for Medicinal Chemists—A Practical Guide, Eds Tsaioun, K., Kates, S. A., Wiley-Interscience, Hoboken, NJ, USA pp. 1–28.
2. Lipinski, C. A., Lombardo, F., Dominy, B. W., Feeny, P. J. (1997) Experimental and computational approaches to estimate solubility and permeability in drug discovery and development settings. *Advanced Drug Delivery Reviews*, *23*, 3–25.

3. Camenish, G., Folkers, G., van de Waterbeemd, H. (1996) Review of theoretical passive drug absorption models: Historical background, recent developments and limitations. *Pharmaceutica Acta Helvetiae*, *71*, 309–327.
4. Wenlock, M. C., Austin, R. P., Barton, P., Davis, A. M., Leeson, P., D. (2003) A comparison of physiochemical property profiles of development and marketed oral drugs. *Journal of Medicinal Chemistry*, *46*, 1250–1256.
5. Egan, W. J., Merz Jnr, K. M., Baldwin, J. J. (2000) Prediction of drug absorption using multivariate statistics. *Journal of Medicinal Chemistry*, *43*, 3867–3877.
6. Teague, S., Davis, A. M., Leeson, P. D., Oprea, T. I. (1999) The design of leadlike combinatorial libraries. *Angewandte Chemie, International Edition*, *38*, 3743–3748.
7. Rishton, G. M. (2003) Non-leadlikeness and leadlikeness in biochemical screening. *Drug Discovery Today*, *8*, 86–96.
8. Veber, D. F., Johnson, S. R., Cheng, H-Y., Smith, B. R., Ward, K. W., Kopple, K. D. (2002) Molecular properties that influence the oral bioavailability of drug candidates. *Journal of Medicinal Chemistry*, *45*, 2615–2623.
9. Bhal, S. K., Kassam, K., Peirson, I. G., Pearl, G. M. (2007) The rule of five revisited: Applying log D in place of log P in drug-likeness filters. *Molecular Pharmaceutics*, *5*, 556–560.
10. Van de Waterbeemd, H., Smith, D. A., Beaumont, K., Walker, D. K. (2001) Property-based design: Optimization of drug absorption and pharmacokinetics. *Journal of Medicinal Chemistry*, *44*, 1313–1333.
11. Collander, R., Bärlund, H. (1932) Permeabilitätstudien an Chara eratophylla. *Acta Botanica Fennica*, *11*, 72–114.
12. Sha'afi, R. I., Gary-bobo, C. M., Solomon, A. K. (1971) Permeability of red cell membranes to small hydrophilic and lipophilic solutes. *Journal of General Physiology*, *58*, 238–258.
13. Hill, A. P., Hyde, R. M., Robertson, A. D., Wollard, P. M., Glen, R., Martin, G. R. (1994) Oral delivery of 5-HT1D receptor agonists: Towards the discovery of 311c90, a novel anti-migraine agent. *Headache*, *34*, 308–309.
14. Hersey, A. (2002) ADME in-silico models – Tools for drug discovery. Presentation at UK-QSAR Autumn Meeting (http://www.ukqsar.org/2011/07/12/autumn-meeting-2002/). Accessed 2013 Aug 16.
15. Young, R. (2008) Physical Chemistry: A medicinal chemist's perspective. Physchem Forum 5 (http://physchem.net/Symp05/symp05.htm). Accessed 2013 Aug 16.
16. Borthwick, A. D., Davies, D. E., Exall, A. M., Hatley, R. J. D., Hughes, J. A., Irving, W. R., Livermore, D. G., Sollis, S. L., Nerozzi, F., Valkó, K. L., Allen, M. J., Perren, M., Shabbir, S., Wollards, P. M., Rice, M. A. (2006) 2,5-diketopiperazines as potent, selective, and orally bioavailable oxytocin antagonists. 3. Synthesis, pharmacokinetics and *in vivo* potency. *Journal of Medicinal Chemistry*, *49*, 4159–4170.
17. Valko, K. (2002) Measurements and predictions of physicochemical properties. In: High-throughput ADMETox Estimation: *In Vitro* and *In Silico* Approaches. Eds Darvas, F., Dorman, Gy., Biotechnologies Press, Westborough, MA, pp. 1–24.

18. Amidon, G. L., Lennernäs, H., Shah, V. P., Crison, J. R. (1995) A theoretical basis for a biopharmaceutic drug classification: The correlation of *in vitro* drug product dissolution and *in vivo* bioavailability. *Pharmaceutical Research, 12*, 413–420.
19. Lindenberg, M., Kopp, S., Dressman, J. B. (2004) Classification of orally administered drugs on the World Health Organization model list of essential medicines according to the Biopharmaceutics Classification System. *European Journal of Pharmaceutics and Biopharmaceutics, 58*, 265–278.
20. Butler, J., Dressman, J. (2010) The Developability Classification System: Application of biopharmaceutics concepts to formulation development. *Journal of Pharmaceutical Sciences, 99*, 4940–4954.
21. Van der Meer, J. W. M., Keuning, J. J., Scheijgrond, H. W., Heykants, J., Van Cutsem, J., Brugmans, J. (1980) The influence of gastric acidity on the bioavailability of ketoconazole. *Journal of Antimicrobial Chemotherapy, 6*, 552–554.
22. Johnson, T. W., Dress, K. R., Edwards, M. (2009) Using the Golden Triangle to optimize clearance and oral absorption. *Bioorganic Medicinal Chemistry Letter, 19*, 5560–5564.
23. Dressman, J. B., Amidon, G. L., Fleisher, D. (1985) Absorption potential: Estimating the fraction absorbed for orally administered compounds. *Journal of Pharmaceutical Sciences, 74*, 588–589.
24. Macheras, P. E., Symillides, M. Y. (1989) Toward a quantitative approach for the prediction of the fraction of dose absorbed using the absorption potential concept. *Biopharmaceutical Drug Disposition, 10*, 43–53.
25. Sanghvi, T., Ni, N., Yalkowski, S. H. (2001) A simple modified absorption potential. *Pharmaceutical Research, 18*, 1794–1796.
26. Ni, N., Sanghvi, T., Yalkowski, S. H. (2002) Independence of the product of solubility and distribution coefficient of pH. *Pharmaceutical Research, 19*, 1862–1866.
27. Johnson, K. C., Swindell, A. C. (1996) Guidance in the setting of drug particle size specifications to minimize variability in absorption. *Pharmaceutical Research, 13*, 1795–1798.
28. Lipinski, C. A. (2000) Drug-like properties and the causes of poor solubility and poor permeability. *Journal of Pharmacological and Toxicological Methods, 44*, 235–249.
29. Sun, D., Yu, L. X., Hussain, M. A., Wall, D. A., Smith, R. L., Amidon, G. L. (2004) *In vitro* testing of drug absorption for drug "developability" assessment: Forming an interface between *in vitro* preclinical data and clinical outcome. *Current Opinion in Drug Discovery and Development, 7*, 75–85.
30. Zhao, Y., Le, J., Abraham, M. H., Hersey, A., Eddershaw P. J., Luscombe, C. N., Butina, D., Beck, G., Sherborne, B., Cooper, I., Platts, J. A. (2001) Evaluation of human intestinal absorption data and subsequent derivation of a quantitative structure–activity relationship (QSAR) with the Abraham descriptors. *Journal of Pharmaceutical Sciences, 90*, 749–784.
31. Zhao, Y. H., Abraham, M. H., Hersey, A., Luscombe, C. N. (2003) Quantitative relationship between rat intestinal absorption and Abraham descriptors. *European Journal of Medicinal Chemistry, 38*, 939–947.

32. Klopman, G., Stefan, L. R., Saiakhov, R. D. (2002) ADME evaluation 2. A computer model for prediction of intestinal absorption in humans. *European Journal of Pharmaceutical Sciences*, *17*, 253–263.

33. Martin, Y. C. (2005) A bioavailability score. *Journal of Medicinal Chemistry*, *48*, 3164–3170.

34. Shanker, L. S., Tocco, D. J., Brodie, B. B. Hogben, C. A. M. (1958) Absorption of drugs from the rat small intestine. *Journal of American Chemical Society*, *123*, 81–88.

35. Dressman, J. B., Thelen, K., Willmann, S. (2011) An update on computational oral absorption simulation. *Expert Opinion in Drug Metabolism and Toxicology*, *7(11)*, 1345–1364.

36. Lewis, D. F., Ito, Y. (2009) Human P450s involved in drug metabolism and the use of structural modelling for understanding substrate selectivity and binding affinity. *Xenobiotica*, *39*, 625–635.

10

MODELS WITH MEASURED PHYSICOCHEMICAL AND BIOMIMETIC CHROMATOGRAPHIC DESCRIPTORS—DISTRIBUTION

The distribution process of an oral drug molecule starts immediately after successful absorption with the help of the moving bloodstream. It is a reasonably quick process as the major function of the circulatory system is to transport compounds to and from the cells. The general circulation helps in distributing hormones and other essential small molecules to the cells while maintaining the pH and ionic strength. The exchange of substances between blood and the body cells occurs in the capillaries. The capillary wall is also specialized for exchange of substances with the interstitial fluid. The total length of capillaries in humans is estimated to be more than 50,000 miles. The diameter of the capillaries is under neural and endocrinal controls. The capillaries are so small that the blood cells travel in single file. No cell in the body is more than 100 μm away from the capillaries. The speed of the flow is slow in the capillaries. On the basis of these features, it is easy to imagine that a dynamic equilibrium process develops quickly between the moving plasma compartment and the stationary tissue compartment. In this respect, the situation is very similar to a chromatographic dynamic distribution process of a sample compound between the mobile and the stationary phases. Both the dynamic equilibrium processes take place on large surfaces between a moving and a stationary phase. The transport processes between the plasma and the tissue compartments are also governed by physicochemical and thermodynamic processes.

The pharmacologically active compound is designed to interact with the target enzyme or receptor. However, only very small fractions of the administered

Physicochemical and Biomimetic Properties in Drug Discovery: Chromatographic Techniques for Lead Optimization, First Edition. Klara Valko.

drug actually reach the target *in vivo*. The distribution of the compound in the body depends on its affinity to many other components, such as proteins, nucleotides, and phospholipids. As they are present in much larger volumes than the target enzyme or receptor, they significantly reduce the amount of the active molecule at the site of action. The distribution of the drug molecule *in vivo* depends on its properties. It may have higher local concentrations in certain tissues or organs. The elimination process also starts at the same time as the distribution process. The elimination process takes place in the liver microsomes via metabolic transformation or in the kidney via the renal excretion process. Monitoring the drug distribution starts with measuring the compound concentration in the plasma (nowadays, it is more common to use whole blood) as a function of time. The plasma concentration versus the time profile provides us with information about the distribution and the elimination processes. The simplest profile can be described by one- or two-compartment models, assuming that the drug distributes between the plasma and the tissue compartments. In order to achieve a constant pharmacologically active concentration of the drug molecule *in vivo*, repeated administration is necessary to compensate for the elimination process. Drugs that have a higher elimination rate have to be administered more frequently to ensure the necessary concentration for the pharmacological activity. Up to the 1970s, the half-life was the most often used term to measure the rate of elimination of the drug from the body. It was supposed that higher volume of distribution means longer half-life as well. It has been demonstrated in Chapter 3 that the steady-state volume of distribution does not correlate very well with the half-life using the data of 85 known drug molecules (see Figure 3.8). In 1973, it was recognized by Rowland et al. [1] that the volume of distribution (V_d), which was obtained from the dose and the average plasma concentration, does not show correlation with the half-life and the concept of clearance was introduced. Equation 10.1 shows the simple concept of volume of distribution (V_d).

$$V_d = \frac{\text{Dose (total amount of drug)}}{\text{Plasma concentration of the drug}} \tag{10.1}$$

It has been observed that the distribution of a drug changes depending on disease state, age, and other parameters; therefore, Benet and Galeazzi [2] defined the so-called steady-state volume of distribution (V_{ss}), which was independent of the half-life or clearance. The steady-state volume of distribution is defined as the product term of clearance and MRT that replaced the half-life.

The steady-state volume of distribution is proportional to the compound's partition coefficient (K) between the plasma and the tissue compartments as described by Equation 10.2.

$$K = \frac{\mathrm{V}_{ss} - V_p}{V_t} \tag{10.2}$$

V_{ss} is the steady-state volume of distribution, whereas V_p and V_t are the volumes of the plasma and tissue compartments, respectively. These volumes can be considered as being constant in each individual.

The steady-state volume of distribution refers to an *in vivo* distribution process and cannot be measured *in vitro*. At early stages of drug discovery, they use various laboratory animals (most frequently rats) for the determination of this important distribution parameter. Therefore, models that can provide an estimate of this *in vivo* parameter without animal experiments are of great importance. The volume of distribution gives us information about the proportion of the administered drug molecules in the plasma and the tissue compartments. It is possible that compounds with similar volumes of distribution concentrate in different tissues; therefore, the volume of distribution can be further divided into various tissue distributions.

Some drugs distribute more into certain types of tissues than into others. Ideally, it would be advantageous if we could design molecules that distribute only into the tissue or organ where the pharmacological target is. Measuring tissue distribution of a drug molecule *in vivo* is quite difficult. One method is to sacrifice laboratory animals after administration of the drug at various time points. We can apply *in vivo* tissue dialysis methods or other methods to measure the amount of compound in various parts of the body. The quantitative whole body autoradiography (QWBA) is an expensive but reliable method to assess the concentration of the administered radio-labeled molecule in various organs at a given time point after administration. It would have a great impact on drug discovery if we were able to estimate compound distribution without using animal or human experiments. Some of the best-known tools are the so-called physiologically based pharmacokinetic (PBPK) modeling method [3–6]. These are based on mathematical models for the estimation of various PK parameters using measured or calculated physicochemical properties such as lipophilicity and acid dissociation constant of the compounds. It is physiologically based as it uses the average physiological volumes of plasma and other tissues and organs that are built in as parameters in the equations. Other input parameters can be the tissue partition coefficients of compounds that can be measured using tissue homogenates. Figure 10.1 shows the schematic distribution and steady-state equilibrium processes in various compartments that serve as the basis of PBPK modeling. When the "free drug" hypothesis is true, we can assume that the free concentration of the drug is the same in the plasma compartment, and the extracellular and intracellular water compartments, which implies that either no permeability barrier exists or no active transport of the compound takes place. However, the drug has potentially very different distribution/partition coefficients between the free and bound compartments depending on their composition, namely, the amount and the types of proteins and phospholipids. When the tissue partition coefficients (K_{pt}), the plasma partition coefficient (K_{pp}), and the volumes of the free and bound compartments are known, we can calculate the amount of compounds in each compartment, and more importantly, we can calculate the free concentration of the drugs administered after a given dose.

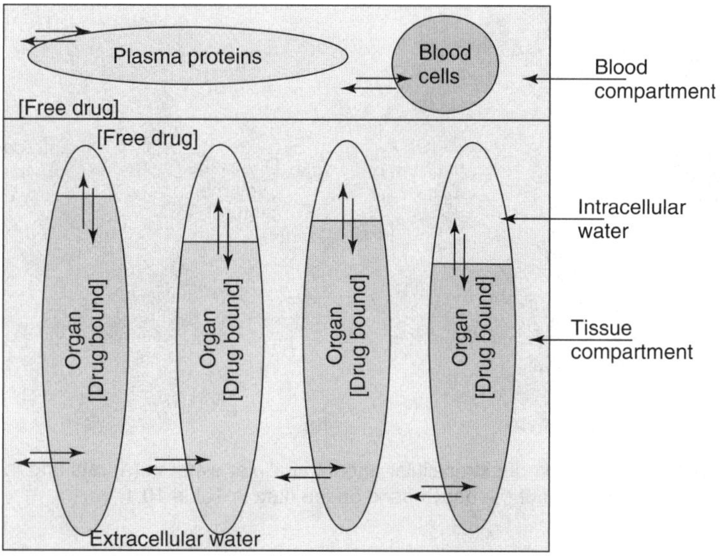

Figure 10.1 The schematic illustration of compounds *in vivo* steady-state distribution in various compartments.

Table 10.1 The average fractional volumes of protein, lipid, and water composition of muscle, plasma, and blood cells in humans and rats[a]

	Muscle		Plasma		Blood cell	
Component	Human	Rat	Human	Rat	Human	Rat
Acidic phospholipid, mg/ml	2.42	1.50	0.041	0.057	0.57	0.50
Neutral phospholipids (fractional volume)	0.0078	0.0072	0.0021	0.0013	0.0033	0.0029
Neutral lipids (fractional volume)	0.022	0.010	0.0032	0.0023	0.0012	0.0017
Total tissue water (fractional volume)	0.745	0.726	0.95	0.96	0.63	0.60
Extracellular water (fractional volume)	0.079	0.118	NA	NA	NA	NA

[a]Taken with permission from Reference [7].

Table 10.1 contains the fractional volumes of lipid, albumin, lipoprotein, and water content of muscle, blood cell, and plasma in humans and in rats found in the literature [7].

Figure 10.2 shows the fraction of rat and human extracellular and intracellular water content based on the data shown in Table 10.1. These data are quite difficult to obtain and are only approximations needed to model compounds distribution

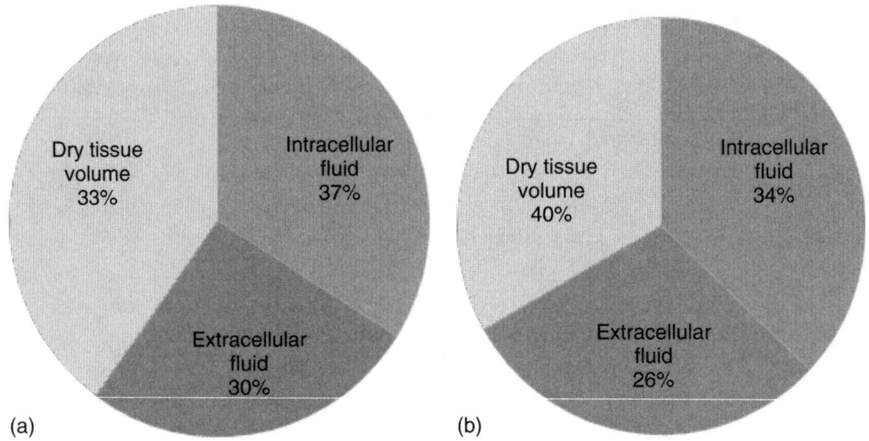

Figure 10.2 The fraction of extracellular and intracellular water in (a) rats and (b) humans relative to the dry content of the body based on the data in Table 10.1.

in various compartment. This is because an unknown amount of structured water molecules surround the proteins and phospholipids that constitute the compartment where compounds "partition" into. This is one of the reasons why the octanol/water partition coefficients are successful in modeling *in vivo* distribution, as the octanol phase also contains a large amount of water. It means that the octanol is a "wet" solvent phase just like the protein and phospholipid compartments in the tissues. When the dry protein and phospholipid content is measured, it usually does not contain the water molecules associated with the proteins and support their tertiary structures, or the water molecules that are associated with the polar head groups of the phospholipid membranes.

Tables 10.2–10.4 show the average weights and volumes of various organs in various species, and the blood content and flow are shown in Table 10.5. Figure 10.3 shows the fractions of various tissue types in rats and humans based on the data shown in Table 10.4. These data are also important as various organs contain different types and amounts of proteins and phospholipids and thus represent the different interactions of the drug molecules. This results in different tissue partition coefficients.

We can see that the percentage of muscle and adipose tissues is quite different in rats and humans. This can cause significant deviations in the distributions of drug molecules in the two species. Various procedures that can be applied to estimate the human drug distribution from various species have been published, such as allometric scaling [9,10]. Some techniques employ scaling factors such as protein binding, brain weight [11], body weight [12], or maximum life span potential [13]. Table 10.3 shows the different plasma compositions of rat and human plasma. This can cause significant differences in the measured plasma protein binding (PPB) in rat or human plasma. The free plasma concentration of drugs that bind strongly

Table 10.2 Species weight, organ weight (% of body weight) in mouse, rat, dog, and human[a]

Parameter/Organ	Mouse	Rat	Dog	Human
Total Body Weight, kg	0.02	0.25	10	70
Adipose tissue				21.42
Bone	10.73		8.10	14.29
Brain	1.65	0.57		1.71
Stomach	0.7	0.46	0.79	0.21
Small intestine	2.53	1.40	2.22	0.91
Large intestine	1.09	0.84	0.67	0.53
Heart	0.50	0.33	0.78	0.47
Kidneys	1.67	0.73	0.55	0.44
Liver	5.49	3.66	3.29	2.57
Lungs	0.73	0.50	0.82	0.76
Muscle	38.4	40.43	45.65	40.0
Skin	16.53	19.03		3.71
Spleen	0.35	0.20	0.27	0.26
Thyroid		0.005	0.008	0.03

[a]Taken with permission from Reference [8].

Table 10.3 Plasma content (in g/100 ml) of mouse, rat, dog, and human[a]

Parameter/Organ	Mouse	Rat	Dog	Human
Total plasma protein, g/100 ml	6.2	6.7	9	7.4
Plasma albumin, g/100 ml	3.27	3.16	2.63	4.18
Plasma AGP, g/100 ml	1.25	1.81	0.37	0.18

[a]Taken with permission from Reference [8].

to AGP, potentially show even greater differences as the AGP concentration in rat plasma is significantly higher than in human plasma. The blood flow in various tissues affect mostly the time versus the concentration profile of the compound distribution. It means that the maximum concentration of the drug is achieved sooner in tissues with higher blood flow. The PBPK modeling approach includes the kinetic aspects of a compound's distribution and elimination *in vivo*; therefore, the blood flow in various tissues are also important input parameters. These models are based on the assumption that electrostatic and nonspecific binding dominate passive distribution in a nonsaturating condition. It means that the compound concentration is always much lower than the available nonspecific binding sites.

Physicochemical parameters such as lipophilicity and acid dissociation constants are used as input for tissue/plasma partition coefficient estimations considering steady-state equilibrium situations. The clearance and elimination rate of the compounds are difficult to model only from the physicochemical properties of the

Table 10.4 Organ volumes (in ml), tissue and fluid volumes (in ml/kg) in different species[a]

Parameter	Mouse	Rat	Dog	Human
Brain		2.2		1,450
Liver	1.3	19.6	480	1,690
Kidneys	0.34	3.7	60	280
Heart	0.095	1.2	120	310
Spleen	0.1	1.3	36	192
Lungs	0.1	2.1	120	1,170
Gut	1.5	11.3	480	1,650
Muscle	10	245	5530	35,000
Adipose		10		10,000
Skin	2.9	40		7,800
Blood	1.7	13.5	900	5,200
Blood, ml/kg	85	54	90	74
Plasma	1	7.8	515	3,000
Plasma, ml/kg	50	31	51.5	43
Total body water	14.5	167	6036	42,000
Total body water, ml/kg	725	668	604	600
Intracellular fluid		92.8	3276	23,800
Intracellular fluid, ml/kg		371	328	340
Extracellular fluid		74.2	2760	18,200
Extracellular fluid, ml/kg		297	276	260

[a]Taken with permission from Reference [7].

Table 10.5 Blood flow (in ml/min) through the organs in mouse, rat, dog, and human[a]

Organ	Mouse	Rat	Dog	Human
Heart	0.28	3.9	54	240
Liver	1.8	13.8	3.9	1450
Kidneys	1.3	9.2	216	1240
Brain		1.3	45	700
Spleen	0.09	0.63	25	77
Gut	1.5	7.5	216	1100
Muscle	0.91	7.5	250	750
Adipose		0.4	35	260
Skin	0.41	5.8	100	300

[a]Taken with permission from Reference [7].

compounds, as the metabolic transformation depends more on the type of functional groups in the molecules than on its overall physicochemical properties. It will be highlighted that the biomimetic chromatographic physicochemical parameters have great potential to serve as input parameters for the estimation of steady-state drug distribution [4].

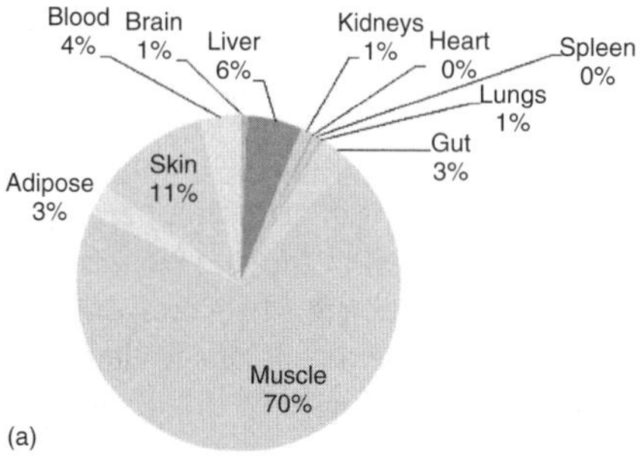

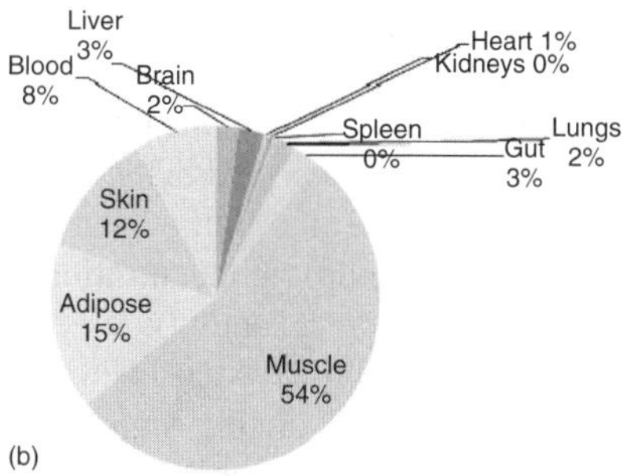

Figure 10.3 The fractions of various tissue types in (a) rats and (b) humans based on the data shown in Table 10.4.

MODELS FOR VOLUME OF DISTRIBUTION

Compounds that interact more strongly with tissue components than with the plasma components will have a higher concentration in tissues and, as a consequence, they have higher volume of distribution. To be able to consider the tissue/plasma distribution as partition equilibrium, we assume that the concentration ratio of the drug in the tissue and plasma compartments is constant and independent of the dose. This is a big assumption as a compound's binding to various tissue and plasma compartments might reach saturation. However, it is reasonable to expect that the drug concentration *in vivo* should not reach such high levels in practice. Some specific binding sites with strong binding affinity might be saturated by the drug molecules in therapeutic concentrations, but as they are present in much smaller quantities than the nonspecific "silent" binding sites, they do not significantly influence the volume of distribution. By ignoring these specific and often saturable binding sites, we do not want to underestimate their importance in drug discovery. Active transporters, saturable metabolizing enzymes, may cause concerns regarding drug–drug interactions.

The aim of using these simplified models presented in this chapter is to establish the overall physicochemical and biomimetic binding profile of a molecule that provides the best estimates for their *in vivo* distribution. If the properties of the molecules do not allow the achievement of the desired volume of distribution or the free steady-state concentration for a given dose, it should be discarded from further development. These calculations are simple and the measurements are inexpensive, yet they are still accurate enough to estimate the approximate dose to achieve the clinically efficacious free concentration at the target.

When the target is in certain tissues or organs, it is preferable to have a high volume of distribution of the compound as the majority of the administered dose will distribute to the tissues and not simply circulate in the plasma compartment. However, when the target is in the plasma compartment, a low volume of distribution is preferred as a high fraction of the administered dose will be where the target enzyme or receptor is. Therefore, the desired volume of distribution value depends on the location and the amount of the target protein.

The steady-state volume of distribution and the clearance can be used to express the MRT of a compound that can be used to determine the required dosing interval of the drug. The aim of understanding these PK and *in vivo* distribution parameters is to relate the drug concentration (the free concentration at the site of action) to the pharmacological activity. When the relationship between the drug concentration and the efficacy *in vivo* has been established, it enables the physicians to decide the dose strength and dosing regimen of the drug for the treatment of the disease and relate the PK parameters to the PD properties.

Several attempts have been published to develop *in vitro* methodology for the estimation of *in vivo* drug distribution. Dal Pozzo et al. [15] described an attempt to design and develop an apparatus for the evaluation of a drug's *in vivo* distribution. They built an apparatus to measure the effect of PPB on tissue distribution using a three-phase system containing water, octanol, and water with dissolved bovine serum albumin (BSA), as is shown in Figure 10.4.

The aqueous phases were separated by a filter and one of the aqueous compartments contained BSA. They injected the drug molecule, and by measuring

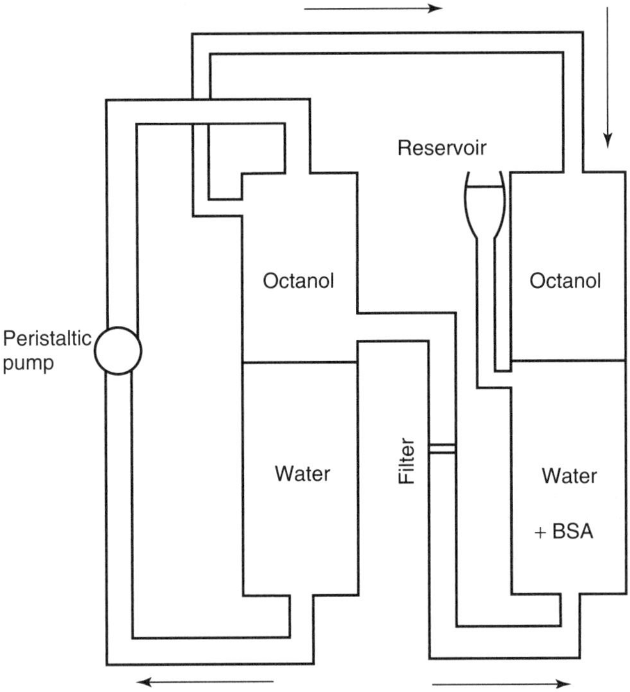

Figure 10.4 Three-phase system water/octanol/water + BSA for *in vitro* evaluation of the compound distribution between the tissue and plasma compartments. (Adopted with permission from Reference [15].)

the concentration of the drug in the aqueous phase as a function of time with and without the addition of serum albumin, they derived a distribution profile for the compound. They investigated the distribution of naproxen, ketoprofen, carbamazepine, and denzimol and concluded that the profiles obtained in the apparatus were very similar to the *in vivo* distribution profiles of these drugs. The octanol phase served to model the compound's tissue partition, whereas the aqueous phase with serum albumin modeled the PPB. The peristaltic pump imitated the blood flow. The setup was quite ambitious, as it also aimed to model the dynamic aspect of the distribution process. The obtained results for the plot of time versus aqueous concentration profile showed great similarity to the *in vivo* time versus the free plasma concentration profile for several investigated drugs. The publication demonstrates the importance of studying three-phase equilibrium distribution. The physicochemical principle of calculating the tissue/plasma partition coefficient of a compound is shown in Figure 10.5.

The tissue/plasma distribution coefficient (K) can be expressed from the tissue and plasma distribution coefficients (K_{pt} and K_{p}, respectively) of the compound according to Equation 10.3.

$$K = \frac{K_{\mathrm{pt}}}{K_{\mathrm{p}}} \tag{10.3}$$

The expression of the tissue/plasma distribution from the plasma and tissue binding is possible when the "free drug hypothesis" (see more details in Chapter 3) is valid and the free aqueous-phase concentration is the same in the tissue and plasma compartments. It is also important that no active transport or permeability barrier disturbs the equilibrium. The difference between the logarithmic value of the tissue and plasma distribution coefficients is equal to the logarithmic values of the tissue/plasma distribution coefficient.

By knowing the plasma binding and the octanol/water partition coefficients of the compounds in the *in vitro* experiment shown in Figure 10.4, the partition coefficient between the plasma and the octanol phase can be calculated. The

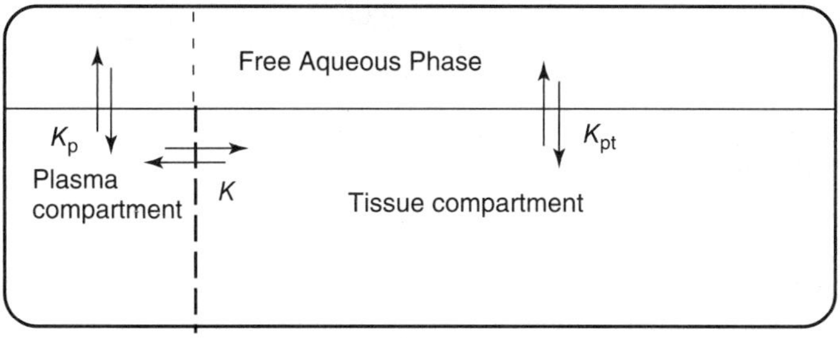

Figure 10.5 The physicochemical principle of the expression of the tissue/plasma distribution coefficient (K) of a compound as the quotient of the tissue distribution coefficient (K_{pt}) and the plasma distribution coefficient (K_{p}).

octanol phase is supposed to model the tissues in this experiment, and the octanol/water partition coefficient models the tissue partition. Shumann et al. [16] suggested using tissue homogenates in the *in vitro* experiment. The tissue-binding data so obtained were used for the estimation of tissue/blood or tissue/plasma ratio of compounds *in vivo*. They used diluted rabbit liver, lung, kidney, and skeletal muscle homogenates and ultrafiltration method for the *in vitro* tissue-binding determination and ultradialysis method for the *in vivo* tissue distribution determination. The results have clearly shown that the *in vitro* determined tissue binding showed good agreement with the *in vivo* tissue-binding data. Highly metabolized drugs in the liver such as propranolol showed the highest *in vitro–in vivo* discrepancy among the investigated drugs. They noticed that cationic and lipophilic drugs provided large tissue-to-plasma ratio and bound strongly to lung tissue. They have expressed the binding as unbound fraction in plasma and tissues. It is very common to express the compound's distribution coefficients in biological tissues, blood, and plasma by unbound fraction (f_u) terms. The unbound fraction is calculated from the total concentration (c_{total}) of the compound added to the tissue homogenates and the measured free unbound concentration (c_{free}) of the compounds shown by Equation 10.4.

$$f_u = \frac{c_{free}}{c_{total}} \tag{10.4}$$

The unbound fraction shows a nonlinear relationship with the distribution coefficient (see Equation 10.5 and Figure 10.6).

$$K_p = \frac{c_{total} - c_{free}}{c_{free}} = \frac{1 - f_u}{f_u} \tag{10.5}$$

Jansson et al. [6] extended the study of tissue bindings for 49 drug molecules by measuring tissue partition (K_{pt}) for the brain, heart, lung, gut, liver, kidney, bone, skin, muscle, and the adipose tissues. They have also tried to substitute the tissue partition coefficients by octanol/water partition coefficient (log P), octanol/water distribution coefficients at pH 7.4 (log D) and oil/water distribution coefficients at pH 7.4. They found that the log D values improved the correlations for most of the tissues, whereas the adipose tissues could be better modeled with oil/water distribution coefficients. The lung tissue binding showed the best correlation with the log P values for neutral and basic compounds. The study clearly demonstrated that using only one liquid/liquid partition coefficient, such as octanol/water log P, is not enough to describe compounds binding to various tissues. The results also revealed that neutral, basic, acidic, and amphiphilic compounds partition very differently for different tissue types, and the charge state has a great impact on the overall tissue/plasma distribution of the compounds. The adipose tissue, which represents 15% of the tissues in humans, has a different behavior from the octanol/water distribution, and it is better modeled by the completely hydrophobic oil and water distribution system where the nonaqueous phase does not contain as much water or polar hydroxyl group as the octanol. Tissue-binding and PPB measurements are time consuming, expensive, and require biological materials in comparison to

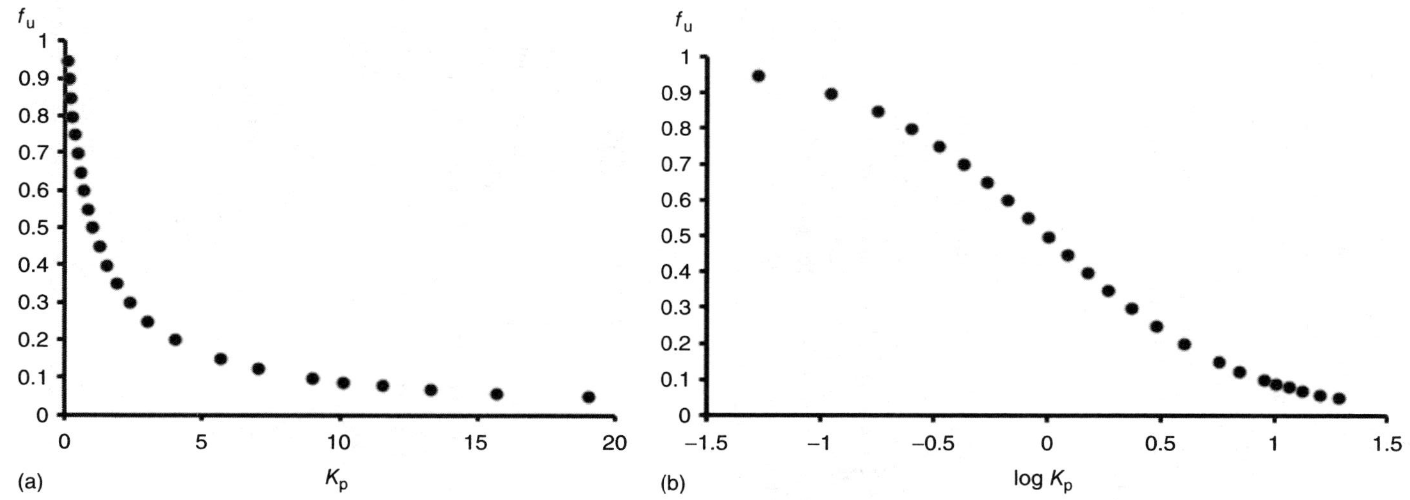

Figure 10.6 The nonlinear relationship between the unbound fraction (f_u) measure of a compound's tissue or plasma protein binding and the compound distribution coefficient ((a) K_p and (b) log K_p).

the determination of liquid/liquid partition coefficients. Therefore, several efforts have been published for the estimation of *in vivo* volume of distribution using only physicochemical property measurements. Lombardo et al. [17] published a method for the prediction of volume of distribution using chromatographically determined lipophilicity ($E \log D$, see more details in Chapter 5), the fraction of ionization of a basic compound using the acid dissociation constant and the Henderson–Hasselbalch equation (see more details in Chapter 8) and the PPB data. They have used the Øie–Tozer equation [18] that expresses the steady-state volume of distribution from the unbound fraction determined in plasma and tissues as shown in Equation 10.6.

$$V_{ss} = V_p(1 + R_{E/I}) + f_u V_p \left(\frac{V_E}{V_p} - R_{E/I} \right) + \frac{V_R f_u}{f_{ut}} \quad (10.6)$$

where V_{ss} is the steady-state volume of distribution, V_p and V_E are the plasma and the extracellular fluid volumes, respectively, and $R_{E/I}$ is the ratio of extravascular to intravascular proteins. The symbols f_u and f_{ut} are the unbound fraction in plasma and tissues. In humans, the average plasma and tissue volumes expressed per body weight (l/kg) (V_p and V_E) are 0.0436 and 0.151, respectively, whereas the ratio of the extra and intravascular proteins is 1.4. Rearranging Equation 10.6, the unbound fraction in tissues (f_{ut}) can be expressed using the constant values of the plasma and tissue volumes, the steady-state volume of distribution, and the unbound fraction in plasma (f_u) as is shown by Equation 10.7.

$$f_{ut} = \frac{V_R f_u}{[V_{ss} - V_p - (f_u V_E)] - [(1 - f_u) R_{E/I} V_p]} \quad (10.7)$$

The literature unbound fraction in tissue (f_{ut}) data of 64 neutral and basic drugs have been modeled using the chromatographically determined lipophilicity ($E \log D$), fraction of ionization at pH 7.4 ($f_{i(7.4)}$) and the unbound fraction in plasma (f_u) measured by equilibrium dialysis as shown by Equation 10.8.

$$\log f_{ut} = -0.0389(\pm 0.1012) - 0.1739(\pm 0.0628) E \log D - 0.8324(\pm 0.1205) f_{i(7.4)} + 1.0400(\pm 0.1376) \log f_u$$

$$n = 64 \quad r^2 = 0.8839 \quad s = 0.3998 \quad F = 152.25 \quad (10.8)$$

The n is the number of compounds, r^2 is the multiple correlation coefficient, s stands for the root-mean-square error, and F is the Fischer test value, which is highly significant. The model shows that the tissue binding is higher for more lipophilic compounds and for compounds for which the fraction of positively charged molecules is high. In principle, the lipophilicity and the fraction of positive charge in the equation describes the tissue partition. The tissue partition contributes positively to the volume of distribution, whereas the PPB reduces it. That is why they have negative and positive regression coefficients. Considering the logarithmic values of the bindings, the volume of distribution has been shown to be proportional to the difference of compounds binding to tissues and plasma proteins. However, the model established by Lombardo et al. [17] could not describe the behavior of acidic compounds.

The biomimetic chromatographic binding data described in detail in Chapter 5 proved to be able to model the volume of distribution of neutral, basic, acidic, and zwitterionic drug molecules as well. Hollósy et al. [19] applied the HSA binding data obtained from the retention times of the compounds on the chemically bonded HSA stationary phase, and the IAM partition data obtained from the retention times on the IAM stationary phase. The mechanistic model is based on the assumption that the major difference between the plasma and tissue compartments is the absence and the presence of phospholipids. Thus, the relative binding strength of the compounds to these major body constituents drives the volume of distribution. They have collected human volume of distribution data of 179 drug molecules and determined the HSA and IAM binding chromatographically. Applying regression analysis, Equation 10.9 was obtained.

$$\log V_{ss} = 0.44(\pm 0.02) \log K(\text{IAM}) - 0.22(\pm 0.02) \log K\ (\text{HSA}) - 0.66$$
$$n = 179 \quad r^2 = 0.76 \quad s = 0.33 \quad F = 272 \qquad (10.9)$$

The letters n, r^2, s, and F denote the same parameters as in Equation 10.8. Although the multiple regression coefficient in Equation 10.8 is smaller than that in the model described by Equation 10.7, the error of the prediction (s) is better and the Fischer significance test value is higher. The model described the volume of distribution of neutral, acidic, basic, and zwitterionic drugs as well, as demonstrated by the plot shown in Figure 10.7.

The question may arise as to how it is possible that compounds with negative charges were not outliers from the model. This happens because the IAM partitioning, that is, the binding of the compounds to the immobilized phosphatidyl choline moiety is stronger for lipophilic and positively charged compounds, whereas the negatively charged compounds bind weakly to phospholipids and more strongly to the chemically bonded albumin stationary phase. The albumin and phospholipid binding of the uncharged molecules at physiological pH showed good correlation. The binding of the uncharged molecules is governed by lipophilicity, but there are probably three-dimensional steric factors that also modulate the binding. The plot of the log K(HSA) and log K(IAM) values for the same set of drug molecules is shown in Figure 10.8.

The difference between the IAM and HSA binding is governed mostly by the charge, the addition of the values for the positive and negative charge (ΔCHI poschg and ΔCHInegchg, respectively) to the equation improved the model (see Equation 10.10).

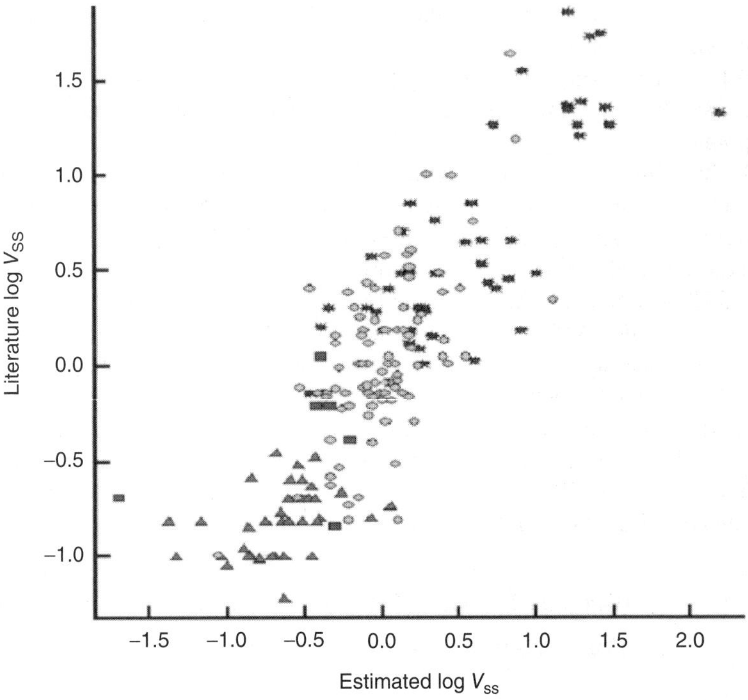

Figure 10.7 The logarithmic values of human volume of distribution data of 179 known drug molecules and the estimated values using the model equation 10.9. The stars, circles, triangles, and square represent, basic, neutral, acidic, and zwitterionic compounds, respectively. (Adopted with permission from Reference [19].)

$$\log V_{ss} = 0.33(\pm 0.03)\log K(\mathrm{IAM}) - 0.11(\pm 0.03)\log K(\mathrm{HSA}) - 0.016(\pm 0.003)\Delta\mathrm{CHInegchg} + 0.005(\pm 0.002)\Delta\mathrm{CHIposchg} - 0.52$$

$$n = 179 \quad r^2 = 0.79 \quad s = 0.31 \quad F = 168 \qquad (10.10)$$

The ΔCHIposchg is higher when a higher percentage of the compound is positively charged, whereas ΔCHInegchg is higher when a higher percentage of the compound is negatively charged at physiological pH. ΔCHIposchg is the measured difference between CHI values of the compounds obtained at pH 10.5 and pH 7.4, whereas the ΔCHInegchg is the difference between the CHI values obtained at pH 2 and pH 7.4. With the recognition of the importance of lipophilicity and the charge state of the drugs in the *in vivo* volume of distribution, it has raised the question whether a simple lipophilicity term such as $c \log P$ and the ionization term calculated from the pK_a values are good enough to use in the model. Therefore, the $c \log P$ and pK_a acid class and pK_a base class parameters have been calculated for the same drug molecules and Equation 10.11 has been obtained.

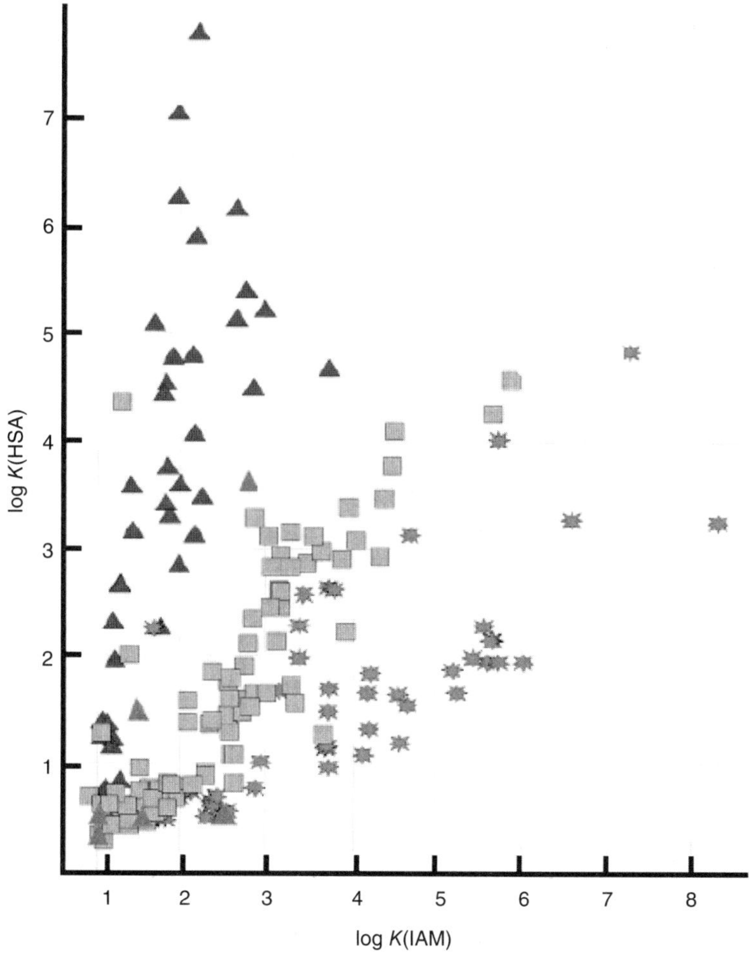

Figure 10.8 The plot of the measured human serum albumin binding (log *K*(HSA) and the immobilized artificial membrane partition (log *K*(IAM)) for the drug molecules used in the model Equation 10.9. The stars, squares, and triangles represent, basic, neutral, and acidic compounds, respectively. (Adopted with permission from Reference [19].)

$$\log V_{ss} = 0.085(\pm 0.02)c\log P - 0.141(\pm 0.011)\mathrm{p}K_{\mathrm{a\ acid\ class}} + 0.078(\pm 0.012)\mathrm{p}K_{\mathrm{a\ base\ class}} - 0.089$$

$$n = 179 \quad r^2 = 0.68 \quad s = 0.38 \quad F = 125 \tag{10.11}$$

The acid class and base class parameters have been calculated using the Henderson–Hasselbalch equation (see more details in Chapter 8) indicating the percentage of ionization. The higher the number, the higher is the percentage of

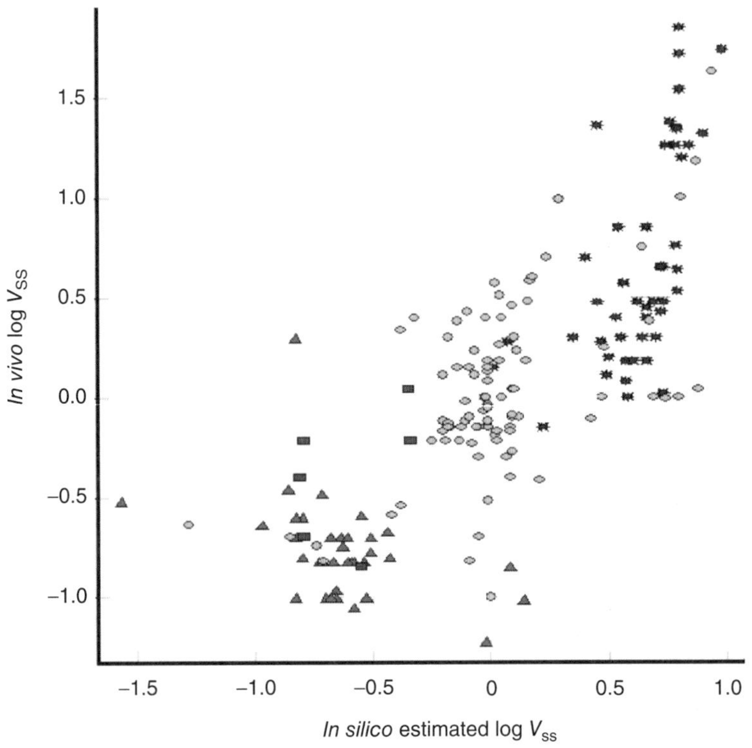

Figure 10.9 The measured and estimated *in vivo* steady-state human volume of distributions including only calculated physicochemical descriptors based on Equation 10.11. The stars, circles, triangles, and square represent basic, neutral, acidic, and zwitterionic compounds, respectively. (Adopted with permission from Reference [19].)

the ionized compound at pH 7.4. The model predicts very well the low, medium, and high volume of distributions for the acidic, neutral, and basic drugs; however, it does not differentiate the volume of distribution very well within the positively and negatively charged compounds as shown in Figure 10.9.

Table 10.6 contains the list of the investigated drug molecules for which the measured IAM and HSA binding data are available in the literature [19]. The CHI obtained at three pHs are also listed in the table, as it was used to assign the drug molecules as bearing positive (basic) or negative (acid) charges and derive the ΔCHI parameters. When the CHI values are the same at all the three pHs, the compounds are assigned as neutral unless they had both the positive and negative charges (zwitterions). The model of the volume of distribution clearly shows the impact of the charge on the molecules, which is one of the major forces to drive them to the tissue compartment (positive charge) or keep them in the plasma compartment (negative charge). Table 10.7 shows literature steady-state volume of distribution, PPB, and calculated physicochemical data of the investigated drug molecules.

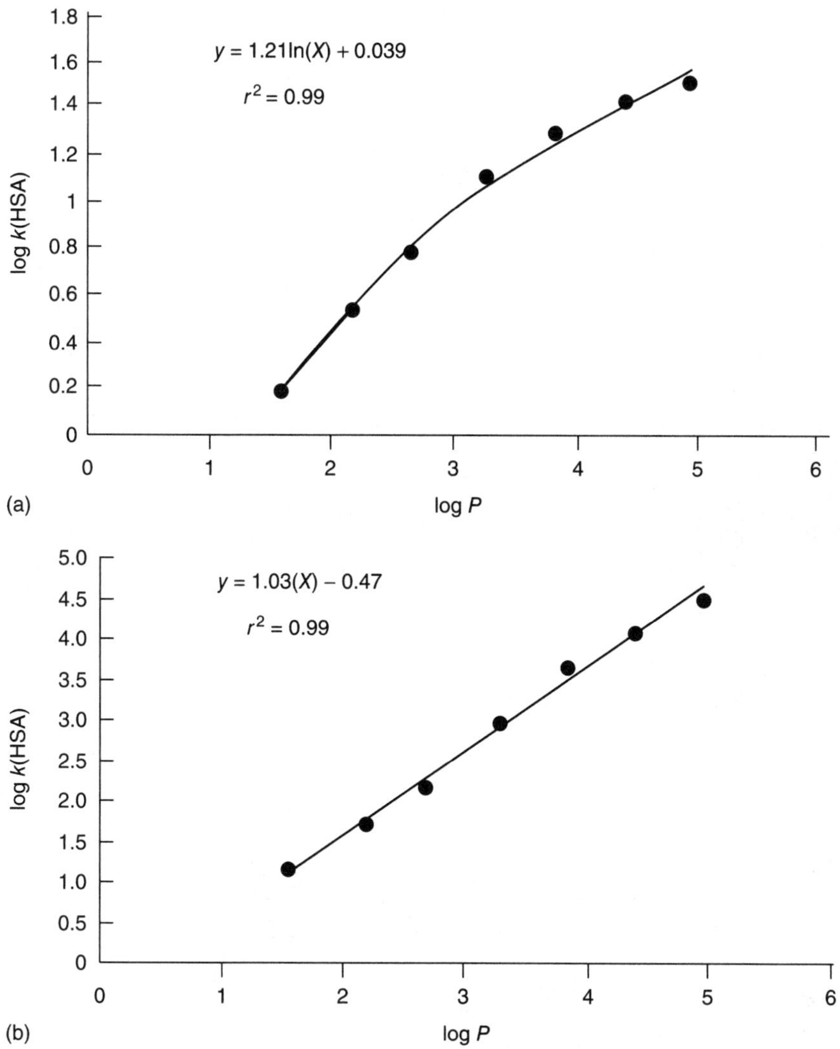

Figure 10.10 (a) The relationship of log *k*(HSA) to log *P* (octanol/water) and (b) the plot of log *K*(HSA) (= $e^{\log k(\text{HSA})}$) versus log *P*(octanol/water) for the acetophenone–octanophenone homolog series. (Adopted with permission from Reference [19].)

It is also important to note that the gradient retention times have been converted to log *k* (logarithmic retention factor) or CHI(IAM) values as described in Chapter 5. These values have then been converted to the logarithm of octanol/water partition coefficient (log *P*) scale using the measured retention data of the acetophenone homolog series up to octanophenone, using the same chromatographic conditions. The plots for linearization of the logarithmic retention factor on the HSA and the IAM columns are shown in Figures 10.10 and 10.11, respectively.

Table 10.6 The measured biomimetic chromatographic binding data of known drug molecules[a]

Drug	CHI pH 2	CHI pH 7.4	CHI pH 10.5	ΔCHInegchg	ΔCHIposchg	CHI (IAM)	Log *K* (IAM)	HSA % Bound	Log *K* (HSA)
Acecainide	13.3	21.3	42.5	0	21.2	23.3	2	29.8	0.69
Acetaminophen	17.4	16.8	−30.6	0	0	4	1.24	14	0.45
Acetanilide	42	41.2	41.8	0	0.6	10.7	1.43	25.3	0.62
Acetazolamide	39.9	33.3	21	6.5	0	−0.5	1.14	71.5	1.47
Alclofenac	72.9	39.4	36.2	33.5	0	17.8	1.71	98.7	5.11
Alosetron HCl	29	45.3	49	0	3.7	32.7	2.69	75.5	1.6
Amantadine	65.9	60.1	61.2	0	1	31.4	2.58	65.5	1.31
Amiloride	12.5	18.6	15.7	0	0	30.7	2.52	20	0.54
Amoxapine	36.4	61.5	89.6	0	28.1	52.7	5.69	86.2	2.15
Amoxicillin	40.9	18.1	8.4	22.8	0	6.3	1.3	16.4	0.49
Ampicillin	33.6	31.5	27	0	0	6.3	1.3	21.2	0.56
Apomorphine HCl	22.9	49.5	78.6	0	29.1	39.7	3.45	90.8	2.58
Aspirin	56.7	19.7	14.9	37	0	−1.7	1.12	37.3	0.79
Bamethan	14.6	24	42.4	0	18.4	18.6	1.75	16.5	0.49
Betamethasone	59.7	57.6	59.7	2.1	2.1	31.7	2.61	55.6	1.09
Bromazepam	44.1	54.2	56.3	0	2.1	28.5	2.34	68.5	1.38
Bromocriptin	49.1	91.6	93	0	1.4	47.9	4.7	94.1	3.13
Budesonide	77.2	75	76.6	2.2	1.6	38.7	3.32	78.7	1.73
Bumetanide	76	49.1	45.2	26.8	0	26.1	2.18	94.3	3.15
Carbamazepine	59.9	59.4	60.1	0	15.4	28.9	3.39	76.5	1.99
Cefazoline	39.3	22	22.8	17.3	0	2.5	1.21	62.7	1.24
Cefixime	30	13.8	10.3	16.2	0	−3.8	1.08	69.1	1.4
Cephalexin	21.6	23.8	23.8	0	0	−2.3	1.11	14	0.45
Chlorpheniramine	28.1	55.3	98.6	0	43.3	47.2	4.58	61	1.2

Chlorpropamide	68.2	34.9	33	33.3	0	5.8	1.29	91.6	2.68
Chlorprothixene	111.2	112.5	113	0	0.6	53.5	5.89	98	4.57
Cinoxacin	46.7	15.6	12.5	31.1	0	1.6	1.19	60.1	1.18
Clonazepam	66.5	66.2	65.1	0.3	0	34.8	2.9	77.1	1.66
Clonidine	15.1	34.3	53.2	0	18.9	28.8	2.37	28	0.66
Cloxacillin	73.4	45.7	43.5	27.7	0	23.1	1.99	92.7	2.86
Colchicine	46.8	44.8	47.5	2	2.6	23.7	2.02	37.1	0.79
Cytarabine	−45.2	−21.7	−27.1	0	0	−15	0.93	29.8	0.69
Diazepam	70.9	80.6	82.3	0	1.7	37.4	3.18	93.2	2.94
Diazoxide	49.2	47.7	27.8	1.5	0	24.7	2.08	75.2	1.59
Diclofenac	89.3	52.1	49.1	37.2	0	33.5	2.77	99	5.42
Diflunisal	91.8	52.9	47.6	39	0	32.5	2.67	98.7	5.15
Diltiazem	40.7	71.7	87.2	0	15.5	41.6	3.71	58.5	1.15
Diphenhydramine	40	59.9	98	0	38.2	44.6	4.14	55.4	1.09
Diprophylline	19.7	18.2	18.1	1.5	−0.1	−4	1.08	25.9	0.63
Dipyridamole	38.2	68.9	72.1	0	3.3	43.2	3.94	87.3	2.23
Doxepin	41.2	64.3	108.2	0	43.9	52.3	5.61	83.2	1.95
Droperidol	36.3	68	76.4	0	8.3	39.3	3.4	88	2.29
Encainide	33	48.2	93.6	0	45.3	41.9	3.74	48.8	0.97
Ethinyl Estradiol	76.9	76.3	77.4	0.6	1.2	46.8	4.52	97.2	4.1
Famotidine	10.7	23.5	25.1	0	1.7	15.7	1.62	14.5	0.46
Felbamate	64.7	63.6	65.8	1.1	2.2	24.6	2.08	68.7	1.39
Felodipine	98.7	99.4	99.5	0	0.1	46.1	4.39	95.5	3.46
Fenclofenac	90.4	55.5	47.7	35	0	35.6	2.98	98.8	5.23

(continued)

Table 10.6 (*Continued*)

Drug	CHI pH 2	CHI pH 7.4	CHI pH 10.5	ΔCHInegchg	ΔCHIposchg	CHI (IAM)	Log *K* (IAM)	HSA % Bound	Log *K* (HSA)
Fenoprofen	83.4	48.7	42.7	34.7	0	26.3	2.19	99.3	5.91
Finasteride	75.9	73.2	75.9	2.7	2.7	38.9	3.35	74.7	1.58
Floxacillin	76.3	48.6	45.4	27.6	0	23.4	2	96	3.62
Flumazenil	49	48.5	50.3	0	1.7	18.4	1.74	20.9	0.56
Flunarizine	55.2	121.2	125.8	0	4.6	52.6	5.67	97.5	4.26
Flunitrazepam	71	71.1	72.3	0	1.2	31.2	2.56	79.7	1.77
Fluoxetine	48.1	67.5	107.3	0	39.8	52.9	5.73	97.1	4.01
Flurazepam	38.2	69.1	93.4	0	24.3	41.8	3.73	59.4	1.17
Flurbiprofen	85.5	47.5	44.8	38	0	26.8	2.22	100	7.81
Furosemide	32.4	−21.7	−31.7	54.1	0	−4.2	1.08	63.8	1.26
Gemfibrozil	51.6	79.7	49.3	0	0	32.9	2.71	96.1	3.64
Glipizide	67.9	46	38.9	21.9	0	21.1	1.87	95	3.32
Griseofulvin	71.4	70.9	72.5	0.5	1.6	33.1	2.73	72.1	1.49
Hexobarbital	63.5	63.3	30.7	0.2	0	18.8	1.76	27.7	0.66
Hydralazine	36.9	31.9	33.2	5.1	1.3	13.3	1.52	48.8	0.97
Hydrochlorothiazide	32	30.8	−30.9	1.2	0	15.9	1.63	35.6	0.77
Hydrocortisone	52.7	50.5	52.4	2.2	2	27.9	2.3	45.8	0.92
Imipramine	46.1	69.1	118.6	0	49.4	54.1	6.04	83.2	1.95
Indomethacin	88.6	54	48.5	34.7	0	32.5	2.68	99.5	6.17
Indoramin	35.2	52.2	71.1	0	18.9	1.9	3.74	72.1	1.49
Isradipine	89	89.8	90.4	−0.8	0.6	40	3.49	92.8	2.86
Ketoconazole	40.9	78.8	81.3	0	2.5	42.9	3.89	93	2.9
Ketoprofen	74.3	41.2	38.1	33.2	0	21.9	1.92	98.4	4.8
Labetalol	34.5	46.3	42.2	0	0	41.4	3.67	64.6	1.28
Levamisol	13.3	44.5	61	0	16.5	34.7	2.89	36.4	0.78

Levonorgestrel	85.6	85	85.8	0.6	0.9	41.2	3.65	93.4	2.98
Lignocaine	22.6	73.9	86.7	0	12.7	29.6	2.43	29.8	0.69
Lorazepam	64.6	63.9	65.4	0.7	1.6	37.3	3.16	91.1	2.62
Maprotiline	45.2	61.1	115.3	0	54.1	52.9	5.74	83.2	1.95
Mebendazole	48.2	61	59.8	0	0	37.5	3.19	89.8	2.46
Methylprednisolone	57.5	56.5	58.2	1	1.6	32.1	2.65	54.9	1.35
Metronidazole	8.5	21	22.4	0	1.5	−3.3	1.09	5.4	0.76
Mianserin	67.1	65.9	68.5	1.2	2.6	33.7	2.79	85.6	2.12
Minoxidil	23.1	33.7	35.9	0	2.2	19	1.77	27.9	0.66
N-Dealkylflurazepam	66.8	68.1	70.6	0	2.5	36.5	3.08	92.6	2.84
Nabumetone	87.9	88.7	89.4	0	0.7	38.4	3.29	94.3	3.16
Nadolol	19.4	25.9	45.2	0	19.3	20.2	1.83	16.5	0.49
Naproxen	75.2	40.2	36.1	35	0	22.9	1.97	99.9	7.06
Neostigmine	5.6	20	37	0	16.9	17.3	1.69	87.6	2.26
Nicardipine	46.3	101.1	103.9	0	2.8	45.9	4.35	93.2	2.93
Nifedipine	76.1	77.3	78.7	0	1.4	29	2.39	69.5	1.41
Nisoldipine	95.8	98	98.7	0	0.7	37.4	3.18	91	2.61
Nitrazepam	59.6	64.2	63.9	0	0	33.3	2.75	82.3	1.91
Nitrendipine	87.9	88.4	89	0	0.6	40.5	3.56	94.2	3.13
Nitrofurantoin	39.9	33.3	21	6.5	0	6.2	1.3	71.5	0.85
Nizatidine	5.6	29.9	33.6	0	3.8	17.9	1.72	20.4	0.55
Nordazepam	55.5	70.3	72.9	0	2.6	38.3	3.28	92.6	2.83
Nortriptyline	44	62.9	118.3	0	55.4	52.6	5.66	86.2	2.15
Ondansetron	31.7	51.8	60.3	0	8.4	37.2	3.15	77.7	1.69
Oxacillin	69.3	42.4	40.4	26.9	0	18.9	1.76	87.8	2.28

(continued)

Table 10.6 (*Continued*)

Drug	CHI pH 2	CHI pH 7.4	CHI pH 10.5	ΔCHInegchg	ΔCHIposchg	CHI (IAM)	Log *K* (IAM)	HSA % Bound	Log *K* (HSA)
Oxazepam	62.3	62	63.7	0.4	1.7	36.2	3.04	94.2	3.12
Papaverine	32.6	65.8	68	0	2.2	34.4	2.86	88.6	2.35
Pentobarbital	62	59.3	31.3	2.7	0	22.4	1.95	28.8	0.67
Pentoxifylline	39.5	37.8	39.2	1.7	1.4	12	1.48	25.9	0.63
Perphenazine	38.8	76	93.7	0	17.7	56.3	6.6	94.8	3.27
Phenytoin	61	60.6	37.3	0.4	0	31.6	2.59	75.5	1.6
Pindolol	20.9	31.5	56.7	0	25.2	30.3	2.49	19.7	0.54
Piperacillin	40	35.1	33.1	4.8	0	9.4	1.39	15.4	0.48
Piroxicam	49.3	38.2	33.8	0	1.4	21.8	1.31	97.7	4.36
Prazosin	30.1	48.7	52.9	0	4.2	31.6	2.6	80.4	1.81
Prednisolone	51.9	49.5	51.5	2.4	2	28	2.31	43.6	0.89
Prednisone	52.2	51.4	53.2	0.8	1.8	25.9	2.16	37.6	0.8
Primidone	37.7	36.1	36.9	1.6	0.8	8.9	1.37	25.3	0.62
Probenecid	79.8	47.4	44	32.5	0	20.1	1.82	95.4	3.44
Procainamide	−34.8	16.6	41.7	0	25.1	19.9	1.81	35.2	0.76
Procyclidine	91.3	90.9	89.3	0.4	0	46.7	4.49	96.5	3.77
Propranolol	37	50.3	83	0	32.7	45.1	4.23	66.5	1.33
Propylthiouracil	30.1	28	8.2	2.1	0	3.9	1.24	33.1	0.73
Protryptyline	43.4	60.6	117.4	0	56.7	51.6	5.45	83.8	1.99
Proxyphylline	26.5	24.9	25.4	1.6	0.5	1.1	1.17	25.9	0.63
Quinidine	14.6	46.4	71.2	0	24.8	45	4.2	77	1.66
Quinine	16.3	46.7	69.4	0	22.7	47.1	4.56	76.7	1.65
Ranitidine	8.9	23.7	41.7	0	18	28	2.31	17.5	0.51
Sulfachlorpyridazine	45.7	23.2	17.5	22.5	0	6	1.29	91.5	2.67
Sulfameter	39	28.2	−26.1	10.8	0	3.7	1.23	83.5	1.97

Sulfamethoxypyridazine	36	33.8	14.4	2.1	0	9.6	1.4	84.5	2.04
Sulfametopyrazine	43.4	22.3	−22.7	21.1	0	−0.4	1.14	68.9	1.39
Sulfapyridine	23.5	29.9	15.4	0	0	11.4	1.46	18.8	0.53
Sulfinpyrazone	77.6	47.8	44	29.8	0	26.1	2.17	97.2	4.09
Sulfisoxazole	51.7	19.5	14.8	32.2	0	2.9	1.21	88.4	2.33
Sulfadimethoxine	56.1	38.4	18.9	17.8	0	10.8	1.44	94.4	3.17
Sulfadimidine	31.7	35.6	19.1	0	0	10.6	1.43	73.6	1.54
Sulpiride	9.3	23.2	48.2	0	25	25.5	2.14	34.3	0.75
Tamoxifen	58.6	107.9	137.2	0	29.3	58.7	7.29	98.4	4.84
Temazepam	70.9	71	73	0	1.9	34.7	2.89	94.9	3.28
Terbutaline	8.8	15.8	18.7	0	2.9	16.8	1.67	28.8	0.67
Theobromine	14.6	13.1	7	1.5	0	−4.1	1.08	64.8	1.29
Tinidazole	33.3	35.6	35.8	0	0.2	0.4	1.16	25.9	0.63
Tolbutamide	71.8	46	35.3	25.8	0	10.2	1.42	96	3.59
Tolfenamic acid	100.7	61.1	52.8	39.6	0	41.9	3.74	98.2	4.68
Tolmetin	71.7	41.5	38.9	30.2	0	20.9	1.87	96.5	3.77
Tramadol	27	41.2	86.4	0	45.2	28.9	2.38	25.3	0.62
Trazodone	34.8	70.9	75.6	0	4.8	36.3	3.05	89.8	2.46
Trimethoprim	21.1	37.1	40.4	0	3.4	20.8	1.86	39.5	0.8
Viloxazine	27.8	42.2	55.4	0	13.2	31.1	2.56	20.9	0.56
Vinblastine	33.8	82.6	93.7	0	11	50.4	5.19	81.7	1.87
Vincristine	31.9	74.1	84.5	0	10.3	47.7	4.67	74.1	1.55
Warfarin	81.5	45.7	34.1	35.8	0	19.9	1.81	97.9	4.45
Zidovudine	31.6	30.1	22.5	1.5	0	1.3	1.18	12.8	0.43
Zolmitriptan	18.1	26.6	49.8	0	23.2	29.7	2.44	30.6	0.7
Zolpidem	32	63.4	66.1	0	2.7	33.9	2.81	73.6	1.54

[a]Taken with permission from Reference [19].

Table 10.7 Literature steady-state volume of distribution, plasma protein binding, calculated log *P*, and pK_a data for the drugs listed in Table 10.6[a]

Drug	CAS Number	VD_{ss} (l/kg)	PPB %	Log *K*(PPB)	*C* log *P*	pK_a Acid Class	pK_a Base Class
Acecainide	32795-44-1	1.5	10	0.38	1.64	0	7
Acetaminophen	103-90-2	0.9	1	0.14	0.49	1	0
Acetanilide	103-84-4	0.7	NA	NA	1.16	0	0
Acetazolamide	59-66-5	0.2	95	3.32	−1.13	2	0
Alclofenac	22131-79-9	0.1	99	5.44	2.73	7	0
Alosetron HCl	122852-69-1	0.9	NA	NA	1.74	0	2
Amantadine	768-94-5	10.4	67	1.34	2	0	7
Amiloride	2609-46-3	5	40	0.83	−0.55	0	5
Amoxapine	14028-44-5	16	NA	NA	4.62	0	6
Amoxicillin	26787-78-0	0.3	18	0.51	−1.87	7	2
Ampicillin	69-53-4	0.4	18	0.51	−1.2	7	3
Apomorphine HCl	314-19-2	2	NA	NA	2.49	1	2
Aspirin	50-78-2	0.2	70	1.42	1.02	7	0
Bamethan	3703-79-5	3.7	NA	NA	1.5	1	7
Betamethasone	378-44-9	1.4	64	1.27	1.79	0	0
Bromazepam	1812-30-2	0.9	52	1.03	1.7	0	1
Bromocriptin	25614-03-3	3	90	2.48	6.59	1	1
Budesonide	51333-22-3	2.9	88	2.29	2.91	0	0
Bumetanide	28395-03-1	0.3	96	3.61	3.37	7	1
Carbamazepine	298-46-4	1.4	75	1.58	1.98	0	0
Cefazoline	25953-19-9	0.1	89	2.39	−1.14	7	1

Cefixime	79350-37-1	0.3	67	1.34	0.45	7	1
Cephalexin	15686-71-2	0.3	14	0.45	−1.64	7	3
Chlorpheniramine	132-22-9	3.4	72	1.48	3.15	0	6
Chlorpromazine	50-53-3	21	98	4.55	5.8	0	7
Chlorpropamide	94-20-2	0.2	90	2.49	2.35	7	0
Chlorprothixene	113-59-7	15	NA	NA	5.48	0	6
Cinoxacin	28657-80-9	0.3	63	1.25	1.5	7	0
Clonazepam	1622-61-3	3	85	2.07	2.38	0	1
Clonidine	4205-90-7	3.5	20	0.54	1.43	0	5
Cloxacillin	61-72-3	0.1	95	3.32	2.52	7	1
Colchicine	64-86-8	1.4	31	0.7	1.2	0	0
Cytarabine	147-94-4	2.5	13	0.44	−2.2	0	1
Diazepam	439-14-5	1.5	98	4.55	3.17	0	1
Diazoxide	364-98-7	0.2	94	3.09	1.2	1	7
Diclofenac	15307-86-5	0.2	99	5.44	4.73	7	0
Diflunisal	22494-42-4	0.1	99	5.44	4.4	7	0
Diltiazem	33286-22-5	5.3	98	4.55	3.65	0	5
Diphenhydramine	58-73-1	4.5	99	5.44	3.54	0	6
Diprophylline	479-18-5	0.8	NA	NA	−1.29	0	1
Dipyridamole	58-32-2	2.5	NA	NA	2.53	0	1
Doxepin	1668-19-5	20	83	1.94	4.09	0	6

(*continued*)

Table 10.7 (*Continued*)

Drug	CAS Number	VD_{ss} (l/kg)	PPB %	Log *K*(PPB)	*C* log *P*	pK_a Acid Class	pK_a Base Class
Droperidol	548-73-2	2	NA	NA	3.26	0	2
Encainide	66778-36-7	2.7	75	1.58	3.52	0	7
Ethinyl Estradiol	57-63-6	3	97	3.99	3.86	1	0
Famotidine	76824-35-6	1.3	17	0.5	−1.2	1	7
Felbamate	25451-15-4	0.8	29	0.67	0.5	0	0
Felodipine	72509-76-3	10	99	5.44	5.58	0	0
Fenclofenac	34645-84-6	0.2	NA	NA	4.71	7	1
Fenoprofen	31879-05-7	0.1	99	5.44	3.82	7	0
Finasteride	98319-26-7	1.1	90	2.49	3.01	0	0
Floxacillin	5250-39-5	0.2	93	2.9	2.66	7	1
Flumazenil	78755-81-4	1	45	0.91	1.09	0	1
Flunarizine	52468-60-7	43	90	2.49	6.34	0	4
Flunitrazepam	1622-62-4	3.7	78	1.7	2.14	0	1
Fluoxetine	54910-89-3	31	99	5.44	4.57	0	7
Flurazepam	17617-23-1	3.4	97	3.99	4.42	0	7
Flurbiprofen	5104-49-4	0.2	99	5.44	3.75	7	0
Furosemide	54-31-9	0.2	97	3.99	1.9	7	0
Gemfibrozil	25812-30-0	0.1	97	3.99	4.16	7	0
Glipizide	29094-61-9	0.2	98	4.55	2.57	7	0
Griseofulvin	126-07-8	1.5	84	2	1.77	0	0
Hexobarbital	56-29-1	1	47	0.94	1.53	0	0
Hydralazine	86-54-4	5	89	2.39	1.02	0	3
Hydrochlorothiazide	58-93-5	3	40	0.83	−0.37	1	1

Hydrocortisone	50-23-7	0.4	85	2.07	1.7	0	0
Imipramine	50-49-7	21	89	2.39	5.04	0	7
Indomethacin	53-86-1	0.2	90	2.49	4.18	7	0
Indoramin	26844-12-2	7	NA	NA	2.84	0	6
Isradipine	75695-93-1	4	95	3.32	4.2	0	1
Ketoconazole	65277-42-1	2.4	99	5.44	2.58	0	2
Ketoprofen	22071-15-4	0.1	94	3.09	2.76	7	0
Labetalol	36894-69-6	7	50	0.99	2.5	2	7
Levamisol	14769-73-4	1.4	NA	NA	1.84	0	2
Levonorgestrel	797-63-7	2.9	94	3.09	3.31	0	0
Lignocaine	137-58-6	1.5	60	1.18	1.95	0	5
Lorazepam	846-49-1	1.5	90	2.49	2.37	0	1
Maprotiline	10262-69-8	52	88	2.29	4.52	0	7
Mebendazole	31431-39-7	2	90	2.49	3.08	4	1
Methylprednisolone	83-43-2	0.7	83	1.91	1.74	0	0
Metronidazole	443-48-1	0.8	20	0.54	−0.46	0	1
Mianserin	24219-97-4	6.3	95	3.32	4.27	0	6
Minoxidil	38304-91-5	3	1	0.14	0.54	0	3
N-Dealkylflurazepam	17617-59-3	0.7	98	4.55	2.81	0	1
Nabumetone	42924-53-8	0.8	99	5.44	2.98	0	0
Nadolol	42200-33-9	2	28	0.66	0.38	0	7
Naproxen	22204-53-1	0.2	99	5.44	2.82	7	0
Neostigmine	59-99-4	0.7	NA	NA	−2.81	0	0

(*continued*)

Table 10.7 (*Continued*)

Drug	CAS Number	VD_{ss} (l/kg)	PPB %	Log *K*(PPB)	*C* log *P*	pK_a Acid Class	pK_a Base Class
Nicardipine	55985-32-5	1.7	95	3.32	5.51	0	3
Nifedipine	21829-25-4	1	96	3.61	3.41	0	1
Nisoldipine	63675-72-9	5	99	5.44	4.86	0	0
Nitrazepam	146-22-5	2.9	87	2.21	2.32	0	1
Nitrendipine	39562-70-4	3.8	99	5.44	4.02	0	0
Nitrofurantoin	67-20-9	0.6	40	0.83	−0.47	2	0
Nizatidine	76963-41-2	1.2	35	0.76	−0.2	0	6
Nordazepam	1088-11-5	0.8	97	3.99	3.02	0	1
Nortriptyline	72-69-5	21	93	2.9	4.32	0	7
Ondansetron	99614-02-5	2.4	73	1.52	2.72	0	2
Oxacillin	66-79-5	0.3	94	3.09	2.05	7	1
Oxazepam	604-75-1	1.3	95	3.32	2.31	0	1
Papaverine	58-74-2	1.5	87	2.21	3.78	0	2
Pentobarbital	76-74-4	1	55	1.08	1.42	0	0
Pentoxifylline	6493-05-6.	2.4	NA	NA	0.12	0	1
Perphenazine	58-39-9	22.5	NA	NA	4.31	0	5
Phenytoin	57-41-0	0.7	90	2.49	2.09	2	0
Pindolol	13523-86-9	1.5	50	0.99	1.67	0	7
Piperacillin	61477-96-1	0.2	22	0.57	1.7	7	1
Piroxicam	36322-90-4	0.1	99	5.44	1.89	0	1
Prazosin	19216-56-9	0.5	94	3.09	1.21	0	4
Prednisolone	50-24-8	1.6	78	1.7	1.42	0	0
Prednisone	53-03-2	0.9	78	1.7	1.66	0	0

Primidone	125-33-7	0.6	20	0.54	0.88	0	0
Probenecid	57-66-9	0.2	89	2.39	3.37	7	0
Procainamide	614-39-1	2.2	15	0.47	1.42	0	7
Procyclidine	77-37-2	1	NA	NA	4.59	0	7
Propranolol	525-66-6	4	93	2.9	2.75	0	7
Propylthiouracil	51-52-5	0.4	80	1.79	2.87	2	1
Protryptyline	438-60-8	22	NA	NA	4.47	0	7
Proxyphylline	603-00-9	0.6	NA	NA	−0.56	0	1
Quinidine	56-54-2	2.5	90	2.49	2.79	0	7
Quinine	130-95-0	1.6	93	2.9	2.79	0	7
Ranitidine	66357-35-5	1.5	15	0.47	0.63	0	6
Sulfachlorpyridazine	80-32-0	0.1	NA	NA	0.56	5	1
Sulfameter	651-06-9	0.3	87	2.21	0.65	2	1
Sulfamethoxypyridazine	80-35-3	0.2	65	1.29	0.41	3	1
Sulfametopyrazine	152-47-6	0.3	NA	NA	1.05	3	1
Sulfapyridine	144-83-2	0.4	NA	NA	0.84	2	1
Sulfinpyrazone	57-96-5	0.1	98	4.55	1.66	7	0
Sulfisoxazole	127-69-5	0.4	88	2.29	0.22	5	1
Sulfadimethoxine	122-11-2	0.2	99	5.44	1.98	5	1
Sulfadimidine	57-68-1	0.6	80	1.79	1.1	3	1
Sulpiride	15676-16-1	2.5	40	0.83	1.11	1	7
Tamoxifen	10540-29-1	55	99	5.44	6.82	0	5
Temazepam	846-50-4	1	97	3.99	2.55	1	1

(*continued*)

Table 10.7 ***(Continued)***

Drug	CAS Number	VD_{ss} (l/kg)	PPB %	Log K(PPB)	C log P	pK_a Acid Class	pK_a Base Class
Terbutaline	23031-32-5	1	25	0.62	0.48	1	7
Theobromine	83-67-0	0.8	NA	NA	−0.67	1	1
Tinidazole	19387-91-8	0.7	12	0.42	−0.32	0	1
Tolbutamide	64-77-7	0.2	95	3.32	2.5	7	0
Tolfenamic acid	13710-19-5	0.2	99	5.44	5.66	7	0
Tolmetin	26171-23-3	0.1	99	5.44	2.21	7	0
Tramadol	27203-92-5	3	4	0.25	3.1	0	7
Trazodone	19794-93-5	1	93	2.9	3.17	0	7
Trimethoprim	738-70-5	1.3	45	0.91	0.98	0	3
Viloxazine	46817-91-8	1	NA	NA	NA	0	6
Vinblastine	865-21-4	35	70	1.42	5.23	1	7
Vincristine	57-22-7	11	70	1.42	4.04	1	7
Warfarin	81-81-2	0.2	99	5.44	2.9	7	0
Zidovudine	30516-87-1	1.4	25	0.62	0.04	1	0
Zolmitriptan	139264-17-8	7	18	0.51	1.29	0	7
Zolpidem	82626-48-0	0.5	93	2.82	2.83	0	2

NA means that the data are not available.

[a]From the supplementary material of Reference [19].

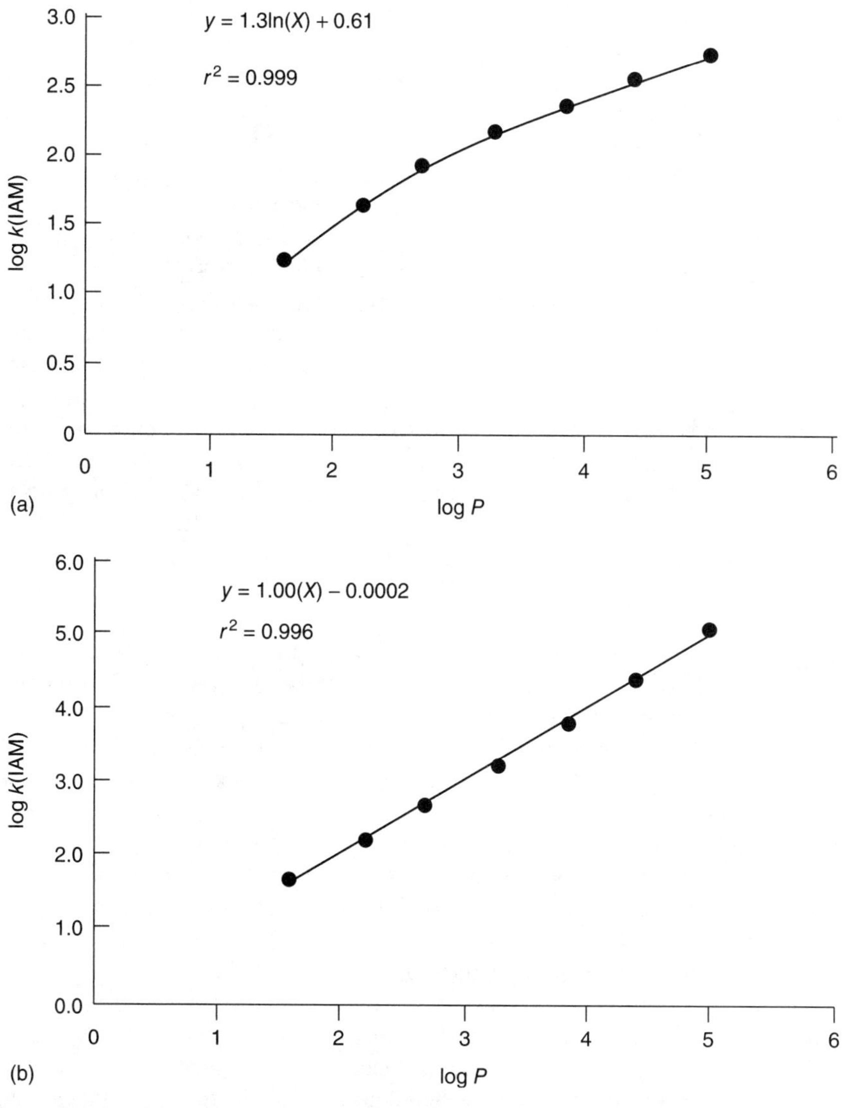

Figure 10.11 (a) The relationship of log k(IAM) to log P(octanol/water) and (b) the plot of log K(IAM) ($= 0.29e^{\log k(\mathrm{IAM})} + 0.70$) versus log P (octanol/water) for the acetophenone–octanophenone homolog series (b). (Adopted with permission from Reference [19].)

The conversion of the scales has two benefits. The first is that the slightly curving chromatographic retention data due to the application of organic phase gradient in the mobile phase are linearized. The second benefit is that we can compare the strength of binding to different biomimetic phases as absolute values expressed in the octanol/water log P scale.

The gradient application is very beneficial, as it reduces the analysis time and makes it possible to elute even strongly bound compounds using the same chromatographic conditions. The octanol/water partition coefficients of neutral molecules showed a similar solvation equation to the HSA and phospholipid binding (log *k* (IAM); see Table 5.5), which means that the size, dipolarity, and H-bond acidity and basicity of the compounds alter the binding to phospholipids and albumin in a manner similar to the octanol/water partition.

Comparing the coefficients of log *K*(IAM) and log *K*(HSA) (0.44 and 0.22, respectively), it is evident that the phospholipid binding has a greater impact on the volume of distribution. It is not surprising, as the tissues represent a much bigger volume than the plasma compartment. Albumin type of proteins that balance out the effect of PPB of compounds can also be found in various cells and tissues.

Another model for the estimation of volume of distribution has been published by Sui et al. [20] using immobilized artificial membrane chromatographic retention data in isocratic mode and PPB data obtained by equilibrium dialysis and the calculated fraction of ionization from the pK_a values. The model is based on the same principles as the model described in Reference [19]. The collection of measured and calculated data from different sources using different methodologies is not very practical when dealing with a large number of research compounds.

In conclusion, there are several models in the literature [18–20] that clearly demonstrate that *in vitro* physicochemical data can be used to estimate compounds *in vivo* distribution between the plasma and tissue compartments. These models revealed that the major driving forces are the positive and negative charges on the drug molecules that pull them to the tissue compartment or keep them in the plasma compartment. The biomimetic chromatographic binding data have also revealed that the effect of charge is the result of the binding preferences of the negatively charged compounds to albumin and the positively charged compounds to phospholipids. Applying directly the phospholipid and the albumin as chromatographic stationary phases makes it possible to detect other binding forces such as steric and hydrophobic interactions between the drugs and biomolecules.

The volume of distribution takes into account an overall tissue binding of the compounds, but it is evident from the literature that drugs bind to various tissues to various extents. When the drug target is concentrated in specific tissues, it is preferred that the majority of the administered dose should be found in that particular tissue. The *in vivo* tissue distribution of a drug is difficult to measure; therefore, several *in vitro* approaches have been investigated using equilibrium dialysis methods and tissue homogenates [16,20].

In principle, the volume of distribution is the sum of the plasma volume and the various tissue volumes multiplied by the tissue partition coefficients as shown by Equation 10.12 [6].

$$V_{\mathrm{ss}} = V_{\mathrm{plasma}} + \sum_{i=1}^{n} V_{\mathrm{tissue,\,i}} \times K_{\mathrm{p\ tissue,\,i}} \qquad (10.12)$$

Equation 10.12 shows that we can calculate the volume of distribution if we know the tissue volumes and the tissue partition coefficients. Björkman [21] made an attempt to identify which types of tissues are the most relevant for the estimation of volume of distribution. The tissue/plasma partition coefficients were collected from the literature for several basic and acidic drugs in various laboratory animals and are shown in Table 10.8. When the *in vivo* total volume of distribution is known, the plasma/tissue partition coefficients can be introduced to Equation 10.12. With the help of regression analysis, we can calculate the volumes of the tissues. Table 10.9 contains the volumes of tissues so obtained from Reference [21].

Table 10.8 Tissue/plasma partition coefficients of acidic and basic drugs in rats[a]

Drug	Brain	Heart	Lung	Gut	Liver	Kidney	Bone	Skin	Muscle	Fat
Biperiden	7	7	61	11		11	2	4	3.1	58
Fentanyl	3.6	4.5	14	8	3.8	12		2.1	3.1	27
Midazolam	3.3	4.2	4.5	4.8	8.8	4.6		1.4	1.3	9
Pentazocine	4.3	5.4	27	4.7	2.3	20	5.4	4.7	5.9	2.5
Procainamide		2.5			3.2	6.4			3.1	0.1
Ceftazidime		0.22	0.44	0.4	0.25	4.8		0.39	0.19	0.16
Ethylbarbital	0.73	0.69	1	0.69	3.7	1.8	0.63	1.2	0.82	0.72
Phenobarbital		0.91	0.77	1.6	1.8	0.73		1.2	0.99	0.30
Phenytoin	0.91	1.2	0.95	2.5	2.3	1.6		1.7	1.1	1.8
Tolbutamide	0.10	0.27	0.25	0.12	0.30	0.22		0.22	0.13	0.13
Valproic acid	0.07	0.43	0.42	0.45	1.8	1.5		0.47	0.16	0.15
Cefazoline		0.1	0.15	0.11	0.79	2.8	0.11	0.30	0.08	
Salicylic acid	0.06	0.19	0.19	0.21	0.23	0.44	0.14	0.24	0.12	

[a]Collated by Björkman [21].

Table 10.9 Estimated tissue volumes in humans (in l)[a]

Organ/Tissue	Tissue Volume, l
Blood	2.44
Brain	1.34
Heart	0.41
Lungs	0.87
Gut	1.21
Liver	2.19
Kidneys	0.34
Bone	9.50
Skin	2.85
Muscle	24.6
Fat	15.3

[a]From Reference [21] with permission.

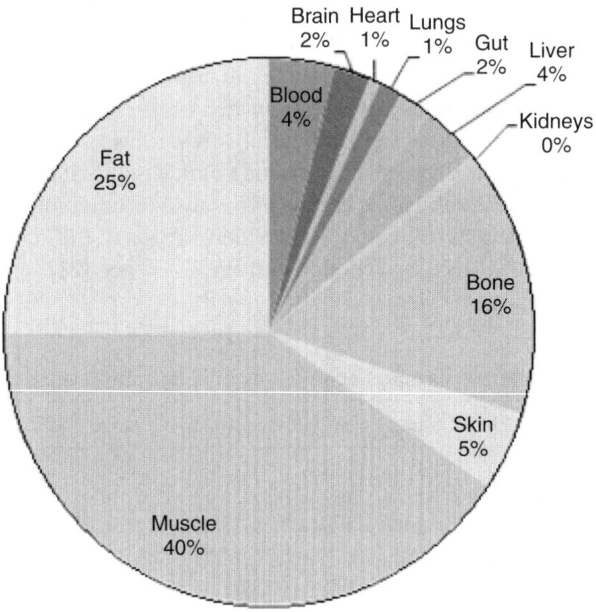

Figure 10.12 The volumes of various tissues that can be used for estimation of drug distribution based on the tissue partition coefficients listed in Table 10.9.

The volumes of various tissues as percentages shown in Figure 10.12 are different from the weight distribution shown in Figure 10.3. These volumes are more relevant when various tissue partition coefficients are considered. The data in Table 10.8 show that basic drugs distribute more to fat than to acidic drugs. Although the blood volume is relatively small (4%) in comparison to the total body volume, acidic drugs stay mostly in the plasma/blood compartment. As the blood/plasma is circulating all over the body, it represents a special compartment. When drugs are targeted to the brain and in general to the CNS, it is crucial to cross the blood–brain barrier. On the other hand, when a drug molecule is targeted outside the CNS, it is better to avoid partitioning into the brain where it may cause side effects. Therefore, the PPB and the brain tissue binding have been studied in great detail and discussed in this chapter separately.

PLASMA PROTEIN BINDING

The PPB of drugs has been of interest for a long time, as drug molecules circulate partly bound to the plasma proteins. The bound drug molecules are not available to interact with the receptors. Plasma proteins are present in much higher concentration than the drug molecule, and, therefore, even weak nonspecific binding to plasma proteins can significantly reduce the free drug concentration. According to the "free drug hypothesis," only the unbound drug molecule can bind to the target protein, or can be cleared from the general circulation. It has been observed that in certain cases, the free plasma concentration is a more relevant parameter monitoring the efficacious concentration of the drug. Trainor [22] reviewed the importance of PPB in drug discovery. On the other hand, Benet et al. [23] have expressed the opinion that changes in PPB have little clinical relevance. Smith et al. [24] have also criticized the strong emphasis on PPB during the lead optimization process in early drug discovery. It is true that there are drug molecules on the market that bind very strongly to plasma proteins (diclofenac, ibuprofen, etc.) and there are drug molecules that show very weak PPB properties. Although PPB alters the free drug concentration, it is not the only binding property that affects it. Phospholipids are also present in large quantities in tissues and these may contribute to a much greater extent to the loss of free concentration of drugs that are available at the site of action. Kratochwil et al. [25] summarized the distribution of the PPB of drugs in various therapeutic areas, such as CNS, anti-inflammatory, renal/cardiovascular, and chemotherapeutics. The PPB of drugs in these areas is evenly distributed, with the exception of anticancer drugs that tend to have lower PPB, as shown in Figure 10.13.

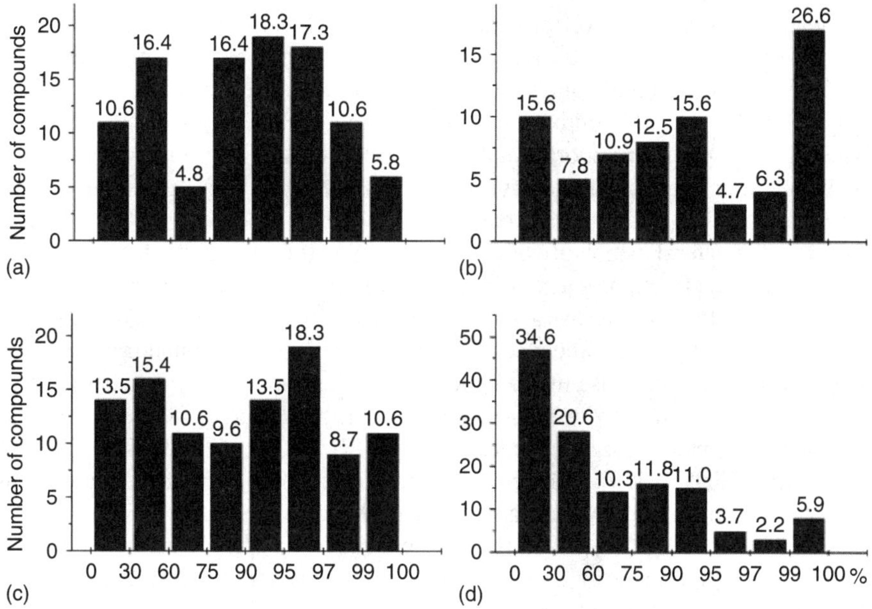

Figure 10.13 The distribution of protein binding of drugs in various therapeutic areas. The plots show the number of drugs with various plasma protein binding ranges in various therapeutic areas. (a) CNS (N = 104), (b) inflammation (N = 64), (c) renal/cardiovascular (N = 104), and (d) chemotherapy (N = 136). N represents the total number of drugs in the particular protein-binding range. (Adopted with permission from Reference [25].)

Liu et al. [26] have also raised the question as to whether we need to optimize the PPB early in the drug discovery process. It is obvious that PPB alone is not responsible for the free concentration of the drug. However, the PPB together with the tissue binding is an important parameter for the estimation of dose, free concentration, efficiency, and PK/PD. In general, we cannot advise the chemistry team what the optimum PPB is for a drug candidate, as it also depends on other properties. For example, the volume of distribution depends on the PPB as well as on the tissue binding.

The estimation of the free concentration *in vivo*, the determination of the unbound AUC (AUC_u) that describes the *in vivo* exposure of the drug, is an important parameter according to Smith et al. [24] and Liu et al. [26]. These parameters cannot be determined without the knowledge of PPB data. The PPB, or more exactly the unbound fraction in plasma, cannot be used on its own as lead optimization criteria. However, together with the clearance and tissue binding, it is used to derive other important PK/PD parameters such as *in vivo* distribution of the unbound compounds.

Herve et al. [27] summarized the specific role and binding characteristics of various plasma proteins. Plasma proteins deliver important endogenous and exogenous compounds such as hormones, vitamins, and fatty acids to which they show specific strong binding. The major component of the plasma is albumin (60%, 600 mM average concentration). The physiological concentration ranges from 40 to 75 g/l. Another important transport protein is the AGP, which represents only 1.5% of the total plasma proteins (0.7 g/l). However, AGP binds basic compounds more strongly, whereas albumin binds more strongly the negatively charged compounds. Immunoglobulins and lipoproteins are the remaining 38.5% of the plasma proteins. They bind lipophilic and poorly soluble molecules. Plasma also contains fibrinogen and other nonprotein small molecules, such as fatty acids, urea, creatinine, ammonium salts, amino acids, and hormones. It also contains inorganic salts in order to maintain the natural osmolality of plasma and cells. It is very difficult to determine an accurate free fraction of a compound in a complex system such as plasma. The composition of the plasma has a wide range of physiological variation (even within the same individual) that influences the measured values of unbound fraction. That is why it is common to use the so-called "pooled" plasma from a number of individuals or animals to provide an average composition. The most frequently applied and accepted methodologies to measure the PPB are the ultrafiltration [28] and equilibrium dialysis [29]. The principle of ultrafiltration is that after equilibrium between the drug and the plasma is reached, the high molecular weight proteins and the bound drug are removed from the plasma using specific filters that do not let through high molecular weight proteins. The concentration of the drug in the filtrate is then determined by an appropriate analytical method, most often HPLC with UV or preferably mass spectrometric detection. The method is very efficient and can be used in a 96-well plate format with 96-well format filter plates. Possible errors may arise from the drug binding to the filter material and the distortion of the equilibrium between the free and bound drug during the filtration process.

The equilibrium dialysis method takes longer, as the drug molecule is equilibrated with the plasma first. Then, a semipermeable membrane that does not let through the large proteins and the protein-bound drug is applied, whereas the small molecules and the unbound drug can permeate through the other side of the membrane. While one side of the membrane contains the plasma and the drug, the other side is filled up with buffer that has the same pH and osmolality as the plasma. The free drug reaches the same concentration on both sides of the membrane after a certain period of time (2–3 h). Then the concentration of the compound is determined using an HPLC/MS analytical method. The volumes of the "bound" and "free" phases are not altered during this measurement; however, we need to take into account the dilution factor, namely, the added "free" volume of the other side of the membrane which contains the buffer solution. The long equilibration process at body temperature may cause decomposition of the drug molecule, which has to be checked. The equilibrium dialysis method also can be used in a 96-well plate format, which makes the methodology reasonably high throughput [29,30]. The equilibrium dialysis method can also be used for the determination of tissue binding. In this case, the tissue (brain, lung, muscle) is homogenized (this usually involves breaking the cells) and diluted down with buffer. In order to accelerate, the equilibration process agitation of the plates is applied. By knowing the total concentration of the drug at the beginning of the experiment and measuring the unbound concentration in the buffer side of the semipermeable membrane, the unbound fraction, or the percentage of bound drug, can be calculated. In general, the ultrafiltration and the equilibrium dialysis methods provide an overall binding data of the compounds to the mixture of proteins in the plasma. The PPB measurements do not reveal which major plasma or tissue component has bound the compound. The accuracy of the method for measuring highly bound compounds is poor, as it depends on the detection sensitivity of the analytical method used for the concentration determination of the unbound molecule. Because of the complexity of the matrix, it is very difficult to derive reliable structure–binding relationships that can be used by the chemistry team during lead optimization. Therefore, the biomimetic HPLC measurements of compounds binding to individual plasma proteins are more useful.

Several applications of chemically bonded proteins on high quality silica supports have been reported and discussed in Chapter 5. These methods are based on the assumption that the chemically bonded protein retains the binding specificity and conformational mobility of the native protein. Comparisons of the chromatographically determined protein binding and binding data obtained by ultrafiltration or equilibrium dialysis methods have been promising, as has been demonstrated in Chapter 5. Here, the models are discussed in more detail where the chromatographically determined HSA and AGP binding have been used together to build a model using regression equations to describe the total PPB. Figure 10.14 shows the percentage PPB, percentage HSA binding, and percentage AGP binding of 80 known drug molecules ranked according to increasing HSA binding.

It can be seen that although the HSA binding dominates the total PPB, strong AGP binding increases the total PPB. It is interesting to note that there are instances

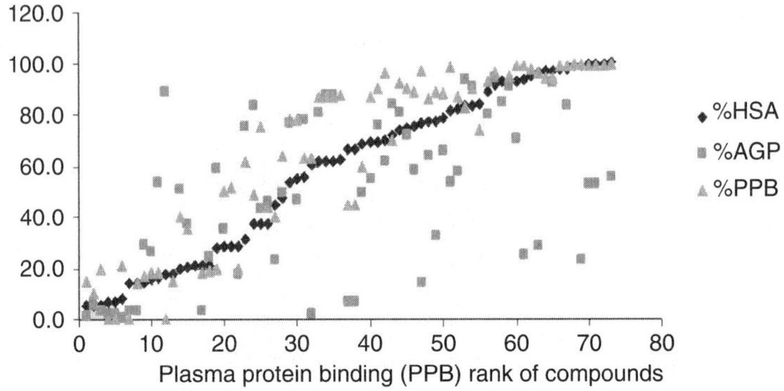

Figure 10.14 The total plasma protein binding (%PPB), HSA binding (%HSA), and AGP binding (%AGP) of 80 known drugs in the order of increasing HSA binding.

when the total PPB is weaker than the HSA binding on its own (see amrinone and bromazepam in Table 10.10). In order to explain these phenomena, we need to understand how the individual binding properties affect the total PPB, which refers to a mixture of proteins. Table 10.10 shows the human PPB and the HSA, AGP binding, and the CHI log D values obtained at pH 7.4 using the HPLC methodology of the studied drug molecules [31].

In order to assess the total PPB from the individual binding constants (that are considered as partition coefficients between the pure aqueous and the protein phases), we need to know the volumes of the free and protein phases. Equation 10.13 shows the expression of total PPB for the situation.

$$K(\text{PPB}) = V_{\text{HSA}} \times K(\text{HSA}) + V_{\text{AGP}} \times K(\text{AGP}) + V_{\text{IG}} \times K(\text{IG}) \qquad (10.13)$$

where IG denotes the other immunoglobulin found in the plasma. The K(PPB), referring to the total plasma partition coefficient of a compound, can be calculated from the percentage binding data shown by Equation 10.14. The percentage binding data show sigmoid relationship with the log K data. However, a significant portion of the sigmoid curve can be considered almost linear, which explains the reasonably good correlation found using the linear combination of percentage binding data in Table 10.10 (Equation 10.14).

$$\%\text{PPB} = 0.86(\pm 0.05)\%\text{HSA} + 0.17(\pm 0.05)\%\text{AGP} + 1.27(\pm 0.43)\ \text{cMR} - 7.03$$

$$n = 80 \quad r^2 = 0.88 \quad s = 11.4 \quad F = 130(p < 0.0001) \qquad (10.14)$$

As we do not have measured immunoglobulin binding constants, the size parameter (calculated molar refractivity, cMR) has been used in the model. The larger size, in general, goes with higher lipophilicity and lower solubility, which results in stronger protein binding. The logarithmic values of the chromatographic binding data can be accurately determined even at the highly bound region [32]. Therefore,

Table 10.10 The human plasma protein binding and the chromatographically determined HSA, AGP binding, and lipophilicity (CHI log *D* at pH 7.4) for 80 known drug molecules[a]

Drug	PPB (%)	log *k*(PPB)	log *k*(HSA)	%HSA Bound	log *k*(AGP)	%AGP Bound	CHI log *D*(pH7.4)	cmr
Acetaminophen	1	−2.00	−0.79	14.0	−1.48	3.2	−0.59	4.2
Acyclovir	15	−0.76	−1.25	5.4	−1.98	1.0	−2.65	5.5
Amiloride	40	−0.18	−0.61	20.0	0.00	50.8	−0.49	5.4
Amoxicillin	18	−0.66	−0.71	16.4	0.05	53.4	−0.52	9.3
Amrinone	45	−0.09	−0.34	31.8	−1.16	6.6	−0.42	5.4
Aspirin	49	−0.03	−0.23	37.3	0.67	83.3	−0.43	4.5
Betamethasone	64	0.24	0.09	55.6	0.74	85.6	1.56	10.1
Bromazepam	60	0.17	0.32	68.5	−0.02	49.4	1.38	7.7
Budesonide	88	0.83	0.55	78.7	0.27	65.7	2.47	11.3
Bumetanide	99	1.69	1.15	94.3	−0.27	35.1	1.11	9.5
Bupivacaine (HCl)	95	1.20	−0.16	41.3	1.35	96.7	2.98	8.9
Ceftazidime	21	−0.58	−1.07	7.9	−2.02	0.9	−0.89	13.9
Cephalexin	14	−0.79	−0.79	14.0	−1.48	3.2	−0.22	9.2
Chlorpheniramine	70	0.35	0.55	79.1	0.58	79.8	1.44	8.2
Chlorpropamide	96	1.28	0.99	91.6	1.15	94.3	0.37	6.7
Cimetidine	19	−0.64	−0.58	21.2	−0.50	24.1	−0.09	6.9
Cinoxacin	63	0.22	0.17	60.1	−1.64	2.3	−0.65	6.3
Ciprofloxacin	40	−0.18	−0.09	45.0	−0.53	22.9	−0.08	8.7
Clonazepam	86	0.76	0.51	77.1	0.23	63.8	2.01	8.3
Clonidine	20	−0.61	−0.42	28.0	0.15	58.9	0.33	5.8

(*continued*)

Table 10.10 ***(Continued)***

Drug	PPB (%)	log k(PPB)	log k(HSA)	%HSA Bound	log k(AGP)	%AGP Bound	CHI log D(pH7.4)	cmr
Cytarabine	12	−0.87	−0.38	29.8	−0.62	19.5	−2.61	5.6
Diazepam	99	1.69	1.08	93.2	0.37	70.7	2.77	8.1
Digitoxin	97	1.38	0.49	76.4	−0.78	14.2	1.11	19.4
Diltiazem	78	0.53	0.14	58.5	1.17	94.6	2.30	11.6
Diphenhydramine	63	0.22	0.08	55.4	0.54	78.2	1.68	8.0
Doxepin	83	0.66	0.67	83.1	1.08	93.3	1.91	8.8
Famotidine	17	−0.69	−0.77	14.5	−0.39	29.4	−0.23	8.1
Finasteride	90	0.91	0.45	74.7	0.39	71.7	2.38	10.7
Flumazenil	50	−0.01	−0.58	20.9	0.42	73.0	1.08	7.7
Fluoxetine	94	1.13	1.39	97.0	1.02	92.1	2.08	8.1
Flurbiprofen	99.96	1.98	2.05	100.1	0.09	55.7	1.03	6.8
Ganciclovir	1	−2.00	−0.84	12.8	−0.91	11.1	−2.64	6.1
Gemfibrozil	97	1.38	1.29	96.1	0.96	91.1	2.72	7.2
Glipizide	98.4	1.58	1.20	95.0	−0.48	25.1	0.95	11.8
Haloperidol	92	1.01	0.61	81.1	1.46	97.6	1.82	10.3
Hydrochlorothiazide	40	−0.18	−0.26	35.6	−0.61	20.0	0.15	6.3
Imipramine	90	0.91	0.67	83.1	0.96	91.1	2.16	9.0
Indomethacin	99	1.69	1.82	99.5	0.04	52.9	1.37	9.5
Isoniazid	1	−2.00	−1.14	6.8	−1.64	2.3	−2.99	3.7
Isradipine	97	1.38	1.05	92.8	1.38	97.0	3.25	9.9

Ketoconazole	99.0	1.69	1.06	93.0	0.72	84.9	2.67	13.9
Ketoprofen	98.7	1.63	1.57	98.3	0.31	67.7	0.69	7.3
Lorazepam	91	0.96	0.96	91.1	1.08	93.3	1.89	8.3
Methylprednisolone	78	0.53	0.08	54.9	−0.06	46.8	1.50	10.1
Metoclopramide	40	−0.18	0.02	51.4	0.87	89.1	0.45	8.2
Metronidazole	10	−0.96	−1.25	5.4	−1.22	5.7	−0.37	4.1
Nabumetone	99	1.69	1.15	94.3	1.40	97.1	3.19	6.9
Naproxen	99.7	1.88	1.95	99.9	0.07	54.7	0.64	6.6
Nicardipine	95	1.20	1.07	93.2	0.94	90.5	3.84	13.3
Nifedipine	96	1.28	0.65	82.6	0.20	62.0	2.59	9.1
Nimodipine	98	1.51	0.62	81.4	0.05	53.3	1.28	11.1
Nitrazepam	87	0.79	0.64	82.3	0.12	57.3	1.90	7.8
Nizatidine	35	−0.28	−0.60	20.4	−0.24	37.1	0.10	9.0
Ondansetron	73	0.42	0.52	77.7	1.30	96.2	1.25	8.6
Phenobarbital	51	0.01	−0.27	35.2	0.31	67.6	1.12	6.2
Phenytoin	89	0.87	0.47	75.5	0.14	58.5	1.72	7.2
Phenytoin	89	0.87	0.61	81.0	0.86	88.8	1.72	7.2
Piperacillin	18.5	−0.65	−0.74	15.4	−0.46	26.2	0.38	13.1
Piroxicam	94.5	1.16	1.47	97.7	0.47	75.4	0.30	8.3
Prazosin	95	1.20	0.59	80.4	1.27	95.9	1.09	10.2

(continued)

Table 10.10 ***(Continued)***

Drug	PPB (%)	log *k*(PPB)	log *k*(HSA)	%HSA Bound	log *k*(AGP)	%AGP Bound	CHI log *D*(pH7.4)	cmr
Prednisolone	70	0.35	−0.12	43.6	0.46	75.0	1.13	9.6
Prednisone	75	0.46	−0.23	37.6	−0.12	43.3	1.23	9.5
Procainamide	16	−0.73	−0.27	35.2	−0.13	43.0	−0.59	7.1
Propranolol	87	0.79	0.29	66.5	0.77	86.5	1.17	7.8
Quinidine	87	0.79	0.51	77.0	0.71	84.5	0.97	9.5
Quinine	94	1.13	0.50	76.7	1.57	98.4	0.99	9.5
Rifampin	89	0.87	0.51	77.2	−0.32	32.5	1.74	21.9
Saquinavir	98	1.51	1.22	95.2	1.59	98.5	3.24	19.0
Spironolactone	90	0.91	0.34	69.1	0.48	75.9	2.81	11.3
Sulfamethoxazole	62	0.20	0.38	71.2	0.46	75.0	0.79	6.3
Terbutaline	20	−0.61	−0.40	28.8	−0.68	17.4	−0.60	6.3
Tolbutamide	96	1.28	1.28	96.0	−0.41	28.3	0.95	7.1
Trazodone	93	1.07	0.90	89.7	1.28	95.9	2.25	10.4
Trimethoprim	44	−0.11	−0.23	37.6	−0.07	46.2	0.49	7.8
Verapamil	90	0.91	0.53	77.9	1.31	96.3	2.32	13.2
Warfarin	99	1.69	1.49	97.9	0.67	83.2	0.93	8.7
Zidovudine	2.5	−1.60	−0.84	12.8	−0.9074	11.1	0.11	6.4
Zolmitriptan	25	−0.48	−0.36	30.6	0.68	83.7	−0.07	8.3
Zolpidem	92	1.01	0.43	73.6	0.60	80.6	1.86	9.3

[a]From Reference [31] with permission.

it is more appropriate to use a model with log k data instead of the percentage binding data as shown by Equation 10.15.

$$\log k(\mathrm{PPB}) = 0.87(\pm 0.05) \log k(\mathrm{HSA}) + 0.17(\pm 0.05) \log k(\mathrm{AGP}) + 0.06(\pm 0.02)\ \mathrm{cMR} - 0.27$$
$$n = 55 \quad r^2 = 0.85 \quad s = 0.36 \quad F = 92 \tag{10.15}$$

A similar model equation can be obtained (Equation 10.16) using the exponential values of the log k binding parameters (denoted as log K), which shows a linear relationship with the logarithmic values of the octanol/water partition coefficients (log P) and puts the HSA and AGP bindings on the same (log P) scale as shown in Figure 10.10.

$$\log K(\mathrm{PPB}) = 0.95(\pm 0.06) \log K(\mathrm{HSA}) + 0.16(\pm 0.05) \log K(\mathrm{AGP}) + 0.10(\pm 0.03)\ \mathrm{cMR} - 0.56$$
$$n = 55 \quad r^2 = 0.86 \quad s = 0.75 \quad F = 104 \tag{10.16}$$

Reduced numbers of compounds have been used in the model described by Equation 10.16 in order to reduce the intercorrelation between the independent variables (the HSA and AGP binding). The plot of the literature and estimated total PPB is shown in Figure 10.15.

Nowadays, whole blood is more and more frequently used than plasma for monitoring drug concentration. Blood is spotted on a filter paper and the water is evaporated; thus, the dry blood spots are easier to transport than the plasma. The blood is reconstituted from the spots by adding water to it. The analysis can be carried out as usual with the reconstituted blood. Therefore, it is interesting to see how the model changes when the blood cells are also present. It is known that compounds can partition into the blood cells, and they have different blood/plasma partition coefficients.

While plasma does not contain phospholipids, the red blood cells are covered by phospholipid membrane. Rapid equilibrium dialysis measurements have been carried out in blood samples of a great number of project compounds at GlaxoSmithKline. The unbound fraction in blood values have been converted to the logarithmic values of the partition coefficient (log k(blood) using the formula by Equation 10.5). The following model equation (Equation 10.17) that has been constructed can be used for the estimation of the free fraction or the bound/free partition of the compounds using the biomimetic HPLC binding data.

$$\log k(\mathrm{blood}) = 0.14\ \mathrm{CHI} \log D7.4 + 0.46 \log K(\mathrm{HSA}) + 0.43 \log k(\mathrm{AGP}) - 0.72$$
$$n = 133 \quad r^2 = 0.79 \quad s = 0.29 \quad F = 164 \tag{10.17}$$

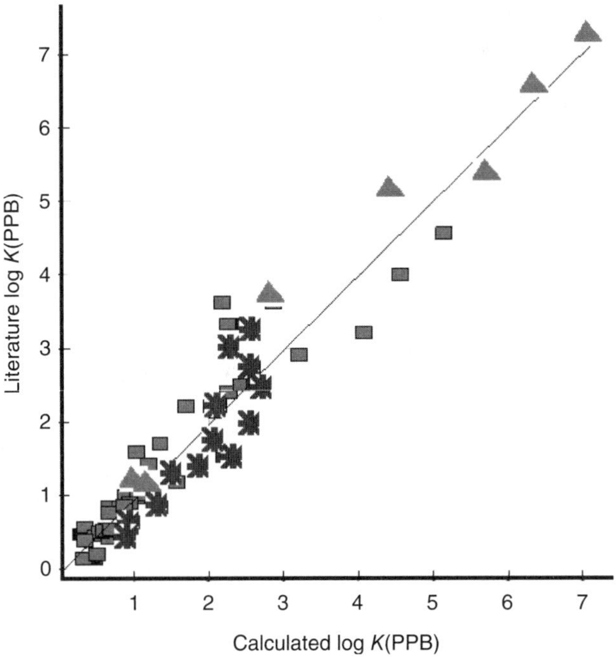

Figure 10.15 The plot of the literature and calculated plasma protein binding (log *k*(PPB)) obtained by Equation 10.16 from Reference [32] (triangles are acids, stars are bases, and squares are neutral molecules).

Note that the HSA binding is expressed by the exponential log *k*(HSA) values (*e* log *k*(HSA) = log *K*(HSA)); that is why the coefficient is small relative to the AGP-binding parameter that is not used as an exponential term.

Figure 10.16 shows the plot of the measured and estimated log *k*(blood) values for various project compounds [32]. There are some outlier compounds, which is not surprising as the blood contains various types of proteins that may specifically and strongly bind certain types of compounds. However, compounds with strong AGP binding were not among the outliers.

It has been observed sometimes that the free concentration of a compound in the PPB or blood-binding measurements by equilibrium dialysis is higher than the free concentration *in vivo*. The explanation in this case may be that we use higher drug concentration in these measurements relative to the plasma proteins for aiding the analytical concentration determination by HPLC with mass spectrometric detection. It may cause saturation of specific binding sites in the plasma or blood, which results in higher free fraction.

In all other cases, we can expect that the equilibrium measurements may give higher binding values when the compound binds strongly to some of the plasma components that are not considered in the model equation. It is always advisable to

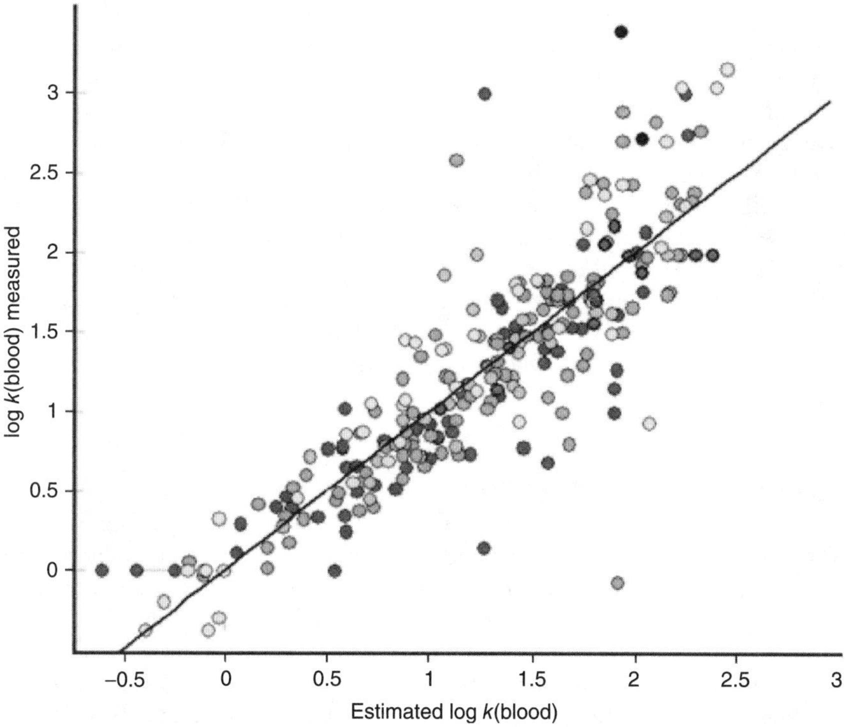

Figure 10.16 The measured and estimated blood partition coefficients for 133 diverse project compounds using Equation 10.16 from Reference [32]. (Various colors represent the compounds from various projects.)

measure the unbound fraction in plasma/blood by equilibrium dialysis using various concentrations of the compound. When the unbound fraction is independent of the compound concentration, we can assume linear binding characteristics. Similarly, we should obtain the same retention times for such compounds when injecting various amounts of drugs on the protein stationary phases. This also means that the compound does not bind to low capacity and high specificity binding sites that can be saturated and may cause drug–drug interactions. Although the serum albumin binding is the dominant factor in both plasma and blood binding, it is important to recognize the importance of AGP binding early in the drug discovery process. Israili and Dayton [33] have provided an excellent review about AGP and its interactions with drug molecules. The AGP concentration can vary significantly in plasma, depending on age, gender, or ethnicity [34–36]. Increased AGP plasma levels have been observed in various disease states, such as inflammation. Therefore, drugs that bind strongly to AGP can cause variable efficacy in the clinical phase. Various genetic variants of AGP may bind certain drugs differently as it has been pointed out by Herve et al. [37]. Elevated AGP levels were reported to cause altered steady-state PKs [38].

During the HSA- and AGP-binding measurements, chiral separation of racemic mixtures has often been observed. It is not surprising, as these stationary phases have been developed for chiral separations. However, significant binding differences of enantiomers imply different PK behavior as well. Levy and Boddy [39] have reviewed the theoretical principles of stereoselective PKs, which had not been investigated in great detail before.

The chiral biomimetic stationary phases and the models derived from them have the ability to explain the stereoselective PK behaviors of racemic mixture. The traditional equilibrium dialysis or ultrafiltration methods cannot detect the difference in the binding of the enantiomers. Using the biomimetic chromatographic technique for the estimation, the differences in the PK behavior and efficacy of enantiomers would be an interesting and important research area for the future generation of drug discovery scientists.

BLOOD/BRAIN DISTRIBUTIONS

When the location of the pharmacological target is known, the preferable distribution is to concentrate the drug in the tissue where the target is. There are several tissues/organs that have been studied in more detail. Targeting the CNS presents additional challenges because of the special properties of the blood–brain barrier layer found between the blood circulation and neural tissue. The CNS drugs need to go to the brain, which has a special barrier and selectivity for certain types of compounds. Another characteristics of the CNS is fluid turnover of the cerebrospinal fluid (CSF) and interstitial fluid (ISF), which provide a continual secretion and drainage [40,41]. As the brain tissue is also networked with blood capillaries, it is very difficult to devise *in vivo* and *in vitro* concentration measurements of compounds in the brain [42]. There are two important aspects of studying drug distribution into the brain. The first aspect is the brain-to-blood distribution ratio in steady state with special emphasis on the unbound concentration of the drug in the brain. The second aspect is the rate of brain permeation of the drug from the blood. These two processes are not completely independent, but they are governed by different physicochemical and physiological processes. The rate of brain permeation studies are very similar in principle to the other permeability studies that have been discussed in more detail in Chapters 7 and 9. In this case, solubility is not an issue, as the compound is dissolved in the plasma; however, the PPB has to be considered, as the bound drug is not able to penetrate through the blood–brain barrier. Similar to all permeability processes, the main driving force is the concentration difference of the compound between the two sides of the membrane. In addition, the blood–brain barrier contains several active transporters, such as the PgP efflux of lipophilic drugs; carrier-mediated influx of small polar compounds such as glucose, amino acids, nucleosides, monocarboxylates, and small peptides; and the tight junction modulation for the permeability of polar solutes. Various *in situ in vivo* methods that are aimed at the determination of the rate of permeation require precise surgical procedure; therefore, the *in vitro* brain cell culture or artificial membrane permeability assays provide a better alternative [43,44]. The measurements of brain permeability [45] for *in situ* microdialysis of the CSF can be carried out using HPLC. The microdialysis technique provides the most relevant brain penetration rate data, including the effects of the active transporters and the rate of blood perfusion [46]. Current and perspective technologies for *in vitro* cell models for the blood–brain barrier have been recently reviewed by Naik and Cucullo [47].

We can expect the brain-to-blood distribution of the molecules *in vivo* to arise from the different plasma and brain tissue binding of the compound. When several molecules have similar brain tissue binding, their differences in the PPB influences the *in vivo* available free drug in the brain. Rowley et al. [48] demonstrated the importance of PPB for the activity of glycine/NMDA receptor antagonist compounds in the brain. We can calculate the brain-to-blood distribution coefficient (K_p) from the unbound fraction in blood and brain ($f_{u,blood}$ and $f_{u,brain}$, respectively) using the assumption that the free concentrations in the blood and brain ($c_{u,blood}$, $c_{u,brain}$) are equal using the formulae described by Equations 10.18–10.21.

$$c_{u,blood} = c_{u,\ brain} \tag{10.18}$$

$$c_u = f_u \times c_{total} \tag{10.19}$$

$$f_{u,\ blood} \times c_{blood(total)} = f_{u,\ brain} \times c_{brain(total)} \tag{10.20}$$

$$k_{brain/blood} = \frac{c_{brain(total)}}{c_{blood(total)}} = \frac{f_{u,\ blood}}{f_{u,\ brain}} \tag{10.21}$$

A study on diverse marketed CNS drugs confirmed [49,50] the validity of Equation 10.21 and highlighted the importance of measuring the unbound fraction of compounds in brain tissue homogenates. Watson et al. [51] have used the unbound fraction in brain tissue data and the total brain concentration obtained by autoradiography using radio-labeled compounds to calculate the free brain concentration. They have found significant correlation between the free brain concentration and the dopamine D2 receptor occupancy of several antipsychotic compounds.

The traditional method for tissue-binding determination is equilibrium dialysis. However, the tissue homogenates are usually very viscous matrix that would require a very long equilibration time. Therefore, it is common to dilute down the tissue homogenate. When the binding ratio is independent of the total concentration of the compounds, the dilution factor should be taken into consideration using Equation 10.22.

$$f_{u,\ undiluted} = \frac{1/D}{((1/f_{u,\ diluted}) - 1) + 1/D} \tag{10.22}$$

where D is the dilution factor in the brain or any other tissue homogenate and the $f_{u,\ diluted}$ of the measured free fraction of diluted brain or tissue homogenate.

The application of HPLC retention data for the determination of molecular descriptors for the Abraham solvation equation (see more details in Chapter 5) can be used to assess the rate of brain permeation ($\log BB_{perm}$) using the published Abraham equation [52] shown by Equation 10.23.

$$\log BB_{perm} = 0.77 \times E - 1.87 \times S - 2.8 \times B + 3.31 \times V - 1.21$$
$$n = 18 \quad r^2 = 0.98 \quad s = 0.481 \quad F = 65 \tag{10.23}$$

where E is the excess molar refraction, S the dipolarity/polarizability, B the H-bond basicity, and V is the McGowan volume of the molecules. Interestingly, the H-bond acidity parameter (A) was not significant in the equation, unlike in the equation describing the blood/brain distribution coefficient. When the equilibrium distribution of 57 drugs was studied, both the H-bond acidity and basicity had negative coefficients (see Equation 10.24), showing that H-bond donor and acceptor groups on the molecules reduce their partition into the brain [52].

$$\log BB_{distribution} = 0.2 \times E - 0.69 \times S - 0.71 \times A - 0.70 \times B + 1 \times V - 0.04$$
$$n = 57 \quad r^2 = 0.95 \quad s = 0.20 \quad F = 99 \tag{10.24}$$

The differences between Equations 10.23 and 10.24 show the different mechanisms in the brain permeability and distribution processes.

The biomimetic HPLC measurements are also very useful for assessing the brain-to-blood distribution coefficient and the brain tissue binding of the compounds. The brain tissue binding obtained by rapid equilibrium dialysis (RED) method for diverse research compounds from various CNS programs have been modeled by the biomimetic chromatographic binding data. Equation 10.25 shows the model for the estimation of brain tissue binding ($\log k$ (BTB)) of compounds, and Figure 10.17 shows the plot of the measured and estimated values [32].

$$\log k(\text{BTB}) = 1.29(\pm 0.10) \times \log k(\text{IAM}) + 1.03(\pm 0.10) \times \log k(\text{HSA}) - 2.37$$
$$n = 135 \quad r^2 = 0.76 \quad s = 0.35 \quad F = 212 \tag{10.25}$$

Note that in this equation, the chromatographic $\log k$(IAM) and $\log k$(HSA) data have not been projected to the $\log P$ scale using the exponential function.

As the brain contains large amounts of phospholipids, the coefficient of the log k(IAM) values are even larger than the coefficient for the albumin binding.

The blood and brain tissue binding data ($\log k$(blood) and $\log k$(brain)) can be converted back to unbound fraction (f_u) values using Equations 10.26 and 10.27.

$$\%\text{Bound} = 100 \times \frac{10^{\log k(\text{Bound})}}{1 + 10^{\log k(\text{Bound})}} \tag{10.26}$$

$$f_u = \frac{100 - \%\text{Bound}}{100} \tag{10.27}$$

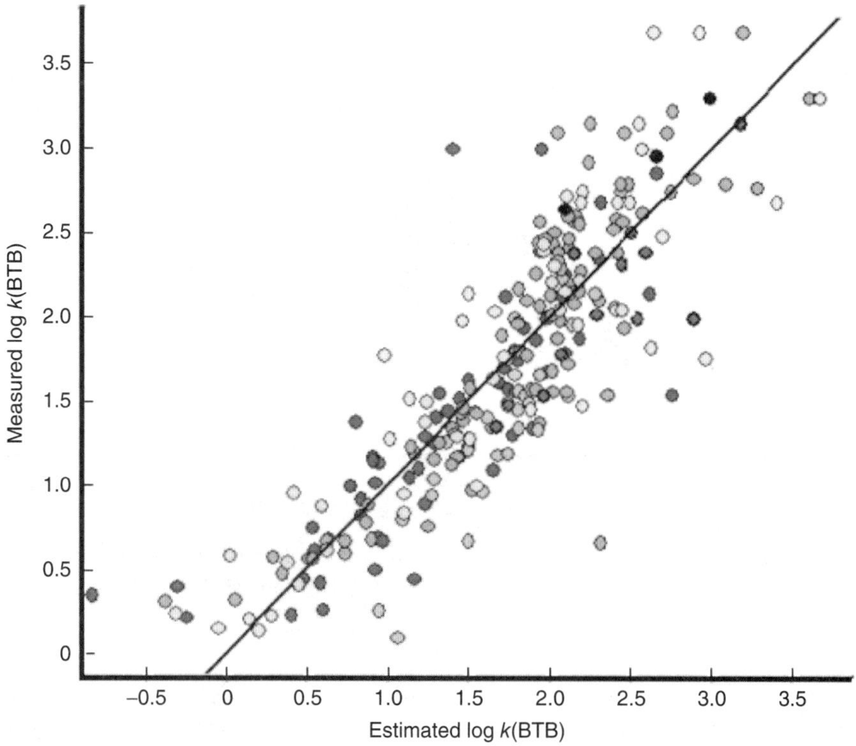

Figure 10.17 The plot of the measured and calculated brain tissue binding data (log *K*(BTB)) using Equation 10.25. The various colors represent the compounds from various programs. (Reproduced from Reference [32].)

where log *k*(bound) can be log *k*(blood), log *k*(brain), or log *k* binding constant for any other tissues. The $f_{\text{u, blood}}$ and $f_{\text{u, brain}}$ values obtained from the biomimetic HPLC HSA, AGP, and IAM binding data can be used to express the brain-to-blood partition data ($k_{\text{brain/blood}}$) according to Equation 10.21. As both the blood and brain tissue binding models have errors, the error of the ratio of the two parameters adds up; thus, the brain to blood distribution coefficient obtained from the equilibrium dialysis method and the biomimetic chromatographic models do not show very good correlation, as demonstrated in Figure 10.18.

Although the correlation between the two measures of brain-to-blood distribution ratio is not very strong, still the distribution coefficients obtained by biomimetic HPLC could be used to differentiate between the CNS penetrant and non-CNS penetrant drugs as shown in Figure 10.19. Some known drug molecules have been categorized as "CNS" because they were specifically designed for CNS targets, drugs that have CNS side effects indicating CNS penetration (that is categorized as "maybe"), and drugs that show no evidence of crossing the blood–brain barrier were assigned as "non-CNS." The box plot shows the average and median values

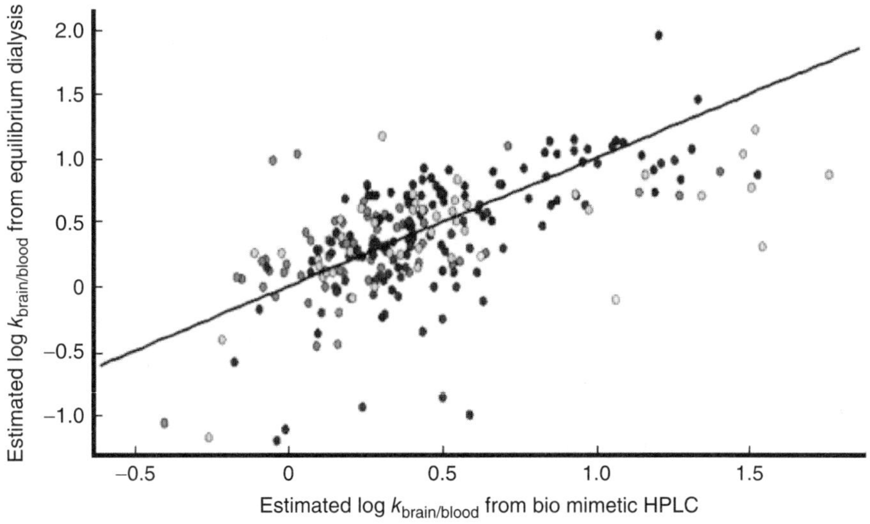

Figure 10.18 The plot of the logarithmic brain-to-blood distribution ratio (log $k_{brain/blood}$) obtained from the equilibrium dialysis and from the chromatographic measurements. (Reproduced from Reference [54].)

of log $k_{brain/blood}$, the variance (standard error) around the average. The circles represent the results of the Tukey–Kramer test in order to establish that the average values are significantly different or not for the different groups of drugs. It can be seen that drugs that are designed for CNS activity and drugs that show CNS side effect have similar average log $k_{brain/blood}$, whereas drugs that do not penetrate into the CNS have significantly lower average log $k_{brain/blood}$. It shows that we can expect a larger brain-to-blood ratio for compounds that have significant potency in the brain, in spite of the fact that it does not guarantee high free concentration in the brain.

If it is true that higher $k_{brain/blood}$ is advantageous for CNS activity, then based on Equation 10.21 the lower unbound fraction in the brain relative to the unbound fraction in blood is advantageous. However, when the unbound fraction in the brain tissue is low, the free concentration of the drug in the brain is compromised. Therefore, the aim is to lower the PPB as much as possible. This optimizing procedure is clearly seen in the early observation that an optimum lipophilicity expressed as octanol/water log P is needed for good brain penetration [53]. Although the lower unbound fraction in the brain is preferred relative to the unbound fraction in the blood for good brain partition of the compounds, in order to obtain efficacious free concentration in the brain, the general protein binding and lipophilicity has to be decreased. Too hydrophilic compounds do not partition into the lipophilic tissue compartments and although the free fraction in plasma is usually high for these compounds because of the low brain penetration, CNS potency cannot be expected.

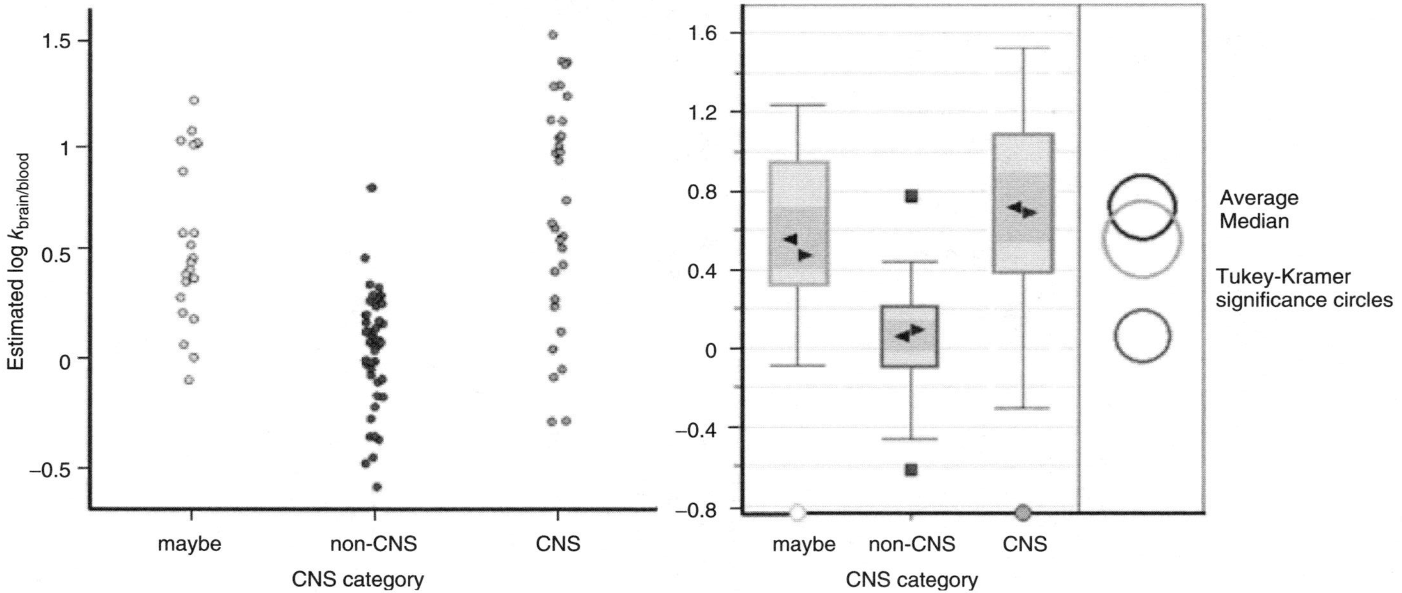

Figure 10.19 The distribution of the estimated log $k_{brain/blood}$ of known drugs categorized according to the CNS penetration (maybe, drugs that are not targeted to CNS but showed CNS side effect; non-CNS, drugs that have no evidence of entering the CNS; CNS, drugs that have CNS activity). (Reproduced from Reference [32].)

TISSUE DISTRIBUTION

The volume of distribution represents the compounds' partition between the tissue and plasma compartments. This is an important aspect for the assessment of the proportion of the administered dose in tissues and in plasma. When the pharmacological target is in the tissues or in a particular organ, it is preferred that a higher proportion of the dose partitions into tissues. However, the free concentration of the drug is more likely to correlate with the clinical outcome; therefore, it is important to know the unbound fraction (f_{ut}), or in another term, the tissue partition coefficients of the compounds. The Øie–Tozer equation [18] shown by Equation 10.6 can be used to express the unbound fraction in tissues, and these data have been modeled for neutral and basic compounds by Lombardo et al. [17]. Using the data of 130 known drugs, Valkó et al. [54] published a model for the estimation of unbound fraction in tissues using the biomimetic HPLC data (HSA and IAM binding). The plot of the measured human clinical unbound fraction in tissues (f_{ut}) and the estimated values using Equation 10.28 is shown in Figure 10.20.

$$\log f_{ut} = -0.66(\pm 0.05) \log K(\text{HSA}) - 0.52(\pm 0.06) \log K(\text{IAM}) + 0.55$$
$$n = 70 \quad r^2 = 0.85 \quad s = 0.58 \quad F = 182 \qquad (10.28)$$

Note that in Equation 10.28, the HSA and IAM binding parameters have been projected to the log P scale. The coefficients of the two biomimetic binding parameters show nearly equal contributions to the reduction of unbound fraction in tissues, indicating that the overall tissue binding is increased by both types of binding properties. It would be valuable if we can characterize the binding properties of

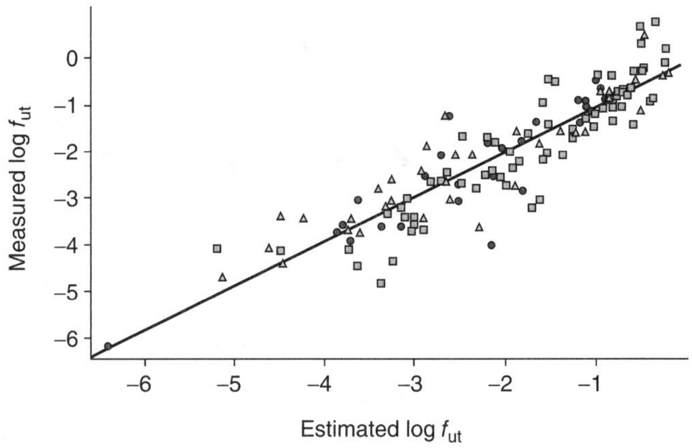

Figure 10.20 Human unbound fractions in tissues (log f_{ut}) and the estimated values using the biomimetic HPLC binding data according to Equation 10.28 with permission from Reference [54] (circle, compound from the training set; star, compound from the test set; various shades represent the acid base character: circle, base; triangle, acid; square, neutral.)

compounds in each type of tissues/organs. Equation 10.25 can be used to estimate the brain tissue binding, showing that the IAM binding is more dominant in brain probably because of the higher phospholipid content than for the tissues overall. It can be supposed that the sum of the two types of binding can be used to estimate various tissue bindings, but the regression coefficients are different for each tissue type depending on the tissue composition.

An interesting tissue binding model for mucin and diluted human lung tissue has been set up for a relatively small set of discovery compounds and known drug molecules for studying inhaled drugs binding. Mucins are produced by the epithelial cells in the lung and they are high molecular weight, heavily glycosylated proteins. In this respect, we can expect similar binding characteristics of compounds to mucins as to AGP. Figures 10.21 and 10.22 show the excellent correlation of AGP binding and mucin binding and the diluted human lung tissue binding of a small but diverse set of molecules [55]. The lung tissue binding could be modeled with the HSA- and AGP-binding data, very similar to the PPB model, but with a much higher regression coefficient for the AGP binding. This indicates that the lung tissue contains higher concentration of highly glycosylated proteins that bind strongly the positively charged compounds.

These findings are in agreement with the observations by Jansson et al. [6] who have noticed that lung tissue binding correlates better with log P than with log D for basic compounds. It is because the lung tissue contains mucus, which contains

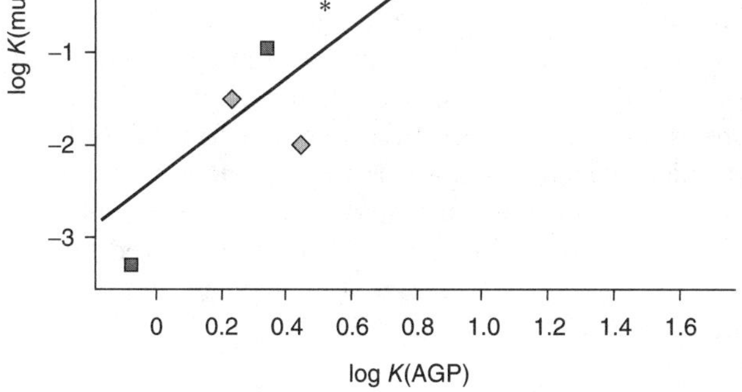

Figure 10.21 The strong correlation found between the mucin binding and α-1-acid glycoprotein (AGP) binding for a small set of diverse compounds from Reference [55]. The correlation coefficient is 0.85. (Known drugs and discovery compounds intended for use in the lung and coded by the acid base character: triangle, square, and circle represent basic, neutral, and acidic molecules, respectively.)

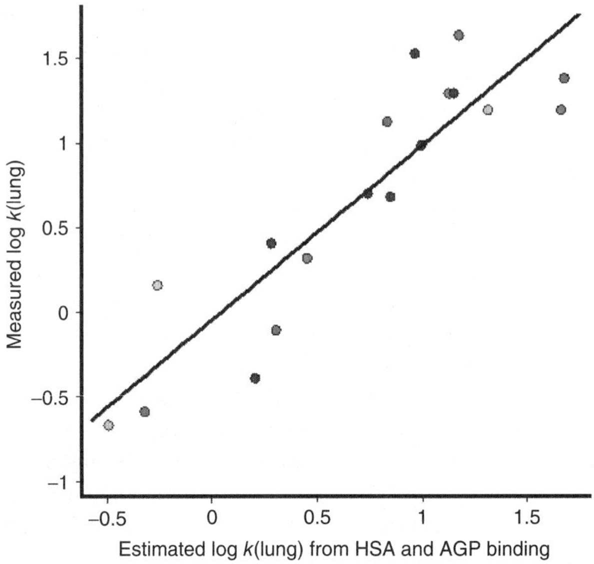

Figure 10.22 The plot of the measured and estimated lung tissue binding data of a diverse set of known drugs and discovery compounds that are targeted to the lung. The estimation is based on the HSA and AGP binding obtained by the biomimetic HPLC measurements. The regression coefficients of HSA and AGP bindings are 0.74 and 0.65, respectively, showing the strong influence of AGP binding on the overall lung tissue binding [55].

highly glycosylated proteins with sialic acid residues that are negatively charged; therefore, they bind strongly the positively charged basic compounds. It has also been reported by Manallack [56] that drugs (especially acting in the CNS) contain at least one basic group (45% of the drugs and 62% of the CNS drug set) that enhances their partition into the tissues. The majority of the basic pK_a values were between 8.5 and 10.5, peaking between 8.5 and 9.5, that means a high percentage of positive ionization of the compounds at physiological pH values.

In principle, when the plasma and tissue partition coefficients of a compound are known, we can calculate the free concentration overall for a given dose. Assuming that 100% absorption and no active transport or permeability barriers exist, it is possible to calculate the required dose of the compounds based on the potency data, which is equivalent to the target concentration at the site of action. The biomimetic HPLC binding data (HSA, AGP, and IAM binding) proved to be appropriate to establish models for *in vivo* distribution of compounds and probably would serve as appropriate input data for PBPK models.

SUMMARY

The distribution of drug molecules *in vivo* is not even; they may accumulate in the plasma or various tissue compartments. It is important to know where the majority of the drug molecule is concentrated *in vivo*. The most often used parameter is the volume of distribution that is equivalent to a hypothetical tissue volume that would be needed to have the same concentration of the drug as it is in plasma. It has been shown that the volume of distribution is proportional to a compound's partition coefficient between the tissue and the plasma compartments. Higher volume of distribution means that higher proportion of the given dose can be found in the tissues relative to the plasma compartment. The steady-state volume of distribution and clearance determines the half-life and the dosing regimen of the compounds. When the pharmacological target is in the tissues, a higher volume of distribution is preferred, whereas when the target is in the plasma, a lower volume of distribution is advantageous. Depending also on the clearance of a particular compound, the desirable volume of distribution can be established. The volume of distribution can be obtained only from *in vivo* measurements; therefore, there is an effort to establish models and *in vitro* measurements for its estimation. The PBPK modeling approach uses various partition coefficients (octanol/water, oil/water) and acid dissociation constants. The biomimetic HPLC binding measurements have been proved very useful to establish *in vitro* models for tissue partition and PPB and volume of distribution of compounds. It has been demonstrated using human clinical volume of distribution data of over 100 known drug molecules that the volume of distribution *in vivo* is governed by the difference in the compounds' binding to phospholipids (IAM) and albumin type of proteins (HSA). It has been observed that while both types of bindings are governed by the compounds' lipophilicity, the presence of positive and negative charges at physiological pH values have a great impact. Positively charged compounds bind much more strongly to phospholipids than to albumin, resulting in strong tissue binding and high volume of distribution, while negatively charged compounds bind strongly to albumin and prefer the plasma compartment, resulting in a low volume of distribution. The importance of the PPB has also been discussed, and it has been shown that PPB on its own is not enough to describe the compounds' *in vivo* distribution. We need to take into account the tissue binding relative to the PPB for the estimation of *in vivo* distribution. The brain tissue binding has been discussed in more detail, as it is an important aspect for designing CNS active drugs, or avoiding the CNS side effects of other drugs. The brain tissue binding is greatly influenced by the phospholipid binding of the compounds, while the lung tissue and mucus binding is highly influenced by the compounds' binding to glycosylated proteins, which could be modeled by HPLC-based AGP-binding data. In principle, the combination of these major biomimetic binding properties (HSA, IAM, AGP binding) can be used to build models for various tissue bindings. When the volumes of these tissues/organs are known, the drug distribution and the available free concentration of the compound administered by a given dose can be calculated. Many scientists' mutual efforts will potentially reveal the real importance of the biomimetic HPLC binding properties of compounds to build more accurate physiologically based models that will dramatically reduce the number of *in vivo* animal experiments during drug discovery and will able more accurately model the actual human distribution of the drugs.

QUESTIONS FOR REVIEW

Q10.1 How can volume of distribution be related to the tissue/plasma distribution coefficients and various tissue partition coefficients?

Q10.2 What are the principles of the *in vitro* estimation of volume of distribution?

Q10.3 What are the major physicochemical properties of compounds that govern their volume of distribution?

Q10.4 Which *in silico* descriptors can be used to estimate the volume of distribution?

Q10.5 Why are volumes of distribution of animals not very good models for volume of distribution of humans?

Q10.6 Describe the principle of a three-phase equilibrium process, for example, PPB, tissue binding, and tissue/plasma partition.

Q10.7 What are the mathematical relationships between unbound fraction, percentage bound, and logarithmic partition coefficients?

Q10.8 How do the HSA and IAM binding correlate for a diverse set of drug molecules?

Q10.9 What methodologies can be used for the PPB determination? What are the advantages and disadvantages of these methodologies?

Q10.10 What is the importance of PPB in lead optimization?

Q10.11 Why is AGP binding important?

Q10.12 How can we model and estimate blood/brain distribution of compounds?

Q10.13 What are the characteristics of unbound fraction in the brain, lung, mucus, and overall tissues?

Q10.14 What are the principles of PBPK modeling and how can the biomimetic HPLC measurements enhance them?

REFERENCES

1. Rowland, M., Benet, L. Z., Graham, G. G. (1973) Clearance concepts in pharmacokinetics. *Journal of Pharmacokinetics and Biopharmaceutics*, *1*, 123–135.
2. Benet, L. Z., Galeazzi, R. L. (1979) Noncompartmental determination of the volume of distribution steady state. *Journal of Pharmaceutical Sciences*, *68*, 1071–1074.
3. Poulin, P., Theil, F. P. (2002) Prediction of pharmacokinetics prior *in vivo* studies. 1. Mechanism-based prediction of volume of distribution. *Journal of Pharmaceutical Sciences*, *91*, 129–156.
4. Rodgers, T., Leahy, D., Rowland, M. (2005) Physiological based pharmacokinetic modelling 1: Predicting the tissue distribution of moderate-to-strong bases. *Journal of Pharmaceutical Sciences*, *94*, 1259–1276.
5. Rodgers, T., Rowland, M. (2007) Mechanistic approaches to volume of distribution predictions: Understanding the processes. *Pharmaceutical Research*, *24*, 918–933.
6. Jansson, R., Bredberg, U., Ashton, M. (2008) Prediction of drug distribution to plasma concentration ratios using a measured volume of distribution in combination with lipophilicity. *Journal of Pharmaceutical Sciences*, *97*, 2324–2339.

7. Kirkovsky, L., Zutshi, A. (2011) Pharmacokinetics for medicinal chemists. In: ADMET for Medicinal Chemists – A Practical Guide. Eds. Tsaioun, K., Kates, S. A., John Wiley & Sons, Inc., Hoboken, NJ, pp. 201–277.

8. Davies, B., Morris, T. (1993) Physiological parameters in laboratory animals and humans. *Pharmaceutical Research, 10(7)*, 1093–1095.

9. Boxenbaum, H. (1983) Interspecies pharmacokinetic scaling and Dedrick plots. *American Journal of Physiology, 245*, 768–775.

10. Hutchaleeha, A., Chow, H. H., Mayershon, M. (1997) Comparative pharmacokinetics and interspecies scaling of amphotericin B in several mammalian species. *Journal of Pharmacy and Pharmacology, 49*, 178–183.

11. Mahmood, I., Balian, J. D. (1996) Interspecies scaling: Predicting pharmacokinetic parameters of antiepileptic drugs in human from animals with special emphasis on clearance. *Journal of Pharmaceutical Sciences, 85*, 411–414.

12. Ritschel, W. A., Vachharajani, N. N., Johnson, R. D., Hussain, A. A. (1992) The allometric approach for interspecies scaling of pharmacokinetics parameters. *Comparative Biochemistry and Physiology, C: Toxicology and Pharmacology, 103*, 249–253.

13. Boxenbaum, H. (1984) Interspecies pharmacokinetic scaling and the evolutionary-comparative paradigm. *Drug Metabolism Reviews, 15*, 1071–1121.

14. Valkó, K. (2004) Application of high-performance liquid chromatography based measurements of lipophilicity to model biological distribution. *Journal of Chromatography, 1037*, 299–310.

15. Dal Pozzo, A., Donzelli, G., Rodriquez, L., Tajana, A. (1989) "*In vitro*" model for the evaluation of drug distribution and plasma protein-binding relationships. *International Journal of Pharmaceutics, 50*, 97–101.

16. Schumann, G., Fichtk, B., Kurz, H. (1987) Prediction of drug distribution *in vivo* on the basis of *in vitro* binding data. *Biopharmaceutics & Drug Distribution, 8*, 73–86.

17. Lombardo, F., Obaxt, R. S., Shalaeva, M. Y., Gao, F. (2002) Prediction of volume of distribution values in humans for neutral and basic drugs using physicochemical measurements and plasma protein binding data. *Journal of Medicinal Chemistry, 45*, 2867–2876.

18. Øie, S., Tozer, T. N. (1979) Effect of altered plasma protein binding on apparent volume of distribution. *Journal of Pharmaceutical Sciences, 68*, 1203–1205.

19. Hollósy, F., Valkó, K., Hersey, A., Nunhuck, S., Kéri, G., Bevan, C. (2006) Estimation of volume of distribution in humans from high throughput HPLC-based measurements of human serum albumin binding and immobilized artificial membrane partitioning. *Journal of Medicinal Chemistry, 49*, 6958–6971.

20. Clausen, J., Bickel, M. H. (1993) Prediction of drug distribution in distribution dialysis and *in vivo* from binding to tissues and blood. *Journal of Pharmaceutical Sciences, 82*, 345–349.

21. Björkman, S. (2002) Prediction of the volume of distribution of a drug: Which tissue-plasma partition coefficients are needed? *Journal of Pharmacy and Pharmacology, 54*, 1237–1245.

22. Trainor, G. L. (2007) The importance of plasma protein binding in drug discovery. *Expert Opinion in Drug Discovery, 2(1)*, 51–64.

23. Benet, L. Z., Hoener, B. (2002) Changes in plasma protein binding have little clinical relevance. *Clinical Pharmacology and Therapeutics, 71(3)*, 115–121.
24. Smith, D. A., Li, D., Kerns, E. H. (2010) The effect of plasma protein binding on *in vivo* efficacy: Misconceptions in drug discovery. *Nature Reviews in Drug Discovery, 9*, 929–929.
25. Kratochwil, N. A., Huber, W., Müller, F., Kansy, M., Gerber, P. R. (2002) Predicting plasma protein binding of drugs: A new approach. *Biochemical Pharmacology, 64*, 1355–1374.
26. Hervè, F., Urien, S., Albengres, E., Duché, J-C., Tillement, J.-P. (1994) Drug binding in plasma. A summary of recent trends in the study of drug and hormone binding. *Clinical Pharmacokinetics, 26*, 44–58.
27. Pacific, G. M., Viani, A. (1992) Methods of determining plasma and tissue binding of drugs. *Clinical Pharmacokinetics, 23*, 449–468.
28. Sebille, B. (1990) Methods of drug protein binding determinations. *Fundamental Clinical Pharmacology, 4(suppl 2)*, 151–161.
29. Banker, M. J., Clark, T. H., Williams, J. A. (2003) Development and validation of a 96-well equilibrium dialysis apparatus for measuring plasma protein binding. *Journal of Pharmaceutical Sciences, 92*, 967–974.
30. Waters, N. J., Jones, R., Williams, G., Sohal, B. (2008) Validation of a rapid equilibrium dialysis approach for the measurement of plasma protein binding. *Journal of Pharmaceutical Sciences, 97*, 4586–4595.
31. Valkó, K., Nunhuck, S., Bevan, C., Abraham, M. H., Reynolds, D. P. (2003) Fast gradient HPLC method to determine compounds binding to human serum albumin. Relationships with octanol/water and immobilized artificial membrane lipophilicity. *Journal of Pharmaceutical Sciences, 92*, 2236–2248.
32. Valkó, K. (2010) Application of bio-mimetic HPLC for protein binding measurements in drug discovery. *Physchem Forum, UK*, Sandwich. (http://physchem.org.uk/symp08/symp08.htm. Accessed 2013 Aug 13.)
33. Israili, Z. H., Dayton, P. G. (2001) Human alpha-1-glycoprotein and its interactions with drugs. *Drug Metabolism Reviews, 33(2)*, 161–235.
34. Fournier, T., Medjoubi-N, N., Porquet, D. (2000) Alpha-1-acid glycoprotein. (Review). *Biochimica at Biophysica Acta, 1482*, 157–171.
35. Johnson, J. A., Livingstone, T. N. (1997) Differences between blacks and whites in plasma protein binding of drugs. *European Journal of Clinical Pharmacology, 51*, 485–588.
36. Kremer, J. M. H., Wilting, A., Janssen, L. H. M. (1988) Drug binding to human alpha-1-acid glycoprotein in health and disease. *Pharmacological Reviews, 40*, 1–47.
37. Hervè, F., Duchè, J.-C., d'Athis, P., Marchè, C., Barrè, J., Tillement, J.-P. (1996) Binding of disopyramide, methadone, dipyridamole, chlorpromazine, lignocaine and progesterone to the two main genetic variants of human α_1-acid glycoprotein: evidence for drug-binding differences between the variants and for the presence of two separate drug-binding sites on α_1-acid glycoprotein. *Pharmacogenetics, 6*, 403–415.

38. Holladay, J. W., Dewey, M. J., Yoo, S. D. (1996) Steady-state kinetics of imipramine in transgenic mice with elevated serum AAG levels. *Pharmaceutical Research, 13(9)*, 1313–1316.
39. Levy, R. H., Boddy, A. V. (1991) Stereoselectivity in pharmacokinetics: A general theory. (Review) *Pharmaceutical Research, 8(5)*, 551–555.
40. Abbott, N. J. (2004) Prediction of blood-brain barrier permeation in drug discovery from *in vivo*, *in vitro* and *in silico* models. *Drug Discovery Today Technology, 1(4)*, 407–416.
41. Abbott, N. J., Rönnbäck, L., Hansson, E. (2006) Astrocyte–endothelial interactions at the blood–brain barrier. *Nature Reviews Neuroscience, 7(1)*, 41–53.
42. Abbott, N. J., Dolman, D. E. M., Patabendige, A. K. (2008) Assays to predict drug permeation across the blood-brain barrier, and distribution to brain. *Current Drug Metabolism, 9*, 901–910.
43. Gumbleton, M., Audus, K. (2001) Progress and limitations in the use of *in vitro* cell cultures to serve as a permeability screen for the blood-brain barrier. *Journal of Pharmaceutical Sciences, 90(11)*, 1681–1698.
44. Di, L., Kerns, E. H., Fan, K., McConnell, O. J., Carte, G. T. (2003) High throughput artificial membrane permeability assay for blood-brain barrier. *European Journal of Medicinal Chemistry, 38(3)*, 223–232.
45. Dash, A. K., Elmquist, W. F. (2003) Separation methods that are capable of revealing blood-brain barrier permeability. (Review). *Journal of Chromatography B, 797*, 241–254.
46. Hammarlund-Udenaes, M. (2000) The use of microdialysis in CNS drug delivery studies. Pharmacokinetic perspectives and results with analgesics and antiepileptics. *Advanced Drug Delivery Reviews, 45*, 283–294.
47. Naik, P., Cucullo, L. (2012) *In vitro* blood-brain barrier models: Current and perspective technologies. *Journal of Pharmaceutical Sciences, 101*, 1337–1354.
48. Rowley, M., Kulagowski, J. J., Watt, A. P., Rathbone, D., Stevenson, G. I., Carling, R. W., Baker, R., Marshall, G. R., Kemp, J. A., Foster, A. C., Grimwood, S., Hargreaves, R., Hurley, C., Daywell, K. L., Tricklebank, M. D., Leeson, P. D. (1997) Effect of plasma protein binding on *in vivo* activity and brain penetration of glycine/NMDA receptor antagonist. *Journal of Medicinal Chemistry, 40*, 4053–4068.
49. Maurer, T. S., DeBartolo, D. B., Tess, D. A., Scott, D. O. (2005) Relationship between exposure and nonspecific binding of thirty-three central nervous system drugs in mice. *Drug Metabolism and Disposition, 33*, 175–181.
50. Kalvass, J. Cory, Maurer, T. S., Pollack, G. M. (2007) Use of plasma and brain unbound fractions to assess the extent of brain distribution of 34 drugs: Comparison of unbound concentration ratios to *in vivo* P-glycoprotein efflux ratios. *Drug Metabolism and Disposition, 35(4)*, 660–666.
51. Watson, J., Wright, S., Lucas, A., Clarke, K. L., Viggers, J., Cheetham, S., Jeffrey, P., Porter, R., Read, K. D. (2009) Receptor occupancy and brain free fraction. *Drug Metabolism and Disposition, 37(4)*, 753–760.
52. Gratton, J. A., Abraham, M. H., Bradbury, M. W., Chadha, H. S. (1997) Molecular factors influencing drug transfer across the blood-brain barrier. *Journal of Pharmacy and Pharmacology, 49*, 1211–1216.

53. Testa, B., Crivori, P., Reist, M., Carrupt, P-A. (2000) The influence of lipophilicity on the pharmacokinetic behaviour of drugs: Concepts and examples. *Perspectives in Drug Discovery and Design*, *19*, 179–211.
54. Valkó, K., Nunhuck, S., Hill, A.P. (2011) Estimating unbound volume of distribution and tissue binding by *in vitro* HPLC based human serum albumin and immobilized artificial membrane-binding measurements. *Journal of Pharmaceutical Sciences*, *100*, 849–862.
55. Valkó, K. (2011) Applications of bio-mimetic HPLC methods for the estimation of *in vivo* distribution of early drug discovery compounds. *2nd World Symposium of Physicochemical Methods in Drug Discovery and Development, Zadar.*
56. Manallack, D. T. (2007) The pKa distribution of drugs: Application to drug discovery. *Perspectives in Medicinal Chemistry*, *1*, 25–38.

11

MODELS WITH MEASURED PHYSICOCHEMICAL AND BIOMIMETIC CHROMATOGRAPHIC DESCRIPTORS—DRUG EFFICIENCY

DRUG EFFICIENCY

Early drug discovery efforts usually concentrate on finding the most potent compounds. However, as a result of increasing potency, the compounds tend to increase in size and lipophilicity [1]. Large and lipophilic molecules usually have poor solubility and permeability, compromising good oral absorption [2]. The *in vivo* distribution profile and pharmacodynamic aspects of discovery compounds are usually taken into account at a later stage of the lead optimization process. It is very difficult, however, to change the structure of potent but large and lipophilic compounds to improve the pharmacokinetic and pharmacodynamic profile, while retaining potency. Drugs in different therapeutic areas usually require different *in vivo* distribution profile depending on the location of the target. On the basis of the free drug hypothesis, only the freely available drug molecules at the site of action are able to interact with the target and exert the required pharmacological effect. Large and lipophilic molecules tend to bind nonspecifically to proteins and lipids in the body, which reduces the overall free concentration. The theoretical basis of the "free drug principle" has been described by Trainor [3], who discusses how the free concentration of the drug *in vivo* influences the efficacy and receptor occupancy. The subsequent optimization of potency and absorption/distribution properties of compounds is time consuming and inefficient. Several new ideas have been suggested and reviewed for the optimization of potency and physicochemical

Physicochemical and Biomimetic Properties in Drug Discovery: Chromatographic Techniques for Lead Optimization, First Edition. Klara Valko.

properties together [4,5]. One of the many approaches is the use of the ligand efficiency (LE) parameter when evaluating the potency. The LE concept was introduced by Hopkins et al. [6]. Rees et al. [7] describes various ways to assess the potency relative to size, the number of nonhydrogen atoms, or the overall polar surface area of the compounds. The most ligand-efficient potent compounds bind to the receptor mostly via stronger polar and steric-specific interactions and not by hydrophobic forces.

The original LE introduced by Hopkins et al. [6] expresses the binding energy relative to the number of heavy atoms, and it can be described by Equation 11.1.

$$\mathrm{LE} = \Delta g = \frac{\Delta G}{N} \tag{11.1}$$

where $\Delta G = -\mathrm{RT} \ln K_\mathrm{d}$ and N is the number of nonhydrogen atoms. The ΔG parameter can be obtained from the measured drug–receptor dissociation constant (K_d). Abad-Zapatero and Metz [4] have suggested simplified LE indices such as percentage efficiency index (PEI) and binding efficiency index (BEI) using percentage inhibition or potency expressed as pIC_{50} values relative to the molecular weight. When we want to reduce lipophilic binding forces, it is important to know the polar surface area of the compound and relate it to the potency. This resulted in the introduction of the surface-binding efficiency index (SEI) that expresses the pIC_{50} relative to the calculated polar surface area [4].

Leeson and Sprinthorpe [8] suggested the ligand lipophilicity efficiency (LLE) to take into account during lead optimization. They defined LLE as the difference between the pIC_{50} and the calculated octanol/water partition coefficient (LLE).

Reynolds et al. [9] have explained that efficient binding is more likely with smaller compounds for two reasons. There is an inevitable reduction in the quality of fit between ligand and receptor when the ligand is big. Larger ligands tend to be more complex; this reduces the accessibility of several atoms in the molecule by the binding site. This results in the reduction of binding ligand surface per atom.

More recently, Mortenson and Murray [10] at Astex Pharmaceuticals proposed a modified LE parameter that includes the heavy atom counts and the lipophilicity (LLE_AT) based on a modified free energy of binding (ΔG^*), which is the difference between the total free energy of binding and the free energy of binding because of lipophilicity. The difference is approximately $\ln(10) \times \mathrm{RT}(\log P - \mathrm{pIC}_{50})$. They have introduced a constant as well and came up with a formula (Equation 11.2) that brings the LLE_AT efficiency parameter to the same scale as the original LE.

$$\mathrm{LLE}_\mathrm{AT} = 0.111 + 1.36 \times \frac{\mathrm{LLE}}{\text{Heavy atom count}} \tag{11.2}$$

This LE parameter has been designed to evaluate the fragment compound's potential to achieve a drug-like lead molecule after optimization. The drug-like end point was estimated using the Lipinksi's rules [2], assuming that a molecule that contains 36 heavy atoms is equivalent of the molecular weight of 500 Da with a 10 nM potency. This profile leads to the commonly used 0.3 kcal/mol per heavy atom LE

Table 11.1 Names, calculations, and reference values of various ligand efficiency indices[a]

Index	Calculation	Reference Value
LE	$\frac{\Delta G}{\text{Heavy atom count}}$	0.3
PEI	$\frac{\%\text{Inhibition}}{\text{MW}}$	1.5
BEI	$\frac{\text{p}K_\text{i},\ \text{p}K_\text{d},\ \text{or}\ \text{pIC}_{50}}{\text{MW}}$	27
SEI	$\frac{\text{p}K_\text{i},\ \text{p}K_\text{d},\ \text{or}\ \text{pIC}_{50}}{\text{PSA}}$	18
LLE	$\text{pIC}_{50} - c \log P$	3-4
LLEAT	$0.111 + 1.36 \times \frac{\text{LLE}}{\text{Heavy atom count}}$	0.3

[a]Taken from References [4,8,10].

value. Table 11.1 shows the most frequently used LE indices and the reference values as minimum required for a potentially developable lead molecule.

Mortenson and Murray [11] suggested using a plot of potency in the function of lipophilicity ($c \log P$). Various lines can be drawn representing the same LLE values combining the potency and $c \log P$. They imply that the same LLE values for compounds with lower potency may have been previously discarded, although their LLE values were the same.

The application of these efficiency indices has already improved fragment selection for lead generation and helps reverse the trend of synthesizing potent but large and lipophilic molecules. This phenomenon has been named "molecular obesity" by Hann [12]. The advantage that these efficiency indices offer is the combination of two or three variables (potency, size, and lipophilicity) into one entity that is much easier to optimize. The sequential optimization of potency and the physicochemical parameters require iterative thinking by adding and removing substituents on the molecules, which is very inefficient. The LE parameters also provide a quantitative way of comparing molecules by giving weighing for size and lipophilicity as well as potency. In this way, we can define more accurately what potency is required for a molecule with a given size and lipophilicity on the basis of binding energies. This approach can be used very early in the drug design process back to the *in silico* docking and scoring [13]. The application of the LE parameters early in the hit to lead optimization ensures good chemical starting points, which are a major determinant of the quality of the final drug [14]. However, beside the optimization of physicochemical properties and the "drug-likeness" features of lead molecules, we need to optimize the *in vivo* absorption, distribution, and metabolism. Avoiding interactions with potential targets that may lead to side effects in the clinic is also paramount. Braggio et al. [15] have introduced the "drug efficiency" concept that takes into account several *in vivo* characteristics of the drug discovery molecules. The drug efficiency ($DRUG_{eff}$) is defined as the percentage of the free biophase concentration relative to the dose that is available at the site of action *in vivo* (Equation 11.3).

$$DRUG_{eff}(\%) = \frac{\text{Free biophase concentration}}{\text{Dose}} \times 100 \tag{11.3}$$

This simple parameter takes into account all the factors that reduce the free biophase concentration *in vivo*. Thus, it includes absorption after oral administration, the clearance, and the distribution of the compounds into tissues, plasma protein binding, tissue binding, permeability barrier, and active transport (efflux) processes as well. During the potency measurements, we can assume that all of the available compounds interact with the target, so the efficiency is 100%. The absorption and clearance can be taken into account as the bioavailability ($F\%$), which can be measured by the total plasma concentration after oral administration. When the target is in the plasma, the free biophase concentration is equivalent to the free plasma concentration that can be measured *in vitro* and calculated by Equation 11.4.

$$\text{Free biophase concentration} = [\text{blood}] \times f_{u,\text{blood}} \tag{11.4}$$

In this case, the unbound fraction in blood ($f_{u,\text{blood}}$) can be obtained from the measurements of plasma protein binding. As the unbound volume of distribution (Vd_u) is defined as the dose divided by the free plasma concentration, its reciprocal value can be considered as the maximum drug efficiency of the compound that can be achieved after intravenous administration or assuming 100% oral absorption (see Equations 11.5 and 11.6).

$$Vd_u = \frac{\text{Dose}}{[\text{blood}] \times f_{u,\text{blood}}} \tag{11.5}$$

$$DRUG_{eff\ max} = \frac{1}{Vd_u} \times 100 \tag{11.6}$$

The unbound volume of distribution already takes into account not only the plasma protein binding of compounds but also the tissue binding. When no permeability barrier or active transport exists between the plasma and the target tissue compartments, we can suppose that the free concentration in plasma is equal to the free concentration in tissues. In these cases, the model that estimates unbound volume of distribution also provides a good estimate for $DRUG_{eff\ max}$. Stepensky [16] has also suggested using the unbound volume of distribution of the drug for pharmacokinetic calculations, as it appropriately describes the drug distribution patterns concerning its free pharmacologically active concentration. Valkó et al. [17] have published a model for the estimation of unbound volume of distribution using biomimetic HPLC binding data. The model has been set up by using the clinical volume of distribution data and literature unbound fraction in plasma values for

70 known drug molecules and used the data of another 50 drug molecules as test set. The literature volume of distribution data of the investigated known drugs have been shown in Table 10.7, the plasma protein binding data have been compiled in Table 10.10, and the measured human serum albumin binding the immobilized artificial membrane-binding data have been shown in Table 10.6. The obtained model equation for the estimation of the unbound volume of distribution is described by Equation 11.7.

$$\log \mathrm{Vd_u} = 0.23(\pm 0.03) \log K(\mathrm{HSA}) + 0.43 \times (\pm 0.03) \log K(\mathrm{IAM}) - 0.72$$

$$n = 70 \quad r^2 = 0.84 \quad s = 0.32 \quad F = 179 \qquad (11.7)$$

The log K(HSA) and log K(IAM) parameters are obtained from calibrated gradient retention times of the compounds on chemically bonded human serum albumin and immobilized artificial membrane stationary phases as described in Chapter 5. Both retention data have been converted to the log P scale, so the regression coefficients of the model are comparable. As is expected, the phospholipid binding of the compounds is a more dominant parameter in reducing the steady-state free plasma concentration. To obtain high free plasma concentration from the possible lowest dose, the unbound volume of distribution should be low. That means it is advantageous when the drug molecule has weak binding to phospholipids and albumin type of proteins. Albumin can be found not only in the plasma but also in tissues; therefore, the regression coefficient of the albumin binding parameter is relatively large. A plot of measured and estimated values of the logarithmic unbound volume of distribution of 130 known drugs is shown in Figure 11.1. The model has been used to predict the unbound volume of distribution for further 60 compounds (test set) which is marked with asterisk in the figure.

Although the maximum drug efficiency can be estimated from the volume of distribution, the actual drug efficiency in the target tissue and organ can be expressed from the unbound fraction of compounds and the total concentration of the drug in various tissues. Equations 11.8 and 11.9 show the formula for calculating the concentration in the brain and lung.

$$\text{Free biophase concentration (brain)} = [\text{brain}] \times f_{\mathrm{u,brain}} \qquad (11.8)$$

$$\text{Free biophase concentration (lung)} = [\text{lung}] \times f_{\mathrm{u,lung}} \qquad (11.9)$$

To be able to obtain the *in vivo* total brain ([brain]) or lung ([lung]) concentrations, it is important to know the brain/blood and the lung/blood distribution coefficients. The tissue partition coefficients ($k_{\mathrm{tissue/blood}}$) can be calculated from the unbound fraction of the drug in plasma or blood and the unbound fraction of the drug in various tissues using the formula described by Equation 11.10 in general.

$$k_{\mathrm{tissue/blood}} = \frac{c_{\mathrm{tissue(total)}}}{c_{\mathrm{blood(total)}}} = \frac{f_{\mathrm{u,blood}}}{f_{\mathrm{u,tissue}}} \qquad (11.10)$$

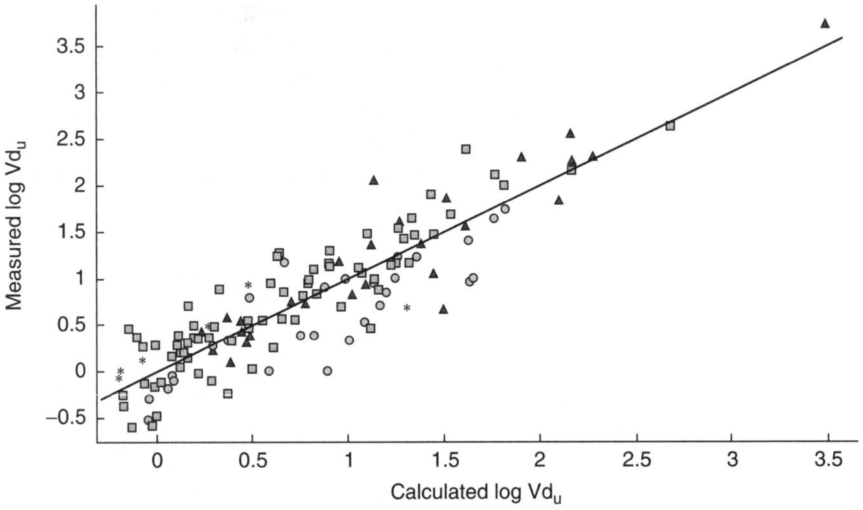

Figure 11.1 The measured and estimated unbound volumes of distribution of 130 known drug molecules using Equation 11.7 from Reference [17] with permission. The circles represent the compounds in the training set, whereas the asterisks represent the compounds in the test set. The colors are based on the acid/base character of the compounds obtained from the determination of the chromatographic hydrophobicity index at three different pH values (circle, acid; square, neutral; triangle, base).

It means that, in principle, the biophase concentration can be estimated from the distribution of the dose in various tissues and in plasma and the particular tissue binding of the drug where the target is.

Valkó et al. [18] have collated the maximum drug efficiency ($DRUG_{eff\ max}$) and the drug efficiency ($DRUG_{eff}$) of more than 100 known drug molecules from various databases [19]. The compiled data are shown in Table 11.2.

As expected, the *in vivo* $DRUG_{eff}$ is usually smaller than the $DRUG_{eff\ max}$. The *in vivo* maximum drug efficiency could be estimated well from the model that has been described for the unbound volume of distribution taking its reciprocal value. The plot of the *in vivo* $DRUG_{eff}$ data and the estimated values from the biomimetic log K(HSA) and log K(IAM) binding data is shown in Figure 11.2.

The agreement between the *in vivo* and *in vitro* $DRUG_{eff\ max}$ data is acceptable if we take into account the fact that the *in vitro* data are based only on two HPLC retention time measurements on biomimetic stationary phases, HSA, and IAM. The *in vitro* $DRUG_{eff\ max}$ data have been derived from the statistically validated unbound volume of distribution model. The model shows that the albumin binding and the phospholipid binding are the dominant factors in reducing the free concentration of the compounds *in vivo*.

The drug efficiency concept combines the overall *in vivo* absorption and distribution properties of the compounds by estimating the *in vivo* available free concentration of the drug at the site of action. The combination of *in vitro* potency

Table 11.2 The generic name, CAS number, dose, pXC_{50}, and *in vivo* $DRUG_{eff}$ and $DRUG_{eff\ max}$ data of known drugs[a]

Generic Name	CAS Number	Target Class	Dose, mg	pXC_{50}	*in vivo* $DRUG_{eff}$	*in vivo* $DRUG_{eff\ max}$
Abacavir	136470-78-5	Enzyme	600	7.15	24.9	62.5
Acebutolol HCl	34381-68-5	7TM	400	6.76	9.6	37
Acrivastine	87848-99-5	7TM	8	8.46	49.2	76.9
Albuterol sulfate	51022-70-9	7TM	6	7.4	11.8	47.4
Alfuzosin	81403-80-7	7TM	7.5	7.64	1.0	4
Alosetron HCl	122852-69-1	Ion channel	1	7.3	9.5	15.7
Amitriptyline HCl	549-18-8	7TM	50	6.15	0.2	0.6
Amlodipine besylate	111470-99-6	Ion channel	5	8.72	0.9	0.3
Amprenavir	161814-49-9	Enzyme	2400	7.92	2.7	8.3
Anastrozole	120511-73-1	Enzyme	1	7.82	64.4	81.7
Aprepitant	170729-80-3	7TM	40	9.92	4.5	5
Aripiprazole	129722-12-9	7TM	10	9.7	1.1	0.2
Bendroflumethiazide	73-48-3	Undefined	2.5	3.44	3.6	4.1
Bezafibrate	41859-67-0	Nuclear receptor	600	4.52	12.5	20.6
Bicalutamide	90357-06-5	Nuclear receptor	50	6.1	4.1	10
Bupropion HCl	31677-93-7	Undefined	300	6.26	2.0	2
Buspirone HCl	33386-08-2	7TM	5	7.62	0.1	0.9
Caffeine	58-08-2	Undefined	50	4.74	51.2	101.6
Candesartan cilexetil	145040-37-5	7TM	8	10.52	0.7	7.69
Carbamazepine	298-46-4	Ion channel	800	4.6	5.6	17.1
Celecoxib	169590-42-5	Enzyme	200	8.52	0.8	0.4
Cetirizine HCl	83881-52-1	7TM	5	8.52	16.7	17.5
Cilostazol	73963-72-1	Enzyme	100	6.7	1.1	0.7

(continued)

Table 11.2 (*Continued*)

Generic Name	CAS Number	Target Class	Dose, mg	pXC_{50}	*in vivo* $DRUG_{eff}$	*in vivo* $DRUG_{eff\ max}$
Citalopram hydrobromide	59729-32-7	Undefined	20	8.34	1.8	1.4
Clomipramine HCl	17321-77-6	7TM	25	7.86	2.6	0.2
Clonidine HCl	4205-91-8	7TM	0.4	8.46	33.6	28.6
Dapsone	80-08-0	7TM	100	4.82	11.2	12
Desipramine HCl	58-28-6	Undefined	100	6.77	0.3	0.3
Desloratadine	100643-71-8	7TM	5	9.4	0.7	0.1
Diflunisal	22494-42-4	Enzyme	1000	4.24	12.0	7.1
Diphenhydramine HCl	147-24-0	7TM	75	8.12	1.2	3.3
Dofetilide	115256-11-6	Ion channel	1	7.5	8.8	8.3
Domperidone	57808-66-9	7TM	30	9.82	1.0	1.2
Donepezil HCl	120011-70-3	Enzyme	5	8.17	1.7	0.3
Doxazosin mesylate	77883-43-3	7TM	1	10	0.7	0.8
Duloxetine HCl	136434-34-9	7TM	40	7.55	0.2	0.2
Ebastine	90729-43-4	7TM	20	7.65	0.0	0.5
Efavirenz	154598-52-4	Enzyme	200	8.92	0.5	0.3
Eplerenone	107724-20-9	Nuclear receptor	25	7.1	70.0	53.8
Escitalopram oxalate	219861-08-2	Undefined	10	6.53	2.9	3.7
Fenofibrate	49562-28-9	Nuclear receptor	43	4.74	2.0	2.3
Fexofenadine HCl	138452-21-8	7TM	60	7.27	12.5	5.5
Finasteride	98319-26-7	Undefined	1	7.28	5.7	9.1
Fluoxetine HCl	59333-67-4	Undefined	20	8.59	3.1	0.2
Flurbiprofen	5104-49-4	Enzyme	50	6.2	10.6	6.7
Fluvoxamine maleate	61718-82-9	Undefined	50	8.3	0.1	0.9
Furosemide	54-31-9	Ion channel	40	5	1.9	8.3

Glimepiride	93479-97-1	Ion channel	1	5.38	8.4	5.3
Glipizide	29094-61-9	Ion channel	2.5	5.52	6.5	6.3
Granisetron HCl	107007-99-8	Ion channel	1	9.89	8.9	9.5
Guanabenz acetate	23256-50-0	7TM	8	8.24	0.9	7.8
Guanfacine HCl	29110-48-3	7TM	1	7.6	2.6	4.5
Haloperidol	52-86-8	7TM	1	10	0.3	0.5
Hydrochlorothiazide	58-93-5	Undefined	25	4.7	25.2	20
Imipramine HCl	113-52-0	Undefined	75	6.55	0.8	0.5
Indapamide	26807-65-8	Enzyme	1.25	4.24	76.4	63.6
Indometacin	53-86-1	Enzyme	25	6.48	3.4	10.4
Irbesartan	138402-11-6	7TM	150	9.3	18.2	10
Isradipine	75695-93-1	Ion channel	5	6.6	0.2	2
Itraconazole	84625-61-6	Enzyme	200	4.19	0.1	0.03
Ketoconazole	65277-42-1	Enzyme	200	4.65	0.6	0.4
Ketoprofen	22071-15-4	Enzyme	100	7.63	1.1	7.7
Lamotrigine	84057-84-1	Ion channel	25	4.04	39.4	37.5
Lansoprazole	103577-45-3	Undefined	15	7.05	1.2	3.6
Letrozole	112809-51-5	Undefined	2.5	7.89	37.0	21.1
Lorazepam	846-49-1	Ion channel	1	8.85	8.8	6.9
Lovastatin	75330-75-5	Enzyme	10	9.52	2.1	5.7
Maprotiline HCl	10347-81-6	7TM	25	4.39	0.5	0.3
Metoclopramide HCl	7232-21-5	7TM	10	8.38	16.8	17.1
Mexiletine HCl	04-01-70	Ion channel	100	4.29	4.0	3.2

(continued)

Table 11.2 (*Continued*)

Generic Name	CAS Number	Target Class	Dose, mg	pXC_{50}	*in vivo* $DRUG_{eff}$	*in vivo* $DRUG_{eff\ max}$
Mibefradil diHCl	116666-63-8	Ion channel	50	6.15	0.2	0.2
Mifepristone	84371-65-3	Nuclear receptor	600	8.02	0.5	1.4
Moricizine HCl	29560-58-5	Ion channel	600	6.21	1.5	1.3
Nadolol	42200-33-9	7TM	40	8.9	5.5	36.8
Naproxen	22204-53-1	Enzyme	500	5.22	7.7	6.3
Nevirapine	129618-40-2	Enzyme	200	8	5.6	30.8
Nicardipine HCl	54527-84-3	Ion channel	60	5.58	2.2	5
Nimesulide	51803-78-3	Enzyme	100	4.4	3.3	3.7
Nimodipine	66085-59-4	Ion channel	360	9.55	0.5	4.5
Nitrendipine	39562-70-4	Ion channel	10	7.64	78.8	0.5
Nortriptyline HCl	894-71-3	7TM	75	8.98	0.1	0.2
Olanzapine	132539-06-1	7TM	5	7.52	1.0	0.4
Olmesartan medoxomil	144689-63-4	7TM	20	8.11	6.3	16.5
Phenytoin	57-41-0	Ion channel	3	4.15	23.3	15.6
Pioglitazone HCl	112529-15-4	Nuclear receptor	15	4.46	2.5	2.2
Piroxicam	36322-90-4	Enzyme	20	6.7	19.3	7.1
Pravastatin sodium	81131-70-6	Enzyme	40	8.16	5.2	90
Prazosin HCl	19237-84-4	7TM	1.5	8.85	1.5	4.1
Probenecid	57-66-9	Ion channel	500	3.82	24.7	38.5
Propafenone HCl	34183-22-7	Ion channel	450	5.7	0.4	0.8
Propranolol HCl	318-98-9	7TM	80	9.38	0.3	2.3
Quinine sulfate	804-63-7	Undefined	1800	6.3	1.7	2.8
Repaglinide	135062-02-1	Ion channel	1	7.54	1.7	5
Riluzole	1744-22-5	7TM	50	4.37	0.6	0.7

Ritonavir	155213-67-5	Enzyme	100	5.66	0.9	0.7
Roxithromycin	80214-83-1	Enzyme	150	6.83	9.3	10
Saquinavir	127779-20-8	Enzyme	2000	8.3	0.0	0.8
Selegiline HCl	2079-54-1	Enzyme	10	8.1	0.1	10.3
Spironolactone	52-01-7	Undefined	100	4.67	0.7	1
Sumatriptan	103628-46-2	7TM	25	8.07	3.0	46.5
Tadalafil	171596-29-5	Enzyme	5	8.3	6.2	6.7
Tamsulosin HCl	106463-17-6	7TM	0.8	11	2.8	4.8
Telmisartan	144701-48-4	7TM	20	8.04	0.3	0.1
Terfenadine	50679-08-8	7TM	120	7.76	0.0	0.03
Tizanidine HCl	64461-82-1	7TM	2	7.51	6.3	29.2
Toremifene citrate	89778-27-8	Nuclear receptor	60	5.52	1.3	0.12
Trazodone HCl	25332-39-2	Undefined	150	6.25	5.3	9.6
Triamterene	396-01-0	Ion channel	150	5.3	1.2	22.1
Valdecoxib	181695-72-7	Enzyme	10	8.4	2.0	2.2
Venlafaxine HCl	99300-78-4	Undefined	37.5	7.74	4.9	15.9
Verapamil HCl	152-11-4	Undefined	320	6.9	0.9	1.6
Zidovudine	30516-87-1	Enzyme	600	8	19.8	36.5
Zileuton	111406-87-2	Enzyme	2400	6.52	1.0	5.8
Ziprasidone HCl	138982-67-9	7TM	20	8.77	0.2	1
Zolmitriptan	139264-17-8	7TM	2.5	8.1	4.2	41.7

[a]Compiled from Reference [18] with permission.

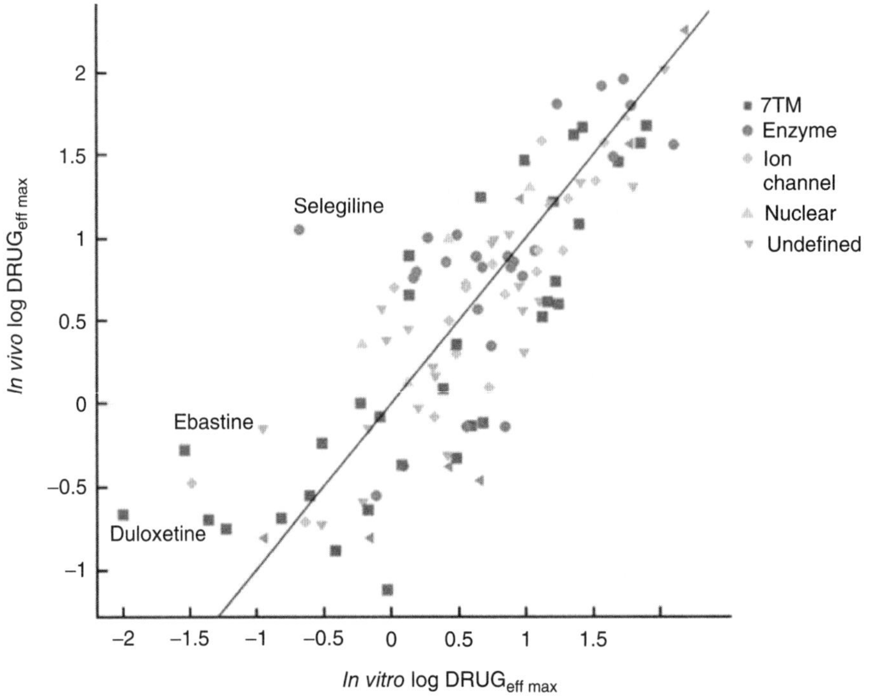

Figure 11.2 The plot of the *in vivo* $DRUG_{eff\ max}$ as a function of the *in vitro* $DRUG_{eff\ max}$ obtained from the reciprocal values of the unbound volume of distribution model described by Equation 11.7. The line is the line of unity. (Reproduced from Reference [18] with permission.)

and the *in vivo* drug efficiency has been described by the drug efficiency index (DEI) that enables chemists to simultaneously optimize potency and *in vivo* properties [20]. The analysis of the data of more than 350 known drug molecules by Montanari et al. [20] revealed that oral drugs very rarely exert nanomolar potency. They are close to the low micromolar range on average. The *in vitro* affinity alone did not show good correlation with the therapeutic dose, suggesting that absorption and distribution properties have a great impact on the therapeutic dose. The low dose is a result of the combination of good potency and ADME properties. The DEI combines these two properties and shows a mathematical relationship with the receptor occupancy (RO%) *in vivo* per unit of dose as shown by Equations 11.11–11.13.

$$\mathrm{DEI} = \log(\mathrm{DRUG_{eff}}) + \mathrm{p}K_\mathrm{d} \tag{11.11}$$

where DEI is the drug efficiency index and $\mathrm{p}K_\mathrm{d}$ is the negative logarithm of the *in vitro* drug–receptor dissociation constant.

$$\frac{\mathrm{RO\%}}{\mathrm{Dose}} = \frac{\mathrm{DRUG_{eff}}}{K_\mathrm{d}+[\mathrm{DRUG}]_\mathrm{Biophase}} \times 100 \tag{11.12}$$

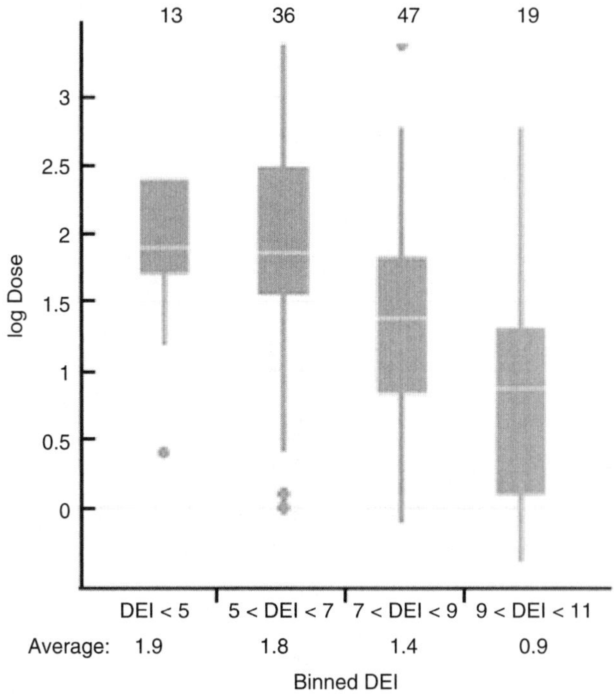

Figure 11.3 The box plot of the binned DEI values (bins: less than 5; between 5 and 7; between 7 and 9; and above 9) and the average dose of the compounds in the bin. Higher DEI values go with lower average dose. (From Reference [18] with permission.)

Combining Equations 11.11 and 11.12, Equation 11.13 is obtained, which describes the relationships between DEI, the proportion of the occupied and unoccupied receptors (RO%/(100 – RO%)), and dose.

$$\mathrm{DEI} = \log \mathrm{DRUG}_{\mathrm{eff}} + \mathrm{pXC}_{50} = \log \frac{\mathrm{RO\%}}{100 - \mathrm{RO\%}} + 2 - \log(\mathrm{Dose}) \qquad (11.13)$$

Table 11.3 shows the *in vivo* DEI and $\mathrm{DEI}_{\mathrm{max}}$ values for the drug molecules listed in Table 11.2.

Higher DEI values can be associated with lower therapeutic dose [18,20] as Equation 11.13 suggests. The therapeutic dose depends on potency, bioavailability, and the number of receptors that needs to be occupied in order to achieve the clinical outcome. Figure 11.3 shows the trend observed between the therapeutic dose and the DEI values of the known drugs listed in Table 11.2 and Table 11.3. Indeed, the trend shows that drugs with higher DEI values have lower dose in average. The DEI values can be estimated only from *in vitro* measurements: *in vitro* potency and two biomimetic HPLC binding parameters describing the compound's protein and phospholipid binding.

Table 11.3 The DEI values and the DEI values derived from the $DRUG_{eff\ max}$(DEI_{max}) for the known drugs listed in Table 11.2[a]

Generic Name	Target Class	Dose, mg	pXC_{50}	*in vivo* DEI	*in vivo* DEI_{max}
Abacavir	Enzyme	600	7.15	8.5	8.9
Acebutolol HCl	7TM	400	6.76	7.7	8.3
Acrivastine	7TM	8	8.46	10.2	10.3
Albuterol sulfate	7TM	6	7.4	8.5	9.1
Alfuzosin	7TM	7.5	7.64	7.6	8.2
Alosetron HCl	Ion channel	1	7.3	8.3	8.5
Amitriptyline HCl	7TM	50	6.15	5.5	5.9
Amlodipine besylate	Ion channel	5	8.72	8.7	8.2
Amprenavir	Enzyme	2400	7.92	8.4	8.8
Anastrozole	Enzyme	1	7.82	9.6	9.7
Aprepitant	7TM	40	9.92	10.6	10.6
Aripiprazole	7TM	10	9.7	9.8	9.0
Bendroflumethiazide	Undefined	2.5	3.44	4.0	4.1
Bezafibrate	Nuclear receptor	600	4.52	5.6	5.8
Bicalutamide	Nuclear receptor	50	6.1	6.7	7.1
Bupropion HCl	Undefined	300	6.26	6.6	6.6
Buspirone HCl	7TM	5	7.62	6.5	7.6
Caffeine	Undefined	50	4.74	6.4	6.7
Candesartan cilexetil	7TM	8	10.52	10.4	11.4
Carbamazepine	Ion channel	800	4.6	5.4	5.8
Celecoxib	Enzyme	200	8.52	8.4	8.1
Cetirizine HCl	7TM	5	8.52	9.7	9.8
Cilostazol	Enzyme	100	6.7	6.7	6.5
Citalopram hydrobromide	Undefined	20	8.34	8.6	8.5

Clomipramine HCl	7TM	25	7.86	8.3	7.2
Clonidine HCl	7TM	0.4	8.46	10.0	9.9
Dapsone	7TM	100	4.82	5.9	5.9
Desipramine HCl	Undefined	100	6.77	6.2	6.2
Desloratadine	7TM	5	9.4	9.3	8.4
Diflunisal	Enzyme	1000	4.24	5.3	5.1
Diphenhydramine HCl	7TM	75	8.12	8.2	8.6
Dofetilide	Ion channel	1	7.5	8.4	8.4
Domperidone	7TM	30	9.82	9.8	9.9
Donepezil HCl	Enzyme	5	8.17	8.4	7.6
Doxazosin mesylate	7TM	1	10	9.8	9.9
Duloxetine HCl	7TM	40	7.55	6.9	6.9
Ebastine	7TM	20	7.65	6.1	7.3
Efavirenz	Enzyme	200	8.92	8.6	8.4
Eplerenone	Nuclear receptor	25	7.1	8.9	8.8
Escitalopram oxalate	Undefined	10	6.53	7.0	7.1
Fenofibrate	Nuclear receptor	43	4.74	5.0	5.1
Fexofenadine HCl	7TM	60	7.27	8.4	8.0
Finasteride	Undefined	1	7.28	8.0	8.2
Fluoxetine HCl	Undefined	20	8.59	9.1	7.9
Flurbiprofen	Enzyme	50	6.2	7.2	7.0
Fluvoxamine maleate	Undefined	50	8.3	7.1	8.3
Furosemide	Ion channel	40	5	5.3	5.9
Glimepiride	Ion channel	1	5.38	6.3	6.1

(continued)

Table 11.3 (*Continued*)

Generic Name	Target Class	Dose, mg	pXC_{50}	*in vivo* DEI	*in vivo* DEI_{max}
Glipizide	Ion channel	2.5	5.52	6.3	6.3
Granisetron HCl	Ion channel	1	9.89	10.8	10.9
Guanabenz acetate	7TM	8	8.24	8.2	9.1
Guanfacine HCl	7TM	1	7.6	8.0	8.3
Haloperidol	7TM	1	10	9.4	9.7
Hydrochlorothiazide	Undefined	25	4.7	6.1	6.0
Imipramine HCl	Undefined	75	6.55	6.4	6.2
Indapamide	Enzyme	1.25	4.24	6.1	6.0
Indometacin	Enzyme	25	6.48	7.0	7.5
Irbesartan	7TM	150	9.3	10.6	10.3
Isradipine	Ion channel	5	6.6	5.9	6.9
Itraconazole	Enzyme	200	4.19	2.9	2.7
Ketoconazole	Enzyme	200	4.65	4.4	4.3
Ketoprofen	Enzyme	100	7.63	7.7	8.5
Lamotrigine	Ion channel	25	4.04	5.6	5.6
Lansoprazole	Undefined	15	7.05	7.1	7.6
Letrozole	Undefined	2.5	7.89	9.5	9.2
Lorazepam	Ion channel	1	8.85	9.8	9.7
Lovastatin	Enzyme	10	9.52	9.8	10.3
Maprotiline HCl	7TM	25	4.39	4.1	3.9
Metoclopramide HCl	7TM	10	8.38	9.6	9.6
Mexiletine HCl	Ion channel	100	4.29	4.9	4.8
Mibefradil diHCl	Ion channel	50	6.15	5.5	5.5
Mifepristone	Nuclear receptor	600	8.02	7.7	8.2
Moricizine HCl	Ion channel	600	6.21	6.4	6.3

Nadolol	7TM	40	8.9	9.6	10.5
Naproxen	Enzyme	500	5.22	6.1	6.0
Nevirapine	Enzyme	200	8	8.7	9.5
Nicardipine HCl	Ion channel	60	5.58	5.9	6.3
Nimesulide	Enzyme	100	4.4	4.9	5.0
Nimodipine	Ion channel	360	9.55	9.2	10.2
Nitrendipine	Ion channel	10	7.64	9.5	7.3
Nortriptyline HCl	7TM	75	8.98	8.1	8.3
Olanzapine	7TM	5	7.52	7.5	7.1
Olmesartan medoxomil	7TM	20	8.11	8.9	9.3
Phenytoin	Ion channel	3	4.15	5.5	5.3
Pioglitazone HCl	Nuclear receptor	15	4.46	4.9	4.8
Piroxicam	Enzyme	20	6.7	8.0	7.6
Pravastatin sodium	Enzyme	40	8.16	8.9	10.1
Prazosin HCl	7TM	1.5	8.85	9.0	9.5
Probenecid	Ion channel	500	3.82	5.2	5.4
Propafenone HCl	Ion channel	450	5.7	5.3	5.6
Propranolol HCl	7TM	80	9.38	8.9	9.7
Quinine sulfate	Undefined	1800	6.3	6.5	6.7
Repaglinide	Ion channel	1	7.54	7.8	8.2
Riluzole	7TM	50	4.37	4.2	4.2
Ritonavir	Enzyme	100	5.66	5.6	5.5
Roxithromycin	Enzyme	150	6.83	7.8	7.8
Saquinavir	Enzyme	2000	8.3	6.6	8.2

(continued)

Table 11.3 (*Continued*)

Generic Name	Target Class	Dose, mg	pXC_{50}	*in vivo* DEI	*in vivo* DEI_{max}
Selegiline HCl	Enzyme	10	8.1	7.2	9.1
Spironolactone	Undefined	100	4.67	4.5	4.7
Sumatriptan	7TM	25	8.07	8.5	9.7
Tadalafil	Enzyme	5	8.3	9.1	9.1
Tamsulosin HCl	7TM	0.8	11	11.4	11.7
Telmisartan	7TM	20	8.04	7.5	7.0
Terfenadine	7TM	120	7.76	5.8	6.2
Tizanidine HCl	7TM	2	7.51	8.3	9.0
Toremifene citrate	Nuclear receptor	60	5.52	5.6	4.6
Trazodone HCl	Undefined	150	6.25	7.0	7.2
Triamterene	Ion channel	150	5.3	5.4	6.6
Valdecoxib	Enzyme	10	8.4	8.7	8.7
Venlafaxine HCl	Undefined	37.5	7.74	8.4	8.9
Verapamil HCl	Undefined	320	6.9	6.9	7.1
Zidovudine	Enzyme	600	8	9.3	9.6
Zileuton	Enzyme	2400	6.52	6.5	7.3
Ziprasidone HCl	7TM	20	8.77	8.2	8.8
Zolmitriptan	7TM	2.5	8.1	8.7	9.7

Abbreviation: DEI, drug efficiency index.
[a]From Reference [18] with permission.

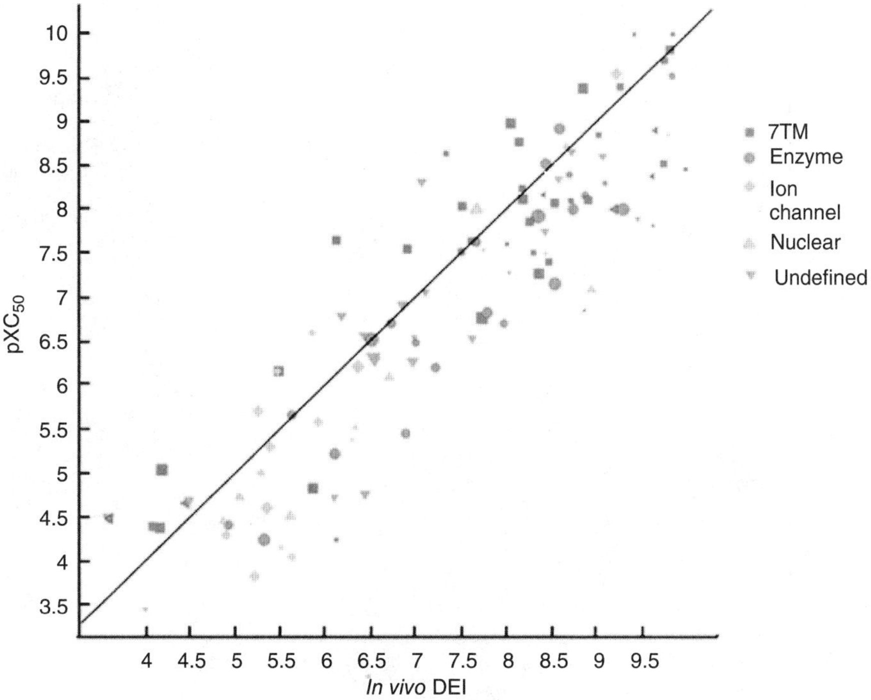

Figure 11.4 The plot of pXC_{50} and *in vivo* DEI values of 115 diverse set of marketed drugs. Markers are sized according to the logarithmic value of the dose; the shape and color is according to the target class. (Reproduced from Reference [18] with permission.)

As the DEI values include the potency and the overall ADME properties of the compounds, it can be used as a single parameter for prioritizing compounds. The low therapeutic dose reduces the risk of toxic side effects [21]. It has been observed that compounds with a therapeutic dose less than 10 mg are less likely to fail in clinical trials [21].

It is also interesting to see how the DEI values can be used for lead optimization. When we plot the *in vitro* potency (pXC_{50}) and the DEI values for the marketed drugs, we see that most of them have at least 1% drug efficiency. They are on the right side of the line of unity, which means pXC_{50} equals the DEI (shown in Figure 11.4). When DEI = pXC_{50}, then the log $DRUG_{eff\ max}$ is zero and the drug efficiency is 1%. The size of the marker is proportional to the therapeutic dose. Drugs acting on enzyme targets tend to have higher dose.

Discovery project compounds usually have very different patterns from known drugs in the relationship between potency and DEI as demonstrated in Figure 11.5. The circles are the marketed drugs, whereas the squares are the project compounds. It is typical that project compounds are within a narrow potency range. It can be seen in Figure 11.5 that the potency of the project compounds are between 7 and 8.5; however, the DEI values show a very wide range from 4.5 to 9. It is also noticeable

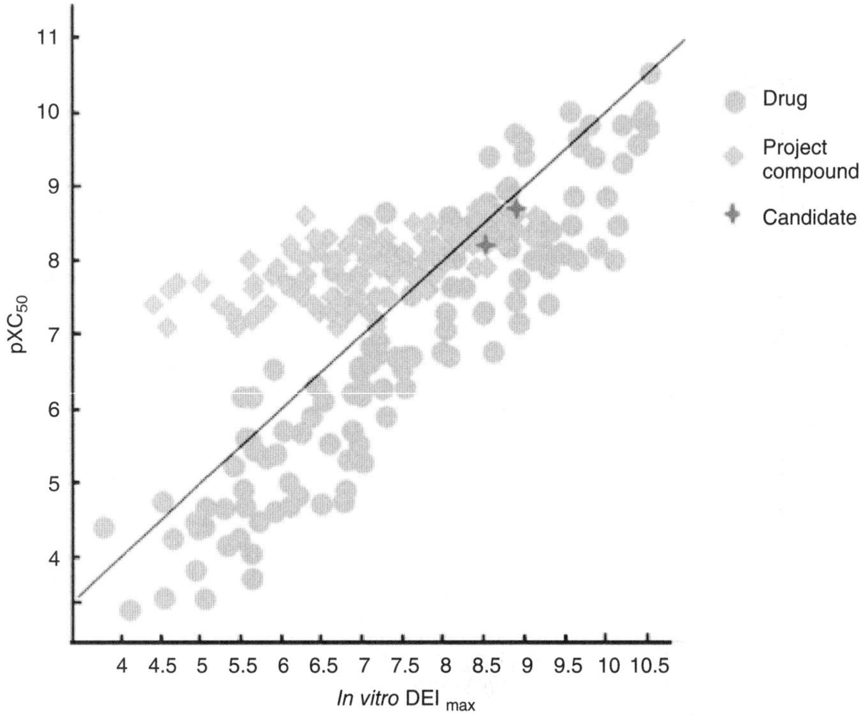

Figure 11.5 The plot of pXC_{50} versus *in vitro* DEI_{max} of known drugs and a set of project compounds. Circles represent the marketed drugs, squares represent the project compounds, and stars represent the selected candidate molecules. (Reproduced from Reference [18] with permission.)

that the majority of the project compounds fall on the left side of the line of unity, which means that the drug efficiency of the compounds is less than 1%.

The candidate molecules, marked as stars, showed better *in vivo* absorption and distribution properties and high DEI values but not necessarily the highest potency. This example shows a retrospective case before program teams started to use the DEI values in lead optimization. If potency alone was considered, several potentially good candidate molecules that showed good drug efficiency would have been discarded. The suggested strategy is that after the *in vitro* potency measurements, the *in vitro* drug efficiency measurements should follow and the DEI_{max} values should be ranked. The compounds should be prioritized by their DEI values for further *in vivo* studies and not based on the potency alone. To be able to guide the optimization process, we need to understand which property of the molecules influences drug efficiency. Drug efficiency has an inverse correlation with a sum of compound's binding to HSA and IAM. Another interesting question is which binding between the HSA or IAM is more influential to the overall drug efficiency. Figure 11.6 shows the plots of the *in vitro* $DRUG_{eff\ max}$ versus the HSA binding (Figure 11.6a) and the IAM binding (Figure 11.6b).

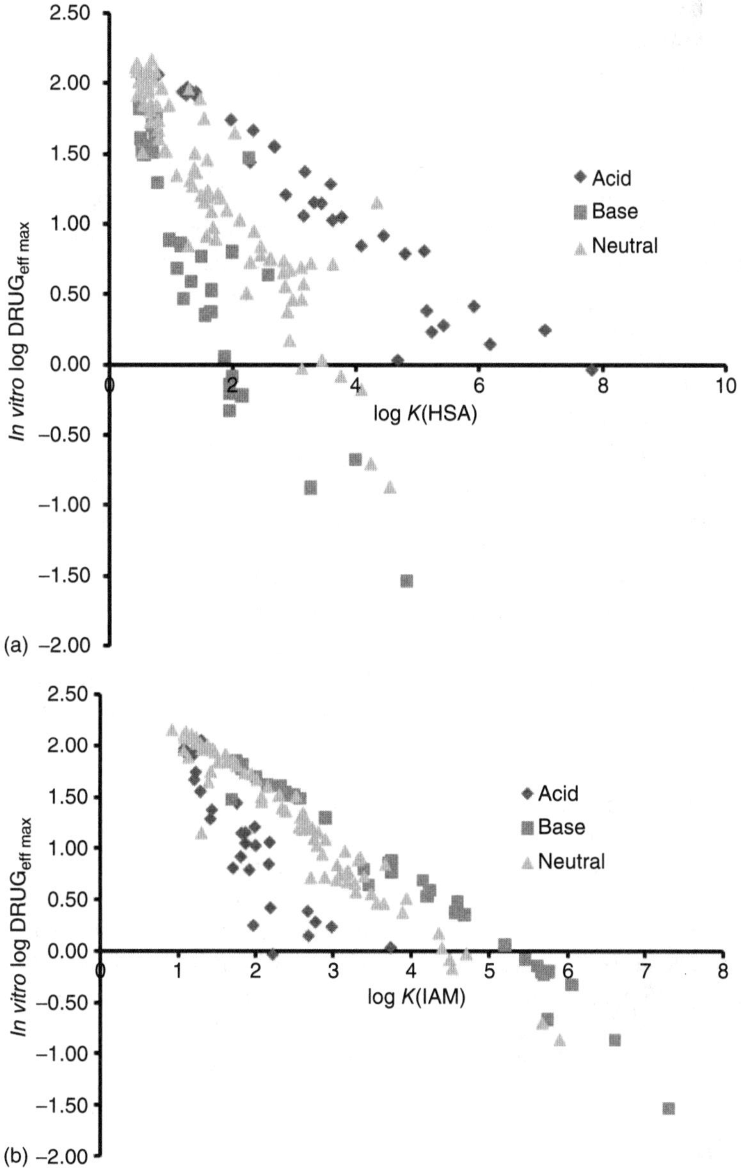

Figure 11.6 The plots of the *in vitro* $DRUG_{eff\ max}$ versus the HSA binding (a) and the IAM binding (b). (Plots are based on the data in Table 10.6 and Table 10.7. Reproduced from Reference [18] with permission.)

It can be seen in Figure 11.6 that the IAM binding, which accounts for the phospholipid binding of the compounds, is more detrimental to drug efficiency. While relatively strong HSA binding does not necessarily decrease the drug efficiency, strong IAM binding causes significant drops. Both types of binding are governed by lipophilicity. The presence of a positive or a negative charge does not reduce the sum of the two types of binding, as the positive charge increases the compounds' binding to IAM, whereas negative charge increases the compounds' binding to HSA. Therefore, the sum of the two types of binding should correlate with the intrinsic lipophilicity (log *P*). Figure 11.7 shows the inverse correlation of the *in vivo* $DRUG_{eff\ max}$ with the *c* log *P* values and much worse correlation with the log *D* values that takes into account ionization. If this is the case, the DEI values should show correlation with the LLE values. Note that $DEI = pXC_{50} + \log(DRUG_{eff})$, whereas $LLE = pXC_{50} - \log P$. Figure 11.8 shows the plots of *in vivo* DEI_{max} versus LLE and *in vitro* DEI_{max}.

A strong correlation is expected between various DEI and LLE values as they all contain the same *in vitro* potency pXC_{50}. It is not surprising that the *in vitro* DEI_{max} shows much better correlation with the *in vivo* DEI_{max} than the LLE (Figure 11.8). It is because in LLE the nonspecific binding of the compounds is modeled by a calculated lipophilicity. The octanol/water partition coefficient is only a model for the albumin and phospholipid binding; it does not take into account any three-dimensional binding interactions of the molecules and also does not model appropriately the charge–charge interactions. Moreover, the lipophilicity in LLE is a calculated parameter, whereas the *in vitro* DEI_{max} is obtained by measured binding properties of the compounds.

The analysis of the data of the marketed drugs suggests that optimizing the compounds on the LLE will help the optimization of DEI too. Reducing lipophilicity without losing the potency is advantageous not only for the efficient ligand binding to the target but also for the good absorption and distribution properties of the compounds *in vivo*. It has been shown that high DEI values in general go together with lower therapeutic dose minimizing the failure of the compounds in the clinic because of toxicity, as the drug efficiency is inversely related to lipophilicity. This encourages chemists to design potent compounds with the lowest possible lipophilicity that increases the chances of good solubility as well.

In conclusion, the LE indices and DEI provide a single point of optimization that is more efficient than the sequential optimizations of various properties of the compounds. It has to be emphasized that an important aspect of the optimization has not been taken into account in DEI: the rate at which compounds are cleared from the body. When the clearance is high, the half-life is short, which means that even if the compound shows high free concentration in the plasma after a low dose, it is cleared rapidly. This requires frequent repeated doses and puts a burden on the body's clearance processes; therefore, good potency and drug efficiency do not necessarily mean good overall exposure.

It has been demonstrated using data for more than 100 marketed drugs that the *in vitro* DEI_{max} data obtained from *in vitro* potency, the HPLC-based HSA and IAM binding measurements, show good correlation with the *in vivo* DEI_{max} values.

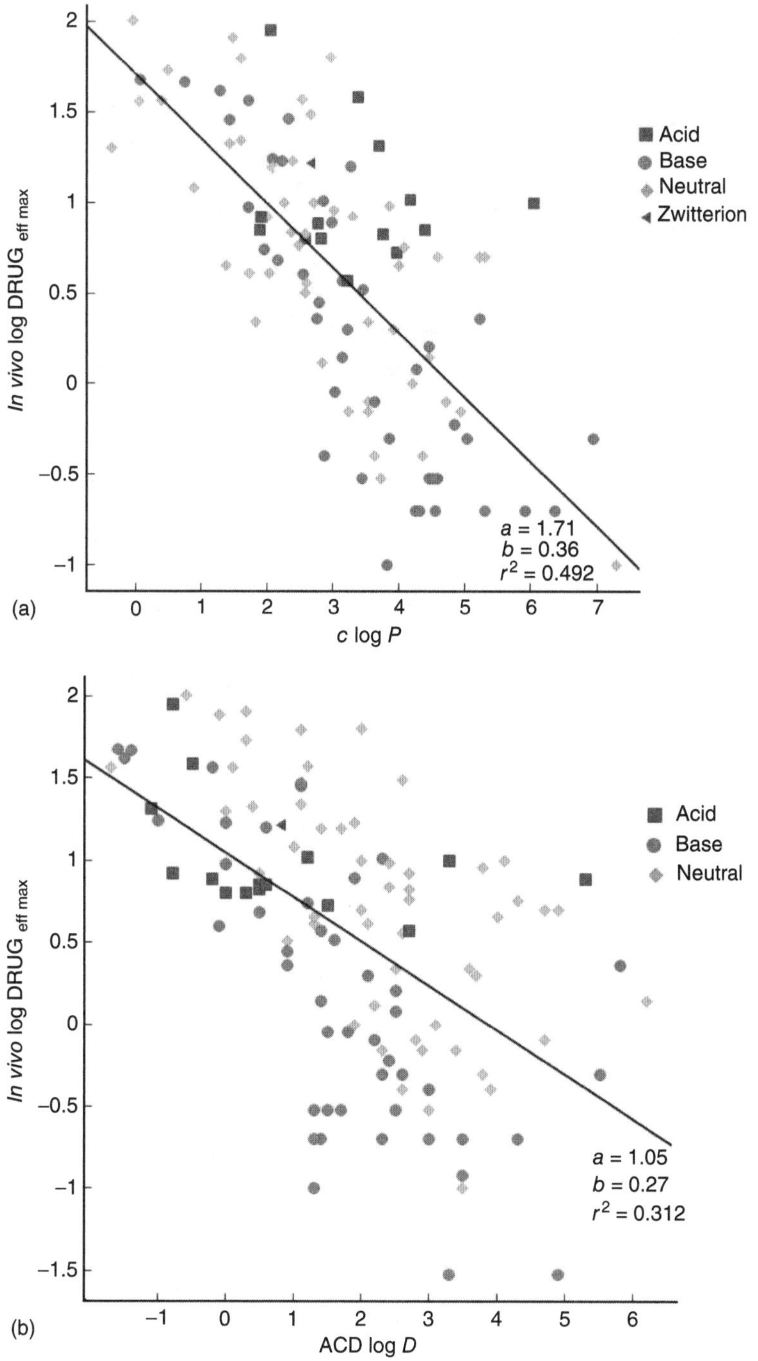

Figure 11.7 The correlation of the *in vivo* $\text{DRUG}_{\text{eff max}}$ with the calculated ACD log *P* (a) and ACD log *D* (b). It is evident that the correlation is much better with the lipophilicity of the uncharged molecules ($r^2 = 0.513$ with ACD log *P*, $r^2 = 0.273$ with ACD log *D*), as charge does not reduce the compounds binding to albumin and phospholipids. (Reproduced from Reference [18] with permission.)

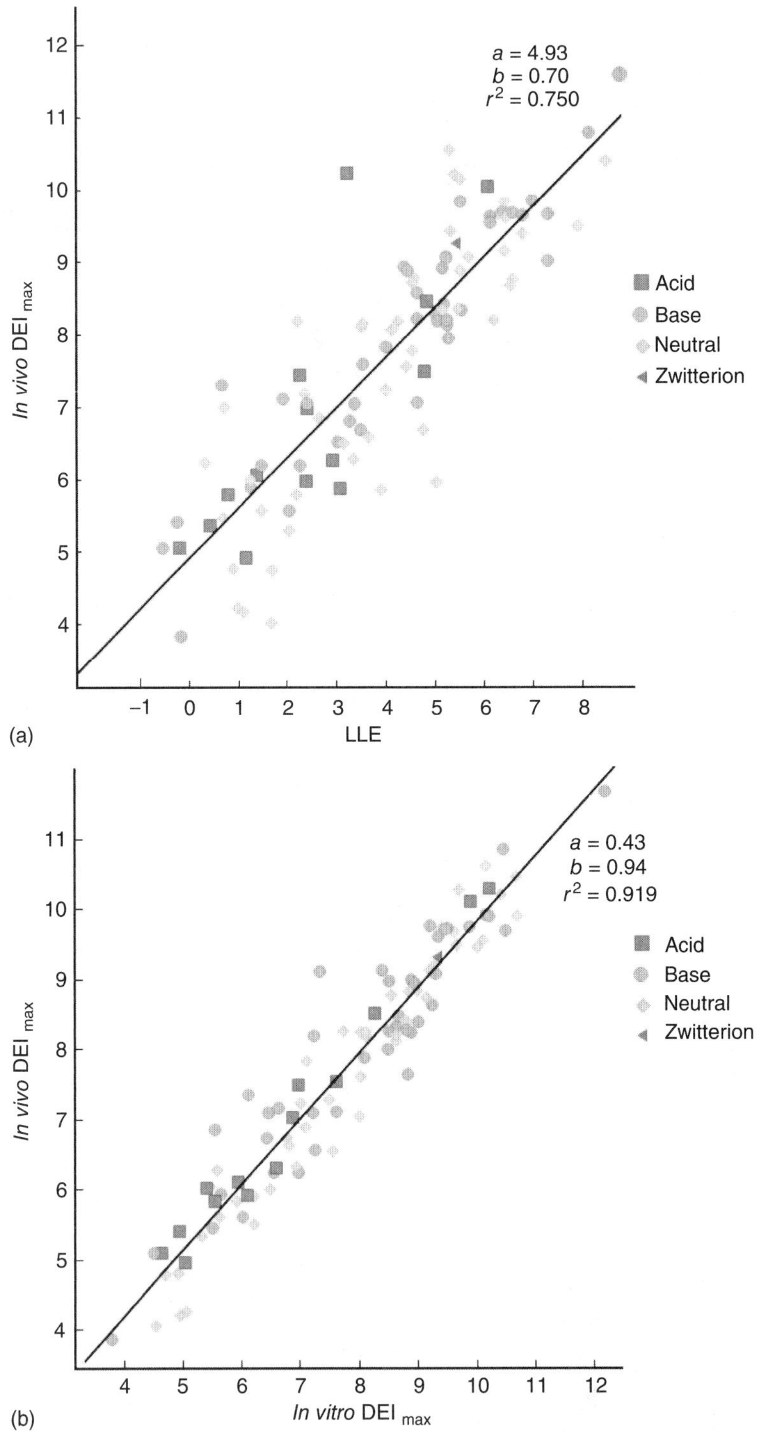

Figure 11.8 The correlations of *in vivo* DEI_{max} values with the LLE ($r^2 = 0.750$) (a) and *in vitro* DEI_{max} ($r^2 = 0.919$) (b). Lines are the regression lines. (Reproduced from Reference [18] with permission.)

This correlation was stronger than that with the LLE values, which incorporate the calculated lipophilicity only. Therefore, the biomimetic HPLC measurements provide additional confidence in a compound's prioritization at early stages of the lead optimization.

SUMMARY

It has been well recognized that the optimization of potency often leads to large and lipophilic compounds that are outside the drug-like property space and fail at later stages of the development process because of poor absorption and distribution properties. The various ligand efficiency indices express the potency relative to the size, polar surface area, or lipophilicity of the compound. Optimizing the ligand efficiency encourages the chemistry efforts to design potent compounds with lower lipophilicity and smaller size. The *in vivo* drug efficiency is defined as the percentage of the free biophase concentration of the drug relative to the dose. It accounts for the losses of the drug through absorption and distribution processes and because of nonspecific binding. It has been demonstrated that the maximum drug efficiency is the reciprocal value of the unbound volume of distribution obtained by intravenous administration. The model for unbound volume of distribution by the biomimetic HPLC HSA and IAM binding measurements can be used to model *in vivo* drug efficiency. The drug efficiency is affected more by strong IAM (phospholipid) binding than by the HSA binding underlying the importance of the tissue binding relative to the plasma protein binding. The two types of binding together show correlation to the intrinsic lipophilicity (log P and not log D). It is due to the fact that the presence of a positive or a negative charge does not reduce the overall nonspecific binding characteristics. The DEI is defined as the sum of *in vitro* potency (pXC_{50}) and the logarithmic value of the drug efficiency. Mathematically, the DEI values are proportional to the expected *in vivo* receptor occupancy of the drug per unit of dose. It has been demonstrated using data for some marketed drugs that with higher DEI values a lower therapeutic dose is required. A low dose is in general advantageous, as it is less likely results in toxic side effects.

The inverse correlation between efficiency and lipophilicity suggests that the DEI and the LLE show the same trend, therefore, optimizing on LLE results in compounds with good DEI values. It has been demonstrated that the biomimetic HSA and IAM binding measurements provide additional confidence in estimating the *in vivo* drug efficiency over the calculated lipophilicity c log P.

QUESTIONS FOR REVIEW

Q11.1 Define the various ligand efficiency parameters, such as LE, BEI, SEI, LLE, and LLE_{AT}.

Q11.2 What is the advantage of early optimization of ligand efficiency instead of potency alone?

Q11.3 What is the definition of drug efficiency?

Q11.4 What is the maximum drug efficiency and how it is related to unbound volume of distribution?

Q11.5 What are the major causes of low drug efficiency?

Q11.6 What is the advantage of using DEI at early lead optimization?

Q11.7 What is the relationship between DEI and the *in vivo* receptor occupancy?

Q11.8 What is the relationship between DEI and LLE?

Q11.9 Why is log P a better model for the *in vivo* nonspecific binding than log D?

Q11.10 Which biomimetic HPLC binding parameter can be used to estimate the drug efficiency of the compounds?

REFERENCES

1. Wenlock, M. C., Austin, R. P., Barton, P., David, A. M., Leeson, P. D. (2003) A comparison of physicochemical property profiles of development and marketed drugs. *Journal of Medicinal Chemistry*, *46*, 1250–1256.
2. Lipinski, C. A. (2002) Drug-like properties and causes of poor solubility and poor permeability. *Journal of Pharmacological and Toxicological Methods*, *44*, 235–249.
3. Trainor, G. L. (2007) The importance of plasma protein binding in drug discovery. *Expert Opinion in Drug Discovery*, *2*, 51–64.
4. Abad-Zapatero, C., Metz, J. T. (2005) Ligand efficiency indices as guidepost for drug discovery. *Drug Discovery Today*, *10*, 464–469.
5. Abad-Zapatero, C. (2007) Ligand efficiency indices for effective drug discovery. *Expert Opinion on Drug Discovery*, *2*, 469–488.
6. Hopkins, A. L., Groom, C. R., Alex, A. (2004) Ligand efficiency: A useful metric for lead selection. *Drug Discovery Today*, *9*, 430–431.
7. Rees, D. C., Congreve, M., Murray, C. W., Carr, R. (2004) Fragment-based lead discovery. *Nature Reviews Drug Discovery*, *3*, 660–672.
8. Leeson, P. D., Sprinthorpe, B. (2007) The influence of drug-like concepts on decision-making in medicinal chemistry. *Nature Reviews Drug Discovery*, *6*, 881–890.
9. Reynolds, C. H., Tounge, B. A., Bembenek, S. D. (2008) Ligand binding efficiency: Trends, physical basis and implications. *Journal of Medicinal Chemistry*, *51*, 2432–2438.
10. Mortenson, P. N., Murray, C. W. (2011) Assessing the lipophilicity of fragments and early hits. *Journal of Computer Aided Molecular Design*, *25*, 663–667.
11. Edwards, M. P., Rice, D. A. (2010) Chapter 23 – Role of physicochemical properties and ligand lipophilicity efficiency in addressing drug safety risks. *Annual Report in Medicinal Chemistry*, *45*, 380–391.
12. Hann, M.M. (2011) Molecular obesity, potency and other addictions in drug discovery. *Medicinal Chemistry Communications*, *2*, 349–355.
13. Garcia-Sosa, A. T., Hetényi, C., Maran, U. (2009) Drug efficiency indices for improvement of molecular docking scoring functions. *Journal of Computational Chemistry*, *31*, 174–184.

14. Hann, M. M., Leach, A. R., Harper, G. (2001) Molecular complexity and its impact on the probability of finding leads for drug discovery. *Journal of Chemical Information and Computer Sciences*, *41*, 856–864.

15. Braggio, S., Montanari, D., Rossi, T., Ratti, E. (2010) Drug efficiency: a new concept to guide lead optimization programs towards the selection of better candidates. *Expert Opinion on Drug Discovery*, *5(7)*, 609–618.

16. Stepensky, D. (2011) Use of unbound volumes of drug distribution in pharmacokinetic calculations. *European Journal of Pharmaceutical Sciences*, *42*, 91–98.

17. Valkó, K. L., Nunhuck, S. B., Hill, A. P. (2011) Estimating unbound volume of distribution and tissue binding *in vitro* HPLC-based human serum albumin and immobilized artificial membrane-binding measurements. *Journal of Pharmaceutical Sciences*, *100(3)*, 849–862.

18. Valkó, K. L., Chiarparin, E., Nunhuck, S. B., Montanari, D. (2012) *In vitro* measurements of drug efficiency index to aid early lead optimization. *Journal of Pharmaceutical Sciences*, *101*, 4155–4169.

19. MICROMEDEX Healthcare Series, DRUGDEX database: http://www.med.yale.edu/eph/phlibrary/uiuctemp/micro.html; PubMed articles: http://www.ncbi.nlm.nih.gov/pubmed; DrugBank Open Data Drug & Drug Target Database: http://www.drugbank.ca/

20. Montanari, D., Chiarparin, E., Gleeson, M. P., Braggio, S., Longhi, R., Valkó, K., Rossi, T. (2011) Application of drug efficiency index in drug discovery: a strategy towards low therapeutic dose. *Expert Opinion on Drug Discovery*, *6(9)*, 913–920.

21. Hughes, J.D., Blagg, J., Price, D. A., Bailey, S., DeCrescenzo, G. A., Devraj, R. V., Ellsworth, E., Fobian, Y. M., Gibbs, M. E., Gilles, R. W., Greene, N., Huang, E., Krieger-Burke, T., Loesel, J., Wager, T., Whiteley, L., Zhang, Y. (2008) Physicochemical drug properties associated with *in vivo* toxicological outcomes. *Bioorganic and Medicinal Chemistry Letters*, *18*, 4872–4875.

12

APPLICATIONS AND EXAMPLES IN DRUG DISCOVERY

STRUCTURE–LIPOPHILICITY RELATIONSHIPS

The importance of lipophilicity in pharmacological responses has been known since Meyer and Overton's work on anesthetic potency at the beginning of the twentieth century. They have described that the anesthetic potency of simple ethers, alcohols, and ketones showed correlation with the oil/water partition coefficients. The widespread application and measurements of lipophilicity via the octanol/water partition coefficients started with the works of Corwin Hansch and Albert Leo [1,2] in 1964. They introduced the quantitative structure–activity relationships (QSAR) approach to drug design. They suggested using lipophilicity measured by the logarithmic value of the octanol/water partition coefficient (log P) in relation to a compound's activity. Since their pioneering works, it has been well recognized that the lipophilicity of a drug candidate molecule can be linked to its *in vivo* absorption, distribution, metabolism, and toxicity (ADMET) properties. The influential work of Lipinski et al. [3] suggested that 90% of orally available drug molecules have log P value less than 5.

The measurements and characterization of a compound's lipophilicity has become widely used in drug discovery. As discussed in detail in Chapter 5, lipophilicity is a composite parameter. It depends on the proportion of polar and nonpolar features in the molecules. There are various types of lipophilicity,

Physicochemical and Biomimetic Properties in Drug Discovery: Chromatographic Techniques for Lead Optimization, First Edition. Klara Valko.

which may be significantly different from the octanol/water lipophilicity, such as hexane/water, or chromatographic C-18 phase, and water lipophilicity. However, it is not enough for the medicinal chemists to know the lipophilicity of the compounds; it is also important to provide guidelines and ideas how to change the structure to achieve a desirable overall property of the drug candidates. In this part of the book, the structure–lipophilicity relationships are presented in more detail, and how lipophilicity affects the ADMET properties of the compounds is explored.

The question is what the ideal lipophilicity value for a drug molecule is. The average log *P* value for synthetic drug molecules is 2.4, whereas natural pharmacologically active compounds are less lipophilic (0.7). Steroids represent the most lipophilic compounds among natural drugs [4]. Wenlock et al. [5] analyzed the average lipophilicity value of drug discovery compounds at various stages in the discovery process. They have found that the lipophilicity of compounds at the end of lead optimization is an average 0.6 log units higher than the marketed oral drugs. They noticed that the lipophilicity of the molecules increases during the lead optimization process. With the aim of increasing the potency, the lipophilicity was also increased.

Gleeson et al. [6] have published a thorough analysis of the effect of lipophilicity on various ADMET properties. They found that 91% of the marketed drugs had log *P* less than 5, 77% had log *P* less than 4, and 58% had log *P* less than 3. They have proposed that log *P* less than 4 provides a safer value for obtaining acceptable ADMET profile. Compounds with lower lipophilicity have greater chance of reaching the market as drugs. Vieth and Sutherland [7] analyzed the lipophilicity of marketed drugs based on the target classes and found that compounds acting on different types of targets may require different optimum lipophilicity. This may provide excuses for chemistry teams to accept compounds with higher lipophilicity. However, this finding has not been confirmed by the studies of Gleeson et al. [6], who showed no significant differences in the average lipophilicity of compounds that achieved good potency on a wide variety of targets. Similarly, Waring [8] compared the lipophilicity of compounds acting on different target classes. He characterized the compounds with the average lipophilicity and the overall lipophilicity distribution for each target class. There were no significant differences in average lipophilicity and lipophilicity distribution of compounds acting on different target classes.

The drug discovery process is by necessity a multiparameter optimization process; there is usually a possibility to trade-off one or two suboptimal properties for the sake of another excellent property. During this process, it is important to be aware of the risks when the compound's lipophilicity is outside the optimum range.

So why does lipophilicity increase during lead optimization?

The energy of drug receptor binding usually increases by adding additional hydrophobic interactions. However, lipophilic or hydrophobic interactions are usually nonspecific, and it often results in a compound's binding not only to the target receptor but also to other targets or proteins and phospholipids. The nonspecific binding to other components in the body reduces the free concentration of the compound at the site of action. In order to evaluate that the potency increases only due to lipophilicity, the ligand lipophilicity efficiency (LLE) described in detail in Chapter 11 can be used. It reveals the difference between potency and lipophilicity. The aim is to increase the potency with reduced lipophilicity, resulting in the increase in the LLE values.

With increased lipophilicity, the aqueous solubility of the compound is reduced. Adequate solubility of a compound is essential for oral absorption. According to the Yalkowski equation [9] discussed in detail in Chapter 6, every log unit increase in lipophilicity will decrease the solubility of the compound by one log unit in molar solubility term. In this respect, chemists should consider a compromise between lipophilicity and solubility. When the required dose is smaller, lower solubility is enough to dissolve the dose.

The permeability of the compounds, which is also essential for oral absorption, has an interesting relationship with lipophilicity. According to the log *D*–cMR plot for modeling absorption, larger molecules need higher lipophilicity for good permeability, as discussed in Chapter 9. Highly polar compounds (log $P < 0$ to -2, depending on the size) show poor permeability without active transport. Permeability increases with increasing lipophilicity. However, very lipophilic compounds may partition into the membrane, but they do not go through the membrane. They prefer staying in the lipophilic environment of the membrane and show poor permeability. Therefore, permeability shows a parabolic relationship with lipophilicity as shown in Figure 7.10.

The clearance and metabolism are complex processes involving various enzymes. Johnson et al. [10] have analyzed the intrinsic clearance with human liver microsomes for more than 47,000 compounds in Pfizer. They have found an inverse correlation with lipophilicity. It was apparent from the analysis that compounds with less than 3 log *D* values (lipophilicity of the compound at pH 7.4) had more likely less than 8.6 ml/(min mg) intrinsic clearance, equivalent to 0.5 extraction ratio. The lipophilicity cutoff has been found to be dependent on molecular size. However, very polar drugs with low lipophilicity may be cleared by the kidney, so it is not advantageous either. Gleeson [11] has studied the cytochrome P450 (CYP) enzyme inhibition data for more than 40,000 GlaxoSmithkline (GSK) compounds. He found that the average CYP pIC_{50} values were significantly lower for less lipophilic compounds (log $P < 3$).

Toxicity in general may occur when the compound binds not only to the target but also to other proteins, receptors, and phospholipids in the body. Compounds' binding affinity because of lipophilicity is not specific. Highly lipophilic compounds often have high binding affinity to several targets, which is known as *promiscuity*. Leeson and Springthorpe [12] found that promiscuity increases dramatically for compounds that have log *P* values higher than 3, based on the data of 2133 drugs and reference compounds. Other types of toxicity, such as phospholipidosis and hERG cardiac potassium channel inhibition, can also be related to high lipophilicity (log $P > 3$). The presence of positive charge provides an additional risk factor for phospholipidosis and hERG inhibition.

The drug distribution characteristics such as the volume of distribution can also be related to lipophilicity, but the ideal distribution is strongly dependent on the target tissue or organ. Therefore, it is difficult to establish even an approximate cutoff value for the volume of distribution. The presence of charge has an even more dramatic effect on the volume of distribution than the lipophilicity as demonstrated in Figure 10.8.

In conclusion, high lipophilicity is generally disadvantageous to the molecule and increases the probability of failure during the drug development process. High lipophilicity (log $P > 5$) causes low solubility, poor absorption, and strong nonspecific binding, which reduces the available free concentration at the site of action. High lipophilicity also increases the chance of microsomal metabolism and CYP inhibition and increases the chances of promiscuous bindings. Higher doses generally increase the risk of unwanted side effects and toxicity. Highly lipophilic compounds may cause phospholipidosis and hERG channel block, especially when it is associated with the presence of positive charge.

Table 12.1 shows the summary of the effect of lipophilicity on various ADME properties. The optimum range of lipophilicity of drug discovery compounds is between 1 and 3 log P. To design compounds with the optimum lipophilicity, we need to understand and estimate how the chemical structure relates to lipophilicity. Hansch and Fujita [1] introduced the substituent constant, π. It relates to the contribution of a given substituent (x) to the log P of the whole molecule when it replaces hydrogen (H). The calculation of a π substituent constant is shown by Equation 12.1.

$$\pi_{\mathrm{x}} = \log P_{\mathrm{x}} - \log P_{\mathrm{H}} \tag{12.1}$$

The π substituent constant can be obtained by measuring the log P values of two compounds, one with the hydrogen and the other one with the x substituent replacing the hydrogen. Fujita et al. [13] have observed that the same substituent contributes to lipophilicity more or less to the same extent in every compound. They have, however, observed that substituents have different contribution to the lipophilicity when they are aromatic or aliphatic.

Table 12.1 The summary of the effect of lipophilicity on physicochemical and ADME properties

Property	log $P < 1$	$1 <$ log $P < 4$	log $P > 4$
Solubility	Good	Medium	Bad
Permeability	Bad	Good	Medium/bad
Absorption	Medium	Good	Bad
Clearance	Bad	Good	Bad
Drug efficiency	Good	Medium	Bad
Brain penetration	Bad	Good (only between 1.5 and 3)	Bad
Toxicity	Good	Medium/good	Bad
Promiscuity	Good	Medium/good	Bad

Table 12.2 Aromatic substituent constants (π) representing the contribution to the lipophilicity of the molecules[a]

Substituent	π Constant	Substituent	π Constant
2-F	0.01	3-CF_3	1.07
3-F	0.13	3-CH_2OH	−1.02
4-F	0.15	4-CH_2OH	−1.26
2-Cl	0.59	3-CH_2COOH	−0.61
3-Cl	0.76	3-COOH	−0.12
4-Cl	0.70	2-$COCH_3$	0.01
2-Br	0.75	3-$COCH_3$	−0.28
3-Br	0.75	4-$COCH_3$	−0.37
4-Br	1.02	3-CN	−0.30
2-I	0.92	4-CN	−0.32
3-I	1.15	2-OH	−0.54
4-I	1.26	3-OH	−0.49
2-CH_3	0.68	4-OH	−0.61
3-CH_3	0.51	2-OCH_3	−0.33
4-CH_3	0.52	3-OCH_3	0.12
2-CH_2CH_3	1.22	4-OCH_3	−0.04
3-CH_2CH_3	0.97	3-OCF_3	1.21
3-*n*-Propyl	1.43	3-NH_2	−1.20
3-*i*-Propyl	1.30	2-NO_2	−0.23
4-*i*-Propyl	1.40	3- NO_2	0.11
3-*n*-Butyl	1.90	4-NO_2	0.24
4-*sec*-Butyl	1.82	3-$NHCOCH_3$	−0.79
3-*t*-Butyl	1.68	3-$NHCOC_6H_5$	0.72
4-Cyclopentyl	2.14	3-SCH_3	0.62
4-Cyclohexyl	2.51	3-SO_2CH_3	−1.26
3-C_6H_5	1.89	–SO_2NH_3	−1.82

[a]Taken from Reference [13] with permission.

Table 12.3 Aliphatic substituent constants (π) representing the contribution to the lipophilicity of the molecules

CH_3	*t*-Bu	OH	$CONH_2$	CF_3	Cl	Br	F
0.52	1.68	−0.67	−1.49	1.16	0.71	0.86	0.14

Hansch and Leo [14] have compiled the π values for various aromatic and aliphatic substituents, and some of them are listed in Table 12.2 and Table 12.3. The positive values mean that the substituent increases the lipophilicity of the molecule, whereas the negative values show a decrease in lipophilicity by the substituent.

The first *in silico* log *P* calculator (*c* log *P*) was developed at Pomona College, California, and is based on compiled π values of hundreds of fragments. It has been recognized that the neighboring substituents may affect each other's contribution to overall lipophilicity; therefore, interaction effects have to be taken into account too. Since then, many *in silico* log *P* calculators have been developed. Rekker [15] has published a fragmental contribution approach. ACD log *P*, *X* log *P*, *A* log *P*, *K* log *P*, and log KOW are all commercially available software based on fragmental and statistical approaches, while *M* log *P*, *V* log *P* are based on topological indices as reviewed by Mannhold [16]. In principle, all of them are based on some sort of fragmentation of the molecule and associating the delta log *P* value with each fragment and interaction. The whole molecule log *P* is then obtained by adding these fragments together. Some of them use statistical approaches; some of them use more systematic assignments of lipophilicity to molecular fragments. The Chemaxon lipophilicity calculator allows using several types of log *P* calculations and allows the user to select a weighted average of the predicted values. The *in silico* calculation of known drug molecules tend to be reasonably good, as these compounds and their experimentally determined log *P*/*D* values serve as a training set for many *in silico* prediction methods. However, the different calculation methods may result in two to three log unit differences in the predicted log *P* for new drug discovery compounds that may contain previously not considered fragments or interactions. Most of the models work well within the same series; therefore, measurements of a few compounds in each series are advisable for checking the predictions.

The addition of only one methyl group may cause dramatic changes in lipophilicity, especially when it is on a nitrogen and it changes the basicity and ionization state of the molecule. Waterbeemed et al. [17] have provided an overview of the effect of lipophilicity in various ADME properties and found that in spite of the obvious trends, it is very difficult to assess the change in lipophilicity to the overall PK behavior of the compounds. Some examples show that changing the lipophilicity of a compound resulted in the change in the characteristics of the drug distribution and PKs. One example is the design of sulmazole. Figure 12.1 shows sulmazole, a cardiotonic drug that was obtained by reducing the lipophilicity of its dimethoxy derivative which showed serious CNS side effects [18,19]. The log *P* of dimethoxy sulmazol is in the ideal range for brain penetration (log $P = 2.4$), whereas the log *P* of sulmazole is only 1.17. The dimethoxy compound caused vision problems

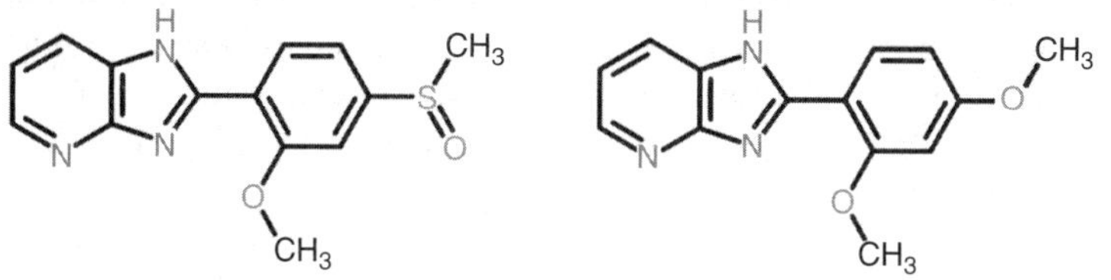

log $P = 1.17$
Non-CNS penetrant

log $P = 2.49$
CNS side effect

Figure 12.1 Sulmazol and its dimethoxy analog. The dimethoxy analog has a side effect of bright vision, because of its brain penetration.

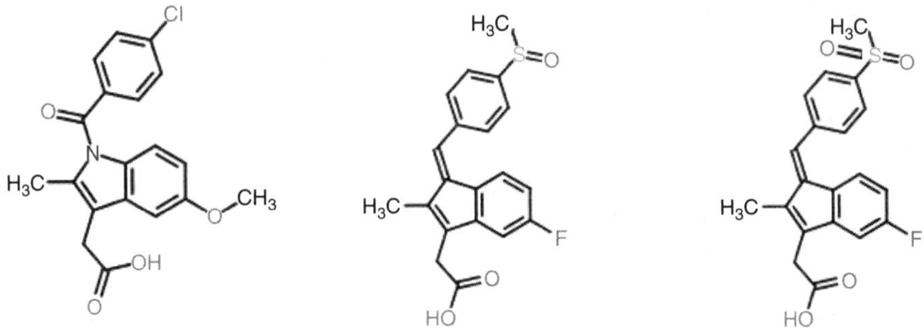

Indomethacine	Sulindac	Metabolite of sulindac
log $P = 4.25$	log $P = 3.29$	log $P = 3.29$

Figure 12.2 The structures and lipophilicity of indomethacine, sulindac, and the metabolite of sulindac.

because of its CNS penetration and activity; this was reduced by simply decreasing the lipophilicity. It is interesting to note that the methylsulfinyl group, instead of the methoxy group, decreased the lipophilicity of the compound by more than one log unit.

Another example of the effect of reduced lipophilicity on plasma protein binding is shown in Figure 12.2. The structure of indomethacin was modified to reduce its lipophilicity and as a consequence plasma protein binding.

Caron et al. [20] have studied in detail the lipophilicity behavior of sulfide, sulfoxide, and sulfon moiety. Figure 12.2 shows the structure and lipophilicity of indomethacin, sulindac, and its metabolite. The plasma protein binding of indomethacine is above 97%, whereas the plasma protein binding of sulindac is 93% because of its reduced lipophilicity.

Another example is ketoconazole and fluconazole, which show markedly different lipophilicity and ADME properties. Figure 12.3 shows the structure and log *P*/*D* values of the two antimycotic agents. Ketoconazole is used as a topical medicine because of poor oral absorption and liver toxicity. Fluconazole, which has much lower lipophilicity and lower molecular weight, can be used as an oral antifungal drug. In this case, not only the reduced lipophilicity but also the reduced molecular size has contributed to the more advantageous absorption properties, reduced the first-pass metabolism and toxicity.

Changing the lipophilicity of barbiturates, for example, not only caused the change of the ADMET properties of the compounds but also has changed the pharmacological effects as well. Figure 12.4 shows the structures and lipophilicity of thiobarbital, barbital, and phenobarbital (Luminal). Thiobarbital is an intravenous narcotic; barbital is a sedative drug; while phenobarbital has a strong anticonvulsive and muscle relaxing activity as well. The volume of distribution of phenobarbital

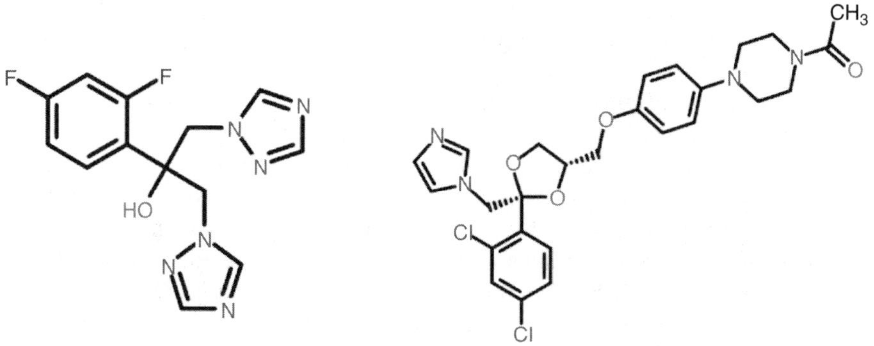

Fluconazole

log P = 0.45
log D = 0.45

Ketoconazole

log P = 4.04
log D = 3.93

Figure 12.3 The structures and lipophilicity of fluconazole and ketoconazole.

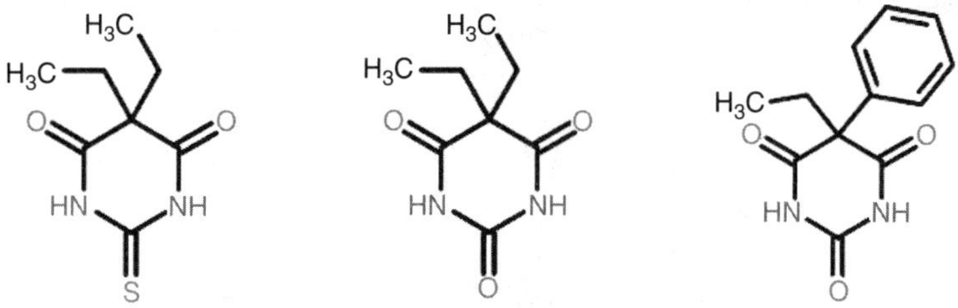

Thiobarbital
log P = 1.63

Barbital
log P = 0.65

Luminal
log P = 1.42

Figure 12.4 The structures and lipophilicity of thiobarbital, barbital, and phenobarbital (Luminal).

(0.55 l/kg) is higher and the half-life is longer than the more hydrophilic barbital (0.15 l/kg).

The Abraham solvation equation model [21] discussed in more detail in Chapter 5 is a useful tool for understanding the structure–lipophilicity relationships. The model clearly shows that the size of the molecule increases the overall lipophilicity, as the required energy to create cavity in water increases. The excess molar refraction parameter that increases with the number of heteroatoms normally has a positive contribution to lipophilicity. Fluorine is the only heteroatom that has a negative contribution to the excess molar refraction, which explains the odd lipophilicity of poly- or perfluorinated molecules. The dipolarity/polarizability, the H-bond donor, and acceptor properties of the substituents reduce the overall lipophilicity of the compounds. As substituents contribute to all five parameters (size, excess molar refraction, dipolarity/polarizability, H-bond acidity and basicity), they all should be taken into account for the assessment of overall lipophilicity contributions.

The Abraham solvation equation model also revealed that there are significant differences in lipophilicity when it is measured in different partitioning solvents. The most significant differences between various solvent partitions and various biological partition and distribution systems are in their sensitivities toward H-bond donor groups. Organic solvents that contain polar groups such as octanol and dissolve relatively large amounts of water are not sensitive to the H-bond acidity. Compounds with H-bond donor functionality are seemingly lipophilic as the H-bond donor group can form hydrogen bonds with H-bound acceptor groups of the solvent, which is energetically favored. In this respect, the biological partitions such as protein binding and phospholipid binding are very similar to the octanol/water system. Figure 12.5 and Figure 12.6 show the plots of the calculated log *P* values and the HSA and IAM binding, respectively, of more than 100 known drug molecules. Both the HSA and IAM binding data have been obtained using the biomimetic chromatographic measurements described in Chapter 5 projected to the log *P* scale. The compounds are marked according to their acid/base character at physiological pH. Acids are negatively charged, bases are positively charged, while neutral compounds, weak acids and weak bases are unionized at physiological pHs.

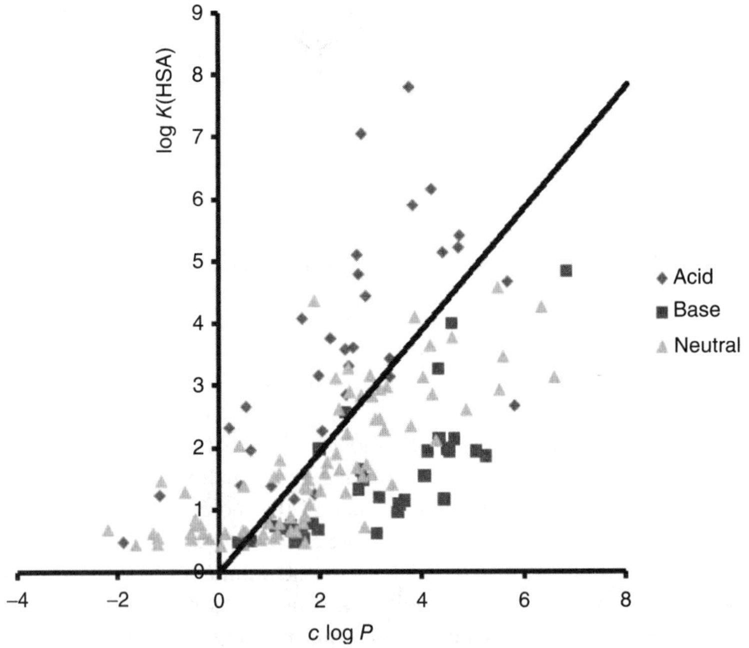

Figure 12.5 The plot of HSA binding (log *K*(HSA)) and ACD log *P* of known drugs. The line represents the line of unity. Plots are based on the data of the known drugs listed in Table 10.6 and Table 10.7.

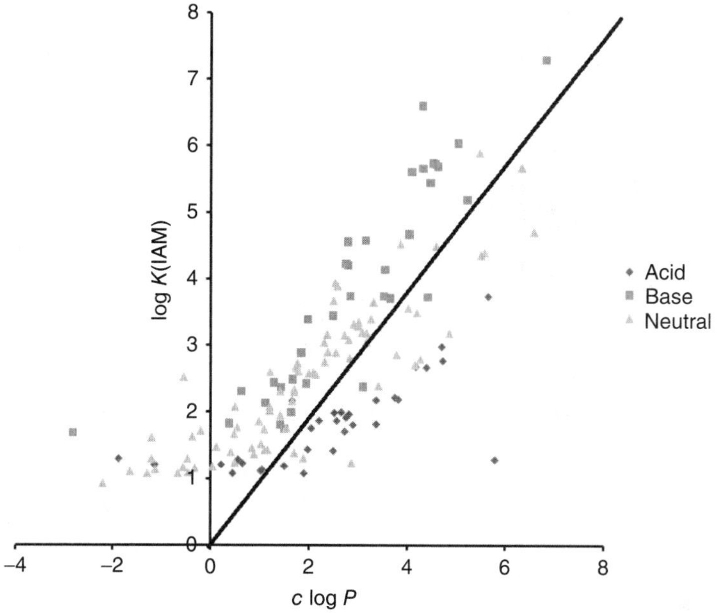

Figure 12.6 The plot of IAM binding (log *K*(IAM)) and ACD log *P* of known drugs. The line represents the equivalent log *K*(IAM) and ACD log *P* values. Plots are based on the data of known drugs listed in Table 10.6 and Table 10.7.

It can be seen in Figure 12.5 and Figure 12.6 that both HSA and IAM binding show a good trend with the calculated ACD log *P*; however, acids bind stronger to HSA and weaker to IAM, whereas bases bind stronger to IAM and weaker to HSA than it is expected from their lipophilicity (ACD log *P*). Plots are based on the data of the known drugs listed in Table 10.6 and Table 10.7.

When the two types of binding are plotted against each other, the differences between the acids and bases are even greater as shown in Figure 10.8.

Neutral and basic compounds are around the line of unity for the HSA binding, while neutral and acidic compounds are around the line for the IAM binding. The plots in Figure 12.5 and Figure 12.6 clearly show that the lipophilicity characterized by the octanol/water partition/distribution coefficients are not ideal for modeling the biomimetic albumin and phospholipid binding strength of the compounds.

We have seen in Chapter 5 that the Abraham solvation equations obtained for octanol/water partition, albumin binding, and phospholipid binding were very similar. The coefficient of the H-bond acidity term is close to zero in the solvation equations established for albumin binding, phospholipid binding, and octanol/water partition coefficients. Therefore, we have expected greater similarities. The discrepancy between octanol/water log *D* and albumin/phospholipid binding is due to different interactions with the charged molecules. The Abraham solvation equations were set up for neutral molecules. The octanol/water distribution coefficients are not able to describe the biomimetic binding of ionized compounds. Ionization,

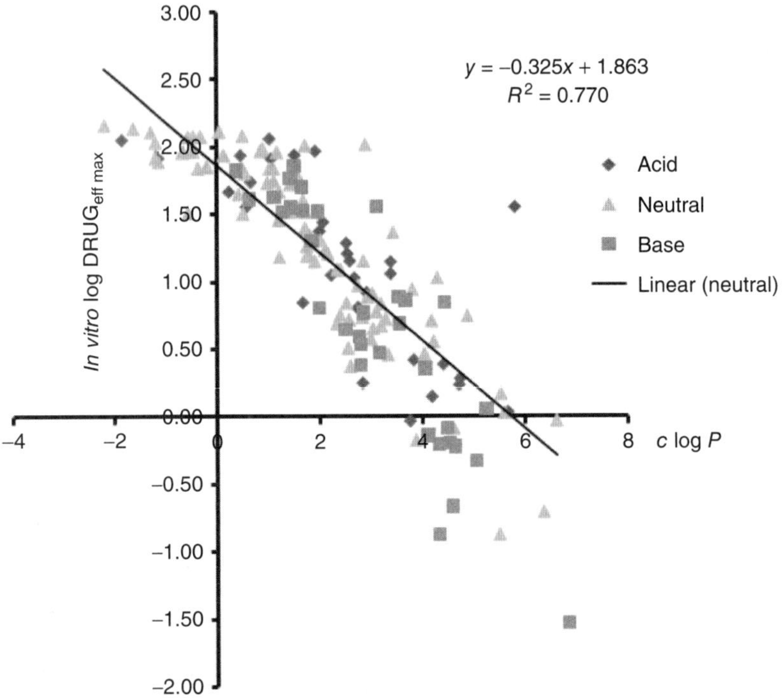

Figure 12.7 The plot of the logarithm of *in vitro* $DRUG_{eff}$ as a function of ACD log *P*. (The regression line is shown with $r^2 = 0.77$.) Plots are based on the data of the known drugs listed in Table 10.7 and Table 11.2.

whether it results in a positive or a negative charge, reduces the octanol/water distribution coefficient of the compounds by three to four orders of magnitude. The albumin and phospholipid binding is affected in different ways by the presence of charge.

The unbound volume of distribution and its reciprocal value, the drug efficiency, are proportional to the sum of the two types of biomimetic binding. Therefore, it is worth investigating how the sum correlates with the octanol/water lipophilicity. Figure 12.7 and Figure 12.8 show the correlation of the logarithm of *in vitro* $DRUG_{eff}$ with the calculated ACD log *P* and log *D*, respectively.

It is clear from Figure 12.7 and Figure 12.8 that the log *P* shows significantly better correlation with the drug efficiency than the log *D*. Acids, bases, and the neutral molecules do not form special clusters in the correlations.

On the basis of this observation, we can answer the long debate among scientists whether log *P* or log *D* is the appropriate lipophilicity parameter. It is evident that the *in vivo* free concentration of the compound relative to dose depends on the lipophilicity of the neutral form of the molecules. So, log *P* is more relevant in this respect than log *D*. It may be just a coincidence that the presence of charge

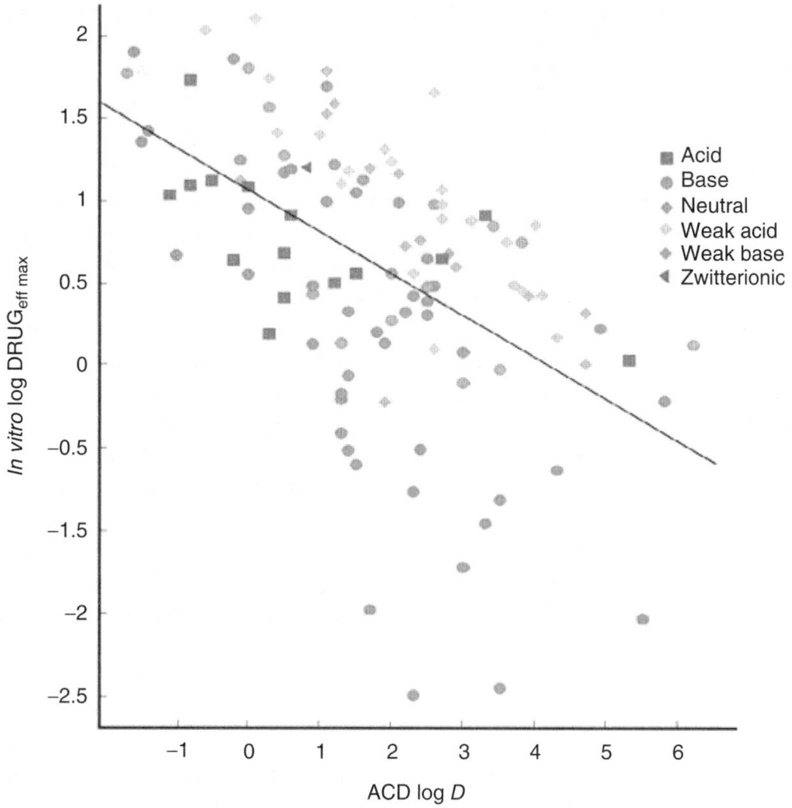

Figure 12.8 The plot of the logarithm of *in vitro* $DRUG_{eff}$ as a function of ACD log *D*. (The regression line is shown with $r^2 = 0.273$). Plots are based on the data of the known drugs listed in Table 10.7 and Table 11.2.

reduces the general lipophilicity to the same extent, as the charge–charge interactions increase the binding to albumin and phospholipids and show correlation with the lipophilicity of the uncharged species. When we want to model general nonspecific binding, the unbound volume of distribution and the drug efficiency of the compounds, the log *P* values are the correct parameters to be used. As a consequence, it is not worth reducing lipophilicity by introducing a charge on the molecules; the strength of nonspecific binding is not reduced. However, as will be discussed later in this chapter, when we want to increase the solubility, the introduction of groups that can be charged is important and the log *D* values are more relevant parameters. As for the permeability, the negative charge, in general, is not advantageous, whereas positively charged compounds may have good permeability.

Lipophilicity plays a special role in the blood–brain distribution and brain penetration of drugs. Hansch and coworkers [22] have found that there is an optimum lipophilicity needed for good CNS penetration and CNS activity: around log $P = 2$. Testa et al. [23] have discussed the role of lipophilicity in CNS penetration in more detail and plotted the brain extraction of labeled methanol, ethanol, isopropanol, *n*-butanol, benzyl alcohol, benzyl methyl ether, hexyl methyl ether, diphenylmethanol, *n*-octanol, benzophenone, benzyl hexyl ether, and triphenylmethanol as a function of their octanol/water lipophilicity (Figure 12.9). The highest brain extraction was obtained for compounds with log *P* values between 1 and 2.5. The Abraham solvation equation for the brain penetration discussed in Chapter 10 indicates that it is different from the octanol/water lipophilicity. It is sensitive to the H-bond acidity of the compounds having a large negative coefficient. It indicates that the brain-to-blood distribution is reduced by the presence of H-bond donor groups unlike the octanol/water partition. These differences are manifested in the so-called Δ log *P* values that can be obtained from the difference of the log *P* values obtained in octanol/water and heptane/water systems [23]. Table 12.4 shows the coefficients of the molecular descriptors (coefficient for excess molar refraction is *e*, dipolarity/polarizability is *s*, H-bond acidity is *a*, H-bond basicity is *b*, McGowan volume is *v*) in the solvation equations obtained for octanol/water, alkane/water, and brain/blood distribution.

Table 12.5 shows the calculated octanol/water partition coefficients, ACD log *P*, free fraction in brain, f_u (brain), and the logarithm of brain tissue partition coefficients for known CNS drugs [24]. The brain tissue partition coefficients have been derived from the free fraction values obtained from brain tissue homogenate and plasma protein binding measurements obtained by equilibrium dialysis at GSK.

By plotting the brain tissue partition as a function of ACD log *P*, the trend is obvious ($r^2 = 0.76$) (see Figure 12.10). However, compounds that show higher ACD log *P* than that expected from the hypothetical line contain H-bond donor functional groups that reduce their binding to brain tissue while they show higher partition

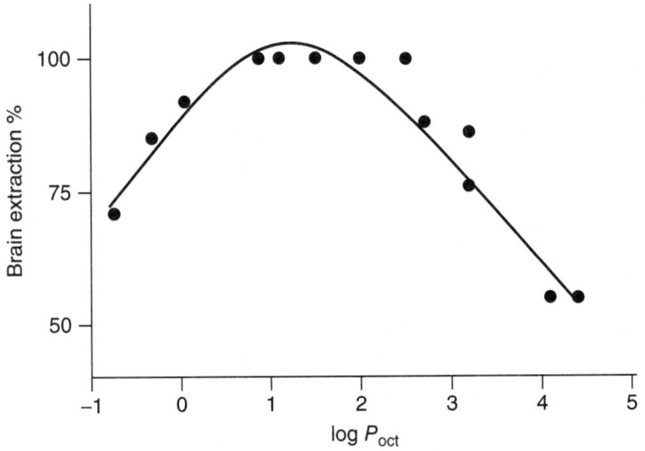

Figure 12.9 The plot of brain extraction obtained from labeled compounds in adult baboons as a function of log *P*. (Adopted from Reference [23] with permission.)

Table 12.4 The coefficients of the Abraham solvation equations for octanol/water, alkane/water, and brain/blood distribution[a]

Distribution	Constant	e	s	a	b	v	R^2
$\log P_{\text{octanol/water}}$	0.09	0.56	−1.05	0.03	−3.46	3.81	0.997
$\log P_{\text{alkane/water}}$	0.29	0.65	−1.66	−3.52	−4.82	4.28	0.998
$\log K_{\text{blood/brain}}$	0.04	0.51	−0.89	−0.72	−0.67	0.86	0.87

e, s, a, b, and v are the coefficients of the excess molar refraction, dipolarity/polarizability, H-bond acidity, H-bond basicity, and McGowan volume (size).
[a]Extracted from Table 5.5.

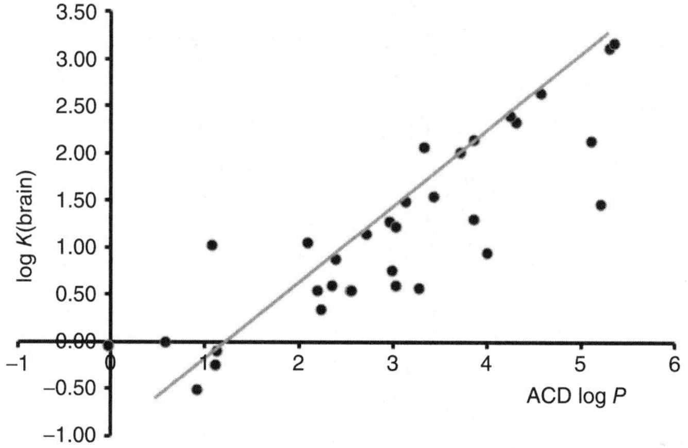

Figure 12.10 The plot of the logarithm of calculated octanol/water partition coefficients (ACD log *P*) and brain tissue binding (log *K*(brain)) for known CNS drugs from the data in Table 12.5. The hypothetical line shows the correlation for compounds without H-bond donor groups.

coefficient in the octanol/water system. Nevertheless, the trend shows that higher lipophilicity results in stronger brain tissue binding and reduced free fraction of the compounds in the brain.

An old example of the modification of lipophilicity in order to avoid the central parasympatholytic effect of atropine is methyl homatropine. While atropine has a log *P* value of 1.83 and has CNS side effects, the quaternary ammonium derivative (methyl homatropine) with a log *P* value of −1.68 is used for peripheral applications (Figure 12.11). The quaternary ammonium compound cannot go through the blood–brain barrier because of its hydrophilic nature and the presence of positive charge.

Large and lipophilic compounds also do not show good brain penetration mostly because of the active PgP efflux that protects the brain of such potentially harmful compounds. An interesting example is the alteration of the PK profile and brain penetration by changing the lipophilicity of two anti-Parkinson compounds. Both

Table 12.5 The ACD log *P* values, the free fraction in brain, and the logarithmic value of the brain tissue partition (log *K*(brain))[a]

Name	ACD log *P*	f_u (brain)	log *K*(brain)
Busprione	2.185	0.22	0.55
Caffeine	−0.04	0.52	−0.03
Carbamazepine	2.38	0.116	0.88
Carisoprodol	2.341	0.2	0.60
Chlorpromazine	5.3	0.00076	3.12
Citaprolam	3.132	0.0306	1.50
Clozapine	3.714	0.0094	2.02
Cyclobenzaprine	5.097	0.0073	2.13
Diazepam	2.961	0.05	1.28
Fluoxetine	4.566	0.0023	2.64
Fluvoxamine	3.321	0.0084	2.07
Haloperidol	3.849	0.0071	2.15
Hydrocodone	1.126	0.55	−0.09
Hydroxyzine	3.995	0.1	0.95
Lamotrigine	2.534	0.22	0.55
Meprobamate	0.915	0.76	−0.50
Methylphenidate	2.556	0.22	0.55
Metoclopramide	2.229	0.31	0.35
Midazolam	3.422	0.027	1.56
Morphine	0.572	0.5	0.00
Nortriptyline	4.315	0.0046	2.34
Paroxetine	4.238	0.0039	2.41
Phenytoin	2.085	0.081	1.05
Propoxyphene	5.21	0.033	1.47
Risperidone	2.711	0.067	1.14
9-OH Risperidone	1.074	0.086	1.03
Selegiline	3.024	0.056	1.23
Sertraline	5.347	0.00066	3.18
Sulpiride	1.11	0.63	−0.23
Thiopental	2.982	0.1463	0.77
Trazodone	3.85	0.047	1.31
Venlafaxine	3.269	0.21	0.58
Zolpidem	3.026	0.2	0.60

[a]Data were kindly provided from Caroline Peet and Phil Jeffrey and also published in Reference [24].

entacapone and tolcapone (see Figure 12.12) inhibit the conversion of L-Dopa into 3-methoxy-4-hydroxy-phenylalanine by blocking the catechol-*O*-methyl transferase (COMT) enzyme. Entacapone cannot cross the blood–brain barrier, whereas tolcapone can, thus inhibiting the metabolism of L-Dopa in the brain as well. The increase in lipophilicity has been achieved by introducing a *p*-methyl phenyl ring instead of the polar diethyl carboxamide group.

In conclusion, the lipophilicity of compounds depends on the proportion of polar and nonpolar groups in the molecules, in an additive manner. H-bond donor and

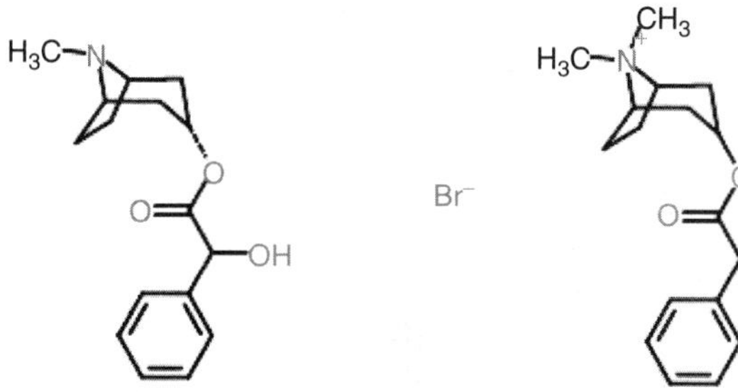

Atropine
log $P = 1.83$

Methyl homatropine Br
log $P = -1.68$

Figure 12.11 The modification of atropine that reduced the lipophilicity resulted in methyl homatropine that has no CNS side effect.

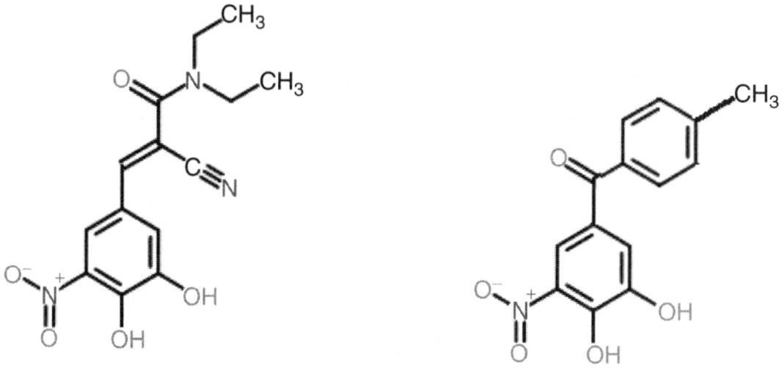

Entacapone
log $P = 0.53$

Tolcapone
log $P = 1.87$

Figure 12.12 Two anti-Parkinson drugs, entacapone and tolcapone: both inhibit the metabolism of L-Dopa, thus increasing its brain concentration. Entacapone acts only in the periphery, whereas tolcapone can go through the blood–brain barrier.

acceptor groups and the dipolarity/polarizability decrease lipophilicity. The contributions of substituents to the overall lipophilicity can be estimated; therefore, *in silico* calculations are available to assess lipophilicity from the chemical structure. Lipophilicity alone is seldom enough to explain the PK and pharmacodynamic properties of the drugs, but it affects several other properties such as solubility, permeability, protein binding, and blood–brain barrier distribution. There are no strict rules about the optimum lipophilicity. However, the drug-like space is limited by lipophilicity and size. Too hydrophilic molecules cannot easily cross the biological membranes; however, very lipophilic compounds (log $P > 5$) are associated with several risks, such as low solubility, high protein binding, poor drug efficiency, and increased toxicity.

STRUCTURE–SOLUBILITY RELATIONSHIPS

Solubility of the compounds is an essential parameter with respect to oral absorption and intravenous administration as well. Drug discovery programs often face the problem of increasing the solubility of the compounds without changing other pharmacological and ADME properties. It is difficult enough to design an appropriate change in lipophilicity, but changing the solubility is a much more challenging task. Aqueous solubility ($\log S_0$) is inversely related to lipophilicity ($\log P$) based on the Yalkowsky equation [25]. The relationship is straightforward for liquids, but an additional parameter, the melting point (T_m), has to be considered too for solids as shown by Equation 12.2.

$$\log S_0 = 0.5 - 0.01(T_m - 25) - \log P \qquad (12.2)$$

The equation is considered as the general solubility equation, which is a semiempirical equation. The melting point is supposed to account for the solid-state properties of the compounds. Before the molecules are solvated by the solvent molecules from the solid form, the intermolecular bonds in the crystal have to be broken. Therefore, the solubility depends not only on the chemical structure of the molecules and its interactions with the solvent molecules but also on the crystal lattice energy in the solid form. The validity of the Yalkowsky equations has been tested on large sets of simple organic molecules, herbicides, and pesticides [26,27]. This semiempirical equation works very well, but it requires the measurement of melting points. While in the past the melting point was used as an indicator of purity and identity of a newly synthesized compound, nowadays it is not measured routinely, as it would decrease the throughput of the discovery chemistry efforts. Discovery compounds are synthesized in small quantities and often purified by HPLC and recovered from the mobile phase in an amorphous form. The estimation of melting point or the enthalpy required for melting is a difficult task [28,29]. Wassvik et al. [30] have investigated the contribution of the solid-state properties to the aqueous solubility on a set of carefully selected drug molecules. They have tried to develop a solubility prediction model for drug molecules by taking into account the solid-phase properties, such as the melting point (T_m), the enthalpy (ΔH), and entropy (ΔS) contributions of melting. The intrinsic solubility (S_0) of the selected drug molecules showed very poor inverse correlation with their lipophilicity ($\log P$); therefore, it was an ideal set to investigate the contribution of the solid-state properties in addition to the lipophilicity of the drugs. Table 12.6 shows the measured intrinsic solubility, melting point, and the enthalpy and entropy of melting of the investigated compounds. The data analysis has revealed that the solid state of the compounds has significantly influenced the intrinsic solubility and explained the majority of the deviations from the solubility–lipophilicity relationship. Their measured data revealed that the constant ΔS_m suggested [26] as being 56.5 J/(mol K) could be as large as 114.1 J/(mol K) (astemizole). Therefore, the authors have suggested including the entropy term in the general solubility

Table 12.6 The intrinsic aqueous solubility (log S_0), calculated c log *P* (Daylight version 4.82), the melting point (in °C), the enthalpy change of melting (ΔH_m in kJ/mol) and entropy change of melting (ΔS_m in J/(mol K)) of drugs with their CAS numbers[a]

Drug	CAS Number	$\log S_0$	$c \log P$	T_m	ΔH_m	ΔS_m
Astemizole	68844-77-9	−7.18 ± 0.14	6.09	174.4 ± 0.10	51.1 ± 0.77	114.1 ± 1.72
Chlorpropamide	94-20-2	−3.30 ± 0.002	2.35	127.8 ± 0.07	25.7 ± 0.41	64.0 ± 1.02
Clozapine	5786-21-0	−4.64 ± 0.03	4.93	183.9 ± 0.07	35.9 ± 0.53	78.4 ± 1.16
Diazepam	439-14-5	−3.85 ± 0.01	3.17	131.6 ± 0.03	24.7 ± 0.10	61.0 ± 0.24
Diazoxide	364-98-7	−3.60 ± 0.01	1.2	327.2 ± 0.24	34.1 ± 1.12	56.8 ± 1.89
Diethylstilbestrol	56-53-1	−5.00	4.96	177.9 ± 0.21	33.4 ± 2.24	73.9 ± 4.96
Fenbufen	36330-85-5	−5.19 ± 0.06	3.14	186.1 ± 0.15	46.2 ± 2.12	100.5 ± 4.64
Gliclazide	21187-98-4	−4.07 ± 0.03	1.09	171.4 ± 0.14	44.2 ± 0.63	99.4 ± 1.45
Glyburide	10238-21-8	−7.05 ± 0.19	4.24	173.6 ± 0.13	46.3 ± 0.09	103.7 ± 0.23
Griseofulvin	126-07-8	−4.83 ± 0.08	1.77	218.0 ± 0.00	44.7 ± 0.78	90.8 ± 1.59
Hydrochlorothiazide	58-93-5	−2.70 ± 0.03	−0.37	267.6 ± 0.41	33.6 ± 0.14	62.2 ± 0.31
(±)-Ibuprofen	15687-27-1	−3.38 ± 0.03	3.68	73.2 ± 0.10	26.6 ± 0.15	76.8 ± 0.42
Indomethacin	53-86-1	−5.95 ± 0.01	4.18	159.8 ± 0.03	37.9 ± 0.18	87.6 ± 0.43
(±)-Indoprofen	31842-01-0	−4.72 ± 0.12	2.74	211.4 ± 0.46	40.3 ± 2.38	83.2 ± 4.84
(±)-Ketoprofen	22071-15-4	−3.52 ± 0.01	2.76	94.8 ± 0.12	37.3 ± 0.33	101.2 ± 0.12
Mifepristone	84371-65-3	−5.75 ± 0.02	4.46	193.9 ± 0.14	31.7 ± 0.61	67.9 ± 1.33
Naproxen	22204-53-1	−4.23 ± 0.02	2.82	155.6 ± 0.12	34.2 ± 0.85	79.7 ± 2.01
Perphenazine	58-39-9	−4.62 ± 0.02	4.31	96.8 ± 0.36	41.8 ± 0.69	113.0 ± 1.91
Phenacetin	62-44-2	−2.48 ± 0.002	1.77	134.2 ± 0.06	34.1 ± 0.92	83.7 ± 2.27
Phenytoin	57-41-0	−4.15 ± 0.04	2.09	295.6 ± 0.28	40.1 ± 0.75	70.4 ± 2.97
Piroxicam	36322-90-4	−4.03 ± 0.01	1.89	200.3 ± 0.45	36.3 ± 0.15	76.7 ± 0.25
Probenecid	57-66-9	−4.90 ± 0.10	3.37	198.9 ± 0.06	40.9 ± 0.20	86.7 ± 0.43
Sulindac	38194-50-2	−4.78 ± 0.03	3.16	187.0 ± 0.06	33.4 ± 0.36	72.5 ± 0.78
Tamoxifen	10540-29-1	−8.49 ± 0.22	6.82	97.8 ± 0.29	34.0 ± 0.40	91.6 ± 1.14
Testosterone	58-22-0	−4.20 ± 0.08	3.22	153.3 ± 0.36	28.2 ± 0.61	66.1 ± 1.48
Trimethoprim	738-70-5	−2.87 ± 0.003	0.98	199.7 ± 0.10	49.8 ± 0.39	105.4 ± 0.85

Abbreviation: CAS, Chemical Abstract Services.

[a]From the study of Wassvik et al. [30] with permission.

equation as shown by Equation 12.3.

$$\log S_0 = 0.5 - \frac{\Delta S_{\rm m}}{5705.85}(T_{\rm m} - 25) - \log P \qquad (12.3)$$

Equations 12.2 and 12.3 show an inverse correlation between log P and solubility with a slope of −1 and intercept of 0.5 that provides an estimate of the maximum solubility expected from a compound without taking into account the effect of the crystal lattice energy. Figure 12.13 shows the maximum solubility line as the plot of measured intrinsic solubility and calculated log P of the 26 compounds in Reference [30].

Perlovich et al. [31,32] have published encouraging results for the calculation of crystal lattice energy using force fields specifically developed for such purposes. However, to date, there are no reliable computational methods for estimating the contribution of the solid-state properties to the intrinsic solubility, purely from the structure. This is not surprising as many drug molecules have several crystal forms in the solid state and the difference in intrinsic solubility of the various forms might differ by several orders of magnitude.

Medicinal chemists have to rely on a few empirical observations of structural motifs that might contribute to higher crystal lattice energy beside the lipophilicity of the molecules, to design soluble drugs. One of these empirical rules can be derived from the general solvation equation to describe intrinsic aqueous solubility by Abraham and Le [33] as has been discussed in Equation 6.3. They have added the product term of H-bond acidity and basicity in the solubility equation for the prediction of possible intermolecular H-bond formation in the crystals. The equation suggests that the presence of H-bond donor and acceptor groups in the molecule potentially decreases the solubility by three log units, as the coefficient of the product term is −3.36. An example for the application of this rule can be seen in Figure 12.14, which shows the structure and the solubility of several steroid derivatives. The 4-pregnane-20a-ol-3one has the lowest solubility. It has got an H-bond donor hydroxyl group and the H-bond acceptor oxo group at the far end of the molecule. In spite of the same substituents, 4-androstene-17b-ol-3one has slightly higher solubility because of the reduced number of methylene groups, which reduce the lipophilicity but possibly did not affect significantly the crystal lattice energy. When the hydroxyl group is replaced by the oxo group in 4-pregnane-3,20-dione the solubility increases. It is much more difficult to rationalize the big difference in solubility of 4-pregnane-17a-ol-3,20-dione (15 μg/ml) and 4-pregnane-21-ol-3,20-dione (191 μg/ml).

The lower solubility compound probably forms intermolecular hydrogen bonds in the solid form, whereas the higher solubility compound might form intramolecular hydrogen bonds. In the logarithmic scale of intrinsic solubility, these are not significant differences: all the steroids can be considered as low solubility compounds. In terms of assessing whether the required dose is soluble or not, these differences can be considered as significant. Steroids are very potent molecules at low doses, which means that a 10 times higher solubility in microgram per milliliter unit can be significant.

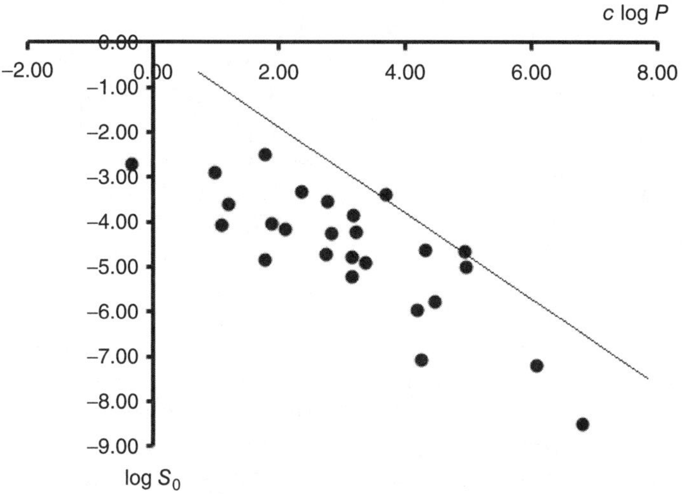

Figure 12.13 The plot of the measured intrinsic aqueous solubility (log S_0) and log *P* values from the 26 drug molecules listed in Table 12.3 from Reference [30]. The line shows the maximum solubility that we can get from a given lipophilicity.

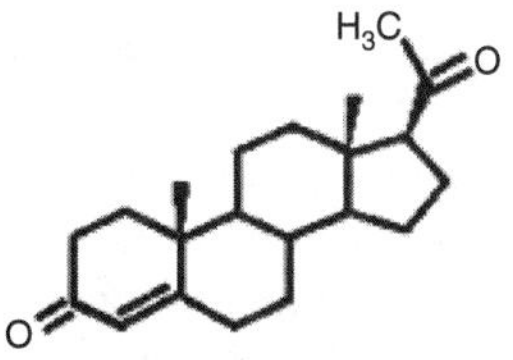

4-Pregnen-3,20-dione

$S_{7,4}$ = 43 μg/ml

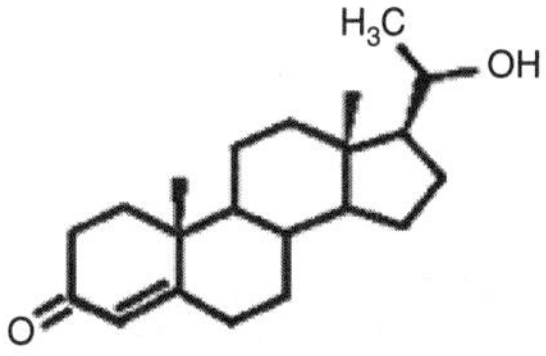

4-Pregnen-20a-ol-3-one

$S_{7,4}$ = 6 μg/ml

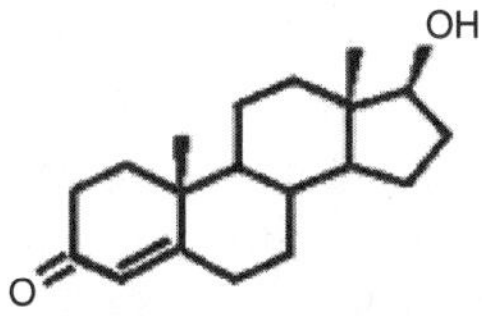

4-Androstene-17b-ol-3-one

$S_{7,4}$ = 30 μg/ml

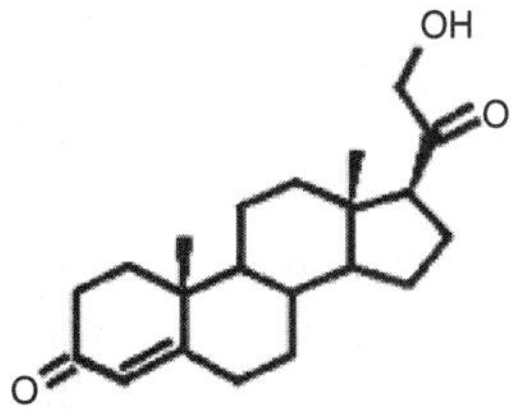

4-Pregnen-21-ol-3,20-dione

$S_{7,4}$ = 191 μg/ml

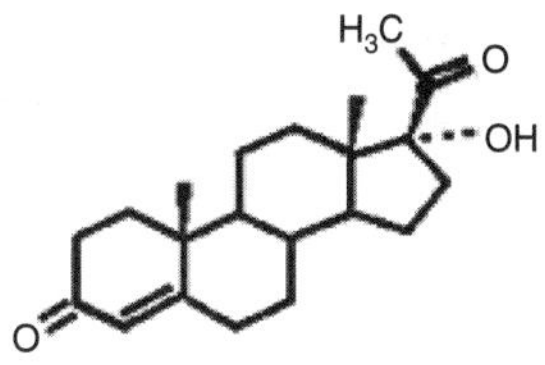

4-Pregnen-17a-ol-3,20-dione

$S_{7,4}$ = 15 μg/ml

Figure 12.14 The structure and the solubility of steroid derivatives.

The influence of the simultaneous presence of the H-bond donor and acceptor groups in the molecules on the melting points has been demonstrated by matched molecular pair analysis by Shultes et al. [34]. An algorithm described by Hussain and Rea [35] can be used to assess the effect of substituents on solubility. The matched molecular pair analysis can be carried out on a large number of molecules with measured data. The computer program pulls pairs of molecules and the measured data together, which differ only by a specified functional group. The average effect of a substituent on solubility is calculated. This powerful technique can be used to identify and quantify the effect of the molecular changes on any measured property when large numbers of measured data are available. The changes are usually averaged and calculations carried out to test whether the substituent significantly contributes to the change in properties. This technique provides a quantitative assignment of the group contributions to a certain property. The work by Shultes et al. [34] analyzed the functional group contributions to the measured melting points of 746 molecules of free bases. Salts were excluded from the analysis as different salts of the same compound usually show different melting points. They assigned simple calculated properties for the structural changes such as H-bond donor, H-bond acceptor, total polar surface area (TPSA), the number of rotatable bonds, $c \log P$, and the number of halogen atoms. They found that H-bond donor and H-bond acceptor groups increased the melting point on an average, 48 and 36 °C, respectively. This is probably due to the increased crystal lattice energy caused by the possible intermolecular H-bond interactions between the donor and acceptor groups. The number of rotatable bonds showed a significant decrease in the melting point. This can be explained by the hypothesis of Dannenfelser and Yalkowsky [28] that the rotatable bonds provide a higher flexibility of the molecule, resulting in higher melting entropy. The matched molecular pair analysis also revealed that halogen atoms significantly influence the melting point of the compounds. Bromine and iodine increase the melting point, whereas the presence of fluorine and chlorine cause a slight decrease. The effect of the hydroxyl group on solubility is quite complex. In general, the addition of a hydroxyl group reduces the lipophilicity by around half a log unit (Table 12.3), so the solubility increases. On the other hand, the hydroxyl group might form hydrogen bonds in the crystal, which reduces solubility. Zhang et al. [36] observed that an addition of a hydroxyl group in general increased the solubility. Their study was based on the intrinsic aqueous solubility of 2794 molecules.

Another interesting observation that can help medicinal chemists design soluble potent compounds is the effect of aromatic rings on the general ADME properties of the molecules. Lovering et al. [37] noticed that saturation of aromatic rings generally improves the clinical success of drug candidates. They have observed that saturation and the introduction of a chiral center in general provide a higher probability of the molecules becoming a drug. They also have shown that the number of aromatic carbons significantly reduce the solubility. Ritche and Macdonald [38] have analyzed the effect of the number of aromatic rings on solubility among other developability characteristics. They have found that more than three aromatic rings in the molecules usually result in low solubility and developability. They have analyzed the precipitative solubility data in pH 7.4 phosphate buffer of more than 30,000 research molecules in GSK. The majority of the molecules had three to four aromatic rings. While the average solubility of compounds with only two aromatic rings was around 100 μg/ml, the average solubility of compounds with four aromatic rings dropped to 25 μg/ml. It is very likely that if they had expressed the solubility in molar term, the drop would have been even more pronounced. Later Ritchie et al. [39] have investigated the effect of heteroatoms in the aromatic rings and the fused aromatic rings. They have found that the heteroaromatic rings showed less detrimental effect on solubility. Table 12.7 shows the rank of various heteroaromatic rings on solubility. The higher number represents bigger enhancement on solubility.

The structure–solubility relationships discussed earlier are related to the intrinsic aqueous solubility of the compounds. We have shown in Chapter 6 that the ionization has a significant impact on solubility. Compounds that have ionizable groups can have increased solubility at pH values where the compounds have charge. The increase in solubility can be three to four orders of magnitude higher at pH values where they are fully ionized. It is difficult to predict the solubility of the fully ionized form because the counterion may affect the crystal lattice energy and may

Table 12.7 The rank of the heteroaromatic rings based on their contribution to good solubility[a]

Heteroaromatic Ring	Prevalence in the Set, %	Solubility Score
Pyrazine	1.90	22
Pyridine	24.40	21
1,3,4-Oxadiazole	1.30	20
Tetrazole	0.80	19
Pyrazole	10.80	18
Pyridazine	1.80	17
Furan	2.50	16
1,2,4-Triazole	2.20	15
Imidazole	8.10	14
Oxazole	2.00	11
1,2,4-Oxadiazole	1.50	10
Pyrrole	7.90	9
1,2,3-Thiadiazole	0.20	8
Isoxazole	1.90	7
1,2,3-Triazole	0.90	6
1,3,4-Thiadiazole	2.10	5
Thiophene	5.60	4
Pyrimidine	15	3
Thiazole	6.90	2
1,3,5-Triazine	2.20	1

The higher the score, the better the ring is in terms of solubility.
[a]Reference [39].

reduce the solubility. Medicinal chemists often consider the solubility of the compounds at physiological pH; therefore, the high throughput solubility measurements often carried out at pH 7.4 buffer. Hill and Young [40] have reported an interesting observation by analyzing the measured solubility data of more than 100,000 research compounds at GSK. They found that in addition to the decrease in solubility because of lipophilicity of the compounds, the number of aromatic rings caused a further decrease in the measured solubility as shown in Figure 12.15. Compounds with the same calculated log *D* (referring to the octanol/water distribution coefficient at pH 7.4) were less likely soluble if they had more aromatic rings. They introduced "solubility forecast index," abbreviated as SFI, which is the sum of the

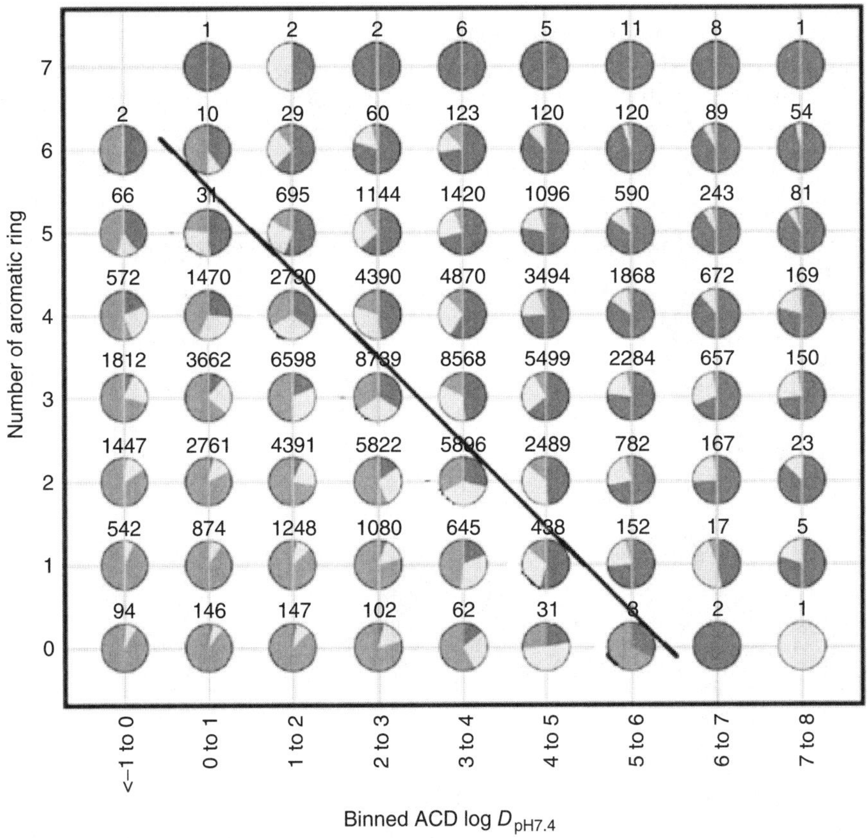

Figure 12.15 The distribution of low (red), medium (yellow), and high (green) solubility compounds with various number of aromatic rings and in various bins of calculated ACD log $D_{pH7.4}$. Compounds with the same lipophilicity bin are less soluble when they had more aromatic rings. The line shows the value when SFI is around 6, where approximately 50% of the compounds are insoluble. The numbers above the circle shows the number of compounds falling into that particular bin. (Adopted from Reference [40] with permission.) *(See insert for color representation of this figure.)*

log *D* value and the number of aromatic rings in the molecules. They found that when the SFI value was around 6, 50% of the compounds had less than 30 μM solubility. Every unit increase in SFI decreased the probability of good solubility (>30 μM) by approximately 10%. Thus, more than 90% of the investigated compounds with SFI value of more than 12 had lower than 30 μM solubility, whereas more than 90% of compounds had solubility of more than 30 μM when their SFI value was below 4. The explanation of the observation is that multiple aromatic rings in the molecules provide a planar structure that can "stack" in the crystals, increasing their melting points and crystal lattice energy.

There are neutral, acidic, and basic compounds included in the study. The effect of ionization is taken into account both in the calculated lipophilicity and in the solubility measurements; thus, they cancel out. On the basis of the empirical observations described, the number of aromatic rings and the presence of both H-bond donor and acceptor groups can decrease the solubility above the contribution of lipophilicity. One of the most popular approaches to increase solubility is the introduction of ionizable groups in the molecule. This makes the molecule partially ionized at the pH of the gastrointestinal tract and the stomach. The introduction of a positive or a negative charge may significantly increase the solubility, but there are other serious consequences of the charge. Negatively charged compounds usually have impaired permeability. The presence of charge significantly influences the *in vivo* distribution of the compounds between the plasma and tissue compartments too. As the pH in the gastrointestinal tract ranges from 2 to 8 or 9, there is a possibility of precipitation during the transit from the stomach to the duodenum. The safest way to increase the solubility of the compounds is to increase the intrinsic aqueous solubility, that is, the solubility of the neutral form. There are some favorite groups that chemists like to introduce for the solubility enhancement, such as the morpholine ring, or the recently introduced oxetane fragment [41] (Figure 12.16).

It has been discussed in Chapter 9 that the developability classification system (DCS) introduced by Butler and Dressman [42] differentiates the low solubility–good permeability class (class II) compounds depending on the cause of the low solubility. Compounds that are poorly soluble because of high lipophilicity are usually more developable than compounds that have poor solubility not only because of high lipophilicity but also because of the large crystal lattice energy. The increase in solubility in simulated intestinal fluids occurs for compounds that have poor aqueous solubility because of the high lipophilicity. These compounds are more likely to partition into the phosphatidyl choline micelles that provide a solubility enhancement (see Figure 6.13). The solubility of such compounds can

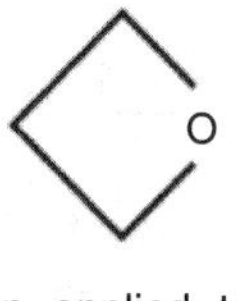

Figure 12.16 The oxetane ring often applied to improve solubility, lipophilicity, and metabolism of drug discovery compounds [41].

be dealt with by formulation scientists more easily than the compounds with low solubility because of high crystal lattice energy.

There are several approaches published for the prediction of solubility from chemical structure [43–47], but, in general, solubility is more difficult to predict than other properties of molecules. When the prediction provided acceptable results in terms of statistics, it would usually be carried out on large sets of compounds covering 6–10 orders of magnitude solubility range. However, in the range of solubility of pharmacological importance, normally between 10 and 1000 μM, the one to two log unit error of prediction is too large to be able to rank compounds and make decisions about their solubility for acceptable absorption. There have been concerns about the quality of the measured solubility data too, which makes model development difficult. As we have discussed in Chapter 6, solubility is condition dependent; even when considering the intrinsic aqueous solubility that refers to the solubility of the neutral form, the measured data showed large discrepancies. It is due to the lack of characterization of the solid form or change in the conditions of the measurements. Many compounds exist as amorphous material or in various types of crystal forms, of which the solubility may differ by several orders of magnitude. In 2008, Llinàs et al. [48] published a paper inviting scientists to a "solubility challenge." They published the intrinsic solubility data of 100 compounds determined by the Sirius "checksol" chasing equilibrium method (see more details in Chapter 6). They stated that the ionic strength, the temperature, and the composition of the buffer and the solubility of all the compounds were measured under identical conditions. They invited the modelers to predict the solubility of 32 molecules for which they withheld the data. Hewitt et al. [49] have published the results of solubility predictions for the challenge, comparing various methods, such as multiple regression analysis using various molecular descriptors, artificial neural network. They compared the root mean square errors of the prediction of several commercially available in *in silico* programs (ChemSilico, Optibrium, Pharma Algorithms, Simulation Plus, etc.). The errors ranged from 0.87 to 1.56 logarithmic molar solubility units. They also suggested dividing the training set of compounds based on the structural similarity and developed local models for the subsets that worked slightly better. They concluded that a simple linear regression approach often was superior to more complex modeling methods. Today, we still rely on chemists' intuitions in designing compounds with better aqueous solubility. The actual measurements and careful characterization of the crystal forms can confirm whether they have reached the required aqueous solubility. Formulation scientists also have a great role to play in increasing the absorption of drugs and new technologies are emerging in this field, which are outside the scope of this book.

STRUCTURE–PERMEABILITY RELATIONSHIPS

The structure–permeability relationships are even more difficult to reveal than the structure–solubility relationships. The permeability of compounds depends on the conditions, such as membrane composition, agitation, and pH of the donor and acceptor compartments. It is important to differentiate between cellular permeability and cell membrane permeability. The most often studied cellular permeability is the permeability of compounds through epithelial cells of the intestinal wall. Permeability together with solubility influences the oral absorption of the compounds through the intestinal wall. There are two major mechanisms of intestinal permeability, the paracellular and transcellular mechanisms. Small hydrophilic molecules prefer the paracellular route, whereas large lipophilic compounds cross the intestinal wall via transcellular mechanism. The log *D*–cMR model for absorption (Figure 9.1) describes this phenomenon in a quantitative way. Large and hydrophilic compounds generally show low oral absorption [50] because of low permeability.

The cell membrane permeability itself plays a significant role for drug molecules that have to get inside the cell where the target enzyme or receptor is. The relationships between the physicochemical (structural) properties and permeability can be expected to hold when the permeability of the compound is governed by passive thermodynamic processes. There is a debate in the literature [51,52] whether the passive permeability exists at all, as more and more transporters are discovered [52] and the active transport processes play a significant role in compounds crossing the biological membranes. The active transport processes play an important role in the permeation of organic acids, amino acids, and sugar molecules. This process helps the permeability of compounds that are essential for the cells, but their physicochemical properties prevent them from passive permeation through the inner hydrophobic core of the membrane. The PgP is an active efflux transporter that protects cells from hydrophobic compounds that are able to cross the cell membrane because of the advantageous physicochemical properties but are potentially harmful to the cells. Here, we only discuss the structural and physicochemical properties of compounds that can be related to passive permeability through a biological membrane bilayer. The general properties that can be related to membrane permeability are the size, lipophilicity, H-bond acidity, H-bond basicity, and charge state. Let us start with the simplest model of the biological membrane bilayer that has been suggested by Wohnsland and Faller [53]. They found that the intrinsic permeability of the compound depends on its partition coefficient between hexadecane and water. It describes the compound's ability to partition into the completely hydrophobic inner layer of the phospholipid bilayer. On the basis of the Abraham solvation equation described in detail in Chapter 5, the alkane/water partition can be described by Equation 12.4 [54].

$$\log P_{\text{alkane–water}} = 0.29 + 0.65E - 1.66S - 3.52A - 4.82B + 4.28V \qquad (12.4)$$

The equation suggests that any polar functionality in the molecule such as *S*, the dipolarity/polarizability; *A*, the H-bond acidity; and *B*, the H-bond basicity decreases the alkane/water partition coefficient and the intrinsic permeability, whereas *E*, the excess molar refraction, and *V*, the size increase them. The solvation equations obtained for the Caco-2 cell permeability and the MDCK cell permeability described by Equations 7.8 and 7.9 also supports the hypothesis.

H-bond donor and acceptor groups in the molecules decrease cellular permeability. The size parameter has an opposing effect, as it increases the alkane/water partition, but it does not have significant effects on the cellular permeability, having a very low coefficient (0.09 and −0.17, respectively). The effect of size on permeability can be explained by the unstirred water layer. This is a water layer close to the membrane where compounds go through via diffusion. The diffusion coefficient depends on the molecular size. While increasing size might help permeability through the hydrophobic core of the membrane, it reduces the diffusion of the compounds in the unstirred water layer, which slows down the permeability. This can be a possible explanation for the insignificance of the size in the cell permeability equations. The effect of the presence of charge on permeability has also been investigated in great detail. On the basis of the pH-partition theory, only neutral molecules can cross the membrane. It has been observed that positively charged compounds, in general, have higher permeability than negatively charged compounds, as they can temporarily dissociate the proton and become uncharged molecules at physiological pH values. Quaternary ammonium salts (which have positive charge not because of an associated proton) are known to be less permeable. Negatively charged and zwitterionic compounds are considered to have much lower permeability [55]. The permeability characteristics of amphoteric drugs have been extensively studied by Tam et al. [54]. They have designed an "in combo" model that predicts the human absolute bioavailability of amphoteric drugs. According to their hypothesis, amphoteric drugs are transported by paracellular routes even when their molecular weight exceeded 350 Da.

A comprehensive study of molecular properties that influence artificial membrane permeability and oral bioavailability has been carried out by Veber et al. [57] based on the data of more than 1100 drug discovery compounds in GSK. The data analysis revealed that the size of the molecules is not influential, whereas reduced flexibility as expressed by the number of rotatable bonds, low polar surface area expressed as the sum of H-bond donor and acceptor groups on the molecules, increased permeability. They suggest compounds with less than 12 rotatable bonds and less than 140 Å polar surface areas are more likely to have good permeability. The $c \log P$ value has been found not to be as significant as the polar surface area, which is supported by the solvation equation theory. Winiwarter et al. [58] carried out similar studies on a diverse set of drug molecules investigating the human *in vivo* intestinal permeability. They used experimentally determined $\Delta \log P$ descriptor [59] for the description of H-bond acidity and other calculated and experimentally derived hydrogen bond strength, lipophilicity, and molecular weight as potentially important properties that may influence the permeability. They found that a combination of H-bond donor descriptors, general H-bonding descriptor, and lipophilicity were sufficient to model the human *in vivo* intestinal permeability of the drugs. We have to remember that sometimes linear correlation with lipophilicity cannot be observed with permeability. As has been shown in the publication of Faller [60], lipophilicity shows a parabolic relationship with the intrinsic permeability. It means that too hydrophilic and too lipophilic compounds have low permeability. Liu et al. [61] recently have reviewed the relationship of lipophilicity with passive drug permeation. They suggest considering the composite property of lipophilicity and highlighting the importance of the separate consideration of hydrophobicity and polarity as the two determinants of lipophilicity.

Gleeson et al. [62,63] have carried out a study on the contribution of substituents to solubility and permeability, based on the measured data on matched molecular pairs. Their results are summarized in Table 12.8, which provides a useful guide to medicinal chemists. It is important to note that the substituent contributions in Table 12.8 to solubility and permeability refer to the physiological pH and includes the possible effect of ionization. Positive numbers show average increase in solubility and permeability by introducing the substituent instead of hydrogen, whereas the negative numbers indicate decrease in solubility and permeability. These results also demonstrate that, in general, solubility and permeability work against each other and it is often not easy to find the optimum compromise between the two properties.

For good bioavailability, we need to have both good solubility and permeability and also have low first-pass elimination. Varma et al. [64] have published a study on the search for the optimal physicochemical property space for good bioavailability. They have analyzed data for 309 known drugs that were characterized by size, lipophilicity, ionization state, polar descriptors, and the number of rotatable bonds. They found that size, the ionization state, lipophilicity, and the number of rotatable bonds significantly affected the bioavailability. The polar descriptors showed parabolic relationships with bioavailability underlining the importance of the optimum polarity for solubility and permeability.

An example of structural change in angiotensin II receptor antagonists and effect on permeability has been published by Ribadeneira et al. [65]. Figure 12.17 shows the general structure of the compounds, and Table 12.9 shows the measured permeability values. The example shown in Figure 12.17 and Table 12.9 clearly demonstrates the effect of the introduction of H-bond donor groups (–OH and –COOH) at position 5: this reduces the cell permeability dramatically. The carboxyl group also introduces negative charge that decreases permeability, even more than the hydroxyl analog.

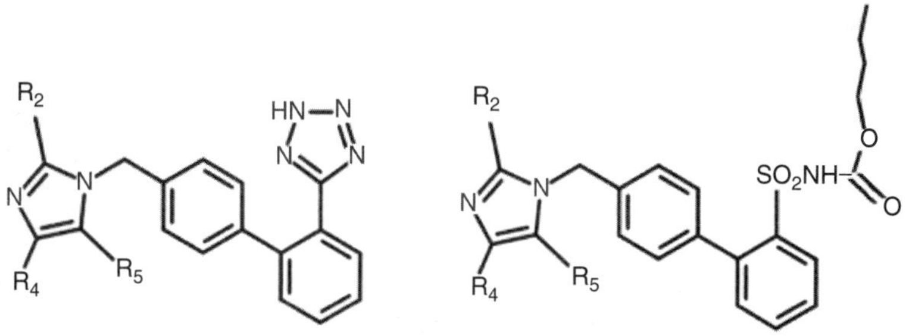

1 2

Figure 12.17 The general structure of losartan analogs (R groups are shown in Table 12.9).

Table 12.8 The average contribution of aliphatic and aromatic substituents to the overall aqueous solubility and permeability of the molecules[a]

Substituent	$\Delta c \log D$	$\Delta c \log P$	ΔMWT	Solubility, mM	Permeability, nm/s
Amine (aliphatic)	−2.4	−1.3	15	0.37	−0.62
Carboxylate	−2.1	−1.3	44	0.57	−0.65
Dimethylamine (aliphatic)	−1.4	−0.4	43	0.11	Not significant
Methylredsulfonamide	−1.3	−1.2	93	0.27	−0.47
Amide	−0.9	−1.2	43	0.15	−0.5
Hydroxy (aliphatic)	−0.9	−1.1	16	0.31	−0.5
Acetamide	−0.8	−0.9	57	0.25	−0.53
Piperidine	−0.8	0.4	83	0.24	−0.71
Morpholine	−0.8	−0.6	85	0.06	−0.5
Methyl alcohol	−0.5	−0.7	30	0.28	−0.45
Methylsulfonyl	−0.4	−0.8	78	0.01	−0.6
Cyano	−0.3	−0.5	25	−0.14	−0.47
Methoxy	0	−0.1	30	−0.03	−0.44
Dimethylether	0.1	0.1	44	0	−0.28
Dimethylamine (aromatic)	0.1	0.3	43	−0.19	−0.48
Fluoro (aromatic)	0.1	0.1	18	−0.1	−0.34
Acetyl	0.4	−0.1	42	−0.05	−0.47
Methyl ester	0.4	0	58	−0.14	−0.39
Methyl	0.5	0.4	14	−0.11	−0.4
Ethoxy	0.5	0.4	44	−0.2	−0.49
Chloro (aromatic)	0.7	0.7	34	−0.35	−0.47
Pyridine	0.7	0.5	77	−0.25	−0.39
Bromo (aromatic)	0.9	0.8	79	−0.53	−0.52
Trifluoromethoxy	1	1	84	−0.72	−0.51
Trifluoromethyl	1.1	0.9	68	−0.54	−0.52
Ethyl	1.2	1.1	28	−0.3	−0.45
Furan	1.2	1	66	−0.65	−0.58
Iso-propyl	1.5	1.3	42		0.23
Propyl	1.6	1.5	42	−0.4	−0.48
tert-Butyl	1.8	1.8	56	−0.47	−0.52
Benzene	1.9	1.8	76	−0.58	−0.59
Thiophene	1.9	1.5	82	−0.6	−0.64
i-Butyl	1.9	1.9	56	−0.3	−0.43
n-Butyl	2	1.9	56	−0.35	−0.51
Benzyl	2.3	2.1	90	−0.49	−0.52

[a]Selected from Reference [63].

The book edited by Fischer and Ganellin, the "Analogue-Based Drug Discovery" [66], shows numerous examples of the effect of substituents not only on the pharmacological behavior of the compounds but also on the absorption and PK change. The book discusses the historical design of drugs from various therapeutic areas by searching for analogs to obtain a better profile. Kerpel-Fronius and Fischer [67]

Table 12.9 The Losartan analogs and their permeability[a]

Compound	Analog	R_2	R_4	R_5	Caco-2 Permeability (cm/s) × 10^6
Losartan	1	nC_4H_9	Cl	CH_2OH	0.35
E-3174	1	nC_4H_9	Cl	COOH	0.069
DMP 581	1	nC_3H_7	C_2H_5	CHO	4.2
DuP 532	1	nC_3H_7	CF_2CF_3	COOH	0.009
DuP 167	1	nC_4H_9	Cl	CHO	9.72
DMP 811	1	nC_3H_7	C_2H_5	COOH	0.021
XM970	2	nC_3H_7	C_2H_5	CH3CO	6.35

[a]Selected from Reference [65].

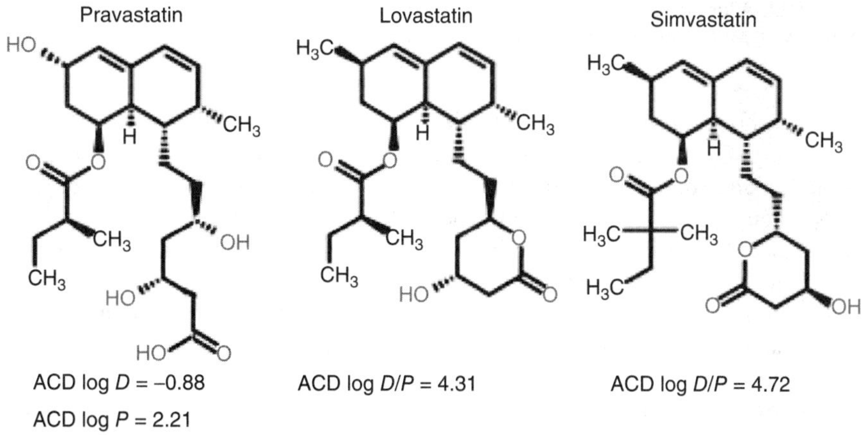

Figure 12.18 The structures of various "statin" drugs that differ in lipophilicity and permeability.

discuss an example from the statin drugs that reduce cholesterol. In this example, they have demonstrated how a prodrug approach helped to overcome the low permeability of pravastatin by forming a lactone ring (lovastatin and simvastin). The lactone ring can open up *in vivo* after successful oral absorption (Figure 12.18).

The higher lipophilicity of lovastatin and simvastin ensures rapid cell uptake in the liver. The lactone ring facilitates the cell penetration by passive diffusion, as it hides the acid functionality. Interestingly, pravastatin should not be cell permeable because of its hydrophilic nature, but it is actively transported into the liver cells by organic acid transporter (OATP) [68,69].

Another interesting study has been carried out on benzamide analog thrombin inhibitors and their intestinal permeability [70]. Figure 12.19 shows the structures of argatroban and inogatran, which were altered by various substitutions changing the lipophilicity (log *P*) of the derivatives from −2.28 to 2.57. The number of H-bond donor and acceptor groups is also altered. Alterations have been carried out

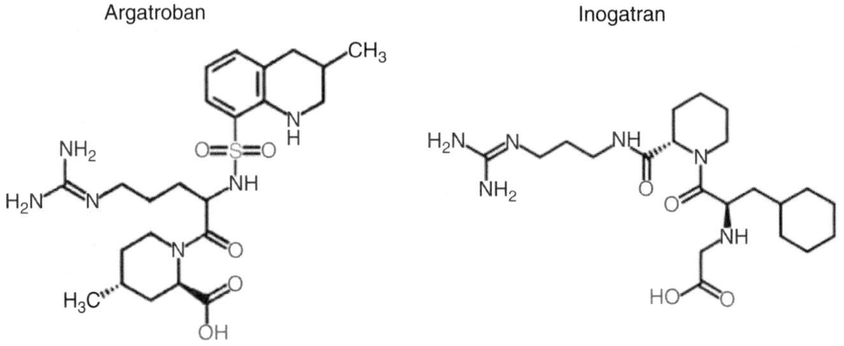

Figure 12.19 The structures of argatroban and inogatran that have been altered for the investigation of the relationships between structure and intestinal permeability by Sugano et al. [68]. Alterations have been carried out on the carboxyl group forming various ester derivatives. More lipophilic and larger derivatives were less permeable.

on the carboxyl group, forming ester derivatives and thus eliminating the presence of the negative charge. They found a negative correlation between intestinal permeability and size. It is surprising that they also found a negative correlation with lipophilicity of the neutral form of the molecules using the calculated Prolog P (a calculated log P by a software from CompuDrug) values as a lipophilicity measure. This is an unexpected result as the compounds are hydrophilic, so the increase in lipophilicity should increase the permeability. However, as the derivatives were large, the increase in lipophilicity was accompanied by a further increase in the size of the molecules, which caused a reduction of permeability.

These examples show that in real drug discovery settings, there is a need to investigate the structure–permeability relationships for the studied analogs and to build a local model using measured data. It is also important to identify the type of permeability we want to optimize and adjust the pH and other conditions accordingly for the permeability measurements. When large discrepancies are observed between the expected and observed permeability behaviors of compounds, it probably means that an active transport process is interfering with the passive permeability.

Permeability through the skin is another interesting topic of research, especially from the environmental toxicology point of view. Skin permeability is also important in drug discovery when the transdermal drug delivery route is considered. The primary role of the skin is to prevent compounds going through, unlike the intestinal wall that is designed for selective absorption of nutrients. It is generally accepted on the basis of the work of Potts and Guy [71] that size and lipophilicity are the main factors that influence skin permeability according to Equation 12.5.

$$\log K_p = 0.71 \log P - 0.0061\ \mathrm{MW} - 6.3 \tag{12.5}$$

where K_p is the skin permeability, log P the logarithm of octanol/water partition coefficient, and MW is the molecular weight. El Tayar et al. [72] carried out a systematic structure–skin permeability relationship study on steroid analogs and found that the H-bond donor properties also influenced the percutaneous penetration of the steroids beside lipophilicity and size. The transdermal delivery of nonsteroid anti-inflammatory drugs has been investigated by Liou et al. [73]. Their study also covered various hydration states of the skin. They found that the solubility of the drugs also influences the transdermal absorption. Because of various systemic side effects of the nonsteroid anti-inflammatory drugs, especially their irritation on the intestinal wall, they are especially convenient for topical applications.

Permeability through the blood–brain barrier is another area of the permeability studies. Sometimes we want to avoid brain penetration to reduce unwanted CNS side effects, whereas drugs that are targeted to the brain need good brain permeability. The blood–brain barrier differs from other biological membranes in its lack of tight junctions. Even small molecules are restricted to go through via paracellular routes. Blood–brain barrier permeability is also heavily influenced by active transporters, especially the PgP that protects the brain from large lipophilic molecules [74]. Table 12.10 shows the descriptors that are generally accepted for brain penetration and their cutoff values. The effect of the ionization state has also been studied, and models to predict brain penetration have been built [75]. The negative effect of H-bond donor groups on CNS penetration has been supported by the large negative coefficient of H-bond acidity in the Abraham solvation equation [74] and the introduction of the $\Delta \log P$ parameter [59] that, in principle, measures the effect of H-bond acidity on organic water/solvent partition.

In conclusion, there are several types of permeability measures that affect the drug distribution properties. Permeability measures the rate by which compounds can go through various biological membrane barriers, such as the intestinal cells, cell membranes, blood–brain barrier, and the skin. In general, the permeability depends on the compound's ability to go through the hydrophobic inner layer of the membranes, and therefore, lipophilicity is a dominant parameter. However,

Table 12.10 Physicochemical properties of compounds and their cutoff values for good CNS penetration

Property	Cutoff for Brain Penetration
Molecular weight	<400 to 500 Da
H-bond donor groups	<3
H-bond acceptors	<7
$c \log P$	<7
$\log D_{pH7.4}$	1–3
Polar surface area	<60 to 70 Å
Rotatable bonds	<8
Flexibility	1.27
pK_a	7.5–10.5
Number of N and O atoms	<6

excessively lipophilic compounds may partition into the membrane but still cannot go through to the other hydrophilic side. The size also plays role, as it controls the diffusion process in the unstirred water layer near the membrane's polar surface. The H-bond donor groups and negative charge slow down the permeability, whereas the positive charge may be temporarily lost during the permeation process when it is acquired via associated proton.

STRUCTURE–CHARGE STATE RELATIONSHIPS

Most drugs contain ionizable groups that may bear charge at physiological pH. The plasma pH is considered to be 7.4, and minor variations in this pH can cause significant pathological effects with serious consequences. However, a wide pH range can be found in the gastrointestinal tract: from pH 2 to pH 9.5. There can also be local pH variations within the cells. For example, the pH around the lysosome (where the acid hydrolyze enzymes work) can be as low as 4.5, whereas the cytosol pH is around 7.2. Manallack [77] has analyzed the distribution of the acid dissociation constants (pK_a) for 582 drugs. On the basis of the World Drug Index, around 63% of drug molecules contain groups that can get charged between pH 2 and pH 12. We have seen that practically all the important physicochemical properties of drug molecules, such as lipophilicity, solubility, and permeability are dependent on the presence of charge and the percentage of charged molecules at a particular pH. The lipophilicity drops by three orders of magnitude because of ionization. The solubility may also be increased a thousand times because of ionization, as a rule of thumb. Not only the physicochemical properties but also other binding properties of the molecules, such as receptor binding affinity, protein binding, and phospholipid binding, are affected by the charge. Medicinal chemists often design compounds by tuning the pK_a values for the optimum receptor binding. We have discussed in Chapter 8 the typical acidic and basic functional groups and also the effect of substituents on aromatic rings on the pK_a based on the Hammett equation.

Table 12.11 shows the distribution of the investigated drug molecules by Manallack [77] according to the number of acidic and basic functionalities. Almost 70% of the CNS subset of the investigated drugs contained at least one basic group. Overall, 50% of the drugs contained a minimum of one basic group, while around 30% contained at least one acidic group.

The range of acidic drug molecules with their pK_a (from 2.8 to 9.7) are shown in Figure 12.20, whereas Figure 12.21 shows a range of basic drugs and their pK_a values (from 2.5 to 10.9).

The acid dissociation constant itself is very rarely used in quantitative structure–property relationship studies: instead, the percentage of the ionized form at physiological pH is used, as it is proportional to other physicochemical and ADME properties of the compounds. Regarding solubility and intestinal permeability, acidic groups are not advantageous. As for the bases, the most advantageous pK_a values for increased solubility and permeability should be just below 7.4. It means that at the low pH in the stomach, the basic compound is in

Table 12.11 The distribution of the drugs, CNS subset, and non-CNS subset containing one or two basic and acidic groups[a]

Set of Drugs	One Acid Group, %	One Basic Group, %	Two Acid Groups, %	Two Basic Groups, %	One Acid and One Basic Group, %	Others, %
Entire set (582)	24.4	45.4	3.8	10.5	11.2	4.8
CNS subset	15.5	62.1	1.1	13.2	7.5	0.6
Non-CNS subset	28.2	38.2	4.9	9.3	12.7	6.6
Word Drug Index [76]	11.6	42.9	3.0	24.6	7.5	10.4

[a]Based on Reference [77].

fully ionized form, which has a positive effect on the dissolution as well. When the pH of the intestine reaches 8 or 9, the compound is in uncharged form, which helps the permeability.

As an example, the pK_a values and percentage of ionized form of a few local anesthetic drugs are shown in Table 12.12. The percentage of ionized form at physiological pH changes the cell penetration rate and the time course of the anesthetic effect. It can be seen in Table 12.12 that small changes in the pK_a values may change the percentage ionized form significantly when the pK_a is close to the physiological pH (see Chapter 8 for the Henderson–Hasselbalch equation).

Charges are often introduced in drug discovery compounds in order to strengthen the drug–receptor interaction. Charge–charge interactions have the highest energy in comparison to hydrogen bonding, dipole–dipole, or simple hydrophobic interactions; thus, they significantly enhance the receptor binding energy. For example, for anticholinergic activity a positive charge is essential. However, most of the positively charged natural product derivatives of atropine can also go to the CNS and cause CNS side effects. The semisynthetic methyl homatropine is a quaternary ammonium compound; thus, it cannot lose its positive charge and cannot go through the blood–brain barrier. By making the positive charge permanent on the molecules, the CNS side effect could be avoided.

The presence of charge has a pivotal effect on the distribution of the compounds between the plasma and tissue compartments and consequently on the half-life and

Table 12.12 The pK_a values of four local anesthetic drugs and the percentage of positively charge molecules at pH 7.4

Compound	pK_a	Percentage of Positively Charged Form at pH 7.4
Procaine	8.9	96.9
Cinchocaine	8.8	96.2
Tetracaine	8.4	90.9
Lidocaine	7.9	76.0

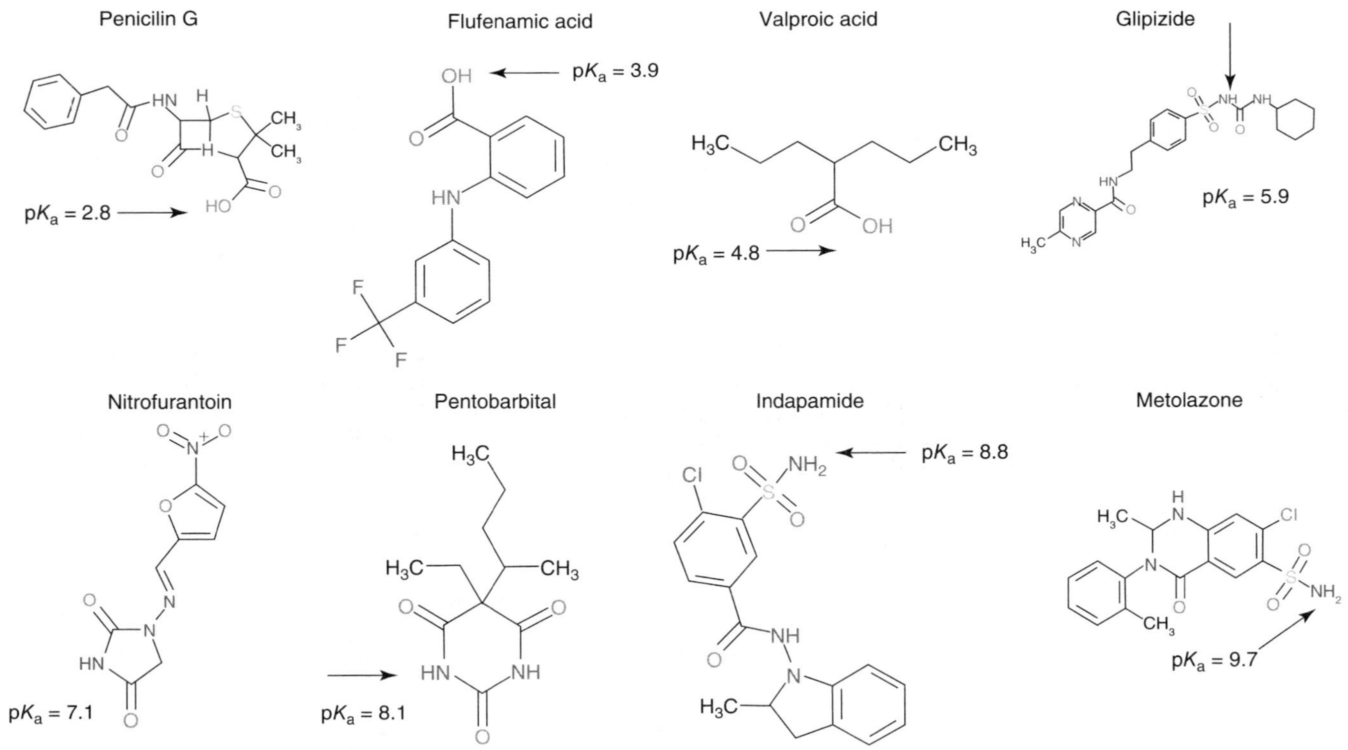

Figure 12.20 Acidic drugs with pK_a values ranging from 2.8 to 9.7. (Adopted from Reference [77].)

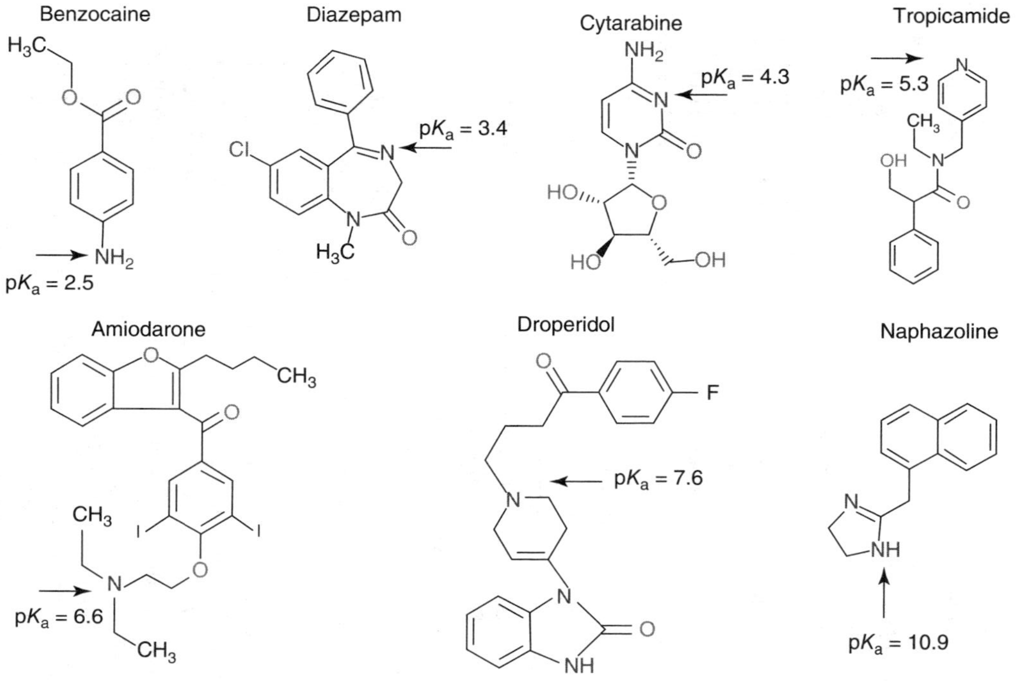

Figure 12.21 Drugs with basic functionality pK_a ranging from 2.5 to 9.6. (Adopted from Reference [77].)

other PK parameters. This is due to the very different protein binding and phospholipid binding characteristics of the positively and negatively charged molecules that are discussed in the following text.

STRUCTURE–PROTEIN BINDING RELATIONSHIPS

Serum albumin is the most abundant plasma protein. One of its main roles is to transport small molecules within the blood stream from various parts of the body. Therefore, it has got promiscuous binding properties, which means that it binds strongly to a wide variety of compounds. There are several binding sites identified, such as the warfarin binding site, the benzodiazepine binding sites, and six fatty acid binding sites [79–82]. The negative charge and nonpolar aromatic substitutions usually increase the albumin binding [83–85]. We have discussed in Chapter 5 that the Abraham equation obtained for the albumin binding for unionized compounds [86] showed strong similarity with the octanol/water partition. It means that H-bond donor groups do not reduce the binding to albumin. The H-bond acceptor descriptor, however, has a negative coefficient, which means that the albumin binding is reduced by H-bond basicity. It is generally accepted that albumin binding increases with increasing lipophilicity and with the presence of negative charge, especially by a carboxyl group. The carboxyl group fits into a cationic pocket of the warfarin binding site. There is another hydrophobic pocket near the positively charged binding area that easily accommodates an aromatic ring. By reducing the hydrophobicity of the substitution that fits into the hydrophobic pocket, the albumin binding can be reduced. One example from a research project was to introduce an ether oxygen in the hydrophobic side chain substitution. A carboxyl group was needed for the potency, but the extremely strong albumin binding reduced the available free concentration of the compound in plasma. Figure 12.22 shows the general structure of approximately 50 derivatives. Basically, by positioning the ether group in the aliphatic side chain, the average albumin binding of the compounds could be reduced from 98% to 92%, regardless of other substitutions in the molecules.

The albumin binding increases with the neutral form lipophilicity of the compounds (log P). The presence of positive charges decreases the albumin. Figure 12.23 shows the relationship of the albumin binding with the distribution coefficient at pH 7.4 (ACD log D) and the lipophilicity of the neutral form (c log P) for acidic, basic and neutral compounds.

The correlation of albumin binding with the lipophilicity of the compounds expressed as octanol/water partition coefficient is not perfect, in spite of the similarities between the two, when comparing the solvation equations. The albumin binding is influenced by the shape of the molecules too and the position of the polar and nonpolar moieties in the molecules. This happens because the binding occurs on the protein surface and does not require full solvation of the molecules by the albumin. In the octanol/water partition system, the molecules are fully solvated in the octanol and the water phase. Actually, we do not know how many molecules are at the interface of the octanol and water because of their amphiphilic nature.

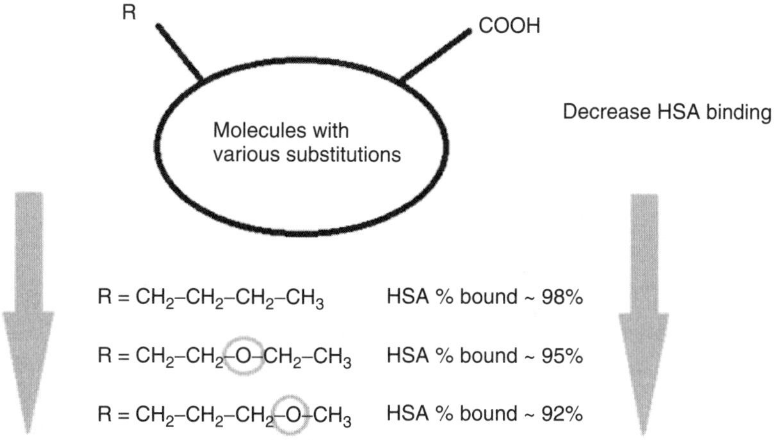

Figure 12.22 The average albumin binding of carboxyl derivatives by changing an alkyl substitution near the carboxyl group. The introduction of a polar ether oxygen, especially toward the end of the alkyl chain, reduced the average HSA binding of the derivatives.

As there are many different binding sites on the albumin surface and the molecular shape also plays an important role in the binding, it is very difficult to build accurate *in silico* calculation models for its estimation. It is also observed that the albumin binding shows good correlation with solubility, specifically the insolubility of the compounds. When a compound is poorly soluble in water, it easily "sticks" to any hydrophobic surfaces, especially to the albumin, which has a large hydrophobic surface area.

The AGP is another important component of the plasma proteins [87]. Its molecular weight ranges from 41 to 46 kDa. The carbohydrate represents 41% of its molecular weight, which consist of five to six highly cialylated *N*-linked glycans. As with other glycoproteins containing the cyalic acid moiety, its isoelectric point is around 2.8–3.8, which means that it is negatively charged at physiological pH. Therefore, its binding properties are very different from albumin. AGP binds the positively charged compounds strongly. The AGP concentration in plasma is much lower than the albumin concentration, 0.75 mg/ml AGP in contrast to 45 mg/ml albumin in human plasma. Israili and Dayton [88] have drawn attention to the importance of the AGP binding of drugs in clinical pharmacology and published the binding characteristics of hundreds of drugs. They also have pointed out that although the AGP concentration is low in plasma, its concentration depends on several factors. Inflammation, age, and gender can all influence AGP levels in plasma, and it can increase four- to fivefold of its normal concentration. Therefore, drug molecules binding strongly to AGP may have different free fraction in plasma depending on these factors, which alter the PK profile of a drug. Drugs that bind strongly to AGP often show variable efficacy in clinic, as the patients have varying AGP levels. Therefore, it is useful to identify early strong AGP binding of discovery compounds, and preferably to avoid it. As we have discussed in Chapter 5,

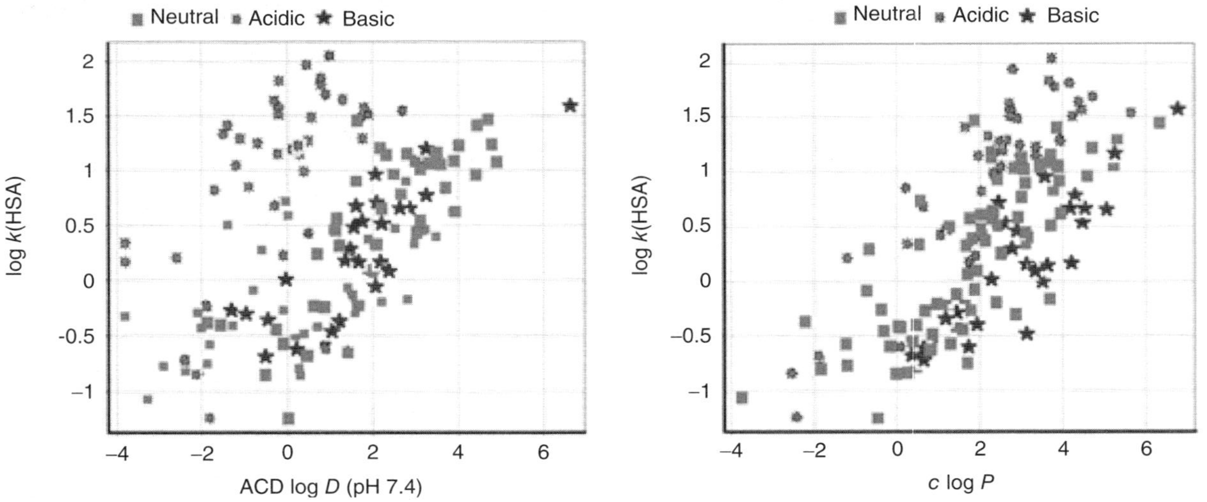

Figure 12.23 The albumin binding expressed as log k(HSA) as a function of the calculated lipophilicity at pH 7.4 (ACD log D) and the lipophilicity of the neutral form (c log P).

the chemically bonded AGP stationary phase can be used to easily measure the compounds' binding to AGP. Kaliszan et al. [89] have analyzed the structure–AGP binding relationships of a diverse set of compounds and hypothesized that the AGP contains several funnel-like binding sites. The negative charge is at the narrow end of the funnel, and compounds that have exposed positive charge that fits to the narrow end of the funnel show especially strong binding as illustrated in Figure 12.24 reproduced from the work of Nasal et al. [90].

The AGP binding of compounds, just like the albumin binding, shows a trend with the lipophilicity of the compounds. Figure 12.25 shows the correlation of AGP binding of 80 known drug molecules listed in Table 10.10, with chromatographically determined CHI log D values. It can be seen that the AGP binding of neutral drugs at physiological pH shows correlation with the lipophilicity, whereas the positively charged compounds show much stronger interaction with AGP than it could be expected from their lipophilicity. Negatively charged compounds have low lipophilicity at pH 7.4, whereas their AGP binding shows a relatively large range. It has been observed that positively charged compounds also show strong binding to IAM; this mimics the phosphatidyl choline binding. Kaliszan et al. [89] have found that the introduction of a charge parameter (the excess atomic charge on aliphatic nitrogen, N_{ch}) and a size parameter (S_t) improves the correlation between AGP and IAM binding, based on the data of 49 known drug molecules. Figure 12.26 shows the plot of AGP binding as a function of IAM binding for 80 known drugs listed in Table 10.10.

This is a very interesting observation, and it may explain one of the roles of elevated AGP binding in inflammation. The explanation may be that it helps to extract positively charged harmful compounds from tissues to the plasma compartment, which helps the body to eliminate them. This idea has been tested by administering phospholipids intravenously in plasma to avoid the lethal consequences of sepsis [91,92]. Figure 12.26 shows that there is a good trend between the IAM

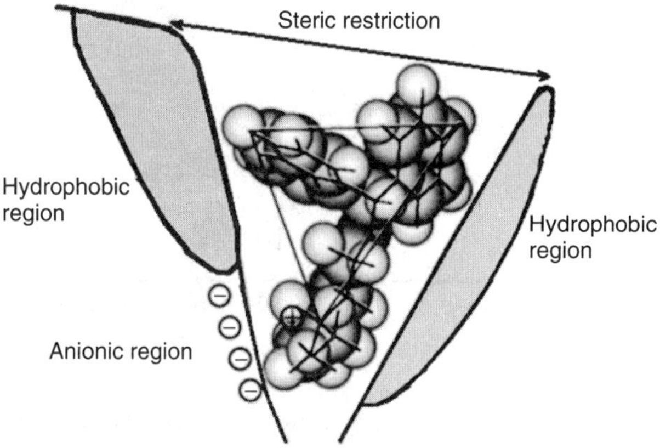

Figure 12.24 The hypothetical binding site of AGP from Kaliszan et al. [89].

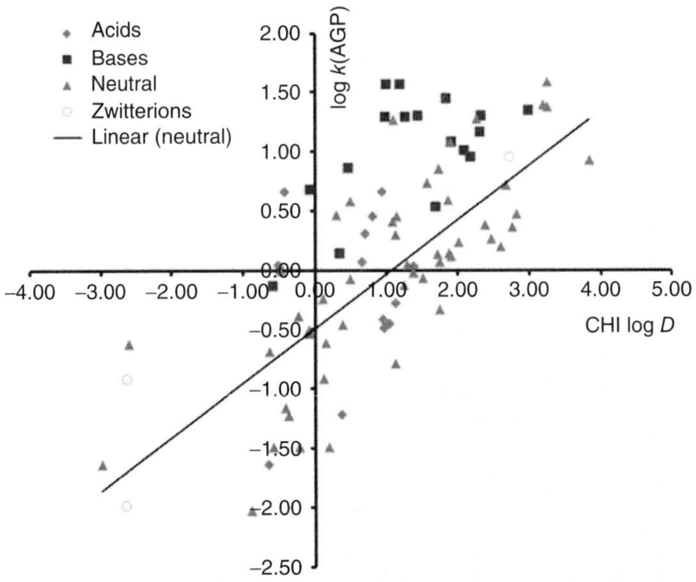

Figure 12.25 A plot of AGP binding as a function of the chromatographically determined lipophilicity (CHI log *D*) at pH 7.4 based on the data of 80 known drugs from Table 10.10. The trend line is based on the data of neutral drugs only.

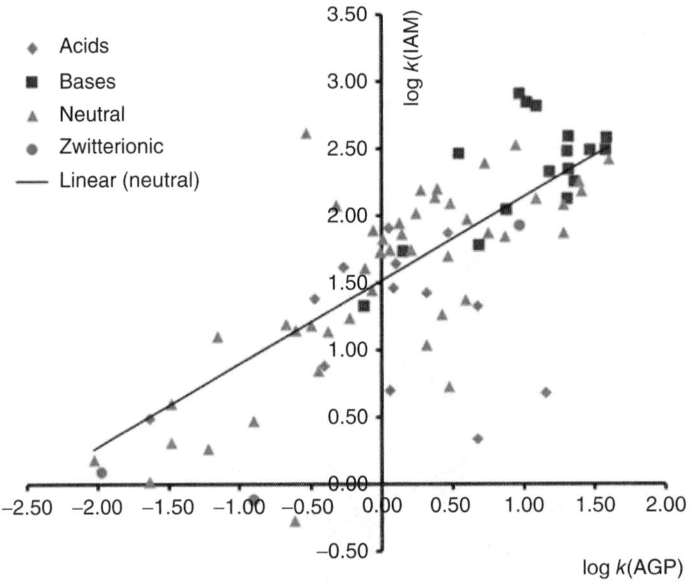

Figure 12.26 The plot of the AGP binding as a function of IAM binding for the 80 known drug molecules listed in Table 10.10.

and AGP binding for neutral and basic molecules, whereas the negatively charged compounds binding is governed mostly by shape and steric factors that are very different and, therefore, no correlation or trend could be observed for acids by Taheri et al. [93].

Holladay et al. [94] reported the effect of elevated AGP binding of saquinavir on the volume of distribution and clearance. We have discussed in Chapter 10 that the volume of distribution is proportional to a compound's partition between the plasma and the tissue compartments; thus, any event that changes the plasma and tissue binding will alter this partition. Compounds that strongly bind to AGP are especially affected by increased AGP concentration in plasma and not only the volume of distribution but also the clearance that depends on the free concentration of the drug in plasma may change significantly. It is also interesting to note that the AGP binding of compounds may show good correlation with the compound binding to other glycoproteins, such as PgP, which is an important active transporter in the gut and in the blood–brain barrier [95]. Similar to lysosomes in the cells, mucus in the lung also contains glycoproteins; compounds that bind strongly to AGP may accumulate in other glycoproteins at higher concentrations. It is shown in Figure 10.21 and Figure 10.22 that the AGP binding data obtained by the HPLC retention time measurements showed good correlation with the mucus binding and lung tissue binding.

The AGP and HSA binding measurements using biomimetic chromatographic retention time measurements can also reveal stereoselective binding of enantiomers.

Actually, the AGP and HSA stationary phases have been developed for chiral separation purposes. Therefore, the biomimetic chromatographic binding measurements to albumin and AGP potentially can be used to predict the stereoselective in vivo distribution of the compounds. An interesting example has been published by Ohmori et al. [94] *showing the different PK behavior of amlodipine in elderly patients. It may well be that the positively charged amlodipine binds strongly and stereospecifically to AGP, which plasma concentration depends on the age. The effect of stereoselective binding of racemates has not been explored yet on the in vivo distribution of drugs. The investigation of stereoselective pharmacokinetics could be the topics of future scientific studies in order to gain better understanding the PK behavior of optically active drugs.*

STRUCTURE–PHOSPHOLIPID BINDING RELATIONSHIPS

Phospholipid binding can be measured by biomimetic HPLC retention times using IAM stationary phases. There is a significant difference between the phospholipid binding and the permeability of compounds through the phospholipid bilayer. Permeability measures rate of transfer, whereas the binding measures the compound's partitioning into phospholipids. For good membrane permeability, a good partitioning into the membrane is essential. Phospholipid binding is also dominated by the lipophilicity of the compounds. Figure 12.27 shows the plot of the IAM binding

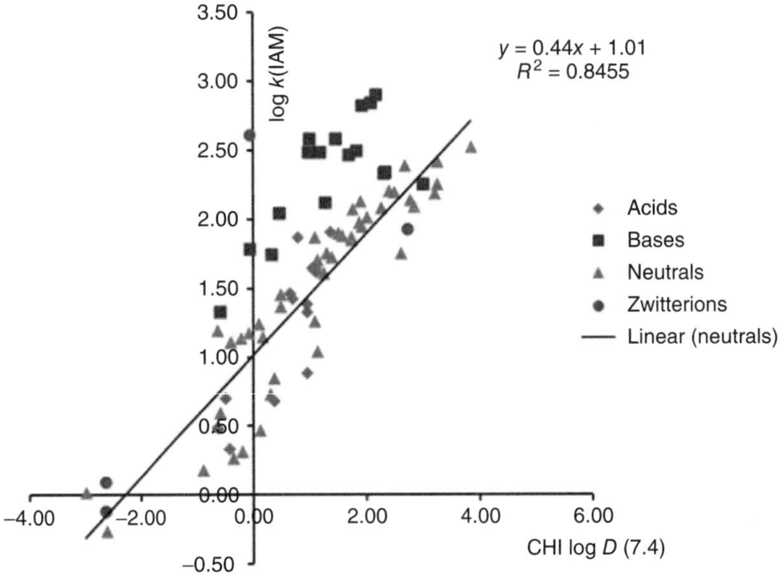

Figure 12.27 The plot of measured phospholipid binding (log *k*(IAM)) as a function of chromatographic log *D* (CHI log *D*) for 80 drug molecules from Table 10.10. Both measurements have been carried out using pH 7.4 mobile phases. The regression line and the correlation equation refer only to the neutral compounds.

(log *k*(IAM)) as a function of the measured chromatographic log *D* (CHI log *D*) obtained on C-18 stationary phase, which is completely nonpolar. It can be seen that while both positive and negative charges reduce the CHI log *D* values, the IAM binding remains strong for positively charged compounds.

The plot in Figure 12.27 shows that there is a correlation between the IAM binding and the C-18 lipophilicity for diverse neutral compounds ($r^2 = 0.84$). Positively charged compounds show stronger binding to immobilized artificial membrane (mimicking phosphatidyl choline), which is due to the negatively charged phosphate groups on the IAM surface. Barbato et al. [97] also observed that protonated β-blockers showed stronger interactions with IAM stationary phase than isolipophilic neutral compounds, in spite of being more than 90% protonated. They have hypothesized that electrostatic interactions take place between the positive charge of the compounds and the negatively charged phospholipid surface of the stationary phase. Similar observations have been published by others as well using more diverse sets of drug molecules [98,99]. Strong binding of positively charged compounds has been observed not only with IAM stationary phases but also using immobilized liposomes prepared from egg phospholipids. On the other hand, other differences have been observed between the IAM partition and the C-18 retention for neutral compounds using the solvation equation model [100–102]. Comparing the Abraham equations obtained on C-18 stationary phase and on IAM, the coefficient of the H-bond acidity parameter is markedly different. H-bond acidity seems to have insignificant effects on IAM partition: the coefficient is close to zero, similar to the octanol/water partition system. The explanation can be that the H-bond basicity of octanol and the IAM stationary phase is close to the H-bond basicity of water. Therefore, H-bond donor groups equally interact with water, octanol, and IAM. The C-18 stationary phase is completely hydrophobic; there are no H-bond acceptor groups present that would attract H-bond donor groups the same way as water molecules. In this respect, the IAM retention, similar to albumin binding, can be best modeled by the octanol/water partition system. However, the presence of charge on drug molecules at physiological pH causes an opposite effect on IAM and albumin. The positive charge increases the phospholipid binding and reduces the albumin binding, as demonstrated in Figure 10.8. These basic differences are the major driving forces for compounds' partitioning between the plasma and the tissue compartments and contribute to the volume of distribution of the drugs *in vivo*. The biomimetic chromatographic measurements on albumin and IAM stationary phases, therefore, can model the distribution of drugs between the plasma and the tissue compartments, and they also explain the dramatic effect of charge. Strong interactions of the positive charge with phosphatidyl choline are also responsible for the solubility enhancement of lipophilic and positively charged compounds in simulated intestinal fluid that contains phosphatidyl choline and taurocholate micelles.

We have observed significant effects of the shape of the molecules on the IAM retention by comparing the retention of structural isomers. Figure 12.28 shows parts of proprietary molecular isomers with the measured log *k*(IAM) values, calculated lipophilicity (*c* log *P*), and the calculated size parameter (cMR). The measured chromatographic lipophilicity values are CHI log $P = 4.2$ for isomer A and CHI log $P = 4.4$ for isomer B. The difference between the lipophilicity of ionized and unionized forms shows that the compounds are ionized at pH 7.4. In spite of the increased lipophilicity of isomer B, isomer A showed much stronger partition into

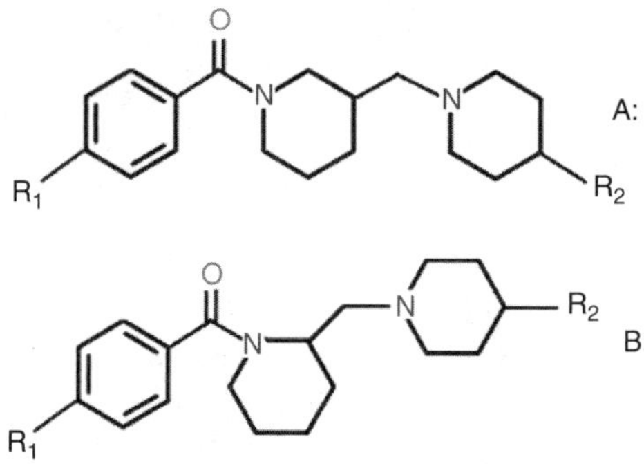

	log *k*(IAM)	*c* log *P*	cMR
A:	3.67	3.8	13.9
B:	3.27	4.6	13.9

Figure 12.28 Measured and calculated log *k*(IAM), *c* log *P*, and cMR values of some structural isomers.

IAM. It is possible that in this case, the structural change affected the pK_a values of the basic nitrogen as well: this contributed to the opposite change in calculated and measured lipophilicity and the IAM retention. The R_1 and R_2 substitutions contained further ring systems, which further enhance the difference in shape between the two structural isomers. While isomer A has an elongated shape, isomer B is more round in shape. The lower IAM retention of isomer B can be explained by the increase in width of the molecule in comparison to isomer A, which made it more difficult to fit between the organized alkyl chains of the immobilized phosphatidyl choline molecules.

Another example of lipophilicity, albumin binding, and phospholipid binding differences between structural isomers is shown in Figure 12.29. The calculated lipophilicity, size, and polar surface area (PSA) are exactly the same for the two molecules shown in Figure 12.29, whereas the measured biomimetic properties show significant differences. The solubility dropped for the more lipophilic compound, albumin binding (%HSA), AGP binding (%AGP), and phospholipid binding (log *k*(IAM)) increased significantly.

These and numerous other examples show the advantages of measured biomimetic properties of drug discovery compounds that could not be predicted using the traditional physicochemical property calculators. This is because *in silico* lipophilicity calculators use only two-dimensional fragment descriptors, which do not include three-dimensional features of the molecules. These unpredictable changes may cause significant differences in the *in vivo* distribution and PK

Compound A

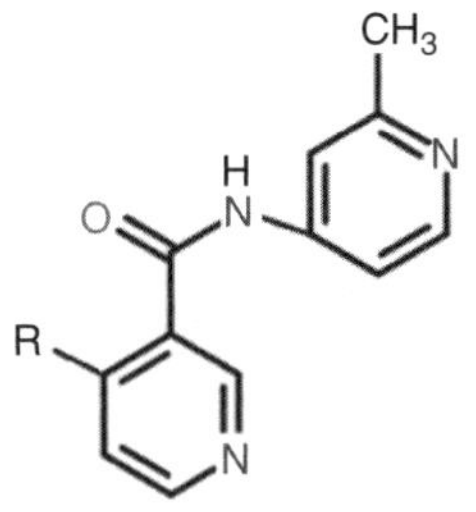

c log *P*	2.19
cMR	9.9
PSA	73.22
CHI log *D*	1.64
Solubility	44 μm
% HSA bound	85.5
% AGP bound	70.9
log *k*(IAM)	2.03

Compound B

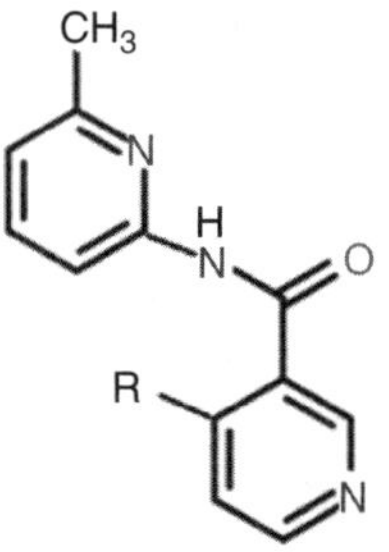

c log *P*	2.19
cMR	9.9
PSA	73.22
CHI log *D*	2.65
Solubility	24 μm
% HSA bound	93.8
% AGP bound	85.7
log *k*(IAM)	2.15

Figure 12.29 Differences between measured biomimetic properties and calculated parameters between positional isomers.

properties of compounds. It has been discussed in Chapter 10 that strong phospholipid binding (IAM retention) increases the compounds' partition into the tissue compartments, which is manifested in high volume of distribution and reduces free concentration of the drug.

Cassartelli et al. [103] have reported that the IAM retention showed strong correlation with the phospholipidogenic potential of the compounds studied. They detected the increased phospholipid accumulation caused by high concentration of drugs by Nile red fluorescence dye. The phospholipid binding of the compounds were measured by the CHI obtained on IAM stationary phase, CHI (IAM). Good correlation has been found between the Nile red fluorescence index and the CHI (IAM) values as shown in Figure 12.30. This finding supports the hypothesis that strong phospholipid binding of drugs may be related to their ability to cause phospholipidosis.

In conclusion, lipophilicity increases protein binding, phospholipid binding, and permeability of compounds. Excessive lipophilicity reduces solubility and permeability. The presence of a positive charge increases the phospholipid binding, tissue binding, volume of distribution, AGP binding, and, in some cases, permeability; on the other hand, the negative charge increases albumin binding and reduces volume of distribution, phospholipid binding, tissue binding, and permeability.

In principle, a few gradient chromatographic retention time measurements can provide valuable information about the physicochemical properties of drug discovery compounds. Measurements with starting aqueous mobile phases at three different pHs provide a measure of lipophilicity at physiological pH, the acid/base character of the compound with the indication of the proportion of ionized form at pH 7.4 and the lipophilicity of the unionized form of the molecules. Calibrated retention time measurements using IAM stationary phase measures a compound's

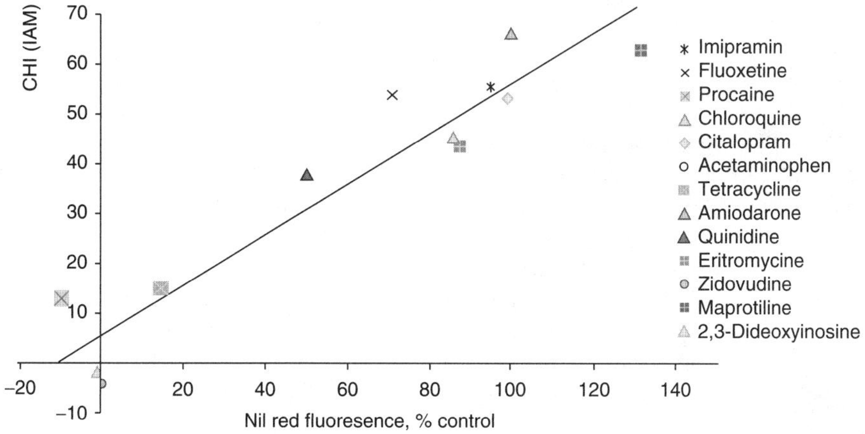

Figure 12.30 The plot of the phospholipid binding, CHI (IAM) in the function of the Nil Red fluorescence index obtained after treating of hepatocyte cells by various known drugs at high concentrations. (With permission from Reference [103].)

partitioning into phospholipids. Further, two gradient retention time measurements using chemically bonded HSA and AGP stationary phase indicate the compound's binding to these two types of proteins. The typical chromatographic conditions and calibration set of compounds for the biomimetic binding measurements used are described in Chapter 5.

The obtained standardized chromatographic retention data can be introduced in various models to estimate *in vivo* distribution properties of drug discovery compounds and accelerate the compound selection and prioritization processes in drug discovery. The measured data can reveal changes unpredictable by traditional *in silico* calculations. These measurements can be fully automated and takes only 4–10 min per compound.

Table 12.13 shows the summary of the models that require measured biomimetic properties as input. It is hoped that based on the methodology described in this book, it is possible to reproduce these measurements in a standard analytical HPLC laboratory. The calculations of the binding data and the estimated *in vivo* distribution data can be carried out in Excel. These models can provide information about the expected *in vivo* steady-state dynamic equilibrium distribution of drug discovery compounds. However, the models cannot provide information about the kinetics of binding: the time needed for establishing the dynamic equilibrium. The *in vivo* steady-state equilibrium processes may be disturbed by permeability barriers and active transport processes that cannot be predicted by the proposed measurements. The measurements of metabolism and clearance are needed for the full description of *in vivo* distribution properties of the compounds. The described biomimetic properties reveal a compound's interactions only with the major constituents of plasma and tissues. Other types of proteins and lipids may specifically and strongly bind the compounds. This may alter the estimated distribution data. Incorporating the proposed methods and models can significantly enhance the success of the drug discovery process.

SUMMARY

Structure–property relationships and some general rules have been discussed in this chapter. Lipophilicity depends on the number of polar and nonpolar fragments in the molecules. The contribution of substituent constants to the overall lipophilicity of the molecules has been shown. Polar heteroatoms and H-bond donor and acceptor groups generally reduce the lipophilicity of compounds. Alkyl substitutions and aromatic and nonaromatic hydrocarbon rings increase lipophilicity. One methyl (or $-CH_2-$) group addition to a molecule increases the octanol/water partition coefficient by approximately half of a log unit. A hydroxyl group reduces lipophilicity over a half log unit. In general, increasing size increases lipophilicity,

Table 12.13 The properties of drugs that can be estimated using biomimetic chromatographic retention data

Property	Biomimetic Data	Model	Comment
Absorption (permeability)	Reversed phase retention (CHI log *D*)	Plot of CHI log *D* and cMR	cMR is the calculated molar refraction for the measure of size
Solubility	Reversed phase retention	Inversely correlates with lipophilicity, including the number of aromatic rings (solubility forecast index)	The model is not able to predict the insolubility because of the crystal lattice energy
Solubility enhancement in biorelevant media	IAM retention	CHI IAM is higher than 40 solubility enhancement because of partitioning into micelles	The solubility enhancement showed good correlation with CHI IAM values
Volume of distribution (log V_{ss})	IAM and HSA binding	0.44 log *K*(IAM) – 0.22 log *K*(HSA) – 0.66	Modeling the difference between tissue binding and plasma binding
Brain tissue binding	IAM and HSA binding	1.29 log *k*(IAM) + 1.03 log *k*(HSA) – 2.37	Note that the logarithmic retention factor is not scaled here to log *P* values
Overall tissue binding	IAM and HSA binding	−0.66 log *K*(HSA) – 0.52 log *K*(IAM) + 0.55	IAM and HSA binding data are scaled to log *P*
Lung tissue binding	HSA and AGP binding	0.74 log *k*(HSA) + 0.65 log *k*(AGP)	Note that the logarithmic retention factor is not scaled here to log *P* values
Mucus binding	AGP binding		

Table 12.13 (*Continued*)

Property	Biomimetic Data	Model	Comment
Plasma protein binding (log *k*(PPB)) = log(%bound/(101 − %bound))	HSA binding, AGP binding	0.87 log *k*(HSA) + 0.17 × log *k*(AGP) + 0.06 cMR–0.27	cMR stands for molar refractivity as size parameter
Unbound volume of distribution (log Vd_u)	IAM and HSA binding	0.23 log K(HSA) + 0.43 × log *K*(IAM)–0.72	IAM and HSA binding data are scaled to log *P*
Drug efficiency max 100 × free plasma concentration/dose	IAM and HSA binding	$100/Vd_u$ = 10^logVd_u	Log Vd_u should be estimated using the unbound distribution model

whereas polarizability/dipolarity reduces lipophilicity together with H-bond donor and acceptor groups. The presence of charge (either positive or negative) decreases lipophilicity by approximately three orders of magnitude. There are several "flavors" of lipophilicity; hence, partition coefficients measured in different solvent systems may not show good correlation. For example, cyclohexane/water partition differs significantly from the octanol/water partition for H-bond donor compounds. The difference between these two partition coefficients ($\Delta \log P$) has been shown to correlate with the blood/brain distribution. The chromatographic lipophilicity measured on C-18 stationary phases is more similar to the alkane/water partition. Lipophilicity inversely correlates with solubility and shows a parabolic optimum relationship with permeability. Solubility is more difficult to predict from structure alone. It depends strongly on the interactions between the molecules in the solid form. Different crystal forms of the same compound may show three orders of magnitude differences in solubility. Taking into account the lipophilicity and the number of aromatic rings that may contribute to the crystal stacking provides the SFI that estimates the probability of low, medium, and high solubility. Compounds that have both H-bond donor and acceptor groups are more likely to form strong H-bonding in the solid form, increasing the crystal lattice energy and decreasing solubility. This effect has been included in the Abraham solvation equation as the product term of H-bond acidity and basicity. The presence of charge increases the solubility by several orders of magnitude.

The permeability of molecules depends on their lipophilicity. Very hydrophilic molecules cannot go through the nonpolar inner core of the lipid bilayer and have poor permeability. In this respect, it is the opposite of aqueous solubility, which requires low lipophilicity. This opposite effect of lipophilicity on solubility and permeability is a major problem for medicinal chemists in optimizing oral absorption.

However, when a molecule is very lipophilic, both solubility and permeability are compromised. Very lipophilic molecules easily partition into the hydrophobic inner part of the membrane, but they cannot go through to the other hydrophilic side. In general, the presence of charge reduces permeability. Positively charged molecules may permeate through the membrane if they are able to lose the charge (proton) temporarily. The permeability of negatively charged compounds is very limited.

Albumin binding is controlled by the lipophilicity of compounds. Aromatic rings and the presence of negative charge significantly increase the albumin binding of the molecules. AGP binding has a significantly different selectivity from the albumin binding. Positively charged molecules bind to the funnel-shaped pockets of AGP, especially when the positive charge can reach the narrow end of the funnel. Lipophilicity plays a major role of AGP binding too.

Phospholipid binding, which is often characterized by the IAM retention of the molecules, correlates with the lipophilicity and is shown to be similar to the octanol/water lipophilicity for neutral molecules. However, positively charged molecules bind very strongly to phosphatidyl choline and IAM. Both protein binding and phospholipid binding are influenced by the shape of the molecules; therefore, it is very difficult to build *in silico* calculation models for their estimation. The measurement of biomimetic binding is therefore providing an additional accuracy for the estimation of compounds *in vivo* distribution, protein binding, and tissue partition.

QUESTIONS FOR REVIEW

Q12.1 List a few substituents that increase lipophilicity and a few that decrease lipophilicity.

Q12.2 Which ADME properties are negatively affected by low lipophilicity ($\log P < 1$)?

Q12.3 Which ADME properties are negatively affected by high lipophilicity ($\log P > 4$)?

Q12.4 How does lipophilicity affect protein and phospholipid binding?

Q12.5 How does lipophilicity affect brain penetration?

Q12.6 How can solubility be related to lipophilicity?

Q12.7 Which solute properties may increase crystal lattice and reduce solubility?

Q12.8 What is the SFI?

Q12.9 Which substituents in general improve the solubility of a compound?

Q12.10 What structural and physicochemical features of the molecules help reduce protein and phospholipid binding?

REFERENCES

1. Hansch, C., Fujita, T. (1964) ρ-σ-π Analysis – A method for correlation of biological activity and chemical structure. *Journal of American Chemical Society*, *86*, 1616–1624.

2. Leo, A. J. (1972) Biological correlations – The Hansch approach. Advances in Chemistry Series, *114*, Ed. Gould R. F., American Chemical Society, Washington D.C., pp. 51–128.
3. Lipinski, C. A., Lombardo, F., Dominy, B. W., Feeney, P. J. (1997) Experimental and computational approaches to estimate solubility and permeability in drug discovery and development settings. *Advanced Drug Delivery Reviews*, *23*, 3–25.
4. Bade, R., Chan, H-F., Reynisson, J. (2010) Characteristics of known drug space. Natural products, their derivatives and synthetic drugs. *European Journal of Medicinal Chemistry*, *45*, 5646–5652.
5. Wenlock, M. C., Austin, R. P., Barton, P., Davis, A. M., Leeson, P. D. (2003) A comparison of physicochemical property profiles of development and marketed oral drugs. *Journal of Medicinal Chemistry*, *46*, 1250–1256.
6. Gleeson, M. P., Hersey, A., Montanari, D., Overington, J. (2011) Probing the links between *in vitro* potency, ADMET and physicochemical parameters. *Nature Reviews Drug discovery*, *10*, 197–208.
7. Vieth, M., Sutherland, J. J. (2006) Dependence of molecular properties on proteomic family for marketed oral drugs. *Journal of Medicinal Chemistry*, *49*, 3451–3453.
8. Waring, M. J. (2010) Lipophilicity in drug discovery. *Expert Opinion on Drug Discovery*, *5(3)*, 235–248.
9. Yalkowsky, S. H., Valvani, S. C. (1980) Solubility and partitioning I: Solubility of non-electrolytes in water. *Journal of Pharmaceutical Sciences*, *69*, 912–922.
10. Johnson, T. W., Dress, K. R., Edwards, M. (2009) Using the Golden Triangle to optimize clearance and oral absorption. *Bioorganic and Medicinal Chemistry Letters*, *19*, 5560–5564.
11. Gleeson, M. P. (2008) Generation of a set of simple, interpretable ADMET rules of thumb. *Journal of Medicinal Chemistry*, *51*, 817–834.
12. Leeson P. D., Springthorpe, B. (2007) The influence of drug-like concepts on decision-making in medicinal chemistry. *Nature Reviews Drug Discovery*, *6*, 881–890.
13. Fujita, T., Iwasa, J., Hansch, C. (1964) A new substituent constant, π derived from partition coefficients. *Journal of American Chemical Society*, *86*, 5175–5180.
14. Hansch, C., Leo, A. (1979) Substituent Constants for Correlation Analysis in Chemistry and Biology. John Wiley & Sons, Inc., New York.
15. Rekker, R. F. (1977) The Hydrophobic Fragmental Constant. Its Derivation and Application. Elsevier, Amsterdam.
16. Mannhold, R. (2007) Calculation of lipophilicity: A classification of methods. In: Pharmacokinetic Profiling in Drug Research: Biological, Physicochemical, and Computational Strategies, Eds. Testa, B., Krämer, S. D., Wunderli-Allenspach, H., Folkers, G., Wiley-VCH Verlag GmbH, Weinheim.
17. van de Waterbeemed, H., Smith, D. A., Jones, B. C. (2001) Lipophilicity in PK design: methyl, ethyl, futile. *Journal of Computer-Aided Molecular Design*, *15*, 273–286.
18. Hashimoto, K., Mitsihashi, H. (1986) Effects of OPC-8212, a new positive inotropic agent on canine ventricular arrhythmias. *British Journal of Pharmacy*, *88*, 915–921.
19. Pattan, S. R., Ahirrao, G. B., Pawar, S. S., Pawar, S. B., Godge, R. K., Kapse, G. K., Bhawar, H. S., Gharate, U., Kavade, S. (2011) A review on quantitative structure activity relationships. *Pharmacologyonline*, *1*, 786–806.

20. Caron, G., Gaillards, P., Carrupt, P.-A., Testa, B. (1997) Lipophilicity behavior of model and medicinal compounds containing a sulphide, sulfoxide or sulfone moiety. *Helvetica Chimica Acta*, *80*, 449–462.
21. Abraham, M. H. (1993) Scales of hydrogen bonding: Their construction and application to physicochemical and biochemical processes. *Chemical Society Reviews*, *22*, 73–83.
22. Hansch, C., Clayton, J. M. (1973) Lipophilic character and biological activity of drugs: The parabolic case. *Journal of Pharmaceutical Sciences*, *62(1)*, 1–21.
23. Testa, B., Crivori, P., Reist, M., Carrupt, P.-A. (2000) The influence of lipophilicity on the pharmacokinetic behavior of drugs: Concepts and examples. *Perspectives in Drug Discovery and Design*, *19*, 179–211.
24. Summerfield, S. G., Jeffrey, P. (2006) *In vitro* prediction of brain penetration – A case for free thinking? *Expert Opinion on Drug Discovery*, *1(6)*, 595–607.
25. Yalkowsky, S. H., Valvani, S. C. (1980) Solubility and partitioning I: Solubility of non-electrolytes in water. *Journal of Pharmaceutical Sciences*, *69*, 912–922.
26. Ran, Y. Q., Yalkowsky, S. H. (2001) Prediction of drug solubility by the general solubility equation (GSE). *Journal of Chemical Information and Computer Sciences*, *41*, 354–357.
27. Ran, Y. Q., Jain, N., Yalkowsky, S. H. (2001) Prediction of aqueous solubility of organic compounds by the general solubility equation (GSE). *Journal of Chemical Information and Computer Sciences*, *41*, 1208–1217.
28. Dannenfelser, R. M., Yalkovsky, S. H. (1999) Predicting the total entropy of melting: application to pharmaceuticals and environmentally relevant compounds. *Journal of Pharmaceutical Sciences*, *88*, 722–724.
29. Chickos, J. S., Nichols, G., Ruelle, P. (2002) The estimation of melting points and fusion enthalpies using experimental solubilities, estimated total phase change entropies, and mobile order and disorder theory. *Journal of Chemical Information and Computer Sciences*, *42*, 368–374.
30. Wassvik, C. M., Holmén, A. G., Bergström, A. A., Zamora, I., Artursson, P. (2006) Contribution of solid-state properties to the aqueous solubility of drugs. *European Journal of Pharmaceutical Sciences*, *29*, 294–305.
31. Perlovich, G. L., Kurkov, S. V., Hansen, L. K. R., Bauer-Brandl, A. (2004) Thermodynamics of sublimation, crystal lattice energies, and crystal structures of racemates and enantiomers: (+)- and (±)-ibuprofen. *Journal of Pharmaceutical Sciences*, *93*, 654–666.
32. Perlovich, G. L., Kurkov, S. V., Kinchin, A. N., Bauer-Brandl, A. (2004) Thermodynamics of solutions: III: Comparison of the solvation of (+)-naproxen with other NSAIDs. *European Journal of Pharmaceutics and Biopharmaceutics*, *57*, 411–420.
33. Abraham, M. H., Le, J. (1999) The correlation and prediction of the solubility of compounds in water using an amended solvation energy relationship. *Journal of Pharmaceutical Sciences*, *88*, 868–880.
34. Schultes, S., de Graaf, C., Berger, H., Mayer, M., Steffen, A., Kaaksma, E. E. J., de Esch, I. J. P., Leurs, R., Krämer, O. (2012) A medicinal chemistry perspective on melting point: matched molecular pairs analysis of the effects of simple descriptors on the melting point of drug-like compounds. *Medicinal Chemistry Communications*, *3*, 584–591.

35. Hussain, J., Rea, C. (2010) Computationally efficient algorithm to identify matched molecular pairs (MMPs) in large data sets. *Journal of Chemical Information and Modeling*, *50*, 339–348.

36. Zhang, L., Zhu, H., Mathiowetz, A., Gao, H. (2011) Deep understanding of structure–solubility relationship for a diverse set of organic compounds using matched molecular pairs. *Bioorganic and Medicinal Chemistry*, *19*, 5763–5770.

37. Lovering, F., Bikker, J., Humblet, C. (2009) Escape from flatland: Increasing saturation as an approach to improving clinical success. *Journal of Medicinal Chemistry*, *52*, 6752–6756.

38. Ritchie, T. J., MacDonald, S. J. F. (2009) The impact of aromatic ring count on compound developability – Are too many aromatic rings a liability in drug design? *Drug Discovery Today*, *14*, 1011–1020.

39. Ritchie, T. J., Macdonald, S. J. F. (2011) The impact of aromatic ring count on compound developability: further insights by examining carbo- and hetero-aromatic and aliphatic ring types. *Drug Discovery Today*, *16*, 164–171.

40. Hill, A. P., Young, R. J. (2010) Getting physical in drug discovery: a contemporary perspective on solubility and hydrophobicity. *Drug Discovery Today*, *15*, 648–655.

41. Wuitschik, G., Carreira, E. M., Wagner, B., Fischer, H., Parrilla, I., Schuler, F., Rogers-Evans, M., Müller, K. (2010) Oxetanes in drug discovery: Structural and synthetic insights. *Journal of Medicinal Chemistry*, *53*, 3227–3246.

42. Butler, J., Dressman, J. (2010) The developability classification system: Application of biopharmaceutics concepts to formulation development. *Journal of Pharmaceutical Sciences*, *99*, 4940–4954.

43. Delaney, J. S. (2005) Predicting aqueous solubility from structure. *Drug Discovery Today*, *10*, 289–295.

44. Dearden, J. C. (2006) *In silico* predictions of aqueous solubility. *Expert Opinion on Drug Discovery*, *1*, 31–52.

45. Wang, J., Krudy, G., Hou, T., Zhang, W., Holland, G., Xu, X. (2007) Development of reliable aqueous solubility models and their application in druglike analysis. *Journal of Chemical Information and Modeling*, *47*, 1395–1404.

46. Huuskonen, J., Livingstone, D. J., Manallack, D. T. (2008) Prediction of drug solubility from molecular structure using a drug-like training set. *SAR and QSAR in Environmental Research*, *19*, 191–212.

47. Hughes, L. D., Palmer, D. S., Nigsch, F., Mitchell, J. B. O. (2008) Why are some properties more difficult to predict than others? A study of QSPR models of solubility, melting point and log P. *Journal of Chemical Information and Modeling*, *48*, 220–232.

48. Llinàs, A., Glen, R. C., Goodman, J. M. (2008) Solubility challenge: Can you predict solubilities of 32 molecules using a database of 100 reliable measurements? *Journal of Chemical Information and Modeling*, *48*, 1289–1203.

49. Hewitt, M., Cronin, M. T. D., Enoch, S. J., Madden, J. C., Roberts, D. W., Dearden, J. C. (2009) *In silico* prediction of aqueous solubility: The solubility challenge. *Journal of Chemical Information and Modeling*, *49*, 2572–2587.

50. Hill, A. P., Hyde, R. M., Robertson, A. D., Wollard, P. M., Glen, R., Martin, G. R. (1994) Oral delivery of 5-HT1D receptor agonists: towards the discovery of 311c90, a novel anti-migraine agent. *Headache*, *34*, 308–309.

51. Kell, D. B., Dobson, P. D., Oliver, S. G. (2011) Pharmaceutical drug transport: The issues and implications that it is essentially carrier-mediated only. *Drug Discovery Today*, *17*, 704–714.
52. Sugano, K., Kansy, M., Artursson, P., Avdeef, A., Bendels, S., Di, L., Ecker, G. F., Faller, B., Fischer, H., Gerebtzoff, G., Lennernaes, H., Senner, F. (2010) Coexistence of passive and carrier-mediated processes in drug transport. *Nature Reviews on Drug Discovery*, *(2009)* 597–614.
53. Wohnsland, F., Faller, B. (2001) High throughput permeability pH profile and high throughput alkane/water log P with artificial membranes. *Journal of Medicinal Chemistry*, *44*, 923–930.
54. Abraham, M. H., Cadha, H. S., Leiato, R. A. E., Mitchell, R. C., Lambert, W. J., Kaliszan, R., Nasal, A., Haber, P. (1997) Determination of solute lipophilicity, as log *P*(octanol) and log *P*(alkane) using poly(styrene-divinylbenzene) and immobilized artificial membrane stationary phases in reversed-phase high-performance liquid chromatography. *Journal of Chromatography A*, *766*, 35–47.
55. Ottiger, C., Wunderli-Allenspach, H. (1997) Partition behavior of acids and bases in a phosphatidylcholine liposome-buffer equilibrium dialysis system. *European Journal of Pharmaceutical Sciences*, *5*, 223–231.
56. Tam, K. Y., Avdeef, A., Tsinman, O., Sun, N. (2010) The permeation of amphoteric drugs through artificial membranes – An in combo absorption model based on paracellular and transmembrane permeability. *Journal of Medicinal Chemistry*, *53*, 392–401.
57. Veber, D. F., Johnson, S. R., Cheng, H. Y., Smith, B. R., Ward, K. W., Kopple, K. D. (2002) Molecular properties that influence the oral bioavailability of drug candidates. *Journal of Medicinal Chemistry*, *45*, 2615–2623.
58. Winiwarter, S., Ax, F., Lennernäs, H., Hallberg, A., Petterson, C., Karlén, A. (2003) Hydrogen bonding descriptors in the prediction of human *in vivo* intestinal permeability. *Journal of Molecular Graphics and Modelling*, *21*, 273–287.
59. El Tayar, N., Tsai, R.-S., Testa, B., Carrupts, P.-A., Leo, A. (1991) Partitioning of solutes in different solvent systems: the contribution of hydrogen-bonding capacity and polarity. *Journal of Pharmaceutical Sciences*, *80*, 590–598.
60. Faller, B. (2008) Artificial membrane assays to assess permeability. *Current Drug metabolism*, *9*, 886–892.
61. Liu, X., Testa, B., Fahr, A. (2011) Lipophilicity and its relationship with passive drug permeation. *Pharmaceutical research*, *28*, 962–977.
62. Gleeson, M. P., Bravi, G., Modi, S., Lowe, L. (2009) ADME rules of thumb II: A comparison of the effects of common substituents on a range of ADME parameters. *Bioorganic and Medicinal Chemistry Letters*, *17*, 6906–5919.
63. Gleeson, M. P., Hersey, A., Hannongbua, S. (2011) In-silico ADME models: A general assessment of their utility in drug discovery applications. *Current Topics in Medicinal Chemistry*, *11*, 358–381.
64. Varma, M. V. S., Obach, R. S., Rotter, C., Miller, H. R., Chang, G., Steyn, S. J., El-Kattan, A., Troutman, M. D. (2010) Physicochemical space for optimum oral bioavailability: Contribution of human intestinal ansorption and first-pass elimination. *Journal of Medicinal Chemistry*, *53*, 1098–1108.
65. Ribadeneira, M. D., Aungst, B. J., Eyermann, C. J., Huang, S.-M. (1996) Effects of structural modifications on the intestinal permeability of angiotensin II receptor antagonists and the correlation of *in vitro*, *in situ*, and *in vivo* absorption. *Pharmaceutical Research*, *13*, 227–233.

66. Analogue-based Drug Discovery. (IUPAC), Ed. Fischer, J., Ganellin, C. R. (2006) Wiley-VCH Verlag GmbH, Weinheim.

67. Kerpel-Fronius, S., Fischer, J. (2006) Optimizing the clinical pharmacologic properties of the HMG-CoA reductase inhibitors. In: Analogue-based Drug Discovery, Wiley-VCH Verlag GmbH, Weinheim, pp. 137–156.

68. Hamelin, B. A., Turgeon, J. (1998) Hydrophilicity/lipophilicity: relevance for pharmacology and clinical effects of HMG-CoA reductase inhibitors. *Trends in Pharmaceutical Sciences*, *19*, 26–37.

69. Hsiang, B., Zhu, Y., Wang, Z., Wu, Y., Sasseville, V., Yang, W.-P., Kirchgessner, T. G. (1999) A novel human hepatic organic anion transporting polypeptide (OATP2). Identification of a liver-specific human organic anion transporting polypeptide and identification of rat and human hydromethylglutaryl-CoA reductase inhibitor. *Journal of Biological Chemistry*, *274*, 37161–37168.

70. Sugano, K., Yoshida, S., Takaku, M., Haramura, M., Saitoh, R., Nabuchi, Y., Ushio, H. (2000) Quantitative structure–intestinal permeability relationship of benzamide analogue thrombin inhibitor. *Bioorganic and Medicinal Chemistry Letters*, *10*, 1939–1942.

71. Potts, R. O., Guy, R. H. (1992) Predicting skin permeability. *Pharmaceutical Research*, *9*, 663–669.

72. El Tayar, N., Tsai, R.-S., Testa, B., Carrupt, P.-A. (1991) Percutaneous penetration of drugs: A quantitative structure–permeability relationship study. *Journal of Pharmaceutical Sciences*, *80*, 744–749.

73. Liou, Y.-B., Ho, C.-J., Lin, Y.-K., Sheu, M.-T. (2009) Construction of a quantitative structure–permeability relationship (QSPR) for the transdermal delivery of NSAIDs. *Journal of Controlled Release*, *138*, 260–267.

74. Palmer, A. M. (2010) The role of the blood-CNS barrier in CNS disorders and their treatment. *Neurobiology Disease*, *35*, 3–12.

75. Lavenskij, K., Japertas, P., Didziapetris, R., Petrauskas, A. (2009) Ionization-specific QSAR models of blood-brain penetration of drugs. *Chemistry and Biodiversity*, *6*, 2050–2054.

76. Platts, J. A., Abraham, M. H., Zhao, Y. H., Hersey, A., Ijaz, L., Butina, D. (2001) Correlation and prediction of a large blood-brain distribution data set – an LFER study. *European Journal of Medicinal Chemistry*, *36*, 719–730.

77. Manallack, D. T. (2007) The pKa distribution of drugs: Application to drug discovery. *Perspectives in Medicinal Chemistry*, *1*, 25–38.

78. *WDI, The World Drug Index*, Derwent Information, London, UK. (www.derwent.com. Accessed 2013 Aug 13.)

79. Dockal, M., Carter, D.C., Ruker, F. (2000) Conformational transitions of the three recombinant domains of human serum albumin depending on pH. *Journal of Biological Chemistry*, *275*, 3042–3050.

80. Wardell, M., Wang, Z. M., Ho, J. X., Robert, J., Ruker, F., Ruble, J., Carter, D. C. (2002) The atomic structure of human methemalbumin at 1.9 angstrom. *Biochemical and Biophysical Research Communications*, *291*, 813–819.

81. Petitpas, I., Bhattacharya, A. A., Twine, S., East, M., Curry, S. (2001) Crystal structure analysis of warfarin binding to human serum albumin: Anatomy of drug site I. *Journal of Biological Chemistry*, *276*, 22804–22809.

82. Curry, S., Brick, P., Franks, N. P. (1999) Fatty acid binding to human serum albumin: New insights from crystallographic studies. *Biochimica et Biophysica Acta, 1441*, 131–140.

83. Noctor, T. A. G., Diaz-Perez, M. J., Wainer, I. W. (1993) Use of a human serum albumin-based stationary phase for high-performance liquid chromatography as a tool for the rapid determination of drug plasma protein binding. *Journal of Pharmaceutical Sciences, 82*, 675–676.

84. Kaliszan, R., Noctor, T. A. G., Wainer, I. W. (1992) Quantitative structure–enantioselective retention relationships for the chromatography of 1,4-benzodiazepines on a human serum albumin based HPLC chiral stationary phase: An approach to the computational prediction of retention and on a human serum albumin based HPLC chiral stationary phase. *Chromatographia, 33*, 546–550.

85. Colmenarejo, G., Alvarez-Pedraglio, A., Lavandera J-L. (2001) Chemoinformatic models to predict binding affinities to human serum albumin. *Journal of Medicinal Chemistry, 44*, 4370–4378.

86. Valkó, K., Nunhuck, S. B., Bevan, C. D., Abraham, M. H., Reynolds, D. P. (2003) Fast gradient HPLC method to determine compound binding to human serum albumin. Relationships with octanol/water and immobilized artificial membrane lipophilicity. *Journal of Pharmaceutical Sciences, 92*, 2236–2248.

87. Fournier, T., Medjoubi-N, N., Porquet, D. (2000) Alpha-1-acid glycoprotein. (Review). *Biochimica et Biophysica Acta, 1482*, 157–171.

88. Israili, Z. H., Dayton, P. G. (2001) Human alpha-1-glycoprotein and its interactions with drugs. *Drug Metabolism Reviews, 33*, 161–235.

89. Kaliszan, R., Nasal, A., Turowski, M. (1995) Binding sites for basic drugs on α_1-acid glycoprotein as revealed by chemometric analysis of biochromatographic data. *Biomedical Chromatography, 9*, 211–215.

90. Nasal, A., Siluk, D., Kaliszan, R. (2003) Chromatographic retention parameters in medicinal chemistry and molecular pharmacology. *Current Medicinal Chemistry, 10*, 381–426.

91. Taylor, F. B. (1999) Baboon model of *E. coli* sepsis: Role of phospholipid microparticles in DIC. *Sepsis, 3*, 125–134.

92. Dellinger, R. P., Tmayko, J. F., Angus, D. C., Opal, S., Cupo, M. A., McDermotts, S., Ducher, A., Calandra, T., Cohen, J. (2009) Efficacy and safety of a phospholipid emulsion (GR270773) in Gram-negative severe sepsis: results of a phase II multicenter, randomized, placebo-controlled, dose-finding clinical trial. *Critical Care Medicine, 37*, 2929–2938.

93. Taheri, S., Cogswell, L. P., Gent, A., Stichartz, G. R. (2003) Hydrophobic and ionic factors in the binding of local anesthetics to the major variant of human α_1-acid glycoprotein. *The Journal of Pharmacology and Experimental Therapeutics, 304*, 71–80.

94. Holladay, J. W., Dewey, M. J., Michniak, B. B., Wiltshire, H., Halberg, D. L., Weigl, P., Liang, Z., Halifax, K., Lindup, W. E., Back, D. J. (2001) Elevated alpha-1-acid glycoprotein reduces the volume of distribution and systemic clearance of saquinavir. *Drug Metabolism and Disposition, 29*, 299–303.

95. Zsila, F. (2007) Overlapping ligand specificity of P-Glycoprotein and serum alpha(1)-acid glycoprotein: Evidences and potential implications. *Current Drug Metabolism, 8*, 563–593.

96. Ohmori, M., Arakawa, M., Kazuhiro, H., Takasaki, H., Hifumi, S., Miyamori, I., Akio, F. (2003) Stereoselective pharmacokinetics of amlodipine in elderly hypertensive patients. *American Journal of Therapeutics*, *10*, 29–31.

97. Barbato, F., di Martino, G., Grumetto, L. La Rotonda, M.I. (2005) Can protonated β-blockers interact with biomembranes stronger than isolipophilic compounds? A chromatographic study on three different phospholipid stationary phases (IAM-HPLC). *European Journal of Pharmaceutical Sciences*, *25*, 379–386.

98. Österberg, T., Svensson, M., Lundhal, P. (2001) Chromatographic retention of drug molecules on immobilised liposomes prepared from egg phospholipids and from chemically pure phospholipids. *European Journal of Pharmaceutical Sciences*, *12*, 427–439.

99. Taillardat-Bertschinger, A., Martinet, C. A. M., Carrupt, P.-A., Reist, M., Caron, G., Fruttero, R., Testa, B. (2002) Molecular factors influencing retention on immobilized artificial membrane (IAM) compared to partitioning in liposomes and *n*-octanol. *Pharmaceutical Research*, *19*, 729–737.

100. Valkó, K., Du, C. M., Bevan, C. D., Reynolds, D. P., Abraham, M. H. (2000) Rapid-gradient HPLC method for measuring drug interactions with immobilized artificial membrane: Comparison with other lipophilicity measures. *Journal of Pharmaceutical Sciences*, *89*, 1085–1096.

101. Lepont, C., Poole, C. F. (2002) Retention characteristics of an immobilized artificial membrane column in reversed-phase liquid chromatography. *Journal of Chromatography A*, *946*, 107–124.

102. Li, J., Sun, J., Cui, S., He, Z. (2006) Quantitative structure-retention relationship studies using immobilized artificial membrane chromatography I: Amended linear solvation energy relationships with the introduction of a molecular electronic factor. *Journal of Chromatography A*, *1132*, 174–182.

103. Casartelli, A., Bonato, M., Cristofori, P., Crivellente, F., Dal Negro, G., Masotto, I., Mutinelli, C., Valkó, K., Bonfante, V. (2003) A cell-based approach for the early assessment of the phospholipidogenic potential in pharmaceutical research and drug development. *Cell Biology and Toxicology*, *19*, 161–176.

APPENDIX A

ANSWERS TO THE QUESTIONS FOR REVIEW

CHAPTER 1

Q1.1 What are the major steps of the drug discovery process?

The following are the major steps of the drug discovery process.

1. Target identification: this comes about from the knowledge of human genome, biochemistry, and disease mechanism. The target is a macromolecule (enzyme, ion channel, G-protein-coupled receptor, or hormone receptor) that is not functioning properly and causing pathological symptoms. The drug can be a small molecule that binds to the target and alters its function in a positive way.
2. Target validation: it has to be proved that binding a small molecule to the target alters its function *in vivo* in such a way that it improves the pathological symptoms in the patients.
3. High throughput or medium throughput chemistry: chemistry effort that produces a variety of small molecules that are able to bind the target.
4. High throughput or medium throughput screening of the potency of the compounds on a specified target: the potency screening measures the concentration of the compound that binds to the target and produces a pharmacological outcome.
5. Hit identification and confirmation: molecules that showed response in the potency screening have to be identified and the potency measurements are repeated at various concentrations, and a concentration versus potency profile is established.

Physicochemical and Biomimetic Properties in Drug Discovery: Chromatographic Techniques for Lead Optimization, First Edition. Klara Valko.

6. Hit to lead design: the structures of the proven potent molecules are modified in order to establish structure–potency relationships and improve the potency of the hit compound.
7. Lead-to-candidate optimization: the structure of the lead molecule is modified in order to achieve absorption and appropriate distribution of the compounds. Several *in vitro* measurements are carried out at this stage. The potency of the compound is measured on various targets in order to test the selectivity. *In vitro* clearance and metabolism are measured together with the basic physicochemical properties of the compounds. Solubility, lipophilicity, and permeability are measured in order to select the best possible molecules. Preclinical toxicity screening and *in vivo* target engagement assays are also carried out before the candidate molecule can be selected.
8. Development phase: after the successful candidate selection, the compounds are synthesized at a larger scale. The solid forms are characterized and formulation studies are carried out. Further studies on absorption, distribution, pharmacokinetics, and toxicity are carried out before clinical trials under GLP conditions.

CHAPTER 2

Q2.1 What are the most important physicochemical properties that can be related to the ADME properties?

Lipophilicity and solubility are the most important physicochemical properties that can be related to absorption. The drug molecule has to be in solution in the gastrointestinal tract to be able to permeate through the intestinal wall. Lipophilicity of the compound influences the permeability through biological membranes, as the compound has to be able to partition into the hydrophobic core of the biological bilayer. The lipophilicity of the compound also influences its plasma protein binding and tissue binding and, as a consequence, the volume of distribution. Both lipophilicity and solubility depend on the compound's charge state; therefore, the acid dissociation constant is an important property.

Q2.2 What is "drug-likeness"?

Several scientists have investigated the properties of drugs and compared with the properties of research compounds that have not become drugs because of various reasons. They have found that drug molecules occupy a relatively small part of the chemical space. It has been found that synthetic drug molecules have molecular weight less than 500 Da, they are not very lipophilic ($\log P$ is <5), and they contain a limited number of H-bond donor groups (<5) and H-bond acceptor groups (<10). The drug molecules have reasonably good solubility as well. The chemical space occupied by the drugs is called the *drug-like space*.

Q2.3 Why is there a need to measure solubility, lipophilicity, and permeability at early stages of the drug discovery process?

As a result of the optimization of the potency of the compounds, synthetic molecules became large and lipophilic, and, therefore, they became less soluble and permeable and produced poor *in vivo* pharmacokinetic and pharmacodynamic properties. It is very difficult to fix these problems at a later stage of the development process. It is more efficient to design molecules in the drug-like property space and avoid unnecessary resources in synthesizing and testing molecules that have high probability of failure at a later stage.

Q2.4 Which physicochemical properties are the most important in terms of drug-like property space?
On the basis of the Lipinski rule of five, lipophilicity, the number of H-bond donor and acceptor groups, and size are the important physicochemical properties that differentiate the orally administered drugs from other compounds. The drug-like molecules fall into a certain lipophilicity range, and the size of the molecules is generally smaller than 500 Da. The number of H-bond acceptor groups is less than 10 and donor groups are less than 5. Drugs from natural sources are often outside the synthetic drugs' property space. The drug molecules should be soluble and permeable to be able to get absorbed from the gastrointestinal tract. Both solubility and permeability are controlled by the general lipophilicity of the molecules. The number of H-bond donor and acceptor groups affects the lipophilicity and, as a consequence, the solubility and permeability. Lipophilicity also depends on the charge state of the molecules. The most frequently measured physicochemical properties are solubility, lipophilicity, permeability, and the charge state (acid dissociation constant).

CHAPTER 3

Q3.1 Define the following parameters: fraction absorbed, unbound fraction in plasma, volume of distribution, clearance, half-life, and unbound volume of distribution?
Fraction absorbed is the proportion of the administered dose that reaches the systemic circulation. It is determined by the proportion of the plasma concentration obtained after oral and intravenous administration of the same amount of drug.
Unbound fraction in plasma is the proportion of the free plasma concentration and the total plasma concentration of a drug. It is usually independent of the absolute concentration.
Volume of distribution is a hypothetical volume of tissues where the concentration of the drug is the same as in plasma. It can be expressed as the quotient of the dose and the plasma concentration. It can be determined from the dose and the plasma concentration–time profile.
Clearance is a measure of the ability of the body or an organ to eliminate a drug molecule from the systemic circulation. Systemic clearance (Cl) relates to the whole body, whereas the hepatic or renal clearance refers to the particular organ—liver or kidney. The clearance shows the elimination rate of the drug from the entire body relative to the drug concentration in plasma. Clearance is usually measured by the apparent volume of a reference fluid (usually the plasma or blood) that is cleared of drug in a unit time. So it means the volume of blood or plasma that is cleared by the drug in 1 min. Similar to the volume of distribution, the clearance is also related to the body mass and expressed in ml/(min kg). Clearance can be calculated from the dose and the AUC (area under the curve) values obtained from the plasma concentration–time profile. The organ clearance of the drug is dependent on the blood flow rate in the clearing organ (liver or kidney); it also depends on the free plasma concentration.
The half-life ($t_{1/2}$) is the time when the drug concentration drops to half of the maximum concentration in the plasma after one single dose. It can be estimated from the drug plasma concentration–time profile. The half-life can be short because of the high rate of clearance/metabolism.

The unbound volume of distribution can be derived from the volume of distribution by taking into account the plasma protein binding. The difference with volume of distribution is that instead of the total plasma concentration, we measure the free plasma concentration. Therefore, the unbound volume of distribution can also be calculated by dividing the volume of distribution by the unbound fraction in plasma.

Q3.2 What is AUC, and how it can be measured?
AUC is the abbreviation of the "area under the curve," which is obtained by plotting the plasma concentration in the function of time elapsed after the administration of the drug. It can be obtained by integrating the plasma concentration–time curve. It measures in principle the exposure of the drug *in vivo*.

Q3.3 Why is plasma protein binding important?
The absorbed drug reaches the systemic circulation where it can bind to plasma proteins. The bound drug is not available for clearance or interaction with the target receptor. Therefore, it is important to know the proportion of the drug that is free in plasma. Drug molecules that bind to the same plasma protein binding sites specifically may influence each other's free concentration that may cause drug–drug interactions. The stronger binding drug replaces the weaker binder from the plasma protein binding site and thus may increase the free concentration of the weaker binder, which may reach toxic levels.

Q3.4 What is the "free drug hypothesis" and when are these assumptions not true?
The free drug hypothesis means that the free concentration of the drug in all compartments in the body is the same. It also supposes that only the free drug molecules at the site of action are able to bind the receptor and show the desired pharmacological effect. Thus, the free concentration of the drug in plasma is the same as the free concentration of the drug in the tissue when the steady-state equilibrium is reached. This hypothesis is not true when active transport process or permeability barrier exists, which works against the thermodynamic equilibrium process in steady state.

CHAPTER 4

Q4.1 Explain how chromatographic separation takes place.
Chromatography is a common name for the methods of separation of substances by differences in rates of migration that arise from their different distribution ratios between a flowing mobile phase (gas or liquid) and a stationary phase (liquid or solid). Compounds are separated as a result of differential migration in the chromatographic system, which depends on their interactions between the stationary phase and the mobile phase. Components of a mixture that is injected into the system will move with the mobile phase, but the speed of the component zones is different for each type of molecule, depending on their average distribution between the mobile and the stationary phases.

Q4.2 What is the retention factor and how does it relate to a compound's distribution coefficient between the mobile and the stationary phases?
The peak retention time, the time elapsed from the injection until the appearance of the peak maximum, is characteristic of the chemical nature of the component. The retention factor (k) is the retention time expressed relative to the dead time, which is equal to the time needed for a compound to go through the chromatographic system with the speed of the mobile phase. It means that there is no interaction with the

stationary phase. The retention factor, therefore, is proportional to the average number of molecules of a compound in the stationary phase relative to the mobile phase. The longer is the retention factor, the higher proportion of the molecules stays in the stationary phase relative to the mobile phase. The retention factor is proportional to the compound's dynamic distribution coefficient between the stationary and the mobile phases. The volume ratio of the mobile and stationary phases has to be taken into account for the calculation of the distribution coefficient from the retention factor.

Q4.3 How can we measure the efficiency of a chromatographic system and how does it relate to the mobile phase flow rate in isocratic and gradient elutions?

The efficiency of a chromatographic separation can be expressed by the theoretical plate number (*N*).

$$N = 16\left(\frac{t_R}{t_w}\right)^2$$

t_R is the retention time and t_w is the peak width. An HPLC system is more efficient: the peaks are narrower with longer retention time. It also implies that more components can be separated in a given time frame. The theoretical plate number increases with the length of the column. Therefore, for the comparison of the efficiency of HPLC columns with various lengths, the efficiency can be described by the height of the theoretical plate (*H*), which can be calculated from the column length (*L*) and the theoretical plate number.

$$H = \frac{L}{N}$$

In gradient reversed-phase chromatography, the efficiency of the system can be described by the peak capacity, which is proportional to the number of peaks fully separated in unit of time.

The efficiency of an HPLC system expressed by the height of the theoretical plate (*H*) depends on the mobile- phase flow rate as described by the Van Deempter equation:

$$H = A + \frac{B}{u} + C \times u$$

A is the eddy diffusion term, *B* the longitudinal diffusion, and *C* the mass transfer speed.

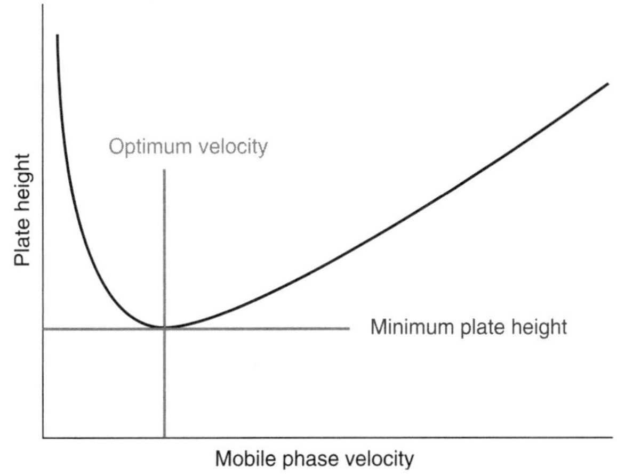

The "*A*" eddy diffusion term does not depend on the flow rate; it depends only on the packing material and packing of the stationary phase particles into the column. Smaller spherical particles packed under constant pressure reduce the *A* term.

The "*B*" term is the longitudinal diffusion term, which is inversely related to the mobile phase flow rate. It means that at very slow mobile phase velocity, the molecules can move backwards as well because of the longitudinal diffusion.

The "*C*" term refers to the mass transfer resistance, which is causing peak broadening at higher flow rates.

The Van Deempter equation describes the dependence of efficiency on the mobile phase flow rate in isocratic mode.

In gradient reversed-phase chromatography, we increase the elution strength of the mobile phase by increasing the organic phase concentration during the chromatographic run. Under these conditions, the later eluting peaks will not become automatically wider. The efficiency that is expressed by the peak capacity increases with increasing mobile phase flow rate. It is not because the same peak broadening factors are not present in gradient mode. It is because certain components injected on the top of the column are not moving till the proper mobile phase strength reaches the column. Thus, the "retention time" in gradient mode cannot be counted from the injection time. The faster flow rate using the same gradient time means shallower gradient, that is, the change of the mobile phase composition is smaller per unit of mobile phase volume, which increases the separation efficiency.

Q4.4 Explain the differences between isocratic and gradient elution.

Isocratic elution

In isocratic mode of HPLC, the mobile phase composition is constant during the analysis. Therefore, the relative lipophilicity of the stationary and the mobile phases is constant. In isocratic reversed-phase chromatography, the retention factor can be directly related to the distribution coefficient of the compound between the mobile and the stationary phases. The peak broadening is dependent on the eddy diffusion (related to efficient column packing, small particle size, and spherical shape), longitudinal diffusion, and mass transfer retention. The peaks are broader when they have a longer retention time. The peak broadening affects efficiency, which can be described by the theoretical plate number (*N*).

$N = 16(t_R/t_w)^2$ t_R is the retention time and t_w is the peak width.

The equation shows that peaks with longer retention times are proportionally wider. The van Deempter equation describes the effect of flow rate on efficiency that expressed as the height of the theoretical plate (*H*), as a measure of efficiency that is independent from the column length. ($H = L/N$, where L is the column length and *N* is the theoretical plate number).

$$H = A + \frac{B}{u} + C \times u$$

where the *A* terms relate to the eddy diffusion, the *B* term relates to the longitudinal diffusion, the *C* term relates to the mass transfer speed, and *u* is the flow rate. Too high a flow rate causes peak broadening because of the *C* term, that is, there is not enough time to develop local equilibrium between the stationary phase and the mobile phase. At slow mobile phase flow, the longitudinal diffusion will cause peak broadening. In isocratic elution, the logarithmic retention factor (log *k*) linearly relates to the organic phase concentration of the mobile phase. It means that, for example, the retention time differences of a homolog series of compounds will be bigger and bigger as more CH_2 groups are added to the molecule. The drawback of the isocratic elution is that it cannot separate compounds with a wide range of lipophilicity. Hydrophilic compounds will elute with the solvent front, and the most lipophilic compounds will not elute from the column. Also, we need preliminary information about the mixture we want to separate to start the method development.

Gradient elution

During gradient reversed-phase chromatography, the mobile phase composition continuously changes during the analysis. Usually, we start with low organic phase concentration. We identify the starting and finishing mobile phase composition and define a gradient time during which the mobile phase composition will change. The steepness of the gradient depends on the concentration change in a given time and the flow rate. In gradient elution, the peak width does not increase with longer retention time. The gradient retention time differences of homologous series of compounds are equal for each additional CH_2 groups. Higher flow rate and shorter columns do not decrease the resolution. That is why we can develop the so-called generic gradient methods. For example, a 40-min generic high efficiency gradient analysis could have been changed to a 5-min analysis without a significant loss of resolution simply by increasing the flow rate and reducing the column length and by applying a fast gradient. The other advantage of using gradient elution is that every compound can be analyzed with the same method. Hydrophilic and lipophilic compounds can be analyzed simultaneously.

CHAPTER 5

Q5.1 What is lipophilicity and how can we measure it?

Lipophilicity or hydrophobicity of the compound refers to the compound affinity for lipophilic environment or the compound's tendency to avoid the aqueous environment. Lipophilicity can be measured by the compound partition coefficient between two nonmiscible solvents or two compartments which are in equilibrium. The partition coefficient is expressed as the equilibrium concentration ratio of the compound in an aqueous and a lipophilic solvent or compartment. The concentration ratio is constant, regardless of the amount of compound (when the solute concentration is low and no interaction between the solute molecules occur). Lipophilicity of the compound can be measured by two potentiometric titrations, one in water and the other one in the presence of organic phase such as octanol. Reversed-phase chromatographic retention can also be related to compound lipophilicity. Chromatographic retention time measurement can be carried out in various reversed-phase systems using aqueous organic phase as the mobile phase. Using a calibration set of compounds, the retention parameters can be converted to chromatographic hydrophobicity index that showed good correlation with partition coefficients.

Q5.2 How can we measure octanol/water partition coefficients?

The measurements of octanol/water partition coefficients for drug discovery compounds are very important. As we discussed earlier, the *in vivo* distribution of drugs between the plasma compartment and various tissue compartments are predominantly dependent on the compound's lipophilicity. The octanol/water partition proposed to be a good model for biological partition has been studied extensively and several experimental protocols have been developed and published for its measurements. In principle, the following procedure can be followed:

Step 1. Prepare an aqueous buffer solution with the required pH. For the determination of the partition coefficient (P) of the unionized form of the compound, the pH of the aqueous buffer should be such that the compound is in its nonionized form (i.e., basic compounds are measured at high pH, acidic compounds are measured at low pH).

Step 2. Equilibrate the octanol with the above-mentioned buffer solution and the buffer with the octanol. Mix and leave it for 24 h at room temperature. Then separate the two layers. (Note: water also dissolves around 5% of octanol!)

Step 3. Dissolve the compound in a known amount of octanol-saturated buffer or the buffer-saturated octanol depending on its expected lipophilicity. Put a known amount of water-saturated octanol and octanol-saturated water with the dissolved compound in to a shake-flask device. Shake and equilibrate (temperature can be controlled for more precise measurements).

Step 4. Separate the two phases (centrifuging the layers can help), and determine the compound concentration in the two layers, using standard concentration determination techniques (e.g., UV spectrophotometry, HPLC).

The octanol/water partition coefficients can be determined using chromatography with octanol-coated stationary phase and aqueous mobile phase.

The "gold standard" determination of octanol/water partition coefficients was introduced by Sirius Analytical Ltd. (http://www.sirius-analytical.com/) with their GLpK_a and T3 instruments that provide a highly accurate determination of octanol/water partition coefficient under a "good laboratory practice" (GLP) environment. The instrument is based on a proprietary Ag/AgCl double junction electrode for accurate pH measurement. It carries out two potentiometric titrations of the compound with and without the presence of octanol. The partition coefficients are calculated using the Bjerrum plots to estimate the pK_a shift caused by the compound partitioning into the octanol phase (see Reference [6] from Chapter 5). From the two titration plots, the pK_a and the log P values are calculated. Sirius pH-metric methods can measure log P values of ionizable compounds between -1 and $+7$, which is the widest range of any log P measurement method. Unlike indirect chromatographic methods, they study partition directly between water and octanol, dodecane, or other partition solvents. They do not suffer from problems of microemulsion formation and lack of dynamic range that affect shake-flask methods.

Q5.3 What is isocratic CHI and how can we determine it?

The isocratic chromatographic retention factor can be measured using various organic modifier concentrations in the mobile phase. Actually, we need to vary the organic phase concentration depending on the lipophilicity of the compound to be able to obtain well-measurable retention times (not too short and not to long). There is a linear relationship between the logarithmic retention factors (log k) and the organic phase concentrations (φ) within a certain range. Several authors suggested using the extrapolated logarithmic retention factors as the measures of lipophilicity that are extrapolated to zero organic phase concentration. This is the intercept of the $\log k$ versus φ straight line, assigned as $\log k_0$ or $\log k_w$. However, these straight lines may cross each other, meaning that different orders of lipophilicity can be obtained for a series of compounds at zero or any other organic phase concentrations. The $\log k$ versus φ lines are not straight within a large range, and different $\log k_w$ values may be obtained using different set of measurements or different types of organic phases (methanol or acetonitrile). The isocratic chromatographic hydrophobicity index (φ_0) is defined as the quotient of the intercept ($\log k_w$) and the slope of the $\log k$ versus φ straight line.

$$\varphi_0 = \frac{\log k_w}{\text{Slope}}$$

The quotient (φ_0) is equivalent to the organic solvent concentration in the mobile phase that is required to get the compound retention time exactly twice that of the dead time, that is, $\log k = 0$. The φ_0 value of a compound is defined as the isocratic chromatographic hydrophobicity index (see Reference [24] from Chapter 5). It ranges from 0% to 100%, which indicates also an appropriate organic phase concentration by which the compound can be eluted from the column with a reasonably short time (retention time is double that of the dead time). It can be calculated from the logarithmic retention factor of a compound measured using three or more organic phase concentration. By fitting a straight line for the points obtained by plotting the measured φ and $\log k$ data, the isocratic chromatographic hydrophobicity index (φ_0) can be calculated.

Q5.4 Why are gradient retention times directly proportional to the logarithmic values of partition coefficients, unlike in isocratic mode, where we need to take the logarithmic value of the retention factor?

During gradient reversed-phase chromatography, the mobile phase composition continuously changes. Usually, we start with low organic phase concentration. We identify the starting and finishing mobile phase composition and define a gradient time during which the mobile phase composition will change. The steepness of the gradient depends on the concentration change in a given time and the flow rate. In gradient elution, the peak width does not increase with longer retention time. The gradient retention time differences of a homologous series of compounds are equal for each additional CH_2 groups. In isocratic mode, the retention time or the retention factor increases exponentially with the addition of one CH_2 group to the homolog series. Therefore, in isocratic mode, the logarithmic retention factor shows a linear relationship with the logarithmic partition coefficients; in gradient mode, the gradient retention time is directly proportional to the logarithmic partition coefficients. This is due to the constantly changing organic phase concentration in the mobile phase.

Q5.5 Approximately how much acetonitrile and methanol do we need to use in isocratic reversed-phase chromatography to elute a compound that has a lipophilicity ($\log P$) of 3 to have a retention time approximately double that of the dead time?

On the basis of the data for more than 500 compounds (see Reference [24] of Chapter 5), an approximation can be made for the organic phase concentration in the mobile phase using acetonitrile and methanol based on calculated octanol/water partition coefficients (log P) of a compound using Equations 5.7 and 5.8.

$$\varphi_{0(\mathrm{AcN})} = 16 \log P + 26 \quad (5.7)$$

$$\varphi_{0(\mathrm{MeOH})} = 13 \log P + 42 \quad (5.8)$$

Introducing log $P = 3$ into Equations 5.7 and 5.8, the compound needs 74% acetonitrile ($16 \times 3 + 26$) and 81% methanol ($13 \times 3 + 42$) to be able to elute with the retention time double the dead time.

Q5.6 Describe the relationship of lipophilicity of ionized compounds as a function of pH.

Lipophilicity, for example, the octanol/water distribution coefficients (log D), drops three to four log units when the compound is ionized. The pH dependence of ionization can be described by the Henderson–Hasselbalch equation. Similar to the pH dependence of ionization, the pH dependence of log D can be described by the following equations:

For Acids

$$\log D_{\mathrm{pH}} = \log P - \log[1 + 10^{(\mathrm{pH}-\mathrm{p}K_a)}] + \log D_{\text{fully ionized form}}$$

For the log D–pH profile plot for acids, see Figure 5.15.

Lipophilicity of acids drops at high pH when higher percentage of ionized form is present, whereas basic compounds lipophilicity drops at low pH values when the positively charged form is present.

For Bases

$$\log D_{\mathrm{pH}} = \log P - \log[1 + 10^{(\mathrm{p}K_a-\mathrm{pH})}] + \log D_{\text{fully ionized form}}$$

For the log D– pH profile plot for bases, see Figure 5.16.

We have to note that the log D of the 50% ionized form of the molecules is only 0.3, less than the lipophilicity of the completely nonionized form (i.e., log P).

Q5.7 We have log D and log P values of a compound with an ionizable group. Which value must be the highest log P or log D?

The lipophilicity of the neutral form of molecule is described by log P. The log D value refers to the logarithmic distribution coefficient of a compound at a particular pH (usually, pH 7.4), and it refers to the sum of ionized and unionized species' distribution. As ionization makes the molecule less lipophilic, log D is always less or equal to log P.

Q5.8 At which pH is a basic compound more lipophilic? At which pH is an acidic compound more lipophilic? Why?

Basic compounds are positively charged at lower pH than their pK_a values. When the proton concentration of the solvent is high enough, the basic compounds are protonated. Charged molecules are always less lipophilic than the uncharged ones; therefore, basic compounds are more lipophilic at high pH. Acidic compounds dissociate a proton at pH values that are higher than their pK_a values. They are negatively charged at high pH and less lipophilic. Therefore, acids are more lipophilic at low pH where they are uncharged.

Q5.9 Describe a few biomimetic stationary phases. What HPLC conditions have to be used to mimic the biological partition behavior of a compound?
Biomimetic HPLC stationary phases contain proteins and phospholipids. The protein phases are chemically bonded HSA and AGP on silica surface. The IAM stationary phases contain chemically bonded phospholipids. The retention factor measured using the biomimetic phases can be used to determine the compounds' integrations with these biomolecules. In order to mimic the *in vivo* conditions, we use mobile phases at physiological pH (pH 7.4). It is important to use only low concentrations of isopropanol with the protein phases in order to retain the natural structure of the protein. Acetonitrile can be used as an organic modifier with the IAM stationary phases preferably less than 70%.

Q5.10 How can we compare various lipophilicity measures?
In order to be able to explain the differences between various lipophilicity measures, we need to have a good understanding of the molecular driving forces behind compound partitioning between an aqueous and the nonaqueous phases. We need to know which molecular property drives the various partition processes. For this purpose, the Abraham solvation equation model (see Reference [60] from Chapter 5) can provide a quantitative insight. The Abraham solvation equation model starts with the description of the solvation forces/energy from the gas phase to a liquid phase. The Abraham solvation equation model (see Reference [60] from Chapter 5) suggests that molecular size and a small number of polarity descriptors (H-bond donor/acceptor, dipolarity) are sufficient to describe compound partition behavior. By measuring a linear free-energy-related solute property such as the logarithmic value of octanol/water partition coefficient ($\log P$) or logarithmic retention factor ($\log k$) values for a set of probe molecules (minimum 25 compounds, preferably 40–60) with known molecular descriptors, such as excess molar refraction, E; dipolarity/polarizability, S, H-bond acidity, A; H-bond basicity, B, and molecular size, V, the solvation equation characteristic for the partitioning system can be constructed using multiple linear regression equations as shown by Equation 5.17.

$$\text{Log(solute property)} = e \times E + s \times S + a \times A + \text{b} \times B + v \times V + \text{Constant}$$

The regression coefficients of the molecular descriptors (e, s, a, b, and v) are characteristics for the partition system. The generally good statistical fit of the chromatographic retention data and the molecular descriptors derived from bulk organic solvent/water partition coefficients support the comparability of the two types of partition data. Solvation equations have also been published for biological partition/distribution processes.

Q5.11 What are the similarities and differences between biomimetic HPLC and the octanol/water partition systems?
In order to compare various partition systems, the coefficients of each molecular descriptor were divided by the volume coefficient (v); otherwise, the scale or unit of the solute parameter distorts the proper comparison of the systems in terms of their sensitivity toward the molecular descriptors. The relative coefficients reveal that the H-bond basicity (H-bond acceptors) parameter is negative in all the listed solvent partition, chromatographic, and biological partition systems. It means that when we have H-bond acceptor groups on the molecules it restricts partition to the nonaqueous hydrophobic compartment. The dipolarity/polarizability parameter is also negative

in most cases, meaning that it increases the hydrophilic character of the compounds (except human intestinal absorption). For the phospholipid partition, serum albumin binding and the C-18 retention with aqueous acetonitrile mobile phase are insensitive to the dipolarity/polarizability parameter, showing close to zero relative coefficients. The biggest variation can be observed for the relative coefficients of H-bond acidity. In most cases, compounds with H-bond donor groups provide a hydrophilic character and discourage compound partitioning out of water. However, there are several systems where the H-bond acidity coefficient is close to zero; therefore, compounds can partition out from the water regardless of the polar H-bond donor groups. These systems are the octanol/water system, serum albumin binding, and phospholipids membrane partition. In these cases, the nonaqueous hydrophobic phase contains H-bond acceptor groups that help in forming H-bonds with the H-bond donor group of the molecules. The explanation for this is that these systems contain water or other H-bond acceptor moiety that accommodates the H-bond donor groups, forming hydrogen bonding. These findings support the idea that the octanol/water partition system is a good model for biologically relevant partitioning of compounds.

Q5.12 What are the five molecular descriptors that can be used to describe a compound's partition in various solvent systems?

The Abraham solvation equation model requires five molecular descriptors that are adequate to describe compound behaviors in various partition and distribution systems, including biological distribution processes. These are the excess molar refraction, E; dipolarity/polarizability, S; H-bond acidity, A; H-bond basicity, B; and molecular size, V calculated from the McGowan volume. The size parameter is proportional to the energy that is required to create a cavity in the solvent in order to solvate the solute. The polar descriptors describe the interactions between the solute and the solvent molecules that may contribute to regain the energy lost during the cavity formation.

Q5.13 Describe the principles of the determination of the Abraham molecular descriptors.

The five molecular descriptors of a molecule can be measured by measuring the SPs in five different partition systems that show orthogonal selectivity. The size parameter (V) and the excess molar refraction (E) can be easily calculated from the molecular structure. We have to determine the solvation equations in the same five orthogonal systems using a diverse set of compounds with known molecular descriptors or we need to use systems for which the solvation equations have already been established. The measured SPs of the molecules determined in the five systems have to be introduced to the known solvation equations. The unknown variables are the molecular descriptors of the new molecule. Then we have to solve the equation systems with five equations and five unknowns. As the measurements contain experimental errors, it is advisable to use iteration or the Excel Solver approach to calculate the molecular descriptors. Chromatographic retention measurements using various stationary and mobile phases can provide a simple way to determine the Abraham descriptors for a new molecule.

The following procedure can be applied for the descriptor determination of a compound by HPLC. First, select three to five chromatographic systems with different selectivity. This can be C-18 column with acetonitrile gradient, C-18 and CN columns with methanol gradient, or polymer-based and perfluorinated-reversed-phase columns with trifluoroethanol gradient. Determine the solvation equations in the five systems using a diverse set of compounds with known molecular descriptors.

Then determine the gradient retention times of the unknown compounds. Use Excel Solver to solve the five equations for the five unknown molecular descriptors.

CHAPTER 6

Q6.1 Define thermodynamic equilibrium, apparent, kinetic, and precipitative kinetic solubilities.

The *thermodynamic equilibrium solubility* refers to the concentration of the saturated solution of the unionized form of the compound, which is in equilibrium with a well-defined solid form at a given temperature.

Apparent solubility refers to the concentration of a saturated solution, also at a given pH without an accurate characterization of the solid material and without ensuring that the saturation equilibrium has been achieved.

Kinetic solubility refers to an aqueous concentration of the compound in a given time frame after adding the aqueous solvent to a solid form of material, or to a dissolved form, usually in dimethyl sulfoxide (DMSO) solvent. When a DMSO stock solution is diluted down by an aqueous buffer and we measure the concentration that remains in the solution after precipitation, we speak about *precipitative kinetic solubility*.

Q6.2 How does the FDA define good solubility for the BCS?

FDA defines good solubility as the lowest solubility in the pH range of the gastrointestinal tract (pH 1 to pH 8) which is equivalent to or a higher concentration than the dose amount in 250 ml. The definition reflects that a single dose will dissolve in a glass of water even when the pH is minimizing the aqueous solubility. The definition comes from a practical point of view; that is, a drug molecule is soluble enough when the dose can be dissolved in 250 ml of liquid at any pH between 1 and 8. Therefore, good solubility for one compound may be not enough solubility for another compound if the clinical dose is higher.

Q6.3 Why should solubility be expressed in molar concentration at lead optimization stage?

When we express solubility in molar term, we define the maximum number of molecules that can be dissolved under given conditions. Therefore, this measure of solubility is suitable for comparing the molecules and for relating it to structural modifications. If we express the solubility in milligram per milliliter unit, a molecule with larger molecular weight seems to be more soluble, in spite of the smaller number of molecules in solution. The potency of the molecules is also expressed in molar concentration at early drug discovery, stating the number of molecules that cause a 50% effect (inhibition or activation) on a receptor. When we would like to check whether the compound is soluble in the screening solution, we also need to express solubility in molar term. When a drug molecule is prescribed, the quantity is expressed in milligram and not in molar units.

Q6.4 There are two compounds A and B with molecular weight 250 and 500, respectively. Both compounds showed aqueous solubility 10 μg/ml. Which compound is more soluble in terms of the number of molecules dissolved?

Compound A is twice as soluble as compound B. Ten microgram of compound A contains twice as many molecules as 10 μg of compound B, as the molecular weight of B is twice of A.

Q6.5 Which properties of molecules have an effect on their solubility?
In principle, the hydrophobicity and the crystal lattice energy are the most important properties that determine aqueous solubility. As lipophilicity depends on the size, dipolarity/polarizability, H-bond acidity, and H-bond basicity, these are important molecular characteristics that influence aqueous solubility. Larger molecules with less polar functionality are less soluble in water. The crystal lattice energy depends on the shape of the molecule and the orientation of functional groups that can interact with each other in the solid form. Thus, planar shape can cause π–π crystal stacking; appropriate position of H-bond donor and acceptor groups can interact in the solid form and increase the crystal lattice energy.

Q6.6 Why is the MP of the solid material used to estimate solubility in the Yalkowsky equation?
The Yalkowsky equation relates the logarithm of molar aqueous solubility to the logarithm of octanol/water partition coefficient and to the MP. The MP is proportional to the crystal lattice energy. Higher MP indicates the higher energy to break the crystal lattice.

Q6.7 How does the pH affect the solubility of ionizable compounds?
Ionization increases the solubility, as it decreases lipophilicity. Ionized, charged species are easily solvated by water molecules that have high dipole moments. The Hendersson–Hasselbalch equation can be used to calculate the percentage of ionized species at a given pH. The further away is the pH from the pK_a values of the compound, the higher percentage is in ionized (or unionized) state. As solubility is inversely related to lipophilicity, we can expect a three to four orders of magnitude increase in solubility by one ionized group (one charge).
However, we cannot predict accurately the solubility enhancement of the molecule caused by one charge. It also depends on the counterion in the aqueous solution. The counterion may form a strong crystal lattice with the compound in the solid form; thus even if the compound is charged, it may not be soluble in water.

Q6.8 How does the presence of 5% DMSO in the aqueous buffer affect the solubility?
DMSO is, in general, a good solvent for a large variety of organic compounds. Therefore, when it is added to water, it enhances the aqueous solubility of organic compounds. The solubility enhancement depends on the solubilizing factor of the cosolvent and its concentration. In general, the solubility does not increase significantly up to 10% DMSO in water. There is an average of 1.5 times increase in solubility when the DMSO concentration in the test solution is increased from 2% to 5%.

Q6.9 Explain the enhanced solubility in simulated intestinal fluids.
There are two reasons that simulated intestinal fluids may increase the solubility of the compound in comparison to pure water solubility. First, the pH of the fluid may enhance the ionization of the compound, depending on the pK_a value. Second, it contains phosphatidyl choline–taurocholate micelles. Positively charged lipophilic compound may partition into the micelles, thus enhancing the apparent solubility of the compound.

Q6.10 Is solubility enhanced in simulated intestinal fluids for every compound?
No. Compounds that have low solubility because of high crystal lattice energy and no ionizable group will not show solubility enhancement in simulated intestinal fluids.

Q6.11 Describe a simple method for measuring DMSO precipitative and thermodynamic solubilities.

Typical solubility measurements at early drug discovery stages are performed from DMSO solutions by diluting down the compounds with pH 7.4 buffer. After an appropriate equilibration time, filtration or centrifugation is applied to separate the solution from the solid. The concentration of the compound is determined by a suitable analytical method, such as UV, HPLC/UV, or chemiluminescence nitrogen detection (CLND) in comparison with a standard solution containing known amount of dissolved compound.

The classical determination of solubility from solid material is the shake-flask method that allows the solid particles to equilibrate with the solvent, after which a concentration determination of the saturated solution is carried out. In order to obtain a true thermodynamic solubility, the temperature, the pH, and the equilibration time should be well defined. The pH should be adjusted to ensure that the compound is in an unionized state in order to obtain the intrinsic solubility of the compound. The solid form of the material should also be characterized, and the crystal form should be stated with the solubility value.

Q6.12 Which experimental conditions should be taken into consideration for the measurement of thermodynamic solubility?

The thermodynamic solubility of a compound refers to the concentration of the saturated solution that is in equilibrium with the solid material. Therefore, the temperature has to be defined together with the solvent pH and ionic strength, as these conditions also affect the thermodynamic equilibrium between the solid and the dissolved particles.

Q6.13 What is IDR?

The dissolution rate is the amount of compound dissolved in a unit of time, starting from the solid form of the compound. It depends on the solubility of the compound, the surface area between the solid and the solvent, the thickness of the diffusion layer around the solid that contains the saturated concentration of the compound, and of course, the diffusion coefficient of the compound. Thus, the surface area between the solid and solvent should be kept constant; the speed of stirring (or rotating the disk containing the solid material) should be controlled, as it affects the thickness of the diffusion layer. The IDR can be defined as the increase in the concentration of a pure potent molecule per unit time, where the surface area, temperature, agitation, the pH, and ionic strength of the media are controlled and kept constant.

Q6.14 What experimental conditions should be kept constant to determine the IDR of a compound?

The surface area of the solid, which is in contact with the solvent, temperature, agitation, pH, and ionic strength of the aqueous solution, should be kept constant for the determination of IDR.

Q6.15 Why do we need to measure the rate of dissolution as well as solubility?

Although the dissolution rate usually shows good correlation with solubility, it is not always the case. Even if a drug molecule has good solubility, it should fully dissolve during the transit time in the gastrointestinal tract in order to be absorbed. Therefore, it is essential to measure the dissolution time of a drug from a given formulation, in order to ensure maximum absorption.

CHAPTER 7

Q7.1 What is the major constituent of biological membranes?

Biological membranes, cell membranes, consist of phospholipid bilayer. The phospholipids are glycerine derivatives. Two of the hydroxyl groups form esters with fatty acids of various lengths with varying saturation, whereas the third hydroxyl group forms substituted phosphate ester. The phosphate group is negatively charged. When the substituent is choline or ethanolamine, there is an additional positive charge on the molecule. Various amino acids can also form esters with the phosphate group, such as serine, but sugar molecules such as inositol can also be attached. These molecules have a very strong amphiphilic nature, which means that one part of the molecule is polar (containing charges), whereas the other part of the molecule consists of hydrophobic nonpolar alkyl chains. It is difficult to dissolve these molecules in water or nonpolar solvents, as they prefer to be at the interface. When the phospholipid molecules are unable to organize themselves at an interface, they can self-associate forming micelles, liposomes, or bilayers. The most frequently used artificial membrane component is phosphatidyl choline, which is the major component of lecithin most often isolated from egg yolk.

The polar head groups of phospholipids cover the outer sides of the membrane, whereas the inner part consists of completely nonpolar hydrocarbons. The polar head group is often covered with carbohydrates and proteins and hydrated with structured water molecules. The inner part of the membrane contains saturated or unsaturated hydrocarbon chains. The hydrophobic forces between the fatty acid chains hold the membrane firmly, but at the same time, they make it very flexible too. There are many "holes" in the membrane that are usually covered with proteins. There are ion channels in the membranes that restrictedly let through various ions. The mechanical strength of the cell membrane is truly amazing, as the membrane thickness results from the length of the alkyl-chain inside the membrane; therefore, it is extremely thin. The surface characteristics of the membrane depend on the nature of the polar head groups.

Q7.2 What are the two major reasons for permeability measurements in drug discovery?

The drug molecules have to be absorbed and most of them have to get inside the cells. Therefore, the membrane permeability of drug molecules is an important factor for delivering the drug to the target. For orally administered drugs, the first barrier is the intestinal wall. The drug has to go through the intestinal cell wall, which requires permeability. The drug molecules have to permeate through the cells (intracellular permeability). The second reason to design permeable molecules is to get the active compound inside the cell. A drug molecule with impaired permeability shows lower potency in the cellular assays in comparison to the enzyme assays. When the cellular potency drops off, medicinal chemists need to understand the reason, and one of the possible causes is low membrane permeability.

Q7.3 What are the major factors that influence the rate of permeation?

Permeability is expressed as the rate by which compounds go through membranes. The basic principle of permeability is based on Fick's law of diffusion. It describes the spread of the molecules from the high concentration to the lower concentration side via diffusion. The major driving force is the concentration difference. The most important factors that influence the rate of transport through biological membrane are the diffusion coefficient of the compound (D), the membrane partition coefficient (K), the surface area of the membrane (A), the membrane thickness (h), and the concentration of the compound in the donor (apical) and acceptor (basolateral) side of the membrane (c_a and c_b, respectively). The diffusion coefficient of the compound can be described by the Stokes–Einstein equation applied for membranes and valid for small spherical molecules. The diffusion coefficient is smaller for bigger compounds. Temperature increases the diffusion, whereas the viscosity of the solvent (in biological permeability, it is the viscosity of the water with possible dissolved proteins and other small molecules) decreases it. The rate of transport is bigger when the membrane surface (A) is bigger. The flux is the normalized rate of transport per surface area unit. The membrane partition coefficient (K) is an important factor as well, because the compound should be able to partition into the lipid membrane from the aqueous phase to be able to go through. However, when the compound is too lipophilic, it might stay in the membrane, which means that the equation is true only up to a maximum K. At high K values, the rate of transport decreases because of the membrane retention.

Q7.4 Explain the experimental setup for *in vitro* permeability measurements.
We can use 24-well to 96-well plate formats. Therefore, the assay is called *parallel artificial membrane permeability assay* (*PAMPA*). We need special plates that contain an upper layer with a filter bottom part that fits into a lower layer of wells. It is possible to shape the lower layer to add an extra curve that can be accessed from the top by an autosampler needle or the tips of a liquid-handling robot. This makes it possible to sample from both the top and the bottom compartments without dismantling the plates. It is advisable to use a liquid-handling robot to form the membrane and add accurate volumes of buffer in the donor and acceptor compartments. The phospholipid membrane component in PAMPA has to be dissolved in an organic solvent with low boiling point that evaporates quickly after it has been spread onto the filter. Alternatively, the phospholipid, typically egg lecithin, can be dissolved in dodecane. Five or more microliters of the lipid solution is applied to a 125-μm thick PVDF microfilter with 0.45-μm pores by a liquid-handling robot. After a few seconds, the membrane covers the filter evenly, and the membrane has to be wetted by aqueous buffer from both sides. Then the sample has to be introduced to the donor compartment, which is typically the lower part of the assembly. The donor compartment contains typically 200 μl buffer solution that contains 10–100 μM test compound. When the sample is added in a DMSO stock solution, it might precipitate out when diluting down with the buffer. This has to be avoided, and the sample has to be dissolved in the buffer. Precipitation causes low permeability results that have been often observed for low solubility compounds. The acceptor compartment contains another 200 μl buffer solution when the experiment is performed in a 96-well assay format. Other protocols may suggest adding cyclodextrin or other solubilizing agents to the donor and acceptor compartments. The equilibration procedure starts after the donor and acceptor compartments are assembled with the membrane-coated filter placed between them and the sample in the donor compartment. The equilibration takes place for 2–4 h with small membrane thickness and efficiently stirred system. Then we have to take the samples out from both the donor and the acceptor compartments and measure the compound concentration at several time intervals using sensitive HPLC/MS generic analytical methods. Some experimental methods describe the concentration measurement of the initial donor compartment and, at the end of the assay, determine the concentration of the test compound in the acceptor

compartment. It is always advisable to check the mass balance, that is, the sum of the amounts of the test compound in both the donor and the acceptor compartments. When it is less than 100%, it suggests that the test compound has accumulated in the membrane. The other important quality control test is the homogeneity of the artificial membrane surface. The calculation of the apparent permeability is carried out from the accurate volumes of the donor and acceptor compartments (V_d and V_a, respectively), the surface area of the membrane (A), and the peak area (or absorbance) ratio of the test compounds (R). The following equation is used to calculate P_{app}, using the time point (t) when the analysis has been carried out and expressed in seconds.

$$P_{app} = -\frac{V_d \times V_a}{(V_d + V_a)A \times t} \times \ln(1 - R)$$

Q7.5 How is HPLC used for the permeability measurements?

HPLC is used for the concentration determination of the compound in the donor and the acceptor compartments before, during, and at the end of the permeability experiment. The concentration ratio of the compound (R) is used to calculate the apparent permeability of the compound.

Q7.6 How can we calculate the apparent permeability from effective permeability and the permeability through the UWL?

In order to calculate a compound's permeability both in the membrane and in the UWL, we can describe the effective permeability as the sum of the reciprocal values of the permeability in each layer as shown by the following equation.

$$\frac{1}{P_{eff}} = \frac{1}{P_0} + \frac{1}{P_{uwl}}$$

P_0 is the intrinsic permeability, which refers to the permeability of a neutral form of a molecule without its permeability through the UWL. P_{uwl} is the permeability of the compound through the UWL. For ionizable molecules, we need to add the reciprocal value of the permeability of the ionized species.

Q7.7 How does the UWL affect permeability?

As the concentration difference is one of the major driving forces for permeability, it is important to note that when the diffusion process is slow temporarily, the concentration near the other side of the membrane can be high, which reduces the rate of transport. It has been observed during the permeability measurements that shaking or stirring the acceptor compartment significantly influenced the rate of transport. It is due to the UWL, which is the aqueous layer around the polar head groups on the membrane surface. In this layer, the solute concentration is controlled purely by diffusion. It has been observed that stirring the aqueous compartments reduces the thickness of the UWL. Owing to the slower movement of the molecules in the UWL, the apparent permeability of some molecules measured lower than its actual membrane permeability. The thickness of the UWL depends on the agitation. The rate of transport through the UWL is controlled by the size of the molecules (diffusion coefficient).

Q7.8 What is "sink condition" and "double-sink condition"?

When the pH is different in the donor and acceptor sides, the so-called "sink condition" occurs for ionizable compounds. Typical "sink condition" for basic compounds happens when the pH in the gut is higher than that in the blood stream. The bases are less ionized in the gut at pH 8, for example. The percentage of the nonionized form is higher, which drives the compound through the membrane. The pH on the other side of the membrane is lower (pH 7.4), which means that when the neutral form of the base reached the other side, it gains a proton and a positive charge, so the concentration of the uncharged form does not increase as fast as it decreases on the other side. This creates a "sink," which means an extra driving force for permeability. The same happens with acids in the stomach, for example, where the pH is lower than that in the blood stream. The unionized acid is high on one side, but it does not increase on the other side because of the pH difference; the unionized molecules get ionized. When we want to imitate these conditions during the artificial membrane permeability measurements applying "sink condition" because of pH differences, the apparent permeability of the compound is increased. There may be other forces *in vivo* that result in the so-called "double-sink condition," which is due to the plasma protein binding of compounds. When the molecules that permeated through the membrane bind to plasma proteins, the concentration on the other side is reduced. It means that the decrease in concentration in the donor compartment is not equivalent to the increase in the concentration in the acceptor compartment. As the free concentration difference on the two sides of the membrane drives the permeability, the concentration difference does not decrease over time. The acceptor compartment behaves like a sink; the molecules that permeated through the membrane bind to proteins and are not present as free species. These conditions can be imitated in the artificial membrane permeability measurements, too. By applying different pH in the donor and acceptor compartments and using serum albumin in the acceptor compartment only, the "double-sink conditions" can be set up and studied. These "sink" conditions may occur *in vivo*, and these conditions usually accelerate the rate of transport of molecules through biological membranes.

Q7.9 How can we determine the intrinsic permeability?
The intrinsic permeability ($\log P_0$) refers to a speed of the compound when going through a membrane in its neutral form. The *intrinsic permeability* does neither include any sink condition nor the slowing effect of the diffusion in the UWL that covers biological membranes. It can be determined using pH conditions where the molecules permeate through the membrane in their unionized form.

Q7.10 Which of the molecular descriptors reduce permeability?
The Abraham solvation equations established for permeability and human intestinal absorption reveal which molecular descriptors affect permeability in which direction. The equations suggest that H-bond donor and acceptor groups reduce permeability, whereas polarizability slightly favors permeability. The size has no or little influence, which might be misleading. It is possible that the permeability of small lipophilic molecules increases with size, but when the size increases above a certain point the permeability decreases. This would result in an overall low coefficient of the size, but it does not mean that permeability does not depend on it. The solvation equation obtained for cell permeability shows great similarity to the equation obtained for diffusion coefficients in water. It suggests that during the cell permeability measurements, the speed of the molecules is dominated by the diffusion through the UWL. The permeability through the artificial membrane, especially when agitation is applied, is controlled by the membrane partition of the compounds. The ionization, which is pH dependent, has a unique effect on permeability. While ionization in general reduces lipophilicity and increases solubility, it decreases permeability.

Q7.11 How does lipophilicity influence permeability?

When comparing permeability with lipophilicity, the permeability usually refers to the membrane permeability and excludes the permeability through the UWL. The effective membrane permeability should be compared with $\log P$, the lipophilicity of the neutral form of the molecules, whereas the apparent permeability at a given pH should be compared to the $\log D$, the lipophilicity of the compound at that given pH. It is important to remember that while lipophilicity is a dimensionless measure (the equilibrium concentration ratio of the compound in an organic and aqueous solvent system), permeability is a rate. A parabolic relationship between apparent permeability and lipophilicity expressed by octanol/water partition coefficients is often observed. It means that permeability is low for hydrophilic and very hydrophobic compounds too. Compounds that have $\log D$ values around 2 and 3 are highly permeable. Compounds with very high lipophilicity show low permeability, as they partition into the membrane, but it is energetically more favorable for them to stay in the lipophilic membrane environments than coming out to the aqueous polar environment on the other side. The reason for the decreasing permeability of highly lipophilic compounds is due to membrane retention.

Q7.12 How does solubility influence permeability?

Both permeability and solubility are governed by lipophilicity, although in the opposite direction. While lipophilicity is essential for permeability, it reduces the aqueous solubility of the compound. An optimum lipophilicity is required for good permeability. Membrane partitioning decreases with the presence of H-bond donor and acceptor groups in the molecules, which is usually not the case for solubility. While hydrophobicity increases with increasing size, apparent permeability reaches an optimum; after that, the further increase in molecular size will reduce the apparent permeability because of the slow diffusion of the molecules in the UWL. While the presence of charge increases solubility and decreases lipophilicity, it also decreases permeability, as charged molecules cannot go through the membrane. Both solubility and permeability are important for good oral absorption, but these properties are difficult to optimize. Increasing polarity, H-bond acidity and basicity, and the introduction of ionizable group increase solubility, but they are usually detrimental to good permeability.

CHAPTER 8

Q8.1 What is the definition of pK_a?

In a simplified way, we can consider the pK_a value as equivalent to the pH of the solution where a functional group of a compound is in 50% ionized and 50% nonionized form, supposing dilute solutions. The pK_a always refers to one particular ionizable group on the molecule. Thus, a compound may have several pK_a values, both acidic and basic.

Q8.2 What types of techniques can be used for the pK_a determination?

Any technique can be used for the determination of pK_a values that can measure the property of the compound that changes because of the ionization. Thus, lipophilicity, solubility, UV spectrum, and conductivity measurements are suitable for pK_a determination. The general principle is that we measure the compound property at various pH values and plot the property in the function of pH. Then we can read the pH when the property is in between the ionized and unionized form of the functional group and calculate the pH at which the functional group is in 50% ionized form.

Q8.3 How does the measured pK_a depend on the ionic strength of the solution?
The effect of ionic strength on the acid dissociation constant can be calculated using the Debye–Hückel equation. The concentration of the compounds in ionic solutions cannot be considered as ideal even with low ionic strength. The concentration should be multiplied with the so-called activity coefficient (γ) that describes more precisely the activity of various species of molecules in solution, and the activity values should be used instead of concentration when describing chemical equilibrium constants. Thus, the pK_a values of compounds differ in solutions with varying ionic strength, where the activity constant depends on the concentration of the other ions in solution. The activity coefficient takes into account the interaction energy of ions in solutions. Typically, both acids and bases get stronger at higher ionic strength. To be able to compare the pK_a values of various molecules, it is extrapolated to zero ionic strength.

Q8.4 How does the organic cosolvent influence the acid dissociation constants?
The pK_a values refer to aqueous environment; more precisely, it refers to water only. When the water is mixed with organic solvents or the compound is near the membrane or proteins, its pK_a shifts depending on the H-bond donor and acceptor property of the solvent and its dielectric constant, dipolarity/polarizability. The Born equation describes changes in the pK_a values (ΔpK_a) of the compounds because of the electrostatic effects of the solvents, as shown in the following equation:

$$\Delta pK_a = \frac{121.6n}{r} \times \left(\frac{1}{\varepsilon} - 0.0218\right)$$

In the equation, n is the number of charged species in the ionization process, r is the common radius of all the ions, and ε is the dielectric constant of the solvent. Typically, when the water is mixed with organic solvents or proteins and membranes are in the aqueous compartment, the dielectric constant, H-bond acidity, and H-bond basicity of the surrounding are decreased relative to the neat water, which results in the weakening of acids and bases. The pK_a of an acid increases and the pK_a of a base decreases, which means the ionization process is suppressed.
The pK_a measured in a range of water–solvent mixtures (p$K_{a\,s}$) usually gives a straight line when plotting the concentration of the organic solvent and the obtained pK_a values. The slope of the straight line depends on the dielectric constant (ε) of the mixture, and the intercept provides the extrapolated pK_a for the pure water as shown in the following:

$$pK_{a\,s} + \log[H_2O] = \frac{a}{\varepsilon} + b$$

When there are multiple pK_a values in the molecule, especially when both acidic and basic pK_a's are present, the slope of the Yasuda–Shedlovsky plot helps identify which is the acidic and which is the basic pK_a. Acidic groups have a positive slope value, whereas basic groups have negative slope values. However, there are exceptions to this rule. It may happen that some of the compounds with sulfonamide acidic functional group show negative slope values.

Q8.5 What are the limitations of the potentiometric pK_a determination?
A relatively large quantity of pure compound is required for the measurements. The limitation of the potentiometric titration is that it requires a relatively high

concentration of test substance (20–500 μM) in a reasonable large volume, 2–20 ml. Potential drug molecules are often insoluble when they lose ionization; thus, precipitation might occur during titration. It is possible to use cosolvents such as methanol, acetonitrile, and DMSO solution. When a cosolvent is used, the pK_a values have to be determined using various concentrations of the organic solvents. Plotting the obtained pK_a values in the function of organic solvent concentration (Yasuda–Shedlovsky plot), usually a straight line can be obtained for which the intercept provides the extrapolated pK_a value to pure water. The method cannot be considered high throughput.

Q8.6 What are the advantages and disadvantages of the spectrophotometric determination of pK_a?

The spectrophotometric method can be used only when the ionizable group is close to a UV chromophore. In this case, the presence or absence of a proton on the ionizable group will shift the UV absorbance of the compound depending on the pH of the solution. The accurate determination and detection of pK_a is easy with the spectrophotometric method; however, it can neither provide information about the group that is ionized nor can it reveal whether the molecule lost or gained proton (charge) during the titration process. Therefore, an expert has to look up the structure and assign the pK_a values to the appropriate functional group. The expert has to assign the ionization center whether it is basic or acidic.

The major advantage of the spectroscopic method is that it is very sensitive and a low concentration of sample is needed. This might prevent precipitation at pH values where the molecule has the least amount of charge.

The UV spectroscopic method provides an advantage over the potentiometric method in that it requires a smaller amount of sample; therefore, solubility is less likely to disturb the measurement. In general, the method is faster than a potentiometric titration that requires proper equilibration of the pH electrode. It can be automated by continuously monitoring the spectrum under continuously changing pH condition. However, it can be used only when the compounds have UV absorbance and the UV spectrum changes because of the ionization state of the molecule.

Q8.7 How can we increase the speed of the spectroscopic pK_a determination?

It can be automated by continuously monitoring the spectrum under a continuously changing pH condition. This is the principle of the Sirius SGA (Spectral Gradient Analysis) instrument. It applies a fast pH gradient using an HPLC gradient pump and a diode array detector to monitor the spectrum change.

Another possibility of high throughput spectrophotometric pK_a determination is to use a liquid-handling robot equipped with a UV plate reader. The robot can perform parallel titration in a 96-well plate format with 1 or 2 ml volumes per well. The test compounds can be dispensed in DMSO solution, preferably in small amounts of DMSO, as it has high UV absorption below 230 nm. The sample solution is diluted down with a base or an acid, and the UV spectrum of each well is recorded after each addition of small amounts of acid or base titrating solutions. The procedure can be fully automated.

Q8.8 What are the principles of the pK_a determination by CE?

The electrophoretic mobility of the compound depends on the ionization state of the molecule. Therefore, measuring the migration time of the compound at various pH values provides the possibility of estimating the pK_a of the compound. A titration curve can be constructed by plotting the migration time of the compound in the function of pH. Then, using the traditional methods of analyzing the titration curve, we can read the pH value from the inflection point where the compound is 50% ionized. This pH provides the pK_a value.

The CE technique does not require expensive columns similar to HPLC; capillaries can be cut to the desired length and are available in several diameters. The sample requirement for the measurements is very small; 20–100 μM material can be easily detected. The solvent consumption is minimal, and only small percentages of organic solvent are applied, which is environmental friendly. It can be automated; moreover, multiple parallel measurements can be easily performed. It has the advantage that the pK_a measurement is based on the migration time measurements at several pH values, so there is no need for concentration determination of the test compound. However, it is difficult to measure the actual pH in the capillary; the composition of various buffers to adjust the pH in a wide range has to be varied that introduce variability in the electroosmotic flow.

Q8.9 How have big pharma companies increased the throughput and reduced the limitations of the pK_a determination by CE?

Pharmaceutical companies have become interested in the methodology as the previously described methods for pK_a determination were difficult to automate and perform on large amount of sample.

They have introduced miniaturization and checked the applicability of the method for compounds with multiple pK_a values. Later they applied pressure to accelerate the experiment, and a diode array UV detector has been attached to the capillaries to detect the spectral changes of the compound in the function of pH. In this way, they obtained the mobility change of the compound as well as the UV spectrum change caused by the pH variations. The run time has become less than 1 min. They validated their methodology using 77 drug molecules with 82 pK_a values.

AstraZeneca and Pfizer followed to set up fast pK_a measurements. Because of the low solubility of discovery compounds that would cause capillary blockage and stop the automated measurements, they have investigated the applicability of acetonitrile, up to 60% for the analysis of basic compounds. Some authors applied the pressure-assisted capillary electrophoresis with mass spectrometric detection. They applied volatile buffers between the pH range 2.5 and 10.5 that increased the sensitivity of the MS detection and has made it possible to pool samples. Another approach is to use multiplexed capillaries connected with detector equipped with multiple UV cells. In this way, the electrophoretic mobility of one test compound at 24 different pH values can be measured simultaneously.

Q8.10 How does the reversed phase retention factor depend on the mobile phase pH?

The first mathematical description of the pH dependence of reversed phase retention of partially ionizable compounds is shown by the following equation.

$$k = \frac{k_u + k_i^{[H^+]/K_a}}{1 + \dfrac{[H^+]}{K_a}}$$

where (k) is the retention factor, (K_a) is the dissociation constant and (k_u) and (k_i) are the retention factors of the fully unionized and the fully ionized species for monoprotic acids, respectively. In principle, the retention factor–pH relationship is similar to the distribution coefficient–pH relationship, and the curve can be described by the

Henderson–Hasselbalch equation. In the past, the limitation in investigating a large pH range was the pH stability of the silica-based reversed-phase stationary phase. The silica gel dissolved, and the phase quickly decomposed when higher than pH 8 mobile phase was used. It is important to remember that if we plot the logarithmic retention factor in the function of pH, the pK_a is not the pH value at the inflection point of the curve. It is the pH when the logarithmic retention factor is log 2 (0.301) less than the logarithmic retention factor of the unionized form of the molecule. The problem of using HPLC retention factor measurements as a function of pH is that we may need to apply an organic modifier in the mobile phase in order to elute the compound in a reasonable time, and the organic phase shifts the mobile phase pH and the pK_a of the compound. It is tedious to measure the retention factor using several mobile phase pH and the procedure is more time consuming than a potentiometric titration.

Q8.11 How can we use the reversed phase gradient chromatography for the determination of pK_a and the acid/base character of compounds?

A simple generic gradient retention time, or its calibrated form, the chromatographic hydrophobicity index (CHI), obtained at several pH values would be sufficient. The CHI values of acids and bases depend on the starting mobile phase pH. Therefore, to determine the ionization state of the molecules, the gradient retention times or CHI values have to be determined using mobile phases with various pH values. The so-obtained CHI versus pH profiles resemble a titration curve. The CHI values are proportional to the logarithmic partition coefficients of the compounds in the mobile and stationary phases, and the increasing CHI value means increasing acetonitrile phase concentration in the mobile phase. It means that the more lipophilic a compound is, the higher the acetonitrile concentration needed for the elution. At higher acetonitrile concentration, the mobile phase pH is lower than the starting mobile phase pH. The pK_a also shifts higher for acids and lower for bases at high concentration of organic solvent. These conditions contribute to the distortion of the theoretical titration curves constructed from gradient retention time versus mobile phase pH. In order to get the CHI versus pH profile for a large number of compounds, the gradient retention time measurements have to be repeated at least 10 times using 10 different starting mobile phase pH. Therefore, the methodology is not suggested for direct pK_a determination. Instead, the CHI values are determined at three different starting mobile phase pH—acidic, neutral, and basic. From the three CHI values, we can determine whether the compound is neutral, acidic, or basic and can approximate the percentage of ionized form at neutral (physiological pH). A slower but more accurate method for the determination of pK_a using gradient elution is to determine the retention time of the compound using pH gradient only with constant organic phase concentration. From another organic phase gradient that is applied at the pH where the compound is not ionized, the lipophilicity of the neutral form can also be determined. Thus, from the two gradient retention times (pH gradient and organic phase gradient), the pK_a values and the lipophilicity of the neutral form can be calculated.

Q8.12 What are the limitations of the chromatographic determination of the acid/base character?

There are several limitations or disadvantages in the chromatographic method. It is time consuming to measure the retention time of the same compound using several mobile phases with varying pH. For lipophilic compounds, it is inevitable to use an organic modifier in the mobile phase, which alters the pK_a of the compound and the pH of the mobile phase, so corrections have to be made. The silica-based stationary phases are not very stable with alkaline mobile phases.

Q8.13 Can we use the pH dependence of solubility for estimating pK_a?
Yes, in principle, we can use the pH dependence of solubility for the estimation of the acid dissociation constant of the compound. It is because the aqueous solubility of a compound depends highly on the pH and the ionization state of the compound. Ionized forms are much more soluble in water. However, the procedure is very time consuming; we cannot estimate the solubility of the fully ionized form of the compound, as it may form different crystals with different counterions. Thus, the determination of the inflection point, where the 50% ionized compound's solubility is, is not very accurate.

Q8.14 How can we calculate the fraction of the ionized form at any pH from the pK_a values?
The percentage of the ionized molecules depends on the proton concentration (pH) and can be calculated at any pH using the Hederson–Hasselbalch equation. The following equation describes the relationship between the percentage of ionized molecules and pH with given pK_a.

$$\%\ \text{Ionized form} = \frac{100}{1 + 10^{(\text{charge}(\text{pH}-\text{p}K_a))}}$$

The charge is −1 for acid and 1 for basic compounds, and it is assumed that only one charge is present on the molecule.

CHAPTER 9

Q9.1 How can the Lipinski "rule of five" be related to solubility and permeability?
The Lipinksi "rule of five" states that poor absorption or permeation is more likely when

1. there are more than five H-bond donor groups (expressed as the sum of OH and NH groups) in the molecule;
2. the molecular weight is more than 500 Da;
3. the $c \log P$ is more than 5;
4. there are more than 10 H-bond acceptor groups in the molecule (expressed as the number of nitrogen and oxygen);
5. compounds that are substrates for biological transporters are exceptions.

Although none of the rules mention solubility and permeability, the size, lipophilicity, and the number of H-bond donor and acceptor groups can be related to solubility and permeability. Large and lipophilic compounds are not soluble, which means that rule 2 and 3 consider the factors that affect solubility. Compounds containing more than five H-bond donor groups are not permeable. H-bond acceptor groups increase the polarity of the compounds, which again has a negative effect on permeability.

Q9.2 How can we model the absorption using the lipophilicity and the size of the molecules as descriptors?
The $\log D$ versus cMR model for absorption visualizes the effect of lipophilicity and size. It has been found that the larger the molecule, the more lipophilic it should be in order to be absorbed orally. The lipophilicity versus size model can be used as

defining the area on the plot where the properties of the molecules are not advantageous for absorption. There can be a defined discrimination line too, which separates compounds with good and poor absorption. The log D versus cMR model can be considered as similar to the solubility and the permeability requirements for absorption. Large hydrophilic molecules are not permeable, and they may have solubility problems as well. The simultaneous presence of H-bond donor and acceptor groups may increase the crystal lattice energy as H-bonds can be formed in the crystals. Large molecules also require more energy to create a solvation cavity in water with the breaking of the H-bonds between the water molecules.

Q9.3 What are the limitations of the log D–cMR model of absorption?

The limitation of the log D–cMR model is that it does not include the potential loss of absorption due to high lipophilicity. Not too large but very lipophilic compounds have low solubility that prevents oral absorption.

Q9.4 What are the criteria of good absorption based on the definition of Class I compounds by the BCS?

Class I compounds have good solubility and lipophilicity.

Good solubility is defined as follows:

A drug substance is considered *highly soluble* when the highest dose strength is soluble in less than 250 ml of water over a pH range of 1–7.5.

A drug product is considered to be *rapidly dissolving* when greater than 85% of the labeled amount of drug substance dissolves within 30 min using USP apparatus I or II in a volume of less than 900 ml of buffer solutions.

This definition is very different from the thermodynamic definition of solubility. It includes the dose; drugs taken in lower dose require lower solubility than drugs taken in higher dose. Therefore, it is not possible to provide a universal cutoff for good and poor solubility.

The good permeability for Class I compounds is defined as follows:

A drug substance is considered *highly permeable* when the extent of absorption in humans is determined to be greater than 90% of an administered dose, based on the mass balance or in comparison to an intravenous reference dose.

The definition does not contain an absolute number for the permeability cutoff. More than 90% of the administered dose should appear in the blood stream for highly permeable compounds.

Q9.5 How can the MAD be calculated?

For the calculation of the MAD, we need to define first the AP. The AP is a single dimensionless number that excludes the effect of the first-pass metabolism and metabolism in the gut. The fraction of a dose absorbed (F_{abs}) is a function of the permeability of the gut wall (P_w), the aqueous permeability of the drug (P_{aq}), the intrinsic aqueous solubility of the neutral form of the drug (S_0), the dose (X), the unionized fraction of the drug at pH 6.5 (F_{non}), and the volume of the luminal content (V_L) as shown by the following equation:

$$F_{abs} = f(P_w, P_{aq}, S_0, X, F_{non}, V_L)$$

Making a few further assumptions, such as having only the neutral form of the drug being permeable and the average pH of the intestinal lumen where the majority of absorption takes place is pH 6.5, the variables of interest can be arranged in a dimensionless number called *absorption potential* (*AP*) and can be described by the following equation:

$$\text{AP} = \log\left(P \times F_{\text{non}} \times \frac{S_0 \times V_{\text{L}}}{X}\right)$$

The AP shows a sigmoidal relationship with the fraction absorbed. In principle, the AP is proportional to the product term of intrinsic solubility, permeability, and the fraction of nonionized form of the molecule at pH 6.5. It means that if any of the product term is close to zero, that is, either the solubility or the permeability is poor, the absorption will be poor. The V_{L} parameter can be considered constant using the average volume of the intestinal fluid in the lumen. For a given series of drug discovery compounds, we can aim at a given dose (X), so it can also be considered constant. Therefore, the product term of solubility, permeability, and fraction of nonionized form can be used for calculating the AP. There is a qualitative relationship between the AP and the fraction of dose absorbed (FA) by passive mechanism shown as follows:

$$\text{FA} = \frac{(10^{\text{AP}})^2}{(10^{\text{AP}})^2 + F_{\text{non}}(1 - F_{\text{non}})}$$

The AP term was calculated from the logarithmic value of the product term of octanol/water partition coefficient instead of permeability, and intrinsic solubility, fraction of the nonionized form of the molecule at pH 6.5, and the luminal volume divided by the dose. The AP equation can be simplified using the lipophilicity and solubility of the ionized form of the molecule, thus avoiding the use of the fraction of nonionized form. We can substitute the luminal volume with 0.25 l and come up with a modified absorption potential (MAP) equation

$$\text{MAP} = \log\left(\frac{K_{\text{OW}} S_{\text{w}}}{4D}\right)$$

K_{ow} is the octanol/water partition coefficient, S_{w} is the intrinsic solubility of the drug molecule in water (mg/ml), and D is the dose in milligrams.
This equation can be used to calculate the MAD by introducing the transit time and intestinal surface area available for the compound to be absorbed. The absorbable dose also can be expressed from this that gives an estimate based on measured solubility and lipophilicity. The AP equations clearly show that there should be a balance between dose, solubility, and permeability. Lower solubility and/or permeability are acceptable for lower dose compounds, whereas the required solubility is much higher for higher dose drugs.

Q9.6 How does the pH affect the solubility and permeability and ultimately the AP?
The pH affects solubility and permeability of ionizable compounds in an opposite way. When the compound gets ionized, its solubility increases but permeability decreases. As the AP is the product term of solubility and permeability, the pH does not affect the AP. We need to make sure, however, that both solubility and permeability are measured at the same pH.

Q9.7 Which molecular descriptors have the most significant effect on oral absorption?
H-bond acidity and basicity have the most significant effect on oral absorption. The Abraham equation revealed that absorption is generally good for a compound (the constant term is around 90%) unless too many polar groups (H-bond donor and

acceptor) are introduced to the molecule. Polar groups reduce absorption. The size parameter has a small positive coefficient.

Q9.8 What are the limitations of the Abraham solvation equation models for absorption?
There are several limitations of the Abraham equation that model absorption.
The original solvation equation model refers to any solute property that is linear free energy related. However, the percentage or fraction absorbed data are not, strictly speaking, linear free energy related. The solvation equation model can be used only for neutral compounds; therefore, it does not take into account the effect of ionization.
Strictly speaking, the Abraham models make use of *in silico* calculated molecular descriptors as independent variables. The main advantage of the models is that they always use the same molecular property description, so it is possible to compare models without the necessity of using the same set of compounds to check correlations between various dependent variables. It is also not a "black box" model, which means that medicinal chemists can get information about the effect and the magnitude of changing certain molecular properties on absorption.

Q9.9 Which measured physicochemical properties can be used to estimate human intestinal absorption of drug discovery compounds and how?
The measured lipophilicity (log *D* or chromatographic lipophilicity, CHI, CHI log *D*) can be used together with the cMR for modeling absorption via the log *D* versus cMR model.
Solubility and permeability can be measured using the same pH. The product term of the two parameters is proportional to the AP of the compounds. The measured intrinsic solubility and permeability of the compound can also be used for estimating the passive intestinal absorption. Dissolution rate measurements are also useful, but they are used rather for characterizing the dissolution rate of the active ingredients from formulations.
The acid dissociation constant (pK_a) is also useful for assessing absorption of the compounds. Special care has to be taken when the compound is ionized at acidic pH, similar to that in the stomach, but might precipitate out from solution at basic pH values (reaching the duodenum and the small intestine).

Q9.10 What are the limitations of using only physicochemical properties in modeling absorption?
Although it is essential to understand the expected behavior of compounds based solely on their physicochemical properties, it is also necessary to mention at least some of the important biological processes that may change compounds behavior *in vivo*. When the physicochemical principles of the absorption process are investigated, it is inevitable to have an insight into the active transport and metabolic transformation processes. These processes are governed by specific structural features of the molecules, but the overall physicochemical properties of the molecules also play an important role. Thus, several acid transporters (MCT) may help the absorption of negatively charged acids. Hepatic uptake transporters help extract drugs with a molecular weight of more than 400 Da from the blood, and there are hepatic efflux transporters that help to excrete high molecular weight ionizable compounds into the bile. When the compound goes through metabolic transformation in the liver, it also reduces the oral exposure. The CYP isoforms are responsible for most of the biotransformations in the liver (see Reference [36] from Chapter 9). Thus, CYP1A2 (12% of total hepatic CYP) targets compounds with an average log *P* of 2 that contain

poly(hetero)aromatic amines and amides. The CYP3A4 is another major component of the CYP isoforms; it takes 30% of the total hepatic CYP, but it also can be found in the gut. CYP3A4 is responsible for metabolizing large and lipophilic compounds (up to log *P* 7) that are uncharged at physiological pH. Amides and amines are mostly metabolized by CYP2C19, whereas basic positively charged compounds are transformed by CYP2D6. There is an interesting association between some CYP enzymes with the important transmembrane pump protein, PgP expressed by the MDR1 gene. PgP can also be found in the blood–brain barrier protecting the brain from large lipophilic compounds. PgP works in concert with CYP3A4 to eliminate potentially harmful xenobiotics. If PgP is not able to efflux all the potential harmful exogenous compounds, then CYP3A4 transforms them. Interestingly, these processes mostly target large and lipophilic compounds whose properties have already been considered in several absorption models based only on the physicochemical principles.

CHAPTER 10

Q10.1 How can volume of distribution be related to the tissue/plasma distribution coefficients and various tissue partition coefficients?

The volume of distribution can be defined as the dose divided by the plasma concentration (in steady state with repeated dose). Thus, the volume of distribution is proportional to the product term of the dose and the plasma volume (V_p) divided by the amount of drug in plasma. The tissue partition coefficient can be expressed as the quotient of the concentration of compound in the bound tissue phase and the concentration of the compound in the free tissue phase (see Equations 3.7–3.10). Thus, the tissue partition coefficient is proportional to the volume of distribution (V_{ss}) minus the plasma volume (V_p) divided by the tissue volume (V_t):

$$K(\text{tissue/plasma}) = \frac{(V_{ss} - V_p)}{V_t}$$

Expressing V_{ss} from the equation, it can be considered as the sum of the various tissue volumes (V_t) multiplied with the tissue partition coefficient (K_{pt}) plus the plasma volume (V_p):

$$Vd_{ss} = \sum(V_t \times K_{pt}) + V_p$$

On the basis of the given equation, the total volume of distribution can be considered as proportional to the compound's partition coefficient between the tissue and the plasma compartments (the constants are the total tissue volume and plasma volume). The total volume of distribution can be considered as the sum of the volume of distribution of the compound into each tissue compartment. To be able to estimate the volume of distributions, we need to know the tissue volumes, tissue partition coefficient, plasma volume, and plasma partition coefficient of the compounds.

Q10.2 What are the principles of the *in vitro* estimation of volume of distribution?

The physicochemical principle of calculating the tissue/plasma partition coefficient of a compound can be visualized by the various partition coefficients of the compound in plasma and in tissues.

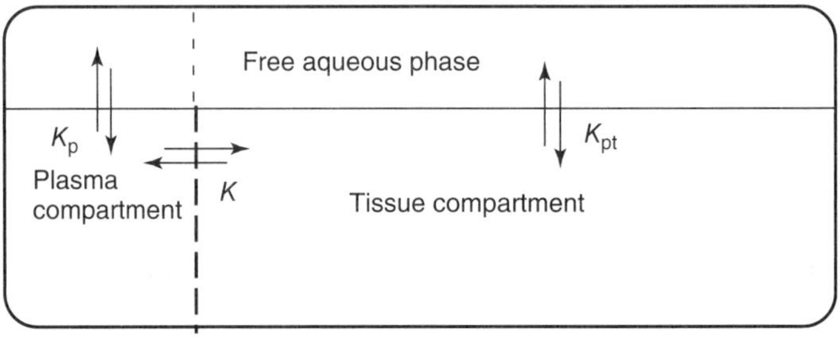

The tissue/plasma distribution coefficient (K) can be expressed from the tissue and plasma distribution coefficients (K_{pt} and K_p, respectively) of the compound according to the following equation:

$$K = \frac{K_{pt}}{K_p}$$

In vitro measurements of plasma protein binding and tissue binding can be used to calculate the *in vivo* volume of distribution. The *in vitro* measurements of tissue and plasma protein binding can be carried out using equilibrium dialysis measurements. We need to know the volumes of various tissues. A simple *in vitro* method for the estimation of volume of distribution is the application of biomimetic retention time measurements and the determination of serum albumin and phospholipid binding of the compounds. The difference between the phospholipid bindings (obtained on IAM stationary phase) and serum albumin binding (obtained on chemically bonded HSA stationary phase) turned out to be proportional to the human volume of distribution.

Q10.3 What are the major physicochemical properties of compounds that govern their volume of distribution?

The volume of distribution can be considered as proportional to the compound's partitioning between the tissue and the plasma compartments. The major difference between these two compartments is that the tissue compartment contains phospholipids, whereas the plasma compartment does not. Therefore, the difference in compounds binding to phospholipids relative to the plasma proteins is proportional to their volume of distribution. It has been observed that the lipophilicity of the compound increases both types of binding. As the tissue compartment is more lipophilic than the plasma compartment, the volume of distribution increases with the lipophilicity of the compound. The presence of a positive charge significantly increases the compound's binding to phospholipids and tissues, whereas it reduces the albumin binding. Therefore, a positive charge usually increases volume of distribution. The opposite is true for negatively charged compounds. They bind strongly to albumin and do not bind strongly to phospholipids. Negatively charged compounds usually have very low volume of distribution and they stay mostly in the plasma compartment. Lipophilicity and the pK_a values of the compounds can be calculated *in silico*.

Q10.4 Which *in silico* descriptors can be used to estimate the volume of distribution?

The volume of distribution can be modeled *in silico* with calculated lipophilicity (c log P) and the calculated percentage of the positively and negatively ionized from of the molecule. The percentage of ionization at pH 7.4 can be calculated from the calculated acid dissociation constant pK_a and the Hendersson–Hasselbalch equation.

Q10.5 Why are volumes of distribution of animals not very good models for volume of distribution of humans?

The volume of distribution depends on the volumes and the composition of various tissues. The volume of distribution is influenced especially by the lipid content of the body. The volumes and lipid contents are very different in humans and in animals, especially the percentage of muscle and adipose tissues. Various procedures that can be applied to estimate the human drug distribution from various species, for example, allometric scaling, have been published. Some techniques employ scaling factors such as protein binding, brain weight, body weight, or maximum life span potential. There are significant differences in the rat and human plasma compositions. This can cause significant differences in the measured plasma protein binding in rat or human plasma. The free plasma concentration of drugs that bind strongly to protein (AGP) potentially show even greater differences, as the AGP concentration in rat plasma is significantly higher than that in human plasma.

Q10.6 Describe the principle of a three-phase equilibrium process, for example, PPB, tissue binding, and tissue/plasma partition.

When there is no permeability barrier or active transport between various compartments in the body, the drug molecules distribute according to their partition coefficients in the plasma and tissue compartments. When the compound's distribution is independent of the dose, the concentration ratios between the free and bound phases in tissues and plasma are constant. The plasma distribution of the compound between the free and the protein-bound phase is constant. It depends only on the plasma protein binding affinity of the compounds. The same is true for the concentration ratio between the free and bound phase in tissues. The tissue components are different from the plasma components, tissue contains lipids, and, possibly, more negatively charged proteins such as in the lysosome. Therefore, the plasma and tissue partition coefficients are different for each compound. However, there is equilibrium between the free phases. On the basis of the "free drug hypothesis," we can assume that the free concentration of the drug is the same in plasma and in tissues. Therefore, the quotient of the tissue and plasma partition coefficients provides the third partition equilibrium constant of the compound, which refers to the tissue/plasma partition. It has been proved experimentally that, for example, the brain/plasma concentration ratio *in vivo* can be calculated from the *in vitro* measured brain tissue binding and plasma protein binding data.

Q10.7 What are the mathematical relationships between unbound fraction, percentage bound, and logarithmic partition coefficients?

The commonly used unbound or bound fraction of compound in plasma and tissues do not show linear correlation to the plasma and tissue partition coefficients. The sigmoidal curve describing the relationship between bound fraction and the logarithmic tissue partition coefficient means that at the strong binding region (above 90%), a small percentage increase in binding is the result in much larger increase in the partition coefficient. When models are built with linear-free-energy-related parameters, such as log *P*, the logarithmic values of the plasma and tissue distribution coefficients should be used.

Q10.8 How do the HSA and IAM binding correlate for a diverse set of drug molecules?
For a diverse set of drug molecules, the HSA and IAM binding does not show good correlation. Positively charged compounds at pH 7.4 bind more strongly to IAM than to HSA. The opposite is true for the negatively charged compounds; they bind more strongly to HSA than to IAM. Neutral compounds usually show the same trend in HSA and IAM binding. However, the correlation is not perfect, as there are different steric effects that influence HSA and IAM binding. In general, long-shaped molecules bind stronger to IAM than round-shaped molecules.

Q10.9 What methodologies can be used for the PPB determination? What are the advantages and disadvantages of these methodologies?
Plasma protein binding can be measured with pooled plasma using the equilibrium dialysis and ultrafiltration technique. The principles of these measurements are that the compound has to be added to the plasma. The unbound molecules can go through a semipermeable membrane until the unbound concentration becomes equal at both sides of the membrane. Molecules bound to the plasma proteins cannot go through the membrane. The total concentration of the drug and the free drug concentration, which is usually measured by HPLC, are used to calculate the unbound fraction. The advantage of the procedure is that it measures the unbound fraction in plasma that contains a mixture of proteins. The disadvantages are that the equilibrium procedure is slow, and decomposition of the drug may occur, the dilution factor by the buffer on the clear side of the membrane has to be calculated, and a sensitive analytical method is needed for the concentration determination. The ultrafiltration procedure starts with spiking the plasma with the compound and with waiting for the equilibration. Then the mixture has to be filtered using ultracentrifuge and a filter that does not let through the plasma proteins together with the bound compound. The compound concentration has to be determined by a sensitive analytical method in the filtrate. The total concentration and the unbound drug concentration are used to calculate the unbound fraction. Both the procedures are slow and difficult to automate, although multiple well plates can be used to make the procedure parallel for several compounds at the same time. A possible source of error is to apply a higher drug concentration than the concentration of drug under physiological conditions.
A higher throughput approach is to use chemically bonded albumin and α-1-acid glycoprotein HPLC phases and measure the retention time of the compounds in order to calculate the binding. This method is very rapid, and the retention time measurements are very accurate. It provides information about the compound's binding of each protein that may have very different structure binding properties; therefore, it is useful for identifying the structural features that can cause strong binding. The peak width and peak shape may provide information about the kinetic aspects of the binding, the onset and offset rates.

Q10.10 What is the importance of PPB in lead optimization?
Plasma protein binding influences the available free drug concentration that is responsible for the pharmacological activity. It also influences the rate of the clearance, metabolism. The free concentration of the drug may be also responsible for the side effects. However, plasma protein binding on its own is not enough to assess the free concentration of a drug available after administering a given dose. Plasma protein binding together with the tissue binding determines the free drug concentration. Tissues are present in much larger volume in the body

than in the plasma; therefore, the tissue binding has a significant effect on the free concentration. Strong plasma protein binding reduces the volume of distribution; thus, it reduces the amount of compound in tissues, which may be disadvantageous when the target is in the tissues. In general, it is advantageous to reduce the plasma protein binding, but we cannot establish a general rule how much it should be. When plasma protein binding causes disadvantageous pharmacokinetic/pharmacodynamic properties, the lead optimization process should focus on reducing it.

Q10.11 Why is AGP binding important?
AGP, α-1-glycoprotein, is a minor component of the plasma proteins. However, it has very different binding characteristic from albumin and globulins. It may specifically and strongly bind positively charged drug molecules. The AGP concentration of the plasma may vary in each individual; moreover, disease state, age, gender, and other factors may influence its concentration. Therefore, compounds that bind strongly to AGP may cause variable efficacy in clinic and variable PK/PD profiles. The AGP concentrations of the plasma of laboratory animals are very different from that of humans; thus it may result in wrong extrapolation of various PK/PD parameters in human. Therefore, it is important to identify strong AGP binding compounds early in the drug discovery process.

Q10.12 How can we model and estimate blood/brain distribution of compounds?
We can expect the brain-to-blood distribution of the molecules *in vivo* to arise from the different plasma and brain tissue binding of the compound. When several molecules have similar brain tissue binding, their differences in the plasma protein binding influences the *in vivo* available free drug in the brain. We can calculate the brain-to-blood distribution coefficient (K_p) from the unbound fraction in the blood and brain ($f_{u,blood}$ and $f_{u,brain}$, respectively, using the assumption that the free concentrations in the blood and brain ($c_{u,blood}$ and $c_{u,brain}$, respectively) are equal). The expected brain-to-blood ratio of a compound can be determined by plasma protein and brain tissue binding data obtained by equilibrium dialysis. The solvation equation approach can also be used for the estimation of the brain-to-plasma distribution of a compound. It was found that both H-bond acidity and basicity reduce the brain penetration of a compound. In this respect, it is very different from the octanol/water lipophilicity, which is not affected by the H-bond donor groups. It is generally accepted that for good penetration, we need an optimum lipophilicity of the compound, which is around $\log P = 2$. Brain/blood distribution model can be built using biomimetic HSA and IAM binding data, as well. The brain is protected by several active efflux and transport systems; thus, the models based on the physicochemical measurements will not be valid for compounds that interact with PgP efflux transporter.

Q10.13 What are the characteristics of unbound fraction in the brain, lung, mucus, and overall tissues?
Brain, lung, mucus, and tissue binding could be estimated using biomimetic HPLC binding data obtained on HSA, AGP, and IAM stationary phases. The coefficients of these binding data are very different in various tissues. The brain tissue binding is dominated by the IAM binding of the compound. Lung tissue binding is dominated by HSA and AGP binding. As the mucus contains highly glycosylated proteins, the mucus binding showed excellent correlation with the AGP binding obtained by HPLC.

Q10.14 What are the principles of PBPK modeling and how can the biomimetic HPLC measurements enhance them?

PBPK modeling methods are based on mathematical models for the estimation of various pharmacokinetic parameters using measured or calculated physicochemical parameters such as lipophilicity and acid dissociation constant of the compounds. It is physiological based, as it uses the average physiological volumes of plasma and various tissues and organs that are built-in as parameters in the equations. Other input parameters can be the tissue partition coefficients of compounds that can be measured using tissue homogenates. For example, a PBPK model for volume of distribution uses the sum of various tissue volumes multiplied by the tissue partition coefficients plus the plasma volumes. When the tissue partition coefficients are not available for a compound, we can use octanol/water partition coefficients, olive oil/water partition coefficients, albumin binding, and calculated parameters for the ionization state of the molecules. The biomimetic HPLC binding data may also be used in PBPK models, and, potentially, they improve their prediction power in comparison to other calculated and measured physicochemical parameters.

CHAPTER 11

Q11.1 Define the various ligand efficiency parameters, such as LE, BEI, SEI, LLE, and LLE_{AT}.

Ligand efficiency parameters describe various ways to assess the potency relative to size, the number of nonhydrogen atoms, or the overall polar surface area of the compounds. They were introduced for simultaneous optimization of the compound potency and physicochemical properties. The idea of the ligand efficiency parameters is not only providing help in the compound's design and progression based on the potency but also finding the most potent compound that is the least lipophilic and small. The most ligand-efficient potent compounds bind to the receptor mostly via stronger polar and steric-specific interactions and not by hydrophobic forces.

The original LE expresses the binding energy relative to the number of heavy atoms as shown in the following equation:

$$\text{LE} = \Delta g = \frac{\Delta G}{N}$$

where $\Delta G = -\text{RT}\ln K_\text{d}$ and N is the number of nonhydrogen atoms. The ΔG parameter can be obtained from the measured drug–receptor dissociation constant (K_d).

A simplified LE index is the BEI that can be calculated using percentage inhibition or potency expressed as pIC_{50} values relative to the molecular weight.

The importance of reducing lipophilic binding forces required taking into account the polar surface area of the compound together with the potency. This resulted in the introduction of the SEI that expresses the pIC_{50} relative to the calculated polar surface area.

The LLE takes into account the difference between the pIC_{50} and the calculated octanol/water partition coefficient.

Efficient binding is more likely with smaller compounds for two reasons. There is an inevitable reduction in the quality of fit between ligand and receptor when the ligand is big. Larger ligands tend to be more complex; this reduces the accessibility of ligand surface per atom bases.

Astex Pharmaceuticals proposed a modified ligand efficiency parameter that includes the heavy atom counts and the lipophilicity (LLE_{AT}) based on the modified free energy of binding (ΔG^*), which is the difference between the total free energy of binding and the free energy of binding because of lipophilicity. They have introduced a constant as well and came up with a formula (see the following) that brings the LLE_{AT} efficiency parameter to the same scale as the original LE.

$$LLE_{AT} = 0.111 + 1.36 \times \frac{LLE}{\text{Heavy atom count}}$$

This ligand efficiency parameter has been designed to evaluate the fragment compound's potential to achieve a drug-like lead molecule after optimization. The drug-like end point was estimated using the Lipinksi's rules assuming that a molecule that contains 36 heavy atoms is the equivalent of molecular weight of 500 Da with a 10 nM potency. This profile leads to the commonly used 0.3 kcal/mol per heavy atom LE value.

Q11.2 What is the advantage of early optimization of LE instead of the potency?

The application of these efficiency indices has already improved fragment selection for lead generation and helps reverse the trend of synthesizing potent but large and lipophilic molecules. The advantage that these efficiency indices offer is the combination of two or three variables (potency, size, and lipophilicity) into one entity that is much easier to optimize. The sequential optimization of potency and the physicochemical parameters requires going forwards and backwards by adding and removing substituents on the molecules, which is very inefficient. The ligand efficiency parameters also provide a quantitative way of comparing molecules by giving certain amounts of weights for size and lipophilicity as well as potency. In this way, we can define more accurately what potency is required for a molecule with a given size and lipophilicity on the bases of binding energies.

Q11.3 What is the definition of drug efficiency?

Drug efficiency ($DRUG_{eff}$) is defined as the percentage of the free biophase concentration relative to dose that is available at the site of action *in vivo*:

$$DRUG_{eff}(\%) = \frac{\text{Free biophase concentration}}{\text{Dose}} \times 100$$

This simple parameter takes into account all the factors that reduce the free biophase concentration *in vivo*. Thus, it includes the percentage of absorption after oral administration, the clearance, and the distribution of the compounds into tissues, plasma protein binding, tissue binding, permeability barrier, and active transport (efflux) processes as well. More efficient drugs produce higher free concentration at the site of action with the smallest possible dose. Drugs that have good absorption and smallest possible nonspecific binding to plasma proteins and tissues (proteins and phospholipids) have the highest $DRUG_{eff}$.

The maximum $DRUG_{eff}$ refers to the free concentration of the compound in plasma relative to dose assuming 100% absorption, no permeability barrier from the plasma to the target, and no active transport process take place. By definition, the drug efficiency refers to an *in vivo* situation, but it is possible to estimate it by measuring plasma protein binding and tissue binding *in vitro*.

Q11.4 What is the maximum drug efficiency and how is it related to unbound volume of distribution?

The maximum drug efficiency is the percentage of the dose that can be found free in plasma after intravenous administration or after oral administration assuming 100% bioavailability of the compound. During the potency measurements, we can assume that the total amount of compound is available to interact with the target, so the efficiency is 100%. The absorption and clearance can be taken into account as the bioavailability (*F*%), which can be measured by the total plasma concentration after oral administration. When the target is in the plasma, the free biophase concentration is equivalent to the free plasma concentration that can be measured *in vitro* and expressed by the following equation:

$$\text{Free biophase concentration} = [\text{blood}] \times f_{\text{u,blood}}$$

In this case, the unbound fraction in blood ($f_{\text{u,blood}}$) can be obtained from the measurements of plasma protein binding. As the unbound volume of distribution (Vd_u) is defined as the dose divided by the free plasma concentration, its reciprocal value can be considered as the maximum drug efficiency of the compound that can be achieved after intravenous administration or assuming 100% oral absorption as shown by the following equations:

$$\text{Vd}_\text{u} = \frac{\text{Dose}}{[\text{blood}] \times f_{\text{u,blood}}}$$

$$\text{DRUG}_{\text{eff max}} = \frac{1}{\text{Vd}_\text{u}} \times 100$$

The unbound volume of distribution already takes into account not only the plasma protein binding of compounds but also the tissue binding. When no permeability barrier or active transport exists between the plasma and the target tissue compartments, we can assume that the free concentration in plasma equals the free concentration in tissues. In these cases, the model that estimates unbound volume of distribution also provides a good estimate for $\text{DRUG}_{\text{eff max}}$.

Q11.5 What are the major causes of low drug efficiency?

The drug efficiency is defined as the percentage of the dose that is available as free concentration at the site of action. We cannot expect 100% of the dose to just go to the target and not interact with other components in the body. Only a portion of the orally administered drug is absorbed and goes to the systemic circulations. A significant portion of the dose immediately may be eliminated by the first-pass metabolism in the liver. The other portion of the administered and absorbed drug may bind to various plasma proteins, distribute into the tissues, and bind nonspecifically to tissue components, proteins, and lipids. These events cause the loss of the free concentration at the site of action and, as a consequence, reduce drug efficiency. Therefore, low drug efficiency can be caused by low absorption and strong protein and phospholipid binding.

Q11.6 What is the advantage of using DEI at early lead optimization?

The combination of *in vitro* potency and the *in vivo* drug efficiency has been described by the DEI that enables chemists to simultaneously optimize potency and *in vivo* properties. The advantage of ranking compounds by their DEI value is that we do not discard compounds only because of low potency if they have excellent drug efficiency. When the drug efficiency is high (above 1%), it means that the compounds require a lower dose in order to achieve the same free concentration. The *in vitro* potency value provides information about the required free concentration at the site of action to achieve a 50% binding or response. Thus, high drug efficiency compounds require a lower dose to achieve the required free concentration at the site of action. Optimizing DEI means that we can optimize simultaneously the potency and the drug efficiency, which is more efficient than first optimizing potency and then drug efficiency.

Q11.7 What is the relationship between DEI and the *in vivo* receptor occupancy?
The combination of *in vitro* potency and the *in vivo* drug efficiency has been described by the DEI that enables chemists to simultaneously optimize potency and *in vivo* properties. The DEI combines these two properties and shows a mathematical relationship with the receptor occupancy (RO%) *in vivo* per unit of dose as shown in the following:

$$\mathrm{DEI} = \log(\mathrm{DRUG_{eff}}) + \mathrm{p}K_\mathrm{d}$$

where DEI is the drug efficiency index and $\mathrm{p}K_\mathrm{d}$ is the negative logarithm of the *in vitro* drug–receptor dissociation constant.

$$\frac{\mathrm{RO\%}}{\mathrm{Dose}} = \frac{\mathrm{DRUG_{eff}}}{K_\mathrm{d} + [\mathrm{DRUG}]_\mathrm{biophase}} \times 100$$

Combining these equations, we can describe the relationship between DEI and receptor occupancy (RO%) or, more exactly, the proportion of the occupied and unoccupied receptors (RO%/(100 – RO%)) and dose.

$$\mathrm{DEI} = \log \mathrm{DRUG_{eff}} + \mathrm{pXC_{50}} = \log \frac{\mathrm{RO\%}}{100 - \mathrm{RO\%}} + 2 - \log(\mathrm{dose})$$

Q11.8 What is the relationship between DEI and LLE?
When the drug efficiency is modeled by *in vitro* measured properties of the compounds, it was found that the sum of the albumin and phospholipid binding of the compounds explains the main portion of the loss of drug efficiency. Both phospholipid and albumin binding depend on the lipophilicity of the compounds. More lipophilic compounds bind more to albumin and phospholipids. When the compound has a positive charge, it binds to phospholipids strongly. When the compound has a negative charge, it binds to albumin very strongly. It means that the sum of the two types of binding is not affected by the charge, and it increases approximately with the lipophilicity of the neutral form of the molecules. Therefore, LLE, which is the difference between the potency and lipophilicity, shows good correlation with the DEI, which is the sum of the potency and drug efficiency.

Q11.9 Why is log *P* a better model for the *in vivo* nonspecific binding than log *D*?
Log *P* refers to the lipophilicity of the neutral form of compounds, whereas log D is usually much lower when the compound has an ionizable group. The *in vivo* nonspecific binding is the sum of albumin types of binding and phospholipid binding. Both types of binding are governed by lipophilicity. The presence of positive or negative charge does not reduce the sum of the two types of bindings, as positively

charged compounds bind strongly to IAM, whereas negatively charged compounds bind strongly to HSA. Therefore, the sum of the two types of binding should correlate with the intrinsic lipophilicity (log *P*) and not with the lipophilicity of the charged molecule (log *D*).

Q11.10 Which biomimetic HPLC binding parameter can be used to estimate the drug efficiency of the compounds?
The human serum albumin binding obtained on chemically bonded HPLC biomimetic phase can be used to measure compound's binding to albumin. The immobilized artificial membrane stationary phase can be used to measure the phospholipid binding of the molecules. The weighted sum of the two binding parameters has been used to model the human clinical unbound volume of distribution drugs. The drug efficiency can be calculated as the reciprocal value of the estimated unbound volume of distribution.

CHAPTER 12

Q12.1 List a few substituents that increase lipophilicity and a few that decrease lipophilicity.
Substituents that increase lipophilicity are fluoro (aromatic), acetyl, methylester, methyl, alkyl, ethoxy, chloro, pyridine, bromo, trifluoromethoxy, furan, *i*-propyl, thiophene, benzyl, etc.
Substituents that decrease lipophilicity are amine, carboxylate, hydroxyl, amide, acetamide, morpholine, methylsulfonile, methyl alcohol, cyano, etc.

Q12.2 Which ADME properties are affected negatively by low lipophilicity (log $P < 1$)?
Permeability gets affected when the compound is too hydrophilic (log $P < 1$). As a consequence of low permeability, the absorption will be low. When the compound is hydrophilic and contains both H-bond donor and acceptor groups, the solubility may be low because of high crystal lattice energy. This also affects absorption.
Low lipophilicity of the compound may result in low brain penetration and weak tissue partition.

Q12.3 Which ADME properties are affected negatively by high lipophilicity (log $P > 4$)?
High lipophilicity of the compound causes low solubility and low absorption. It also may cause high plasma protein binding, high tissue binding, low drug efficiency, and low free concentration at the site of action. High lipophilicity causes the increase of nonspecific binding of the compound resulting in the need of higher efficacious dose. High lipophilicity may reduce permeability if the compound accumulates in the membrane phospholipids. Compounds with high lipophilicity together with a positive charge may cause phospholipidosis.

Q12.4 How does lipophilicity affect protein and phospholipid binding?
Lipophilicity of the compound affects protein and phospholipid binding. Lipophilic compounds try to avoid the aqueous environment and bind to the hydrophobic part of proteins and phospholipids. The effect of charge on lipophilicity (expressed by partition coefficient) and the protein and phospholipid binding is very different. The charge (whether it is positive or negative) reduces the lipophilicity of the compound; however, it strengthens the interaction with the albumin (negative charge) with phospholipids (positive charge).

Q12.5 How does lipophilicity affect brain penetration?
Brain penetration of the compound is affected by lipophilicity. There is an optimum lipophilicity for brain penetration, which is equivalent to $\log P = 2$. Higher and lower lipophilicities reduce the chances of brain penetration. It is important to note that lipophilicity measured by the octanol/water partition coefficients are not very good models for the brain penetration. It was found that brain-to-blood partition is affected by the presence of H-bond donor groups (on the contrary to octanol/water lipophilicity). The effect of the H-bond acidity can be measured by the so-called $\Delta \log P$, which is the difference between the octanol/water $\log P$ and hexane/water $\log P$.

Q12.6 How can solubility be related to lipophilicity?
Aqueous solubility ($\log S_0$) is inversely related to lipophilicity ($\log P$) based on the Yalkowsky equation. The relationship is straightforward for liquids, but an additional parameter, the MP (T_m), has to be considered too for solids as shown in the following.

$$\log S_0 = 0.5 - 0.01(T_m - 25) - \log P$$

The equation is considered as the general solubility equation, which is a semiempirical equation. The MP accounts for the solid-state properties of the compounds. Before the molecules are solvated by the solvent molecules from the solid form, the intermolecular bonds in the crystal have to be broken. Therefore, the solubility depends not only on the lipophilicity, which describes the interactions with the solvent molecules, but also on the crystal lattice energy in the solid form. We can expect an inverse relationship between solubility and lipophilicity for a series of compounds in which the crystal lattice energy is similar. When the solubility of the ionized form of the molecule is of interest, the type of the counterion may affect solubility because of very different crystal lattice or crystal forms.

Q12.7 Which solute properties may increase crystal lattice and reduce solubility?
The two main interactions that increase the crystal lattice energy significantly and, therefore, reduce solubility are intermolecular hydrogen bonds in the crystal and the so-called crystal packing of flat aromatic molecules. The flat aromatic molecules may interact in the solid form via π–π interaction. Therefore, the simultaneous presence of H-bond donor and acceptor groups and the presence of aromatic rings generally reduce solubility.

Q12.8 What is the SFI?
The SFI is the sum of $\log D$ and the number of aromatic rings in the molecule. Every unit increase in SFI decreased the probability of good solubility (>30 μM) by approximately 10%. Thus, more than 90% of compounds with SFI value of more than 12 have lower than 30 μM solubility, whereas more than 90% of compounds have solubility of more than 30 μM when their SFI value is below 4. The explanation for the observation is that multiple aromatic rings in the molecules provide a planar structure that can "stack" in the crystals, increasing their MPs and crystal lattice energy, thus reducing solubility.

Q12.9 Which substituents in general improve the solubility of a compound?
Substituents that reduce lipophilicity are expected to increase solubility. Therefore, carboxylate, hydroxyl, amide, acetamide, morpholine, methylsulfonile, methyl alcohol, and cyano groups are expected to increase solubility. There are special

hetero ring systems that increase solubility, such as pyrazine, pyridine, oxadiazole, tetrazole, pyrazole, pyridazine, furane, morpholine, and oxetene.

Q12.10 What structural and physicochemical features of the molecules help reduce protein and phospholipid binding?

Albumin binding is increasing with increasing lipophilicity. Negative charge and an aromatic ring system near the charge may increase albumin binding dramatically. Reducing lipophilicity and building polarity in the molecule will reduce albumin binding. Exclude carboxyl groups or other negatively charge functional groups, or at least reduce their ionization by increasing the pK_a values. AGP binding is also related to lipophilicity, but positively charged compounds prefer to bind to AGP, and they show weaker binding to albumin. AGP has a funnel-shaped binding pocket, and at the end of the funnel, there are negatively charged sialic acids. Thus, when the positively charged group is not surrounded by bulky substituents, the AGP binding may get very strong. The phospholipid binding is also governed by lipophilicity. More lipophilic compounds bind more strongly to phospholipids. Positively charged compounds have high affinity toward phosphatidyl choline. Long-shaped lipophilic molecules may bind more strongly to phospholipids, as they fit into the arranged hydrocarbon inner core of the membrane. Generally, we can say that lipophilicity, shape, and charge affect protein and phospholipid binding.

APPENDIX B

LIST OF ABBREVIATIONS AND SYMBOLS

%Abs	percentage of absorption
%HSA bound	percentage of compound bound to human serum albumin
RO%	percentage of receptor occupancy
$[H^+]$	proton concentration
5-HT	5-hydroxy-trypatime (serotonine) receptor
7TM	7 transmembrane receptor
A	area or H-bond acidity descriptor in the solvation equation
ACD	Advanced Chemical Development Laboratories, Incorporation
AcN	acetonitrile
ADME	absorption, distribution, and metabolism
AGP	α-1-acid glycoprotein
AMP	Artificial Membrane Permeability
AP	absorption potential
API	active pharmaceutical ingredient
AUC	area under the curve (of the time plasma concentration profile)
B	H-bond basicity descriptor in the solvation equation
BCS	biopharmaceutical classification system
BEI	binding efficiency index
BTB	brain tissue binding
c	concentration

Physicochemical and Biomimetic Properties in Drug Discovery: Chromatographic Techniques for Lead Optimization, First Edition. Klara Valko.

C-18	chemically bonded hydrocarbon stationary phase containing 18 carbon atoms in the chain
C-8	chemically bonded hydrocarbon stationary phase containing 18 carbon atoms in the chain
Caco-2	human epithelial colorectal adenocarcinoma cells for permeability measurements
CE	capillary electrophoresis
Ch	charge
CHI IAM	Chromatographic Hydrophobicity Index obtained on IAM stationary phase
CHI	Chromatographic Hydrophobicity Index
CHI log *D*	CHI values projected to the log *D* scale
CHIN	Chromatographic Hydrophobicity Index of the neutral form of the molecule
Chrom log *D*	CHI values projected to the log *D* scale (using different constants from the CHI log *D*)
CLND	chemiluminescence nitrogen detector
C log *P*	calculated logarithm of octanol/water partition coefficient
c log *D*	calculated log *D* using *c* log *P* and pK_a
Cl_s	systemic clearance
cMR	calculated molar refractivity (proportional to the size of the molecules)
CN	–CN, cyano, group on chemically bonded HPLC stationary phase
CNS	central nervous system
CSF	Cerebrospinal fluid
c_u	unbound concentration
CYP	cytochrome *P*450
D	diffusion coefficient or dose
Da	Dalton, the measure of molecular weight (mass)
DCS	developability classification system
DEI	drug efficiency index
DMPK	drug metabolism and pharmacokinetic
DMSO	Dimethyl sulfoxide
$DRUG_{eff\ max}$	maximum drug efficiency
$DRUG_{eff}$	drug efficiency
E	excess molar refraction descriptor in the solvation equation
E log *D*	logarithm of distribution coefficients determined by chromatographic method (Pfizer)
E log *P*	logarithm of partition coefficients determined by chromatographic method (Pfizer)
est	Estimated
F	Fischer-test value
FA	fraction absorbed
FaSSIF	fasted-state simulated intestinal fluid
FDA	Food and Drug Administration

FeSSIF	fed-state simulated intestinal fluid
f_u	unbound fraction of the drug in plasma
f_{ut}	unbound fraction of the drug in tissue
GLpK_a	Sirius instrument for measuring log P and pK_a
GSK	GlaxoSmithKline
H	theoretical plate height
H-bond	hydrogen bond between two electronegative atoms
HBC	hydrogen bond donor count
HCl	hydrochloric acid
HIA	human intestinal absorption
HPLC	high performance liquid chromatography
HSA	human Serum Albumin
HTS	high throughput screening
IAM	immobilized artificial membrane (HPLC stationary phase)
IC_{50}	inhibitory concentration of a compound that causes 50% inhibition
IDR	intrinsic dissolution rate
ISF	interstitial fluid
IV	intravenous
j	dissolution flow
K	equilibrium constant (partition coefficient)
k	retention factor or rate constant
KOH	potassium hydroxide
K_p	plasma partition coefficient
K_{pt}	tissue partition coefficient
L	column length or length; thickness of the diffusion layer
LE	ligand efficiency
LLE	ligand lipophilicity efficiency
log D	logarithm of octanol/water distribution coefficient at pH 7.4
log k_w	logarithm of extrapolated retention factor to 0% organic phase concentration
log P	logarithm of octanol/water partition coefficient (unionized molecules)
log P_e	logarithm of effective permeability
MAD	maximum absorbable dose
MAP	modified absorption potential
MCT	monocarboxylate transporter
MDCK	Madin-Darby canine kidney cells
MeOH	methanol
MP	melting point
MR	molar refraction
MRT	mean residence time
MS	mass spectrometer
n	number of observations
N	theoretical plate number

negchg	negative charge
NMR	nuclear magnetic resonance
Obs	observed
$OP\%_0$	organic phase concentration causing retention time double of the dead time
P_0	intrinsic permeability
PAMPA	parallel artificial membrane permeability assay
P_{app}	apparent permeability
PC	principal component
PCTFE	polycarbonate track-edge filter
P_{eff}	effective permeability
PEI	percentage efficiency index
PerfluoroC8	perfluorinated hydrocarbon bonded stationary phase with 8 carbon atoms
PgP	p-glycoprotein
pH	negative logarithm of the proton (hydrogen ion) concentration
pIC_{50}	negative logarithm of the inhibitory concentration of a compound causing 50% inhibition
PK	pharmacokinetic
pK_a	acid dissociation constant
PO	per os (oral)
poschg	positive charge
PPB	plasma protein binding
PTFE	polytetrafluoroethylene filter
PVDF	polyvinylidene difluoride membrane
QSAR	quantitative structure–activity relationship
QSPR	quantitative structure–property relationship
R	concentration ratio in the donor and acceptor compartments
r	radius
r^2	regression coefficient
R_s	resolution factor
S	excess molar refraction descriptor in the solvation equation
s	standard error of the estimate
S_0	intrinsic solubility expressed in molar term
S_{app}	molar solubility measured under a given condition
SEI	surface-binding efficiency index
SFI	Solubility forecast index, Chrom log D plus number of aromatic rings
SGA	spectral gradient analysis (Sirius SGA for high throughput pK_a determination)
SGF	simulated gastric fluid
SLF	simulated lung fluid
SP	solute property (linear free energy related)
S_{pH}	molar solubility at a given pH
S_w	aqueous solubility in molar term

t	time
t_0	dead time
$t_{1/2}$	half-life (the time till the max plasma concentration drops to half)
TFE	trifluoroethanol
t_g	gradient retention time
T_m	melting point
t_R	retention time
u	mobile phase velocity
USP	Pharmacopeia in the United States
UV	ultraviolet
UWL	unstirred water layer
V	volume or McGowan volume descriptor in the solvation equation
Vd	volume of distribution
Vd_{ss}	steady-state volume of distribution
V_p	plasma volume
V_t	tissue volume
γ	activity coefficient
ΔCHI	the difference of chromatographic hydrophobicity index obtained at different pH
ΔG	free energy change
$\Delta\log P$	the difference between octanol/water and cyclohexane/water log P
ε	dielectric constant
π	Hansch substituent constant expressing lipophilicity contribution
σ	Hammett substituent constant
φ_0	organic phase concentration in the mobile phase that produce $k = 1$ retention

INDEX

Physicochemical and Biomimetic Properties in Drug Discovery: Chromatographic Techniques for Lead Optimization, First Edition. Klara Valko.